The Geography of South Africa in a Changing World

Edited by RODDY FOX and
KATE ROWNTREE

OXFORD
UNIVERSITY PRESS

OXFORD
UNIVERSITY PRESS

Great Clarendon Street, Oxford OX2 6DP

Oxford University Press is a department of the University of Oxford.
It furthers the University's objective of excellence in research, scholarship,
and education by publishing worldwide in

Oxford New York

Athens Auckland Bangkok Bogotá Buenos Aires Calcutta
Cape Town Chennai Dar es Salaam Delhi Florence Hong Kong Istanbul
Karachi Kuala Lumpur Madrid Melbourne Mexico City Mumbai
Nairobi Paris São Paulo Singapore Taipei Tokyo Toronto Warsaw

with associated companies in Berlin Ibadan

Oxford is a registered trade mark of Oxford University Press
in the UK and certain other countries

Published in South Africa
by Oxford University Press Southern Africa, Cape Town

The Geography of South Africa in a Changing World

ISBN 0 19 571682 5

First published 2000

Editor: Robert Plummer
Designer: Mark Standley
Indexer: Susan Sayers

Published by Oxford University Press Southern Africa
PO Box 12119, N1 City, 7463, Cape Town, South Africa

Set in 9.5pt on 13pt Century Schoolbook by RHT desktop publishing
Reproduction by RHT desktop publishing
Cover reproduction by The Image Bureau
Printed and bound by Creda Communications

The authors and publishers gratefully acknowledge permission to reproduce
material in this book. Every effort has been made to trace copyright holders,
but where this has proved impossible, the publishers would be grateful for
information which would enable them to amend any omissions in future
editions.

The Geography of South Africa in a Changing World

Contents

Colour plates between pages 254 and 255

List of contributors

RODDY FOX
Associate Professor, Geography Department, Rhodes University

KATE ROWNTREE
Associate Professor, Geography Department, Rhodes University

BELINDA DODSON
Research Associate, Queen's University, Kingston, Canada

EVAN DOLLAR
Research Officer, Geography Department, Rhodes University

PATRICK ERIKSSON
Professor, Geology Department, University of Pretoria

ANDREW GOUDIE
Professor, School of Geography, University of Oxford

RICHARD GRIGGS
Head of Research, Independent Projects Trust, Durban

MICHIEL C LAKER
Professor, Department of Plant Production and Soil Science, University of Pretoria

ANTHONY LEMON
Lecturer, School of Geography, University of Oxford

ALAN LESTER
Senior Lecturer, St Mary's College, University of Surrey

LINDISIZWE M MAGI
Vice Rector, Umlazi Campus, University of Zululand

MICHAEL E MEADOWS
Associate Professor, Department of Environmental and Geographical Science, University of Cape Town.

IAN MEIKLEJOHN
Senior Lecturer, Department of Geography, University of Pretoria

ETIENNE NEL
Senior Lecturer, Geography Department, Rhodes University

PETER OMARA-OJUNGU
Dean, School of Environmental Sciences, University of Venda

BRIAN PECKHAM
formerly Senior Lecturer, Law Faculty, Rhodes University

MAANO RAMUTSINDELA
Lecturer, Geography Department, University of the North

DAVID SIMON
Professor, Geography Department, Royal Holloway College, University of London

PAUL SUMNER
Lecturer, Department of Geography, University of Pretoria

HANNES VAN DER MERWE
Senior Lecturer, Department of Geography and Environmental Studies, University of Stellenbosch

IZAK VAN DER MERWE
Professor, Department of Geography and Environmental Studies, University of Stellenbosch

COLEEN VOGEL
Senior Lecturer, Department of Geography and Environmental Studies, University of the Witwatersrand

Preface

This volume has been the result of some two-and-a-half years' work and reflects collaboration between the authors, ourselves, and our editor at Oxford University Press. The editorial process required a number of amendments and excisions and we are grateful to the authors concerned for the good grace with which they accepted our suggestions. We are very pleased to have worked with the many co-operative people who have responded at short notice to our requests, and we would also like to express our thanks to those who stepped into the breach to author or co-author chapters as deadlines came and went.

A multi-authored volume of this kind presents a number of problems with the standardization of language and technical terms, which we hope have been overcome. One of the biggest problems is the use of the term 'black', which can be used generically to describe people who are 'black', 'coloured', or 'Indian/Asian'. As a general rule we have used the term 'black' in the narrowest sense, so that it does not include 'coloured' or 'Indian/Asian' people, as this enables more specific discussion of the differential impact of policies on South African society. Similarly, 'African' is a term which can be used interchangeably with 'black', thus excluding 'coloured', 'Indian/Asian', and 'white' people. Authors were much more sparing in their use of 'African', and in fact its use is highlighted in only one chapter, in which 'African' is taken to mean the indigenous people of Africa.

This book would not have been possible without the input of many people. At Oxford University Press, Linette Downes-Webb, the commissioning editor, was responsible for much early encouragement and lent a sympathetic ear to us when we needed to extend deadlines. Following her departure we were placed into the hands of OUP's academic editor Robert Plummer, and we are indebted to him for his painstaking work on the text. A great deal of time and effort has also been spent on the illustrations in the book. A number of authors sent photographs and to them were added several from our own collections; all photographs are acknowledged in the text. The Graphics Services Unit at Rhodes University provided the bulk of the black-and-white artwork for the figures, and we would also like to acknowledge Paul Sumner and Ian Meiklejohn for the figures in Chapter 13. We produced a number of the colour plates ourselves, Plates 1, 10, 16–18, 31–32, 48, 50–51, and 53, as well as the figures in Chapters 9, 16, and 19.

Finally, we must thank our children, Helen and Jeannie, for their tolerance over the last two years when we have been busy with this book, particularly for the lost Christmas of 1998.

Roddy Fox and Kate Rowntree
September 1999

Introduction

RODDY FOX AND KATE ROWNTREE

The appropriate place to start this book is with a definition of geography, followed by a brief discussion of why it is important to maintain a disciplinary focus in South Africa's rapidly changing educational, developmental, and natural environment. This should facilitate an understanding of the purpose of this volume.

In the USA, geography has been rediscovered and identified as a core subject for schools. It has also become a growth area for tertiary-level students and a major focus of interest in the research community. Indeed, the National Research Council (NRC) recently sponsored a Rediscovering Geography Committee and produced *Rediscovering Geography: New Relevance for Science and Society* (National Research Council 1997). The following quotation is taken from the NRC report, and it is worth reflecting that the conditions described in it apply to South Africa as much as they do to the USA:

> Research at the frontiers of fields as diverse as planning, economics, finance, social theory, epidemiology, anthropology, ecology, environmental history, conservation biology, and international relations has highlighted the importance of geographic perspectives. In particular, the importance of spatial perspectives – through such notions as *place* and *scale* – is being recognized in many fields, extending the influence of geography well beyond its relatively small group of professional practitioners. The increased use of perspectives, knowledge, and techniques associated with a relatively small academic discipline raises several questions for the scientific community. Most directly, what is geography, and how does it connect with broad concerns of society and science? If geography is to play a more prominent role in education and decision making, do its scientific foundations need to be strengthened in order to support its expanded responsibilities? (National Research Council 1997: 10)

The most telling difference between South Africa and the United States in this context is that our educational system is moving in the opposite direction. The adoption of nine 'learning areas' in the South African school curriculum is intended to reduce the importance of a disciplinary base to learning. Geography, for example, will no longer be taught as a discipline but subsumed as part of the Human and Social Sciences and Natural Sciences learning areas. At the tertiary level, universities are developing 'programmes' that interlink disciplines to produce qualifications in, for example, Sustainable Development or Planning and the Environment. The conundrum for South Africa is how to produce the future geographers, economists, biologists, or whatever, on which educational and scientific developments depend, if there is no longer any disciplinary base.

The Association of American Geographers' World Wide Web site defines geography as follows: 'Geography is the science of place and space. Geographers ask where things are located on the surface of the earth, why they are located where they are, how places differ from one another, and how people interact with the environment' (Association of American Geographers 1997). Some twenty-five years ago, Bird (1973: 201) expressed this even more succinctly when he said that geography is the study of 'the changing spatial patterns and relationships of terrestrial phenomena viewed as the world of man'. The common geographical concerns of these three quotations are place, space, scale, and people–environment interactions. The NRC report takes these themes a little further:

> A central tenet of geography is that 'location matters' for understanding a wide variety of processes and phenomena. Indeed, geography's focus on location provides a cross-cutting way of looking at processes and phenomena that other disciplines tend to treat in isolation. Geographers focus on 'real-world' relationships and dependencies among the phenomena and processes that give character to a place. Geographers also seek to understand relationships among places, for example the flows of peoples, goods, and ideas that reinforce differentiation or enhance similarities. In other words, geographers study both the 'vertical' integration of characteristics that define place and the 'horizontal' connections between places. Geographers also focus on the importance of scale (in both space and time) in these relationships.
> The study of these relationships has enabled geographers to pay attention to complexities of places and processes that are frequently treated in the abstract, if at all, by other disciplines. (National Research Council 1997: 11)

It is perhaps time for South African geographers to question whether their own research and teaching do in fact address the vertical and horizontal themes identified so clearly above. We would go further and maintain that when our activities do attempt to undertake either of these themes, then we are doing 'strong' geography. There is also 'weak' geography, in which neither the horizontal nor vertical theme is central. Rather, they may be either implicit or incidental, in which case they will await stronger geographical treatment. Where strong geography comes from is the next important topic to consider in this discussion.

The trend in South African universities is now away from identification as a Geography Department (there are nine such departments) and towards a Department of Geography and Environmental Studies, a Department of Environmental and Geographical Science, or some such designation (twelve departments). If the curricula and research endeavours mirror the joint designation of the majority of departments, then they will tend to produce weak geography, because environmental science differs from geography in a number of ways:

- it is not necessarily terrestrially focused, but also encompasses oceanography;
- it does not need to address either the vertical integration of or horizontal connections between places; and
- it need not be interested in scale or the defining characteristics of places.

By addressing environmental issues, problems, and concerns, however, environmental scientists have to deal with the simple fact that all phenomena are grounded in space and time and are the product of interrelationships between people and the environment. This, of course, means that environmental science is dependent on geography for the techniques and approaches developed to understand why 'location matters'. Environmental science is also dependent on other disciplines such as zoology, botany, geology, and chemistry for their own particular knowledge, approaches, and skills. Environmental science undertakings are therefore likely to have at least some elements of weak geography, just as geographical research and teaching will have elements of weak botany or geology. What is critical, therefore, is that geography continues to produce work from within its own disciplinary genre. If it does not, South Africa's educational system, research community, and society will suffer from the lack of geographical input which Americans are now trying to redress.

It is into this context that we have produced this volume – the first 'Geography of South Africa' since Cole's *South Africa* (1960) and Wellington's *Southern Africa: A Geographical Study* (1955). We are attempting to produce a piece of strong geography in that the book has an explicitly spatial focus: in each chapter there is specific consideration of conditions at the global, southern African, and South African scales. In this way we hope that the reader can appreciate that phenomena and processes occur at different scales and may or may not produce outcomes that help define the character of this place known as South Africa. Thus the material addresses both the vertical and horizontal components of geography identified by the NRC above. The author of each chapter was asked to examine a certain theme's characteristics, distributions, and processes since approximately 1980 – what the characteristics are, where they occur, why they have formed in the way they have, and how our understanding of them has developed.

A brief review of many of the relevant books on South African geography or by South African geographers through time shows us that there has been no recent attempt of this kind. This listing is not intended to be exhaustive, but it does serve to support the point made here. Three comprehensive geographies were written before 1980 by Wellington (1955), Cole (1960), and Lemon (1976). In this period, too, King (1963) produced his classic text on South Africa's landscapes. In the 1980s, many aspects of geography were addressed. In the field of human geography, Fair (1982) assessed South Africa's development geography. Smith (1982) examined urbanization, as did McCarthy and Smit (1984). Apartheid's changing geography came under scrutiny by Smith (1985) and later by Lemon (1987). South Africa's landscapes and historical geography were described and analysed by Christopher (1982, 1984). Physical geographers and environmentalists were equally active. The initial edition of Fuggle and Rabie's ground-breaking *Environmental Concerns in South Africa* appeared in 1983. Meadows (1985) gave biogeographers their first South African university text. Tyson (1986) and Preston-Whyte and Tyson (1988) comprehensively addressed southern Africa's climatology at much the same time that its geomorphology was analysed by Dardis and Moon (1988) or Moon and Dardis (1988). Finally, from the Department of Water Affairs (1986) came the standard work on the country's hydrological resources. In the early 1990s, geographers produced similar types of works. Urbanization was again addressed by Lemon (1991) and Smith (1992), and the reality of a changing South Africa (Rogerson and McCarthy 1992) and themes surrounding the geography of change (Lemon 1995) were also covered. Environmentalists were again dealing with the environmental challenges the country faced (Clarke 1991).

Southern Africa is a term which needs consideration. The human geographers who contributed to this book were encouraged to consider the countries of the Southern African Development Community (discussed in Chapter 4). Plate 1 (colour section) portrays the distribution of countries which the authors incorporated: Lesotho, Swaziland, Namibia, Botswana, Zimbabwe, Mozambique, Angola, Zambia, and Malawi. The physical geographers were advised to treat southern Africa as the areas to the north of South Africa which share common biomes, drainage basins, and geological or landscape formations with it, depending on the theme of the chapter. The geographical area of interest therefore varies from chapter to chapter. A succinct overview of the physical geography of southern Africa is presented in Gregory (1990).

The themes tackled in this book have been placed into fairly traditional groupings. We start with a section called 'Geography in time', which begins with a chapter covering the philosophical and methodological evolution of the discipline of geography, a discussion which is necessary for the reader to appreciate geographical thought and endeavours. Following that are two chapters on southern and South Africa's environmental change and historical geography, which provide the long-term temporal component fundamental to understanding the processes and phenomena of the recent past which are then discussed in the following chapters.

There are seven chapters in the 'Human geography' section and six in 'Physical geography'. In each case the authors have followed their own philosophical standpoints, sometimes explicitly but at other times not, and have identified what they felt to be the characteristics and processes that needed examination. The scaled approach has been easily adopted in most chapters, but there are cases where its success has been more limited. Some authors found that there was very little literature on southern Africa given the themes they were tackling; others admitted that one of the products of South African isolation has been their ignorance of southern African trends. We hope that one of the outcomes of this book will be the stimulus to fill these gaps in our understanding. The chapters cover most of the important features that we felt needed to be understood: economic development, agriculture, landscape evolution, soil resources, and so on. We are aware that there could have been more chapters; the development of Geographical Information Systems, for example, would have been a valuable addition. In this context it is pertinent that we have made considerable use of the electronic databases of Corbett and O'Brien (1997), the Water Research Commission (1997), and Enpat (1998) to produce the maps in the book. We were, however, both restricted by the practical consideration of size and guided by the market for the book. It is intended for undergraduate use, particularly at first- and second-year level, but not exclusively so.

The last section of the book is entitled 'People and the environment', and its three chapters move away, to greater or lesser extents, from the strong geographical focus of the rest of the book. The first of these presents us with an overview of the relationships between environmental resources and development, which is probably the dominant overarching theme that concerns environmental scientists. The following chapter demonstrates the need for more studies of how black people perceive and use the environment, and the final chapter gives the all-important legal basis underpinning environmental law and management. This is an appropriate moment to note our respect for Brian Peckham of Rhodes University, who died while co-authoring the chapter.

Finally, it is our hope that this book stimulates a much better understanding of South Africa by South Africans and all those interested in this wonderful, challenging, and frustrating country. If we have succeeded, that understanding will be based on a enhanced appreciation of the geography of South Africa and also a deeper awareness of the geographer's craft.

SECTION I

Geography in time

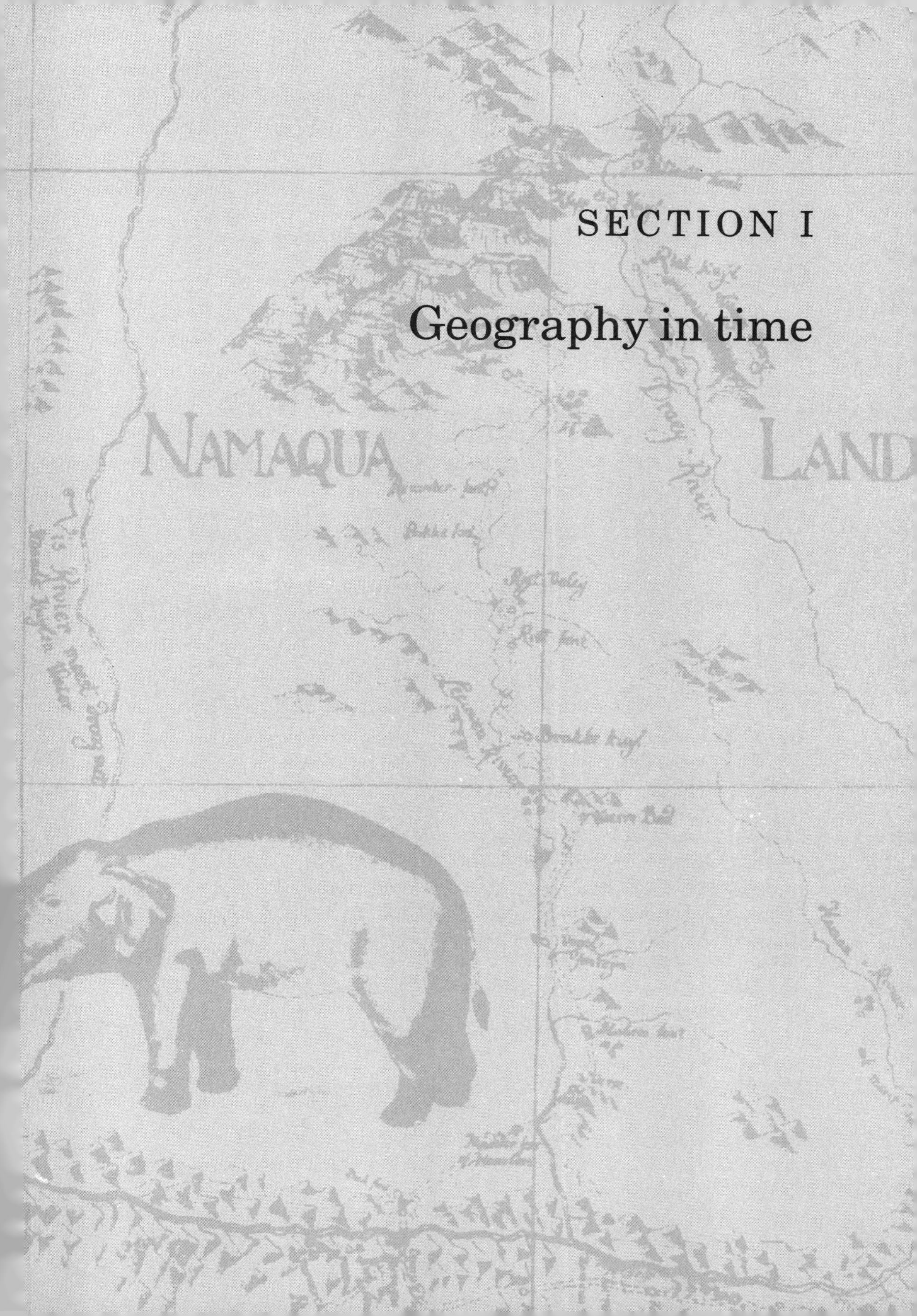

1 Philosophy and methodology in geography

RICHARD GRIGGS

The purpose of this chapter is to introduce new geography students to the philosophical schools that form a foundation for work in the discipline. This 'Cook's tour' answers three questions in as many sections:

1 Why should geographers care about philosophy?
2 What are the major schools of philosophy affecting geography?
3 What is the easiest way to learn more about the subject?

To accomplish this task, philosophical terms and vocabulary will be minimized and some complex debates simplified in order to furnish a broad and comparative overview. Ontology and epistemology are the two most basic philosophical terms you will have to understand. *Ontology* refers to philosophical assumptions and ideas about the nature of existence. What phenomena exist in the world that we can know about? It may surprise you to learn that most philosophies upon which science is based assume we can know only the concepts and ideas contained in our own minds and can say little about what lies outside that cognitive dimension. *Epistemology* refers to the philosophical assumptions and ideas about knowledge. What can we know and how can we know it? All philosophies must address these questions because they affect the selection of data, the methods of research employed, and the mode of explanation.

Now armed with the most basic of terms you can embark on a basic introduction, with just one proviso: do not take this account to be exhaustive. It is aimed at stimulating your intellectual appetite to read and learn more about the philosophies undergirding geography as an academic discipline. Many students who start off with weighty texts of philosophy become so overwhelmed that they turn a blind eye to the subject after only a brief encounter. This is a danger to the future of any academic discipline, for philosophy is the foundation of all knowledge. So please recognize and enjoy this chapter as a friendly bridge connecting you to a much wider world of ideas.

Why should geographers care about philosophy?

The role of philosophy in geography and science is much greater than many university students might suppose. Philosophy offers a carefully formulated framework for making observations and appraising what we observe. Without it there would be:

- no tests and standards for scientific knowledge;
- no agreement regarding the subject matter of various disciplines; and
- no professional standards for geographers and other academics.

Let us understand each of these contributions of philosophy in turn.

Firstly, the accepted tests and standards for scientific knowledge have been established through ontological and epistemological debates. Unless we make some ontological assumptions about the nature of existence, we have no guide to our selection of objects to study. If we cannot address the epistemological question of what is knowable and how we can know it, we have no guidelines for our observations, methods, and modes of explanation. Thus these two great questions of philosophy create sets of assumptions, methods, and modes of explanation that a community of scholars agrees to follow. Each set is referred to as a *paradigm* (Kuhn 1970), and different 'schools' of geographers base their work on different paradigms.

An example of a contrast between two philosophical schools is that between *realism* and *idealism*. Realists make the ontological assumption that there is a real, existing world whether humans perceive it or not. Idealists begin with the opposite assumption: that there is no 'reality' independent of our ideas or consciousness. These antithetical ontological positions produce different paradigms that include divergent approaches and methods. The realists strive for some correspondence between observation, explanation, and an independent 'reality', while the idealists want to examine and explain how different 'realities' are socially constructed. This has resulted in a very lively exchange across the sciences in the 1990s, known as the realism/anti-realism debate.

One might be tempted to ignore such philosophical debate as meaningless, but the irony of philosophy is that all attempts to ignore its questions simply result in another philosophy. This is because philosophy is intrinsic to any claim to knowledge. For example, logical positivism was established by a group of thinkers who decided that any discussion of ontology should be avoided. The result of numerous meetings of this 'Vienna Circle' in the 1920s and 1930s was a system of enquiry some refer to as the 'scientific method'. This method was an attempt to use logical propositions in languages (including mathematics) to avoid speculation on what actually exists beyond our reasoning minds. Most scientists, including South African geographers, claim this as their philosophy, but, as we shall see, few manage to stay within the tight confines of the method that it prescribes.

Once one realizes that any work in geography is based on some paradigm or other, the next step must be to assess each school and understand its strengths and weaknesses. This allows one to choose a system of research that best fits one's own logic and personal philosophy, rather than being indoctrinated into a system of thought. The types of questions one might ask, the standards and tests by which one's work should be judged, and the paradigm that offers the most rigour in a chosen sub-discipline are all born of philosophical enquiry. Only the naive or arrogant would attempt to arrive at their own philosophy of science without first understanding the philosophical debates that have occupied some of the greatest minds in history.

A second role of philosophy is the definition of the subject matter of a particular discipline. Logic, semantics, epistemology, ontology, ethics, aesthetics, and other sub-disciplines of philosophy are used to justify the academic division of labour. What does it mean to be a geographer? What separates history, geology, or any other field from geography? How does the individual geographer account for his or her role within this enterprise? Mission statements, course offerings, the very existence of departments, and the demarcation of geography's sub-disciplines are decided within philosophical debate (see Unwin 1992). To maintain the profession in the competitive academic arena, one must master this debate to justify the very existence of the discipline (some universities have eliminated geography in times of budget shortages) and defend one's own role within geography. Each philosophy proposes the question 'what is geography?', and the different answers can have an integrating or disintegrating effect on the discipline.

Thirdly, geographers, like all scientists, make oral and written presentations based upon research. Research is built upon certain methodologies. Underpinning these methodologies are specific theories generated by certain philosophies. This makes philosophy fundamental to the development and presentation of logical argument, the capacity to identify, analyse, and interpret published research results, and the ability to discuss method and theory in the context of the philosophical schools in which they originate. In other words, philosophy facilitates credible geographic research by structuring our observations and analyses, as well as the forms of presenting them.

Since the study of philosophy is really the business of training our minds and fine-tuning our thinking, it has pay-offs that reach beyond academic pursuits. It challenges unquestioned assumptions, prejudices, and past indoctrination that interfere with sound logic. Knowing ourselves makes us less arrogant about our claims to knowledge and more open to other ways of 'seeing'. Given the original meaning of the term (*philo* = love and *sophy* = wisdom), such wisdom may be the end point of philosophical study.

What are the major schools of philosophy in geography?

We shall now address seven schools of thought affecting the discipline. These are:

- naive realism;
- logical positivism;
- behaviouralism;
- humanism;
- Marxism;
- postmodernism; and
- neorealism.

These seven constitute a manageable approach as long as the reader recognizes that such categories are tools of explanatory convenience. There can be as many philosophies as there are people, and there are many internal disagreements or 'wings' within any school. Nevertheless, discussing schools rather than the ideas of millions of individuals allows us to engage in meaningful debate on central issues. Since these categories are a social construction, other writers might reduce them to two (positivist and post-positivist schools, for example) or expand the number of categories to twenty or more (to include 'offshoots' of Marxism, humanistic geography, and neorealism, and so on). Compare, for example, the seven schools listed above with the branches of the discipline identified by Peet (1998), which are shown in Figure 1.1. This chapter provides explanatory tables in each of the seven sections, so that you can more readily grasp and compare the different philosophical systems according to their ontologies, epistemologies, methods, and claims. These tables list some of the numerous terms and divisions by which each school may be defined, along with the most common critiques.

FIGURE 1.1 Schools of human geographical thought

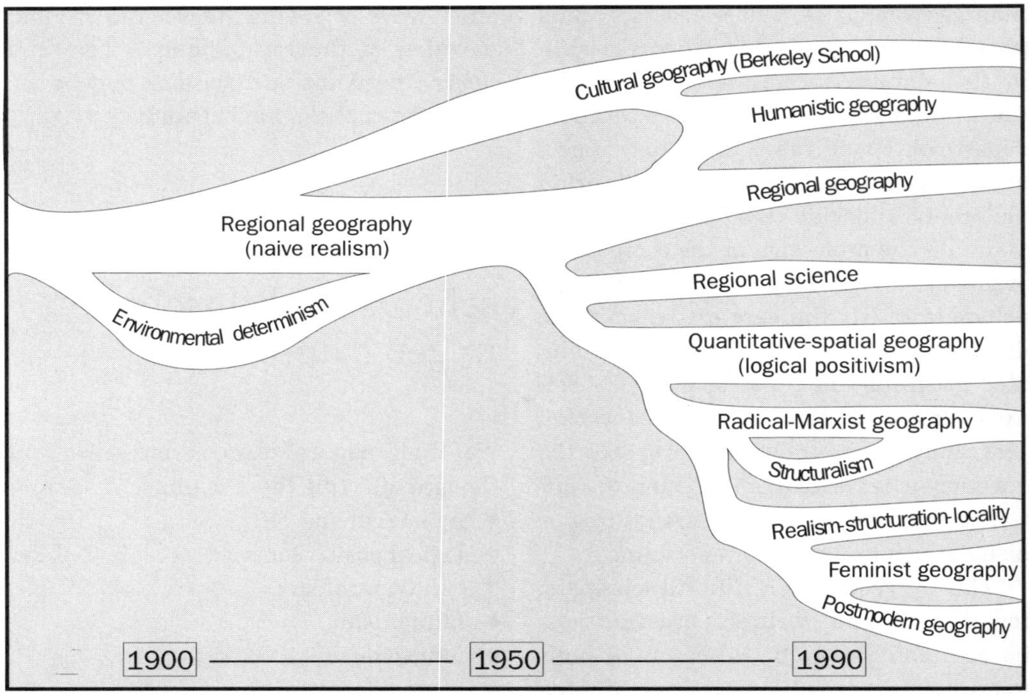

SOURCE: Peet 1998.

It would be convenient to categorize the various chapters of this book according to the approaches the authors have followed in presenting their ideas and material. Unfortunately, the reality of academic writing and research means that this is not particularly easy to do, especially where the authors have attempted to synthesize material from across a number of philosophical divides. One good indicator, however, is the authors' use of language and technical terms; Chapter 3, for example, acknowledges the influence of Marxism and poststructuralism, and Chapter 18 has clear connections to behaviouralism.

Naive realism

This category includes classical realists, positivists, and empiricists, who share the assumption that knowledge of the world can be accessed directly through human experience (see Table 1.1). This is an epistemology that posits no need to account for how our minds process and select data (hence 'naive'). For example, the nineteenth-century founder of positivism, Auguste Comte (1798–1857), argued that knowledge began with experience and was destined to move through several higher orders (such as religion, mathematics, astronomy, and eventually sociology) until a perfect understanding was achieved (Chalmers 1982). All that was required was to categorize and systematize knowledge. Hence the term 'positive knowledge' or 'positivism'.

This school was attacked by the majority of late-nineteenth-century philosophers and formally rejected by the Vienna Circle of the 1930s, whose thinkers argued that naive realism mistakenly equates data with knowledge and therefore does not

TABLE 1.1 Naive realism

OTHER TERMS	empiricism; realism; positivism; the encyclopaedists; 'old' regional geographers
ONTOLOGY	What we sense and experience is what exists
EPISTEMOLOGY	Knowledge is gained through lived experience. The cumulative observations of individual events leads to encyclopaedic knowledge of the world and its content
METHOD	observation; induction; classification; ideographic (focuses on the particular or unique)
CLAIMS	grounded; real-world geography; policy-orientated; practical; not ideological
CRITIQUES	authoritarian; lacks any framework for observation; biased; naive; atomistic (reduces events to their parts rather than understanding the mechanisms and structures behind all similar events); atheoretical; arrogant (my experience = the event); subjective; commonsensical

account for the processing of information. Karl Popper (1968, 1969) derided it as the bucket theory of knowledge: we simply keep filling up the bucket eventually to achieve a unitary scientific view of the world. It is no wonder that the earliest naive realists were known as the 'encyclopaedists'. If a lecturer or professor refers to one of your essays or research papers as 'encyclopaedic', he or she means that you have accumulated 'facts' without analysing how and why you selected your data or how they might fit into a theory of knowledge.

The central problem of naive realism is that our impressions of life do not lead automatically to concepts and descriptions that are 100 per cent accurate. There are many things in the world that we must 'see' by inference (for example the effects of gravity), and some things we see may bear no relationship to what is in the real world (such as delusions from disturbances in brain chemistry). Methodical enquiry must mediate between our senses and the outside world. The abstract theories of modern science are more reliable than our senses when it comes to constructing the knowledge by which we solve human problems (for example, our senses might indicate that the sun 'rises', but abstract thought processes teach a more practical theory of time and planetary motion).

By the 1930s, naive realism had been thoroughly discredited as a philosophy of science. Ironically, it was information accumulated by the naive realists themselves that led people away from their philosophy. New details about retinas, electric impulses, and neurotransmitters demonstrated that there was a complex of mediating processes between 'us' and the outside world. What we 'see' is limited by our very human form (for example, the human eye can perceive only a small band along the entire spectrum of electromagnetic radiation). Theories built upon mental abstractions are required to comprehend 'reality'. The outside world cannot simply be observed and recorded.

It is somewhat unfortunate that geography achieved little or no philosophical awareness until the late 1960s (Harvey 1969). Naive realism reigned in the form of the 'regional concept', whereby geographers defined regions as objective spatial realities and accumulated data about them. Richard Hartshorne (1939: 383) defined the 'ultimate purpose of geography as the study of the areal differentiation of the world'. Symbolic of this period were large regional textbooks with endless accumulation of factual data seldom bound by any theory.

Studies that focus on the uniqueness of regions are termed *ideographic*, while systematic studies seeking general laws are known as *nomothetic*. The regionalists recognized that nomothetic studies were important, but primarily as an adjunct to the explanation of individual regions. It was not until 1953 that this emphasis on 'exceptionalism' was challenged (Schaefer 1953). The critique did not emerge out of a concern for the philosophical underpinnings of the discipline, however. Rather, geographers sought prestige by association with the natural sciences and sought to emulate its methods (Unwin 1992: 122).

Geography's late entry into philosophical debate contributed to perceptions that the discipline employed discredited approaches and outdated methods The staid regional geographies were far outside the currents of mainstream academia. Throughout the 1960s, 1970s, and 1980s, geography departments were vulnerable to closure or amalgamation with other disciplines as geographers struggled to find philosophical justifications for the discipline. In hard economic times, it was easy for administrators and departments outside geography to see that there was little philosophical cohesion between the various practitioners of geography. Why not place biogeographers in botany, geomorphologists in geology, political geographers in political science, and so on?

Despite the dangers that accompany a lack of philosophical sophistication, many geographers still treat philosophy with contempt, as a 'spooky' encumbrance to engaging with the 'real world' of geography. Nevertheless, it is not possible simply to 'explain the facts' without addressing the basic problems of philosophy, namely:

- What can we know?
- How can we know it?
- How can we explain what we know?

Instead of dismissing the subject, it is more challenging, rigorous, and rewarding for geographers to explore it in a disciplined way. After all, geography's advances since the 1980s and the renewed cross-disciplinary interest in the subject during the 1990s have resulted in large part from geographers who engaged the subject of philosophy with vigour. To them the renewed vitality of the discipline owes a great debt.

Logical positivism

The contradictions of naive realism led scientists and philosophers to understand that information must be placed into pre-existing forms. Just as a computer must be programmed prior to data being inputted, a philosophy exists prior to data collection. The empiricists did not recognize this bimodal relationship and therefore advocated what is now called the *monadic* view of scientific knowledge.

Logical positivism was an attempt to overcome these problems by accounting for the relationship between meta-theory and data collection (see Table 1.2). Scientists such as Bertrand Russell (1872–1970) and Ludwig Wittgenstein (1889–1951) argued that since we do not have direct access to knowledge through experience, we must resort to logically connected propositions in language that correspond with our experiences. We can then test the propositions. If our logical forms can accurately predict what we experience, then we can establish laws of science that fit the world in which we live.

Interestingly, logical positivism abandons any effort to describe the essential nature of the world. It limits one to making empirical statements that require verification (testing) in order to be accepted as true or analytical statements that can be accepted in logic. Thus metaphysics, the sub-discipline that examines the nature of being and existence, is abandoned as meaningless. We can structure a description of the world only through propositions in language or mathematics that correspond to our experience of the world.

The result of this philosophy is a method of research described as the *hypothetico-deductive* method. We arrive at a proposition or hypothesis by formal logic and then deduce a law through repeated testing. If a sufficient number of experiments show a constant correspondence

TABLE 1.2 Logical positivism

OTHER TERMS	positivism; naturalism; quantitative geography; spatial science
ONTOLOGY	Logical positivism abandons metaphysics (of which ontology is a part). The true nature of existence is beyond what science should address
EPISTEMOLOGY	We cannot really know anything about the objective world and must confine knowledge to the experience of reality. Practical knowledge is achieved through abstract logic based on propositions in language. We can verify a correspondence between our propositions and our experiences to produce practical theories
METHOD	hypothetico-deductive; nomothetic (the search for general laws)
CLAIMS	scientific; not woolly; predictable; quantifiable; aimed at technically useful knowledge; neutral; not ideological; tested
CRITIQUES	mechanistic; scientistic (appears to be scientific but does not take full account of all phenomena, particularly human subjectivity); ahistorical; makes false claims of neutrality; technocratic; deterministic (suggests that human behaviour and activities are predictable); reductionist; physicalistic; unable to explain human behaviour

between our logical forms and what we 'see' (verification), we establish a universal law that is the necessary explanation for all phenomena of that type. By contrast, the naive empiricists were *inductive*: they attempted to observe the laws of nature directly and then record them logically. Figure 1.2 illustrates the differences between induction and deduction in two flow charts.

The beginning student of philosophy should be alerted to two major problems with logical positivism, although there are many more to investigate. First, it makes an ontological assumption: one cannot not postulate universal laws without assuming that the universe is uniform and law-like, constant and unchanging. Thus the attempt to escape metaphysics failed. Second, is it possible to verify such laws by experiment and observation? David Hume (1711–76) had already shown two centuries earlier that observation and experiment are capable of establishing only what might happen, but not what must happen for all times in the future. Amassing instances in which a swan is found to be white cannot lead us to the conclusion that all swans are white now and forever (there are, in fact, black swans). The hypothetico-deductive method fails to sidestep this classic problem of induction (reasoning from experience to the statement of general laws).

Most physical geographers describe themselves as logical positivists because

FIGURE 1.2 Inductive and deductive routes to scientific explanation

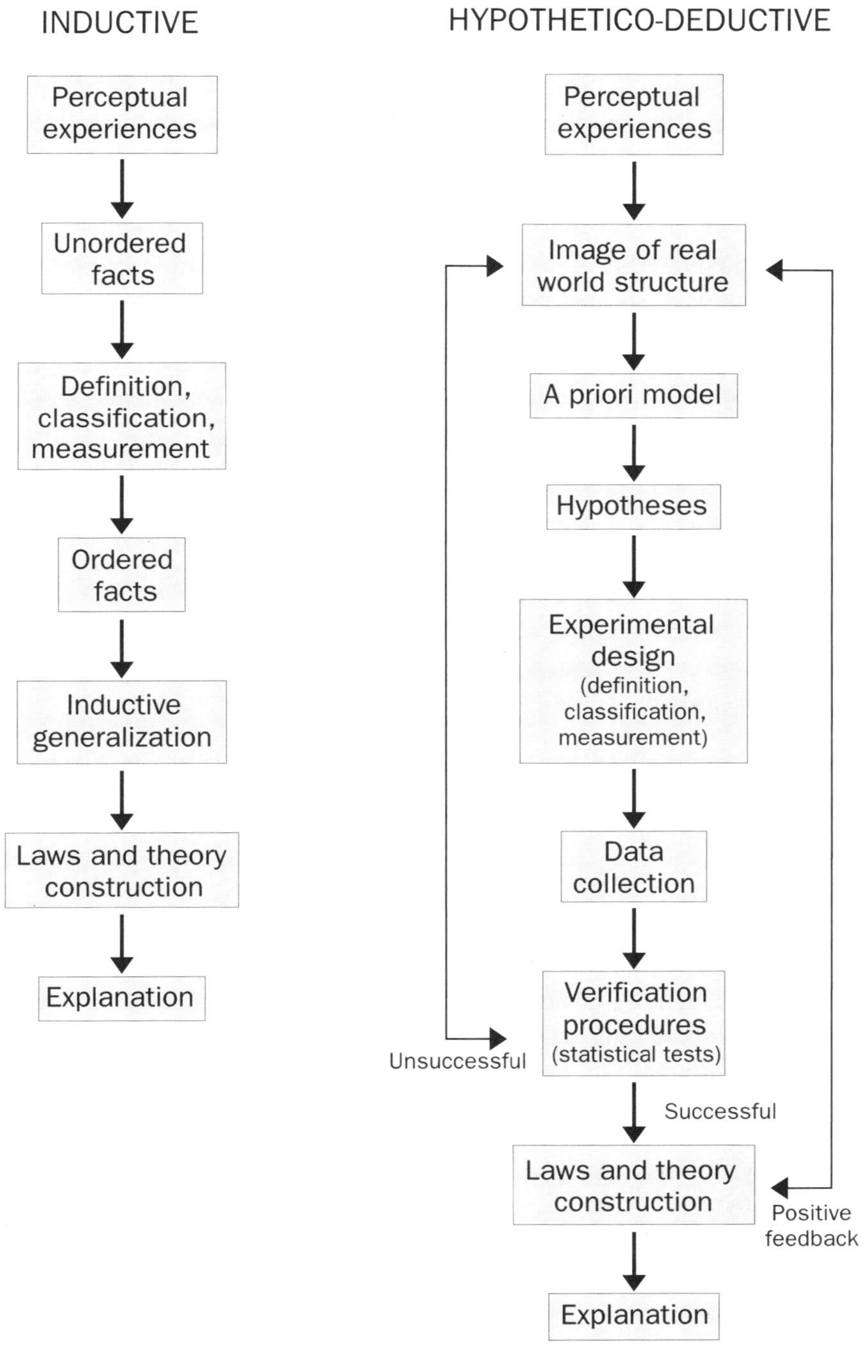

the philosophy works reasonably well in relation to mathematics and measurement. Physical forms have formal properties that can be measured and described. However, the philosophy becomes problematic in practice. Physical geographers are not always focused on testing hypotheses to verify a necessary truth arrived at by deduction. Many reverse this direction and use fieldwork, observation, induction, and synthesis to arrive at an explanation (Hay 1985). (This is well illustrated in Chapter 13, in which landscape evolution is unravelled using various strands of evidence derived from fieldwork, observation, induction, and synthesis.) Histories of science have shown that many scientific advances such as the discovery of penicillin or the theory of tectonic plates were arrived at by accident or as a result of the doggedly determined opinion of a scientist that he or she was 'right' (Chalmers 1982)! There is, therefore, a very tenuous relationship between logical positivism and how scientists actually work.

Human geographers adjusted less easily to logical positivism, although for a time bold attempts were made to overthrow the old guard of regional geographers with 'spatial scientists'. The period between the end of World War II and the 1970s saw a 'quantitative revolution' as numerous human geographers declared that they would make geography no different in method than biology or physics (Abler et al. 1971). Mathematics and scientific models would describe the 'laws' that governed locations: these are nomothetic studies. By contrast, others who practised the older empirical geography aimed at what was unique or special about a place: theirs were ideographic studies.

The quantitative revolution had both positive and negative effects on the discipline. It did raise the standing of geography within the academic community by bringing to an end a long era in which the main geographic product was a tedious and encyclopaedic text on a region. It also led to more methodological rigour. Most importantly, it stirred up philosophical debate so that geography grew increasingly multi-paradigmatic.

On the negative side, the philosophical break with the past created a sharp divide between human and physical geographers that has widened over time. Physical geographers found a home in logical positivism despite its contradictions, while human geographers were unsettled by its lack of utility. Geography as a spatial science was too reductionist, mechanistic, deterministic, and ahistorical to explain the human-made patterns on the earth. Social, cultural, and political geographers must examine many phenomena that cannot be quantified: values, attitudes, beliefs, emotions, intentions, and historical circumstances.

Behaviouralism

Like many schools, behaviouralism developed out of a critique of earlier systems of thought (see Table 1.3). Wolpert (1964), Brookfield (1969), and several others (see the edited volume by Cox and Golledge, 1969) criticized the failure of theories derived from logical positivism to predict spatial outcomes (such as retail locations). People do not simply react to larger economic or political laws but make daily decisions based on incomplete knowledge,

TABLE 1.3 Behaviouralism

OTHER TERMS	behaviourism (critics of behavioural geography often liken it to the stimulus–response psychology of B. F. Skinner known as behaviourism); positivism
ONTOLOGY	What exists is beyond human measurement
EPISTEMOLOGY	We can know only human experience and not the 'real' objective world. By observing the outward behaviour of human beings we can infer what is perceived and even measure human actions that shape the world around us
METHOD	largely inductive; quantitative; nomothetic; employs interdisciplinary models that explain human behaviour; micro-scale studies into human perception; mental mapping
CLAIMS	overcomes the pitfalls of logical positivism to create a true spatial science; accounts for human decision-making in measurable ways; examines the perceptual world in which humans operate
CRITIQUES	reductionist; formulaic; studies at the micro scale fail to support macro-level statements; assumes human rationality

sentiment, habit, and other irrational factors. The behaviouralists proposed that humans do not operate as perfect 'optimizers' of their environment but as 'satisficers' who do the best they can within a bounded rationality (Wolpert 1964; Pred 1967).

This discovery shifted the questions asked and the methods employed to answer them. Unlike the logical positivists, who studied dehumanized and static spatial patterns, the behaviouralists purported to examine the active decision-makers who produced such patterns. This required cognitive, attitudinal, and perceptual studies using methods often drawn from psychology. Behavioural geographers tested and analysed individuals and small groups using mental maps, attitudinal tests, and many forms of statistical measurement (see, for example, Gould and White 1974).

Philosophically, the behavioural school of geography did not seriously challenge the epistemology and ontology on which logical positivism was based. It was in fact a methodological advance aimed at creating 'better' formulas and models for explaining spatial outcomes. In this sense it was, as David Harvey once argued, merely an appendage of the quantitative revolution. Behaviouralists analysed quantitative data just as the logical positivists did, to discover the general laws affecting behaviour and cognition. They also reproduced the same philosophical weaknesses. One example is the classic problem of induction outlined earlier. How can analysing small groups lead to generalizations about all human behaviour?

The credit owed the behavioural geographers is not for their philosophical innovation but for their critiques. At an early stage they recognized that humans do not fit neatly into law-like equations. Although they did not find their way clear of that quagmire, this understanding provided a 'bridge' leading to the development of humanistic geography (Cloke et al. 1991: 67). They also deserve credit for some good empirical work, such as influencing US government policy on natural hazards (White 1973) and producing geographical tools such as mental maps that aid development planning (Lynch 1960) or time–geography prisms that assist with siting public institutions such as hospitals or schools (Pred 1973, 1977).

Perhaps the school's greatest weakness was its failure to attract many followers. The behaviouralists drew so heavily upon cross-disciplinary discussions of perception and cognition that many geographers felt out of their depth. Why should environmental and geographical scientists do what psychologists can do better? On the other hand, the philosophy seems to have attracted physical and human geographers in equal numbers, an achievement surpassed only by the naive realists in their efforts to produce regional geographies.

Humanistic geography

Humanistic geography first appeared in the early 1970s as a critique of the approaches found in both logical positivism and behaviouralism (Buttimer 1976; Tuan 1974a, 1977; see Table 1.4). All the basic assumptions of these philosophies were attacked, including the notions of objectivity, abstraction, generalization, law-like mechanisms, and the division between subject and object. Instead, the humanistic geographers came to focus on the 'lived world' of human experience, the particular or unique, and the connection between our subjective world and the spaces and places we create.

The basic ontology and epistemology of the humanistic approach suggest that knowledge must be acquired by engaging with a world of meaning, values, and emotions. Existence is the lived world we know and therefore the only real 'facts' are what people experience. The result is a broad spectrum of approaches and methods that raise questions about the cohesion of the category 'humanistic geography'. Instead of concentrating on this diversity, however, I can suggest four common features of humanistic geographical approaches:

- their links to methods in the humanities;
- their effort to understand places in terms of their meaning for human existence;
- their emphasis on particular places; and
- their concern for a holistic understanding of places that includes the subjective experience of them.

These features stand in direct contrast to logical positivism, which is characterized by:

- links to methods in the natural sciences;
- a dismissal of metaphysics as meaningless;
- an effort to deduce general laws that apply to all places; and

TABLE 1.4 Humanistic geography

OTHER TERMS	existentialism; phenomenology; idealism
ONTOLOGY	All we can know of existence is the limited world that we experience and not the objective 'real' world
EPISTEMOLOGY	Knowledge can be acquired only through inter-subjective understanding: engaging with a world of meanings, emotions, and values
METHOD	empathy; feeling; understanding; participant observation; in-depth interviews; hermeneutics; ideographic studies
CLAIMS	understands the ideas upon which our behaviour is based; bridges the philosophical division between subject and object; holistic rather than atomistic
CRITIQUES	woolly; lacks a sound methodology; eclectic; self-indulgent; anthropocentric (makes humans the centre of the universe)

- an emphasis on the objective description of geographic localities.

The three major strands of humanistic geography are idealism, existentialism, and phenomenology. The focus differs for each but the methods are similar. *Idealism* aims at reconstructing the ideas and ways of thinking that lead to patterns on the ground (Guelke 1974, 1982; Harris 1978). Both historical scholarship and empathy are required to present a plausible theory regarding the complex of emotions, motivations, and thoughts that lead to certain decisions affecting a place. *Existentialism* in geography focuses on how people assign meanings to geographic spaces (Entrikin 1976; Samuels 1978; Tuan 1974b). It is a philosophy that insists that human environments are not given but constructed (Appleton 1994). Since spaces and places reflect attempts to give meaning to our lives, the existential geographer concentrates on this process. *Phenomenology*, like existentialism, is concerned with the meanings people assign to places, but en route to an explanation of what they experience and how they experience it (Relph 1981). For the phenomenologists, places are not landscapes strewn with objects and people but places full of human meanings that may be partly represented in objects and spatial arrangements.

In terms of method, humanistic geography is mostly inductive and interpretive. The humanists also embrace metaphysics in their attempts to understand how individuals and groups construct meaning in geographical places. The techniques for uncovering this meaning include participant observation (experiencing what the subjects experience), archival research, empathetic analysis ('putting yourself in their shoes'), and an interpretive approach known as *hermeneutics*, which involves understanding as opposed to explaining. This approach is named after Hermes, a messenger from the gods in ancient mythology. The first practitioners of hermeneutics were theologians who studied the language of biblical texts in search of God's intended meaning. Today it is also a secular concern that includes the analysis of any text, language, discourse, symbol, spatial arrangement, building, or landscape that conveys and represents human meaning (Pickles 1992).

Humanistic geography has been productive for the discipline in terms of the volume of work it has produced. Moreover, the humanistic critique of both behaviouralism and logical positivism has helped to refine geographic thought, showing that the description and explanation of a static location according to universal laws rather than human experience is reductionist. The humanist geographers probe deeper into an understanding of a place than those who stand far outside to measure, analyse, and explain it in quantitative terms. Finally, the introduction of hermeneutics into geography is a tool of long-standing utility that has been adopted into other philosophical approaches (such as structuration, postmodernism, and some forms of neorealism, which will be explained later in this chapter).

Nevertheless, the humanistic geographers are hardly above critique. Their methods are eclectic, selective, and filtered through interpretation. It must be asked: to what degree is the 'lived experience' of place merely the imaginings of the interpreter? There is no system of verification, a problem which perhaps leads us back to the empiricism of the naive realists. Furthermore, the humanists make abundant use of generalizations in their texts, because any broad description of place requires general data (possibly drawn from the general statements of the logical positivists!). Perhaps the greatest weakness is the lack of attention to the kinds of structural constraints under which communities operate. Oppressive political and economic forces play a role in the creation of the spaces in which we live. Finally, this school fails to unite all geographers in a common set of practices. It attracts hardly any physical, political, or economic geographers, and is principally confined to cultural, historical, and social geographers.

Marxism

The term Marxism covers a variety of philosophical branches influenced by the German philosopher Karl Marx (1818–83) (see Table 1.5). The founding principle that informs both the ontology and epistemology of Marxism is *historical materialism*, which emphasizes the need for people to survive in material ways. Historically, all people have had to find food, shelter, and clothing. The way this is achieved, however, has changed over time, giving rise to a process known as the *historical dialectic*, by which societies evolve though successive stages. Technological and social innovations can change the ways societies organize the production of material goods and meet their needs. For instance, the change from feudalism to capitalism adjusted the complex of social, political, and economic relationships that Marxists call the *mode of production*. Since power relations are usually based on a division of labour, the relationship between classes underlies Marxist explanation.

TABLE 1.5 Marxism

OTHER TERMS	structuralism; radical geography; political economy; historical materialism; critical theory
ONTOLOGY	What exists is nature and matter which humans act upon to meet their needs and desires
EPISTEMOLOGY	We can know about our material relationship with the earth by understanding how people transform nature through labour. This creates powers, structures, and relations of production. Knowing this process moves us beyond idle facts and towards action-orientated knowledge: how to emancipate ourselves from oppressive social forms
METHOD	dialectical materialism; deductive; nomothetic; analysis of class relations with reference to the mode of production
CLAIMS	a socially responsible philosophy (since Marxism aims not just to describe 'what is' but to change things for the better); holistic; grounded in the productive forces of society; integrative (structural forces, economic processes, and human relations are integrated both locally and globally); scientific
CRITIQUES	ahistorical; deterministic; reductionist; selective science; mechanistic; economistic; reifies (discusses abstract entities such as 'capitalism' as if they were human actors); totalizing (asserts meta-theories that ostensibly explain all phenomena); static; repetitive; neo-positivistic; anachronistic; fails to account for human agency

Many geographers have found the premise of historical materialism very attractive to a discipline concerned with human–environment relations. Moreover, the assumption of an historical dialectic provides a useful tool for explaining how the spatial structure of society is affected by the mode of production. Changing geographic patterns and the use of space in a particular area (for example the closing down of an auto plant) can be explained in terms of large social, political, and economic conditions (such as the global movement of capital). In a landmark work, the Marxist geographer David Harvey (1973) showed how the economic injustices of the capitalist system created discriminatory housing markets and ghettos in Baltimore. Similar studies of ways in which uneven spaces were produced by exploitative systems proliferated, because they appeared relevant and replicable at the urban, regional, and global scales. By the early 1990s a disproportionate number of the most frequently cited geographers in the field were Marxist (Bodman 1992).

In terms of explanation, this school differs substantially from the previous schools we have analysed. Marxist geography looks 'behind' both human action and behaviour to examine how structures (large social, economic, and political formations such as capitalism) bind land and society together into certain logical relationships that are expressed spatially. Thus Marxism promotes a more impersonal theory of agency than do the other schools, particularly humanistic geography. The Marxists oppose logical positivism, too, on the grounds of its objects of study (such as quantifying the processes which predict the best site for an industrial location) and its claims to neutrality. The selection of our subject matter is in itself a political act, and the Marxists attempt to encourage a more socially responsible choice. In terms of ontology and explanation, however, there are many similarities between Marxism and logical positivism. Both assume that there are certain necessary internal relations in a society that can be regularly observed, and both make deductions from a basic set of premises to reach their conclusions.

The greatest impact of Marxist geographers on the history of geographic thought has been its stimulation of lively philosophical debate. Marxist geography had to respond to critiques that it left out significant actors by subsuming them under a spectrum of political and economic forces that determine human and environmental spaces (Duncan and Ley 1982). Empirical, humanistic, and positivistic geographers, in turn, had to respond to Marxist critiques that they had not made their theories explicit and that they paid insufficient attention to the structural causes of spatial outcomes such as global trade, capitalism, and class relations. Both critiques appear to have much validity. When one focuses too heavily on structure, the result is a deterministic brand of geography in which the actors are dehumanized and reduced to passive parties. If agency is overemphasized, the actions of individuals and organizations appear completely unconstrained by economics, politics, and even material conditions.

The debate came to be known as the problem of 'structure and agency'. It was a central issue in various brands of Marxist geography (Peet and Thrift 1989) and led to some post-Marxist philosophies. One of these was the concept of 'structuration', which was imported from the sociology of Anthony Giddens (1982). Giddens sees structure and agency as reciprocal and 'interpenetrating' in their relationship. That is, human agents must act within a certain social structure or institution, but at the same time they are also active in shaping those institutions. Giddens also weds the Marxist concern for the effects of macro-scale political, social, and economic systems to the humanistic concern for local meaning by introducing hermeneutics to explain the context in which people act. Few geographers actually took up Giddens's approach, partly because of its sociological content and partly because it seemed a contrived compromise. Nonetheless, all geographers are faced with the dilemma of structure and agency and must find some approach to resolving it.

Feminist geography is another product of the lively debate generated by Marxism (Bowlby et al. 1989; McDowell 1986). Here the assumptions informing the historical dialectic are questioned. According to feminist thinkers, the first set of unequal social relations was not between social classes but between men and women, and therefore the patriarchal system that resulted from this relationship must be examined. There is no point in discussing the relations of production or power if one fails to note that, historically, women have been relegated to positions of relative powerlessness and their labour has been exploited. Feminist geographers also examine how inequality between men and women produces gendered spaces. This includes the domination of many geography departments by men (Rose 1993)!

Aside from stimulating debate that has generated new theories and perspectives, Marxism has also helped advance a better understanding of the interrelationship between the global and the local. For instance, both human and physical geographers better appreciate the links between the global economy and issues such as the destruction of a rainforest. Philosophically, Marxism's greatest weakness has been its basic premise. Similarly to the logical positivists, Marxists assume that there is an order to be discovered and explained. This leads to a search for laws, grand theorizing, and totalizing tendencies (such as the assumption that all workers share the same plight). Individuals and the chaos of life are removed and replaced by a logical, deduced order that helps account for the eloquence and appeal of Marxist explanations. We have seen how Anthony Giddens attempted to bridge the gap and humanize Marxism, but a fresh assault on the philosophy was en route in the form of postmodernism.

Postmodernism

In many ways, it is easier to explain what postmodernism rejects than to define and group its eclectic arrangement of arguments, explanations, and objects of study (see Table 1.6). Empiricist, positivist, and behaviouralist philosophies are rejected because no regular laws can predict human behaviour. Postmodernism stands structural Marxism on its head by positing that we are not so much the passive recipients of our social structures as the inventors of our world. There is a complete rejection of logical positivism because there is no such thing as a single reality to measure: there are many contested realities, which are all in flux. In fact, all grand theoretical frameworks are rejected as 'totalizing' stories that scientists tell either to make sense of the world or to build a position of privilege and authority.

What remains in the wake of the rejection of all grand theories is an epistemology that suggests that whatever is known is relative to the knower. There are no epistemological guides because there are no static categories of knowledge. In terms of method, we must not 'privilege' one storyteller above another but disentangle and deconstruct various texts to reveal the multiple discourses that inform human action and create human landscapes (Duncan 1995). Such a philosophy in geography promotes ideographic study – the search for difference and disorder in the world (Gregory 1989). There can be no attempt to define objective regions as the naive realists sought to do. Instead, hermeneutics is used to understand how people create places through language, ideas, icons, representations, symbols, and discourse.

This leads to some obvious critiques. If postmodernists claim that 'reality' and 'truth' are artificial constructs, how can we

TABLE 1.6 Postmodernism

OTHER TERMS	post-positivism; poststructuralism; idealism; deconstructionism
ONTOLOGY	There is no given state of existence, because all phenomena are in flux
EPISTEMOLOGY	Knowledge is constructed by the human mind and institutionalized through inter-subjective agreement. People are enrolled into different schools of thought
METHOD	hermeneutics; ideographic study; discourse analysis; symbolism; mythology; deconstruction; participant observation; informal interviews
CLAIMS	the only philosophy with an accurate epistemology (that the truth is a social construction); liberates the individual; overcomes the oppression of universal truth and laws
CRITIQUES	idealistic; trivializes morality; makes truth relative; solipsistic (you cannot get outside your own mind); enmeshed in a web of subjective words; narrowly focused; limited policy application

believe what they themselves have to say? This emphasis, however, may lead to an unwarranted dismissal of the postmodernist contribution. Within geography, postmodernism has revived the ideographic study of place and regions, leading to new and better-informed regional studies. The products are more sophisticated than those of the encyclopaedic regionalists of the naive realist school. In addition, postmodernists' writings reflect critically on the role of the geographer and on the ways in which places are represented in texts. This introspection has been a healthy process, because it has brought into question some of the authoritarianism and absolutism that has accompanied much geographic scholarship.

Probably the most significant contribution of postmodernist geography has been its refinement of the hermeneutic method first pioneered by the humanistic geographers. This has reinvigorated certain subdisciplines through the provision of new tools of analysis. For instance, discourse analysis has helped to produce an entirely new branch of political geography known as critical geopolitics (Agnew and Corbridge 1985), which examines the representations, narratives, and metanarratives that are used to manipulate political processes. Representations are the words, symbols, and pictures that a person might use to shift the discourse and affect our use of space (for example, by referring to environmentalists as 'bunny huggers' to limit support for their efforts to stop a development). Narrative is a chronological, ordered text that builds personal authority (such as 'expert' testimony promoting a new development). Metanarratives are grand theoretical efforts to bring order to disorder; these are epitomized by 'scientific' explanations. Metanarratives easily co-opt people because they are constructions which are difficult to escape, such as the unquestioned neutrality of science. A further contribution of the hermeneutic method has been its revival of cultural geography through studies of place.

Despite its useful innovations, postmodernism is troubled by many philosophical problems. Firstly, it views all systems of knowledge as equivalent narratives. Science and its philosophical schools are differentiated from superstition or witchcraft only in terms of the believers that are enrolled. This destroys any basis for assessing or verifying knowledge claims by the use of logic and brings epistemology to a standstill. Secondly, postmodernism confronts us with the problem known as solipsism: if nothing can exist independently of our own minds, then how can we know anything outside the content of our minds? Bodies of theory cannot put us in touch with other peoples' minds, ideas, and experiences. Thirdly, the relativism of postmodernism may be seen as an absurdity. That is, if one says that everything is relative, one is ironically making an absolute statement. Similarly, to posit minds that construct ideas suggests some constant underlying form of reality, which contradicts the postmodernist stance. Finally, postmodernism has attracted few, if any, physical geographers.

Neorealism

Neorealism in all its variations (critical realism, scientific realism, transcendental realism, and others) is distinguished by a relatively complex explanation of ontology (see Table 1.7). By detailing different categories of what exists, it overcomes many problems associated with the other schools. For instance, it is able to defend the common-sense assumption of naive realism that material things exist externally to our ideas about them while it does not attribute a reality independent of humans to social phenomena (Bhaskar 1975, 1986; Outhwaite 1987; Sayer 1992; Werlen 1993). This teaches us to embrace more complexity and to choose our research methods in accordance with the ontological categories of 'reality'. Unfortunately, this also leads to critiques that the epistemology and ontology of neorealism are too complicated to be adapted easily (Sarre 1987; Pratt 1995). Among the principal schools offering theoretical underpinnings for the knowledge claims of geography in the last twenty years, neorealism has been one of the last to be fully operationalized (Cloke et al. 1991; Unwin 1992; Pratt 1995).

Neorealism has a number of appealing features that make the complexity worth penetrating, however. Firstly, it is not limited to 'experienced' phenomena. Instead it posits three ontic classes: the real, the actual, and experienced events. The longest-term project of science is the attempt to explain 'real' events, the underlying structures and mechanisms that underpin the universe. In this case we can hope only to achieve a correspondence between reality and our theories. 'Actual' events are experiences that can be surmised by abstraction for all practical purposes, because their effects are observable (an example is the observation that gases expand under pressure). 'Experienced' events are filtered through human res-

TABLE 1.7 Neorealism

OTHER TERMS	scientific realism; critical realism; transcendental realism; realism; post-positivism
ONTOLOGY	Existence is many-layered. There are real things, experienced things, and abstracted things. Each represents a different ontic direction
EPISTEMOLOGY	Knowledge is not limited to the content of one's own mind. Theoretical abstraction can lead us to approach a real world that exists outside our sense perceptions (the long-term project of science)
METHOD	Many different methods are possible (for both nomothetic and ideographic studies), but one must match the method to the ontic direction of the research. One can build abstract theories to explain the structures and mechanisms that produce events (the real) or the perceived regularity in events (the actual), and one can use a variety of methods, including hermeneutics, to examine human experiences. One may also attempt to find correspondence between real, actual, and experienced events (synthetic research)
CLAIMS	closer to the way science is actually practised than is logical positivism; resolves agency–structure debate; reconnects philosophy and science; not ideological; appeals to both physical and social science; applies to policy; embraces complexity; context-sensitive; includes the most accurate ontology
CRITIQUES	utilitarian (aims at instrumental or practical purposes rather than a true understanding of philosophy); rationalist (aims to justify what scientists are already doing); naive realism in sheep's clothing; postulates an epistemology of objective truths; too inclusive; can mean all things to all people; the school itself is so diverse that 'realists' argue among themselves as to which realism is 'real'

ponses and subjectivity. An understanding of such events requires methodologies similar to those used by humanists and postmodernists. Thus, one must match the method to the ontic direction of the research. In addition, one can conduct intensive (how does something happen?) and extensive (how widespread is the phenomenon?) studies in all three ontic directions, and synthetic research that attempts to link the three (Sayer 1992, Harré 1986: 8–33).

Secondly, neorealism does not insist that every research production is a final description of intransitive mechanisms which cause events. It is, rather, a self-conscious employment of the transitive categories of science (models, theories, and so on) to help us approach a more practical understanding of both the physical and social world (Sayer 1992; Harré 1986). For instance, certain forms of political organization at a certain juncture of space and time can produce regular effects if it is understood that such effects will no longer be felt if there is transformation of agency–structure relations (such as changed policy). Such research then may not be the final 'truth', but it is 'practically adequate' for a contribution to policy formulation.

Thirdly, neorealism insists that it is necessary to be clear about our objects of study (maintaining 'ontic' direction) and to segregate contingent from necessary factors in our explanations. Work in human geography often focuses on the contingently related conditions manifesting in temporal spaces (Griggs 1996). Allen (1983) has shown, however, that it is possible to demonstrate that certain necessary relations obtain within a defined socio-economic formation (such as property relations and landlordism).

Fourthly, neorealism embraces the complex interactions of agency, structure, and discourse when addressing the operations of society. All three interact to produce particular spaces as contingent products (of a certain time and sets of relations). This overcomes the straitjacket of logical positivism, the arrogance, eclecticism, and privileged authoritarianism of empiricism, the determinism of Marxism, the hyper-individualism of humanism, and the solipsism of postmodernism. Moreover, it transcends the objective/subjective divide that concerned the humanists, because categories of human action rather than the ontologically questionable category of 'space' are used to describe geographic conditions and change.

Neorealism has been slow to take hold in geography, but historically geography has been slow to take on new philosophies. Outside geography it has been a growing approach embraced by sociologists, psychologists, engineers, political scientists, historians, physicists, and many others across the spectrum of both the physical and human sciences. In some fields such as international relations it predominates. Its cross-disciplinary appeal suggests that a refined neorealism could finally bring physical and human geographers into a single paradigm. Others are sceptical because its multi-methodological approach tends to justify almost all approaches. While one might think that all philosophical roads lead to realism, another might think that realism is too amorphous to lead anywhere (Sayer 1985; Cloke et al. 1991: 135).

Like all philosophical systems, neorealism is troubled by certain contradictions. The most significant of these is its attempted escape from solipsism. How can neorealism construct the categories of 'real' and 'actual' if we have no direct access to them except by abstraction (our own ideas)? This has established a debate between postmodernists and realists that may continue for a long time. The postmodernists accuse the realists of taking a 'giant leap of faith' into a reality for which they cannot establish an existence. The neorealists see postmodernism as reductionist, anthropocentric (seeing the human mind as the centre of the universe), and contrary to common sense. This debate crosses the full spectrum of the sciences, because one side seeks to defend the project of science and the other seeks a radical restructuring of what constitutes knowledge. This is a good point on which to end the Cook's tour because it beautifully illustrates the foundational role that philosophy plays in justifying our claims to knowledge. In all humility, one must confess that any scientific or disciplinary claim to knowledge is shaky and that we have much philosophical work ahead.

What is the easiest way to learn more about philosophy?

Here are a few pointers on how to follow up on this introduction. First, obtain a good dictionary that covers the philosophical terms employed in geography. A very good start is *The Dictionary of Human Geography* (Johnston et al. 1994). Just perusing the dictionary will give you a healthy appreciation of the extent to which philosophy and geography are integrated. Then there are some very readable texts covering the philosophical schools (Cloke et al. 1991), their historical development (Johnston 1997), and the specific debate on the role of geography as an academic discipline (Unwin 1992). A good complement to these three texts is *Human Geography: An Essential Anthology* (Agnew et al. 1996). Included there are many of the key essays that led to the philosophical development of the discipline.

2 Environmental change

EVAN DOLLAR AND ANDREW GOUDIE

The environment of planet earth has been changing ever since its formation some 4.6 billion years ago. In Chapter 11 we shall see how the configuration of the continents and oceans, the atmosphere and climate, and the fauna and flora have evolved together through geological time. When the human species first made its appearance some two million years ago, a new environmental variable came into play. All species modify their environment to some extent, but humans have undoubtedly had the most intense and pervasive effect on it. Environmental change as it relates to the human species, both in terms of the impact of our activities and our ability to cope with future change, is at the centre of modern environmental research.

It is difficult to evaluate the environmental impact of early humans, but human impact became increasingly significant after the inception of cultivation around 10 000 BC and the development of urban civilization over the last four to five thousand years. In the nineteenth and twentieth centuries the human impact on the environment has undoubtedly far surmounted that of earlier time. Colonialism, industrialization, population growth, and globalization of the economy have together intensified and spread the impact. In the late twentieth century the environment has become a political issue which has galvanized both the richer and poorer countries of the world into wide debate and into developing agendas for action.

Much of the debate on environmental change revolves around the questions of how we measure real change and how much of the observed change is the result of human activity. The landscape that we see around us is in part a reflection of present-day processes controlled by the present-day environment, but it also bares the imprint of past environments. It is therefore important that we recognize the changes that have taken place in the past so that we can interpret what we see in the present day. Environmental change can take place in a number of guises: long-term irreversible change whereby one state is replaced by another, cyclical change whereby a pattern is repeated over time, and seemingly random change whereby different states replace each other but no long-term trend or pattern is evident. It is difficult to evaluate the significance of observed changes outside the context of these long-term trends.

Human-induced environmental change must therefore be evaluated against long-term natural trends. This chapter first examines the natural changes that have taken place through geological time, and then assesses possible human impacts. Environmental change is a global issue, and it is first examined at the global scale, before we turn to southern Africa itself.

Global environmental change

Geological and geomorphological change

Our solar system, together with planet earth, probably originated about 4 600 million years ago. This means that our planet has evolved over an almost inconceivable time span, especially when we recall that our own species – *Homo sapiens* – has inhabited the earth for only a minute fraction of that time. *Homo sapiens* probably emerged in Africa a mere couple of million years ago, and reached the New World and Australia only in the last few tens of thousands of years. The earth has an enormous and complicated history that earth scientists have striven to understand by reading the evidence presented in its rocks.

Earth scientists have developed a widely accepted and unified view of earth history, known as *plate tectonics*, which is based on sea-floor spreading, the movement and collision of lithospheric plates, the building of continents, and the opening and closing of ocean basins. Many of the world's most important features – mountains, volcanoes, island arcs, earthquakes, ocean ridges, deep sea trenches, and so on – can be explained in the context of plate tectonics. The earth's surface is ever-moving both horizontally and vertically, though we find a remarkable contrast between some areas of the land surface that are very ancient and relatively stable (such as the basement shield of large portions of Africa) and other areas that are relatively very young (the ocean basins). The geological evolution of the earth is more fully described in Chapter 11.

The **lithosphere** is the outer, rigid exterior of the earth. It contains the crust, continents, and plates. A **lithospheric plate** is one of the dozen or so segments that make up the lithosphere, which are internally rigid and move independently of one another. Where plates come together, they meet at convergence zones; where they move apart from each other, in a process known as rifting, a divergent zone exists.

Plate tectonics is the theory and study of plate formation, movement, interaction, and destruction. Seismicity, volcanism, and mountain-building can often be the result of plate tectonics.

Climate change

In addition to such major tectonic changes, the earth has throughout its history witnessed major climatic changes over a wide range of time scales (Goudie and Stokes 1998). Table 2.1 summarizes the main orders of change which can be identified, from the minor fluctuations observed within the period of instrumental record (during which weather-recording instruments have been available), with durations of the order of a decade or decades, to the major geological periods, with durations of many millions of years. The impact of climatic change on the environment is dramatic. Changes in temperature and precipitation impact on evapotranspiration, the hydrological cycle, weathering and soil development, erosion, and fauna and flora. Vegetation belts, for example, can shift through several degrees of latitude in response to such change.

Changes in climate are the result of two major factors: the total amount of energy received by the earth–ocean–atmospheric

TABLE 2.1 Orders of climatic variation

TIME-SCALE UNIT		
(1) Minor fluctuations within the instrumental record	10 yr.	Minor fluctuations with somewhat irregular length and amplitude
(2) Post-glacial and historic	10^2 yr.	Variations over intervals of the order of 250–1 000 years, for example the sub-Atlantic recession and others affecting vegetation in Europe and North America
(3) Glacial	10^4 yr.	The phases within an ice age; glacial periods may be of the order of 50 x 100^3 years
(4) Minor geological	10^6 yr.	Duration of ice ages as a whole; periods of evolution of species
(5) Major geological	10^8 yr.	For example ice ages at intervals of 2.5 x 10^8 years

Principal bases of evidence:
(1) Instrumental; behaviour of glaciers; records of river-flow and lake levels; non-instrumental diaries: crop yields, tree-rings (also for dating)
(2) Earlier records of extremes: fossil tree-rings; archaeological finds; lake levels; lake sediments; oceanic core-samples; pollen analysis; radiocarbon dating; ice-cores
(3) Fauna and flora characteristic of interglacial deposits; pollen analysis; variation of height of snowline and extent of frozen ground; oceanic core-samples (dating through latter)
(4 & 5) Geological evidence: character of deposits; fossil fauna and flora; dating largely through radioactivity of rocks

system (external factors) and the storage and transport of this energy within the earth's oceans and atmosphere (internal factors). External factors include variations in solar energy and adjustment of the orbital parameters of the earth. Orbital variations (known as Milankovitch forcing) are the result of the dynamic relationship between the earth and the sun, as explained further in Chapter 11. Variations in the earth's orbit are thought to have played a major role in the glacial/interglacial cycles of the Quaternary period of geological history. Perturbations in sunspot activity appear to play a smaller-scale role in climatic change. It is believed that they are were responsible for the warming period in the Middle Ages known as the Medieval Warm Epoch and the Little Ice Age between about 1500 and 1850, which apparently coincided with a reduction in solar activity known as the 'Maunder minimum' (Lindesay 1990).

The internal mechanisms of climate changes result from the storage and utilization of the energy within the atmosphere and oceans. Changes in the spatial and temporal arrangement of the continents as a result of wandering plates have resulted in alteration of the configuration of the ocean basins, the creation of new oceans and ocean current systems, sea-level variations, and so on. Orogenic (mountain-building) episodes, the development of the polar ice cover, and changes in the composition of the atmosphere as a result of volcanic emissions also impact on the distribution of energy in the oceans and atmosphere. All these factors have affected global and regional climates, with concomitant environmental impacts.

Various curves which show the nature of the different orders of climatic change are illustrated in Figure 2.1. They serve to show the frequency and magnitude of changes during the Quaternary period, and to place such changes in the context of the longer span of earth history. They have recently been reviewed by Moore et al. (1996). The shorter-term changes include such events as the period of warming that took place in the first decades of the twentieth century. The changes with durations of hundreds of years have been

Geological time is divided, in order of descending magnitude, into eras, periods, and epochs.

The **Quaternary** is the period of geological time following the Tertiary. It consists of two epochs, the Pleistocene and the Holocene, and extends from approximately two million years ago to the present.

The **Pleistocene** is the earlier epoch of the Quaternary, extending from the end of the Pliocene epoch (approximately two million years ago) to the beginning of the Holocene epoch (approximately 10 000 years ago). The Pleistocene is often referred to as the 'Ice Age', a time when the world experienced great fluctuations in temperature resulting in colder periods (glacials) being separated by warmer periods (interglacials).

The **Holocene** is the present epoch, covering approximately the last 10 000 years of geological time. Many consider the Holocene to be one of the interglacials of the Pleistocene, as it has been characterized by an increase in global temperature and a rise in relative sea level.

FIGURE 2.1 The different scales of climatic change

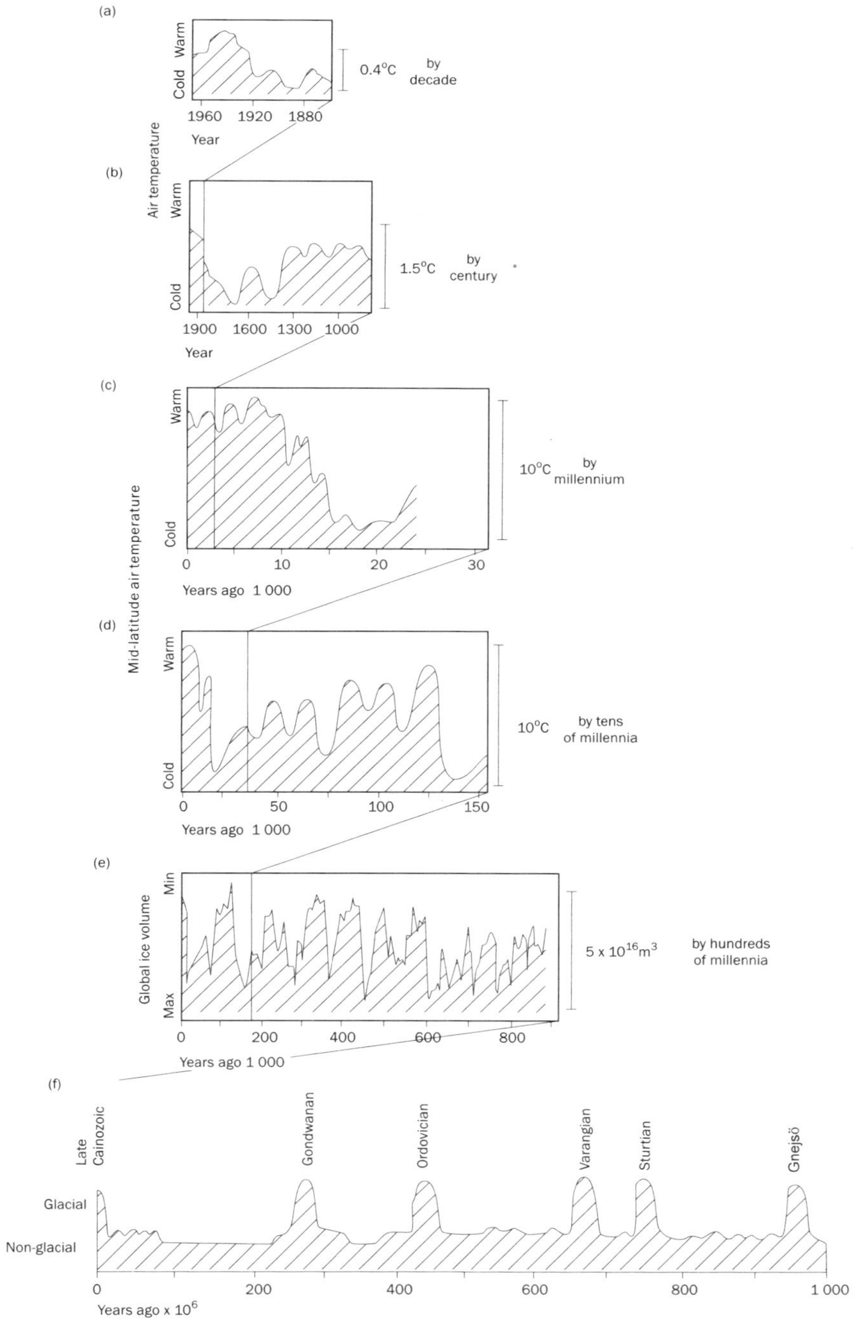

SOURCE: Goudie 1993b.

characteristic of the current Holocene epoch and include various phases of local glacial advance and retreat such as the Little Ice Age. The fluctuations within the previous epoch, the Pleistocene, consisting of major glacials and interglacials, lasted for the order of 10 000 to 100 000 years, while the Pleistocene itself, termed a minor geological variation, consisted of a group of glacial and interglacial events that lasted in total about two million years. Such major phases of ice-age activity appear to have been separated by about 250 million years.

Natural climatic change has been highly significant during the two million years that humans have occupied the earth. In the last 20 000 years alone the area of the earth covered by glaciers has been reduced to one-third of what it was at the glacial maximum; the waters thereby released have raised ocean levels by over one hundred metres, flooding the continental shelves. The land, unburdened of the weight of overlying ice, has risen locally by several hundred metres, permanently frozen ground has retreated from extensive areas of Europe, desert sand fields have advanced and retreated over areas such as the Kalahari, and inland lakes have flooded and shrunk. In response, vegetation belts have swung through the equivalent of tens of degrees of latitude, rainforests have expanded, and many of the largest mammals perished in the catastrophe called 'Pleistocene overkill'. Even at the present time, smaller fluctuations of climate are leading to changes in fish distributions in northern waters, marked fluctuations in valley glaciers, extensive flooding or desiccation of African lakes, and difficulties for agricultural schemes in marginal areas. Africa is noted more for its coral reefs than for its icy mountains, and under present conditions glaciers are of limited extent and area, being confined to Mount Kenya (5 198 m), Kilimanjaro (5 835 m), and the Ruwenzori Mountains (5 089 m). In the Pleistocene, however, glaciers were certainly more widespread, though the chronology is insecure and fragmentary and some of the evidence for past glaciation is controversial.

Sea-level changes

Sea-level changes have been associated with both tectonic and climatic changes. It is possible to categorize these changes according to whether they were of a world-wide nature, related to changes in the volume of water in the oceans (eustatic changes), or of a local nature, involving changes in land levels (tectonic changes). It is not always easy to differentiate between the two on the ground, but the effects of such changes can be seen along most shorelines. Where there are stranded beach deposits, marine shell beds, and platforms backed by steep cliff-like slopes, one has evidence of emerged shorelines. Other signs of submerged coastal features are the drowned mouths of river valleys (rias), submerged dune chains, notches and benches in submarine topography, and remnants of forests or peat layers at or below present sea level. Many coasts show evidence of both emergent and submergent phases in their history.

Human impact

In addition to natural environmental changes brought about by the geological and climatic processes discussed so far, the environment has also been significantly affected by human activities. These changes are discussed at length by Goudie (1993a). As human technology has developed, as human numbers have increased, as humans have diffused over the face of the earth, and as consumption of energy has burgeoned, so human impacts on the environment have been magnified. Even in the earliest stages of human occupation of the land, environmental changes would have occurred. Deliberate manipulation of fire, for example, may have taken place in Africa for over a million years. Fire is a very potent force that is capable of transforming vegetation assemblages, soil characteristics, faunal habitats, and even air quality. Likewise, in later Stone Age times, humans were effective predators and may have, tens of thousands of years ago, caused the extinction of most of the world's megafauna (an event known as the Pleistocene overkill). It is paradoxical that it is only in Africa, the 'cradle of humankind', that a significant number of species of large mammal have survived: elephant, rhinoceros, hippopotamus, giraffe, and water buffalo.

Human-induced environmental changes have become especially significant in the modern era. The ways in which humans affect the environment are proliferating. For example, nearly all the powerful pesticides post-date the Second World War, and the same applies to the increasing construction of nuclear reactors. Environmental issues that were once locally confined have become regional or even global problems. An instance of this is the way in which substances such as lead, sulphates, and DDT, an environmentally damaging insecticide, are found at the poles, far removed from the industrial societies that produced them. Moreover, the complexity, magnitude, and frequency of impacts are probably increasing; a massive, modern dam such as Aswan in Egypt or Kariba in Zimbabwe has a very different impact from that of a small Roman dam. Finally, compounding the effects of rapidly expanding populations, there has been a general increase in per capita consumption of energy, with associated environmental impacts. Energy resources are being developed at an ever-increasing rate, giving humans enormous power to transform the environment. One index of this is world commercial energy consumption, which trebled in size between the 1950s and 1980.

A substantial amount of change has been achieved in recent decades. Table 2.2, based on the work of Kates et al. (1991), attempts to make quantitative comparisons of the human impact on ten 'component indicators of the biosphere'. For each component the authors defined total net change clearly induced by humans to be 0 per cent for 10 000 years ago and 100 per cent for 1985. They then estimated dates by which each component had reached successive quartiles (that is 25, 50, and 75 per cent) of its 1985 total change. They believe that about half of the components have changed more in the single generation since 1950 than in the whole of human history before that date.

As a result of the escalating trajectory of environmental transformation, it is now possible to talk about global environmental change. Turner et al. (1990) distinguish

TABLE 2.2 Quartiles of human-induced change from 10 000 BC to the mid 1980s

FORM OF TRANSFORMATION	DATES OF QUARTILES		
	25%	50%	75%
Deforested area	1700	1850	1915
Terrestrial vertebrate diversity	1790	1880	1910
Water withdrawals	1925	1955	1975
Population size	1850	1950	1970
Carbon releases	1815	1920	1960
Sulphur releases	1940	1960	1970
Phosphorus releases	1955	1975	1980
Nitrogen releases	1970	1975	1980
Lead releases	1920	1950	1965
Carbon tetrachloride production	1950	1960	1970

SOURCE: Kates et al. 1991: Table 1.3.

between two components of this: systemic global change and cumulative global change. In the context of systemic change, 'global' refers to the spatial scale of operation and comprises such issues as global changes in climate brought about by atmospheric pollution. In the context of cumulative change, 'global' refers to the areal or substantive accumulation of localized change, and a change is seen to be 'global' if it occurs on a worldwide scale or represents a significant fraction of the total environmental phenomenon or global resource (see Table 2.3). Both types of change are closely intertwined. For example, the burning of vegetation can lead to systemic change through such mechanisms as carbon dioxide release and alteration of the reflectivity of the earth's surface (albedo change), and to cumulative change through its impact on soil and biotic diversity.

The increasing scale of human impacts has been well summarized by Kates et al.:

> Most of the change of the past 300 years has been at the hands of humankind, intentionally or otherwise. Our ever-growing role in this continuing metamorphosis has

TABLE 2.3 Types of global environmental change

TYPE	CHARACTERISTIC	EXAMPLES
Systemic	Direct impact on globally functioning system	(a) Industrial and land-use emissions of 'greenhouse' gases (b) Industrial and consumer emissions of ozone-depleting gases (c) Land-cover changes in albedo
Cumulative	Impact through worldwide distribution of change	(a) Groundwater pollution and depletion (b) Species depletion/genetic alteration (biodiversity)
	Impact through magnitude of change (share of global resource)	(a) Deforestation (b) Industrial toxic pollutants (c) Soil depletion on prime agricultural lands

SOURCE: Turner et al. 1990: Table 1.

> itself essentially changed. Transformation has escalated through time, and in some instances the scales of change have shifted from the local and regional to the Earth as a whole. Whereas humankind once acted primarily upon the visible 'faces' or 'states' of the Earth, such as forest cover, we are now also altering the fundamental flows of chemicals and energy that sustain life on the only inhabited planet we know. (Kates et al. 1991: 1)

Major types of human-induced environmental change

The range of environmental issues that have arisen in recent years is considerable, and it is possible to categorize them in a wide variety of ways. In this section we shall examine them in the context of each particular component of the environment.

First of all, let us consider changes in vegetation (Meyer and Turner 1994). Fire, fire suppression, clearing, grazing, pollution, urbanization, and other types of land-cover change all contribute to changes in plant communities. Grasslands, heathlands, secondary forests, and shrublands are examples of major vegetation assemblages that owe much of their distribution and character to human influence. Great concern currently attends various aspects of vegetation-related land degradation (Johnson and Lewis 1995), including desertification and the decline in forest health as a result of air pollution. Genetic and species diversity is being modified by tendencies such as domestication and genetic engineering.

Secondly, considerable changes are taking place in animal communities as a result of human impact, and these can be divided into five categories: domestication, dispersal, extinction, expansion, and contraction. Among the most significant changes are the invasion of certain areas by exotic animals (for example the rabbit in Australia), reductions in the health of species due to pollution, deliberate hunting (of whales, for example), and, above all, the removal of habitat. Particular fears have arisen about the very rapid reduction in biodiversity that may be caused by the removal of such habitats as tropical rainforests or areas of fynbos.

In addition to their impact on flora and fauna, humans are also modifying soil characteristics. Of great concern is the spread of soil salinity caused by irrigation or by changes in land cover. Other important issues include laterite crust development as a result of deforestation, soil compaction by heavy agricultural machinery, and reductions in soil organic content produced by cultivation and drainage. In many parts of the world, soil erosion by wind and water is proceeding at unacceptably fast rates because of deforestation, overgrazing, and construction activity.

Laterite is a hard and infertile stratum of tropical ground which forms as a result of weathering.

A further component of the environment that is undergoing change is the hydrological cycle. The quantity and pattern of flow in rivers is being modified on a large scale by the construction of dams, transfers of water from one river basin to another, land drainage, channelization and embanking of streams, and land-cover changes such as urbanization or the removal of forest cover. Elsewhere, groundwater is being mined, lakes are being deprived of their inflow (for example the Aral Sea), and water quality is being impaired by chemicals, sediments, and the spread of organisms.

The land surface is being modified by deliberate excavation, construction, and dumping. Less deliberate but far-reaching geomorphological changes are caused by a wide range of human activities. These include ground subsidence caused by the extraction of minerals, hydrocarbons (coal, oil, and gas), and water, the destabilization of slopes due to forest clearance, the melting of permafrost in tundra areas, and accelerated coast erosion due to sediment starvation.

Perhaps the greatest attention is currently being given to systemic global changes brought about by modification of the atmosphere. Gas emissions (such as carbon dioxide, methane, nitrous oxide, and chlorofluorocarbons, or CFCs), aerosol generation (for example dust and sulphates), thermal pollution (especially in large cities), and albedo changes can all contribute to climatic changes. Equally, some of the substances emitted into the atmosphere can cause severe air pollution (for example ozone depletion and acid precipitation).

Global warming and the 'greenhouse effect' are caused by an increase in the production of the so-called greenhouse gases, mainly as a result of fossil-fuel combustion. These gases – carbon dioxide (CO_2), nitrous oxide (N_2O), and methane (CH_4) – act to retain terrestrial long-wave radiation in the lower part of the atmosphere, resulting in global warming and the greenhouse effect. The recent rapidity of these changes is significant. CO_2 levels have fluctuated over time: during the last interglacial (127 000 years before the present, or BP), for example, CO_2 levels were around 280 parts per million (ppm), during the Last Glacial Maximum (LGM) around 18 000 BP they dropped to 180 ppm, while in the 1850s they were around 270 ppm. Present values of CO_2 are as high as 350 ppm, and are expected to nearly double by the middle of the twenty-first century.

Increased levels of artificial chemical compounds in the atmosphere such as CFCs and halons have created a thinning of the ozone layer. Exposure of the CFCs in the stratosphere to ultraviolet radiation from the sun causes them to break down, releasing a free atom of chlorine which acts as a catalyst, accelerating the conversion of ozone to oxygen, thus depleting ozone levels in the atmosphere. Ozone plays a vital role in filtering out harmful ultraviolet radiation. Increased reports of skin cancer have been shown to be a direct effect of reduced ozone concentrations in the atmosphere.

Recent discoveries of the 'hole' in the ozone layer and increased concentrations of greenhouse gases have led to the development of models designed to predict the impact of these changes. These models, known as general circulation models (GCMs), predict that increased CO_2 levels in the atmosphere are likely to have significant impacts on world climate, including that of South Africa (see Chapter 12).

Environmental changes in southern Africa

Environmental changes in southern Africa are as old as the oldest rocks of the subcontinent. The geological time scale and associated macro-scale changes in the region are summarized in Table 2.4 and are explained in more detail in Chapter 11. On a regional scale the most important geological processes that influence environmental change are lateral plate movements, vertical tectonics, volcanism, and erosional modification of the landscape. Through geological time there have been widespread changes in lithology (the mineral composition and texture of rocks), environment, climate, flora, and fauna. Understanding climate changes in the southern hemisphere is of particular importance, as this hemisphere is more oceanic than the northern hemisphere, and it is the oceans that play a dominant role in redistributing solar energy. The modern floral and faunal assemblage of southern Africa has developed only recently (in the last 5 000 years). Environmental change is therefore in itself a natural process. In southern Africa, however, human intervention in the biosphere during the last 100 000 years through fire, the last 2 000 years through agriculture, and the last 150 years through modernization has had a considerable impact on the environment.

TABLE 2.4 The geological time scale and macro-scale environmental change in southern Africa

ERA	PERIOD	EPOCH	DURATION (millions of years)	AGE (millions of years)	MAJOR EVENTS IN SOUTHERN AFRICA
Cenozoic	Quaternary	Holocene	(10 000 yrs)		Climatic change related to 'glacial' and 'interglacial' cycles
		Pleistocene	2	2	
	Tertiary	Pliocene	3	5	Major uplift
		Miocene	19	24	Moderate uplift
		Oligocene	13	37	
		Eocene	21	58	Development of African erosion surface
		Palaeocene	8	66	
Mesozoic	Cretaceous		78	144	
	Jurassic		64	208	Rifting of Gondwana in late Jurassic; Karoo sediments laid down; evolution of dinosaurs; initiation of Cape Fold Belt
	Triassic		37	245	
Palaeozoic	Permian		41	286	
	Carboniferous		74	360	
	Devonian		48	408	Cape Supergroup laid down
	Silurian		30	438	
	Ordovician		67	505	
	Cambrian		65	570	Gondwana formed; explosion of life forms
Precambrian	Proterozoic		2 000	2 600	Early sedimentary basins
	Archaean		1 400	4 000	Formation of cratons, early continents, and the rocks of the African 'basement'

Compiled by K. M. Rowntree.

Geological change and its impact on geomorphology

Africa was formed as a result of the rifting of Gondwana, the supercontinent comprising the present continents of the southern hemisphere which broke up in phases between approximately 180 and 130 million years ago (see Figure 2.2). The present-day environment of southern Africa owes much of its origin to geological events in the post-Gondwana period.

The pre-rift southern Africa occupied a central position in Gondwana, which was situated over the South Pole some 320 million years ago. Rifting produced new base levels for erosion and rapid denudation of the landscape. Since the fragmentation of Gondwana, large parts of the continent have been exposed to weathering and erosion, resulting in widespread denudation and sedimentation. Kilometres of rock and sediment have been stripped from the landscape (Partridge and Maud 1987) and

FIGURE 2.2 Geomorphic evolution of Africa since the Jurassic

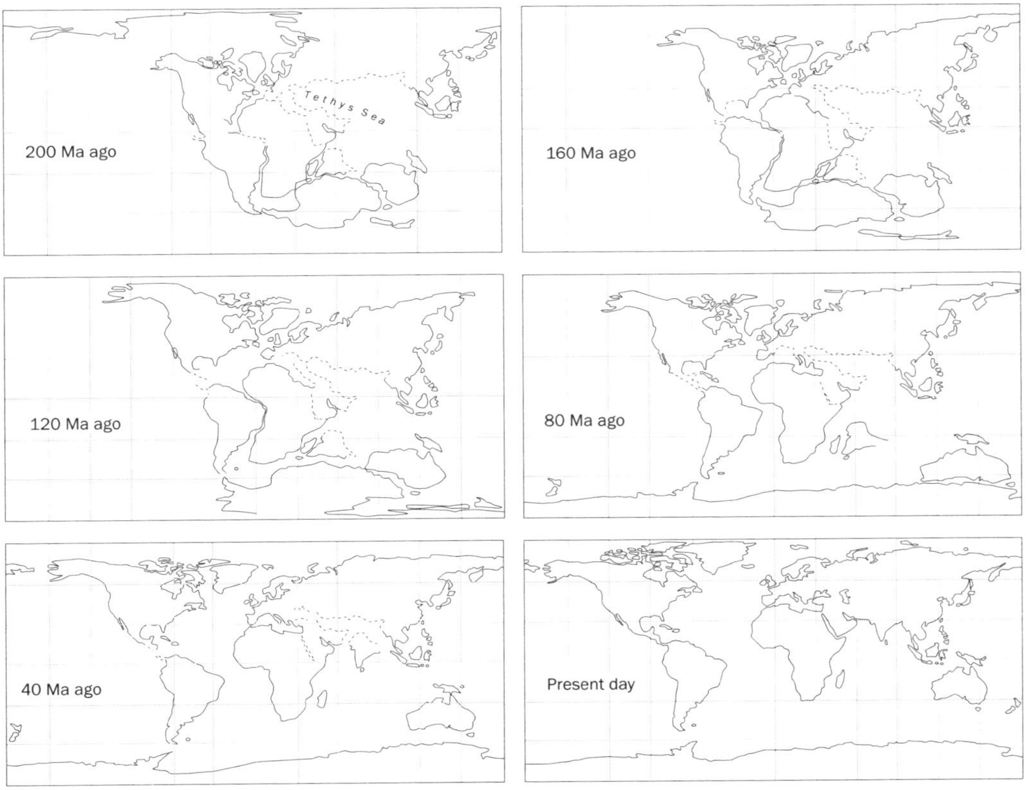

SOURCE: Summerfield 1996.

the Great Escarpment has retreated some 120 km inland from the south-east coast and some 50 km from the west coast.

The African erosion surface evolved over nearly 100 million years after the rifting of Gondwana, during the geological periods known as the Cretaceous and Tertiary. Planation of this surface was terminated by uplift and tilting during the middle Miocene epoch (*c.* 18 Ma BP), which resulted in significant environmental changes. Rivers were incised, erosion rates accelerated, and local climates were changed. Renewed uplift and tilting of between 600 and 900 m in the middle Pliocene epoch (*c.* 4 Ma BP) resulted in further environmental changes.

> The unit **Ma** means mega-annum, or millions of years. Thus *c.* 18 Ma BP means approximately eighteen million years before the present.

Major post-rifting geomorphic events contributed to climatic change. Although southern Africa occupied a position around fourteen degrees further south than at present, the Cretaceous climate of the region was considerably wetter and warmer than its is now, probably due to the configuration of the oceans and continents and the pattern of ocean currents. High offshore sedimentation rates suggest high hillslope runoff and river flows, and high levels of clay in the sedimentary record indicate a warm, wet, topical climate (Partridge and Maud 1987). Evidence suggests that the east–west climatic gradient that presently exists in South Africa may have been in operation then (Partridge 1997). The climatic and environmental effects of late-Tertiary uplift were considerable. Partridge (1997) argues that the combined effect of the middle-Pliocene uplift and late-Pliocene cooling in East and southern Africa had a profound effect on the emergence of early hominids in Africa. These rapidly changing environments may help explain why our ancestors emerged in Africa and not elsewhere.

The rifting of Gondwana and the Miocene and Pliocene tectonic movement along clearly defined axes have defined the macro-level functioning of southern African rivers. Major sedimentation peaks occurred in the Miocene and Pliocene epochs. Imprinted on these tectonic phases have been the impact of climatic change, with associated periods of wetness and dryness, and concomitant changes in vegetation, runoff, erosion, and weathering rates (Dollar 1998). A good example of long-term fluvial change is that of the Orange River. Dingle and Hendey (1984) used the offshore sedimentary record to show that the Orange River, which presently discharges at Oranjemund (28° S), has an episodic mouth. The Orange has on earlier occasions discharged some 500 km further south via the present Krom/Olifants system through the now submerged Cape Canyon during periods of low sea level (see Figure 2.3).

Climatic change

Southern Africa has experienced the impact of at least two great ice sheets. The most recent ice covering, during the Permo-Carboniferous glaciation, lasted over 100 million years, ending roughly 200 million years ago. Evidence of this glaciation can still be found in the glacial

FIGURE 2.3 Evolution of the Orange River system

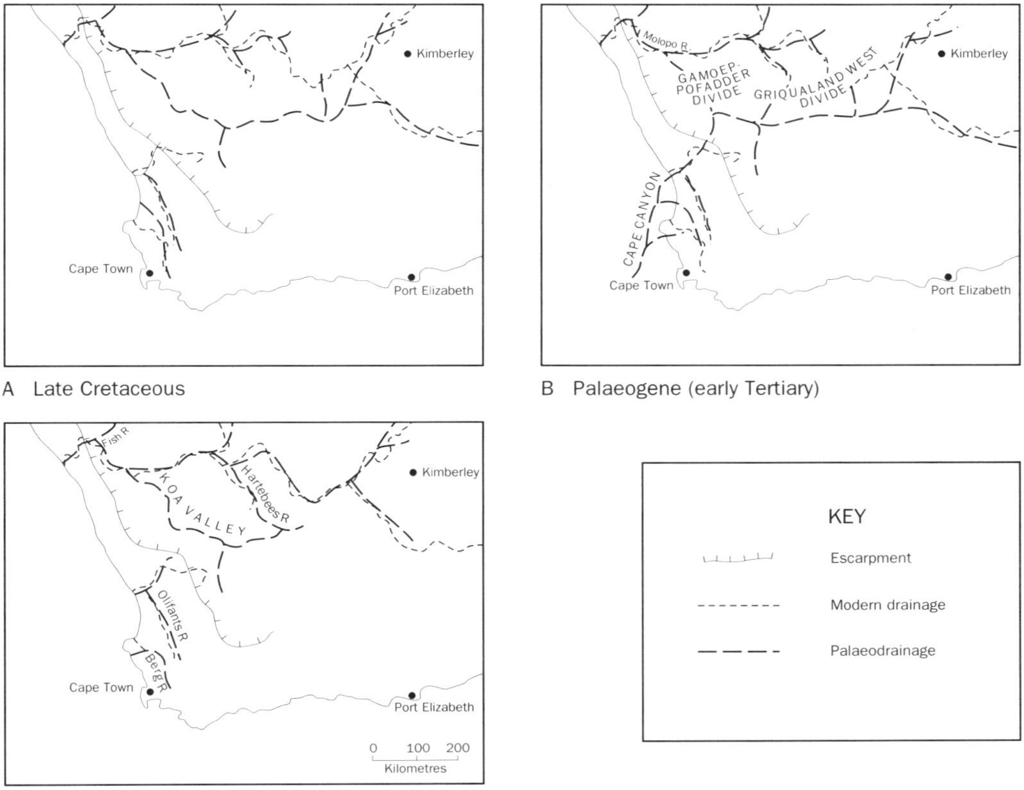

A Late Cretaceous

B Palaeogene (early Tertiary)

C Neogene (late Tertiary)

SOURCE: adapted from Dingle and Hendey 1984.

pavings near Barkly West in the Northern Cape. The Jurassic rifting of Gondwana resulted in the opening up of the proto-Atlantic and proto-Indian Oceans. In Chapter 11 we shall see how the climate changed from glacial to desert as the continent moved from the South Pole to a more tropical position. Since the Cretaceous period the African continent has moved no more than fourteen degrees of latitude. The stability of the continent indicates that plate migration has not been a significant factor affecting environmental change since then, and the major currents washing the southern African coastline, the Agulhas and Benguela, were probably in their present position by the end of the Miocene epoch.

The late-Pliocene global warming was terminated by the growth of the east Antarctic ice sheet and the onset of ice formation in the North Atlantic around 2.4 Ma BP (Partridge 1990). Quaternary climate changes consisted of alternating glacial and interglacial periods, with a quasi-periodicity of approximately 100 000 years. (A quasi-periodicity is an oscillation in a time series of a climatic variable that is found to recur at approximately equal time periods.) Although no true glaciation occurred in southern Africa during the Quaternary, the glacial episodes of the northern hemisphere were reflected in depressed temperatures in southern Africa. Prell et al. (1979) have shown that the region has experienced at least six major cold (glacial) phases and seven major warmer (interglacial) phases during the Quaternary (see Figure 2.4). The two coldest phases occurred around 420 000 BP and during the Last Glacial Maximum around 18 000 BP. The warmest period was probably around 125 000 BP (Tyson 1990). Talma and Vogel (1988) argue that the glacial periods probably represent an air temperature depression of around 3–6°C, while a sea-surface temperature depression of around 5°C probably occurred.

The most reliable palaeoclimatic reconstructions have been achieved for the last 30 000 years. This is primarily as a result of the preservation potential of palaeoclimatic indicators and the reliability of radiocarbon dating. Partridge et al. (1990), utilizing a variety of palaeoenvironmental data including fossil pollen records, environmental isotopes, charcoal, micromammalian remains, and sedimentary and geomorphic evidence, have shown that climatic change has not been coeval (occurring at the same time) in southern Africa. They divide southern Africa into five palaeoclimatic regions: the Southern and Western Cape, Karoo, Namib, Kalahari, and Eastern regions (see Figure 2.5). The major climatic changes experienced in these regions over the last 40 000 years are summarized below.

FIGURE 2.4 Oxygen isotope record for the deep-sea core RC17-69 off KwaZulu-Natal

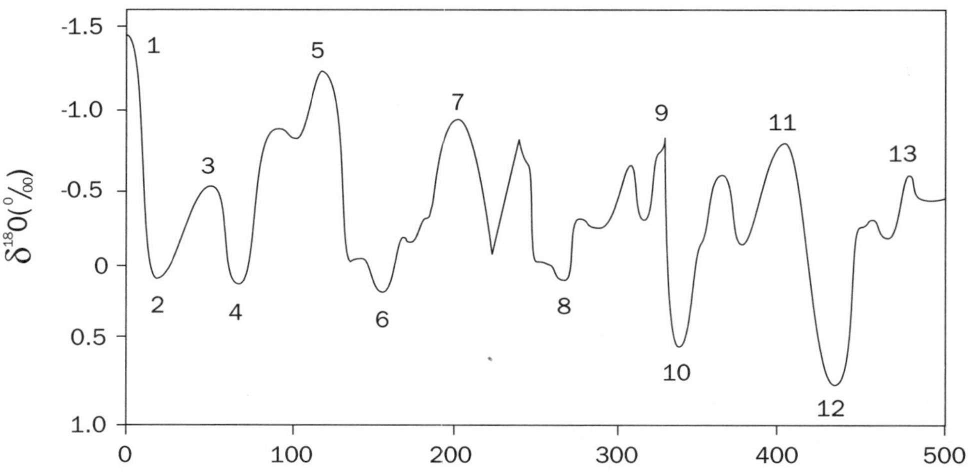

SOURCE: Prell et al. 1979.

FIGURE 2.5 Palaeoclimate regions of southern Africa

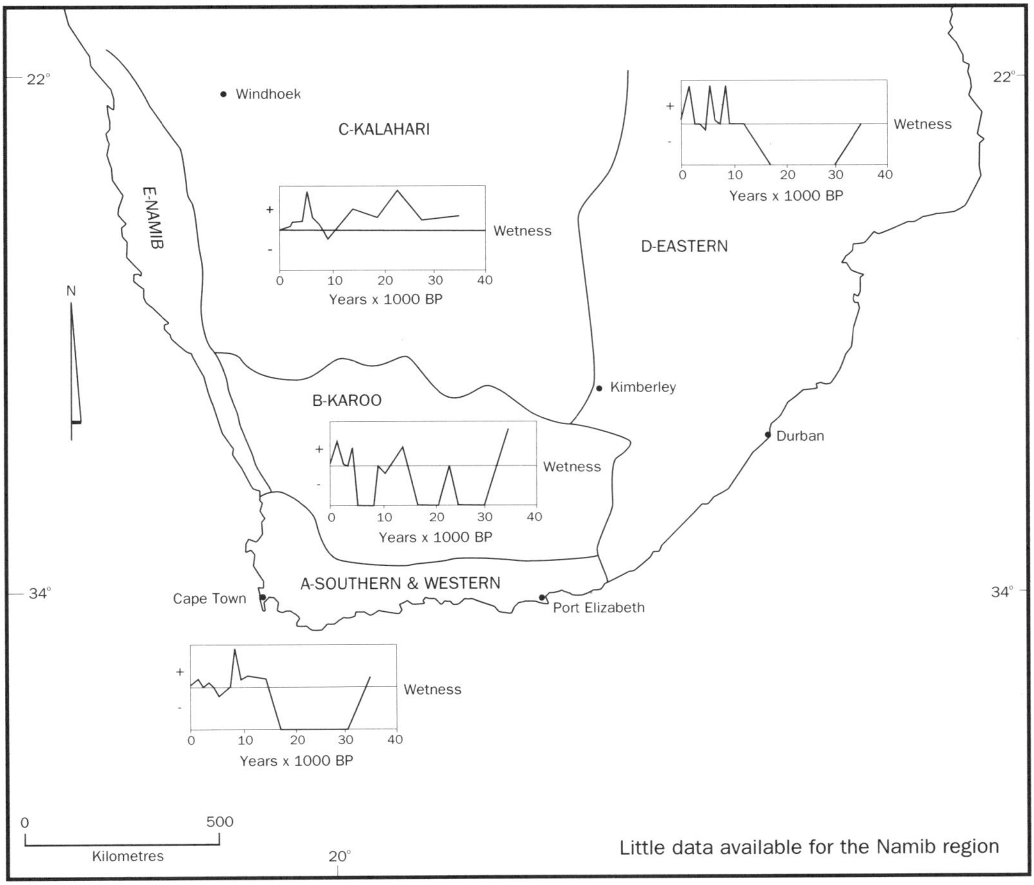

SOURCE: adapted from Partridge et al. 1990 and Dollar 1998.

> The prefix **palaeo-** means ancient, of prehistoric times. Thus 'palaeoclimatic' refers to ancient climates, and 'palaeoenvironmental' to ancient environments.

40 000–30 000 BP

Temperatures were generally lower during this period. With the exception of the Eastern region, biological and geomorphic evidence points to wetter conditions than at present. Oxygen isotopic data from the Cango Caves suggests that the temperature was some 3–4°C cooler around 30 000 years BP.

30 000–25 000 BP

Lower temperatures were experienced in all regions, probably 4–5°C lower than at present. Conditions were wetter in the Namib and Kalahari regions, and drier in the Southern and Western Cape, Karoo, and Eastern regions.

25 000–21 000 BP

Generally cooler conditions were experienced. The Karoo region was probably slightly wetter, as was the Kalahari. Drier conditions were experienced in the Namib region.

21 000–17 000 BP

This period, which includes the Last Glacial Maximum (*c.* 18 000 BP), was characterized by widespread dry and cool conditions. Temperatures may have been 5–6°C cooler than at present, while the sea level was probably some 130 m below the present level.

17 000–12 000 BP

This period was one of temperature amelioration. Widespread warmer and wetter conditions prevailed.

12 000–10 000 BP

This period is known as the terminal Pleistocene. Generally warmer and wetter conditions existed. 10 000 BP marks the boundary between the Pleistocene and the present interglacial, the Holocene.

10 000 BP–present

It appears that the post-glacial warming continued into the early Holocene, although there is evidence for a brief cooling period between 9 000 and 8 000 BP. The period from 8 000 to 5 000 BP was significantly warmer and wetter than present conditions in the Southern and Western Cape and the Eastern region. This period, termed the mid-Holocene hyperthermal, was characterized by a high sea stand of approximately +2 m above present mean sea level and a rise in temperature of around 2°C. Following the Holocene hypothermal, several fluctuations of relatively small amplitudes were experienced both in temperature and wetness. Tyson and Lindesay (1992) have reviewed the evidence for late-Holocene climatic change.

Two significant events during the last 2 000 years were the Medieval Warm Epoch (MWE) between *c.* AD 900 and 1300 and the Little Ice Age (LIA) between *c.* AD 1500 and 1850. There is widespread evidence for the LIA, while the MWE appears to have been less extensive. The period from 1850 to the present has been one of recovery and amelioration onto which human-induced climatic forcing has been superimposed. Although climate change has occurred, there is evidence to suggest that there have been regional variations in climatic response. For example, while the interior of the subcontinent experienced a cooler and more arid LGM, the Western Cape experienced a cooler and wetter climate.

Short-term changes in the recent past have been the subject of much speculation. Although there is evidence from many North African countries that rainfall is declining, this does not appear to be the case in southern Africa. What is clear is the eighteen-year quasi-periodicity of rainfall in the summer rainfall region of the subcontinent (Tyson 1986). The documentary record generally supports the physical evidence of numerous periods of wetter and drier conditions, but it would seem that the temporal characteristics of rainfall in southern Africa have remained essentially the same for the past two centuries (Nicholson 1996). Nicholson (1996) has argued that the major environmental changes that have occurred in the last few decades in Africa can be shown to have historical counterparts, thus casting doubt on the extent of presumed human-induced desertification.

Geomorphology and climatic change

Climatic changes during the Pleistocene were accompanied by worldwide changes in sea level. The Last Glacial Maximum of 18 000 BP was characterized by low sea-level stands up to +130 m below present mean sea level. Rivers were therefore rejuvenated in their lower courses and were incised into their valley floors. Post-glacial climatic amelioration resulted in rising sea levels. The high sea-level stand of +2 m above the present mean sea level some 5 000 years ago resulted in the drowning of many incised rivers and deposition of up to 40 m of sediment on their valley floors. Relative sea-level changes have undoubtedly had a major impact on the ecology and distribution of animals and plant life in coastal areas.

All African deserts have been affected by Quaternary climatic fluctuations. During the Pleistocene, repeated expansion and contraction of desert margins occurred (Goudie 1996). In southern Africa the Kalahari sandveld, a fossil desert which has now been covered with dense vegetation, indicates earlier, more arid phases. Extensive palaeodunes in the Kalahari occur in areas with a present mean annual rainfall of up to 1 000 mm. Under present-day conditions, these dunes exist only where sand supply conditions permit and where mean annual rainfall is less than 200 mm. The dunes thus provide unequivocal indications of earlier aridity. The expansion and contraction of coastal dunefields provide further evidence of environmental change, as does the existence of fossil vegetated dunefields tens of kilometres inland from the present coastline. Many coastal dunefields, such as the Alexandria dunefield east of Port Elizabeth, are thought to be recent. They started to form around 6 500 BP, when the sea level returned to approximately its present position following the LGM (Illenberger and Verhagen 1990).

Climatic change has had a considerable impact on the rates of weathering and the distribution of soils. Widespread evidence for colluviation (the accumulation of weathered debris on a slope) exists in central and southern Africa. This colluvial material, which covers up to 20 per cent of the land surface, is thought to have been laid down over a period starting approximately 30 000 years ago. Pollen analysis indicates that the ground cover was reduced and that dry vegetation typical of the present-day Karoo predominated (Botha et al. 1992). The high sodium content of colluvium makes it highly erodible, resulting in deeply incised dongas (gulleys) (see Plate 2, colour section). Many of these dongas are ancient landscape features, related to a series of cut-and-fill events due to climatic change. The present extensive geographical distribution of these dongas, however, suggests an interplay between climatic forces and human-induced vegetation disturbance.

A further effect of climatic change is its impact on the magnitude and frequency of flooding in South African rivers (Zawada et al. 1996). While the largest gauged flood on the Orange River was recorded as 8 330 cubic metres per second ($m^3 s^{-1}$) at Vioolsdrift in 1974, Zawada et al. (1996) have used the techniques of palaeoflood hydrology (the study of ancient floods) to show that the Orange River has experienced at least thirteen floods with discharges in the range of 10 200 $m^3 s^{-1}$ to 14 660 $m^3 s^{-1}$ in the last 500 years. Similarly, a palaeoflood of 28 000 $m^3 s^{-1}$ has been identified for the Mfolozi River, compared to the highest recorded flood of 6 000 $m^3 s^{-1}$ in 1984.

The effect of climate change on lake levels has also been shown conclusively. Shaw (1988) argues that the Okavango Delta is a recent feature, as the middle Kalahari shows evidence of three major palaeolake basins linking the Okavango to the Chobe and Zambezi Rivers. He identifies two stages of lake development. The first was tectonic in origin and lasted to approximately 35 000 BP (Lake Palaeo-Makgadikgadi, covering 60 000 km^2). In the second stage an episodic lake (Lake Thamalakane) formed and dried in response to oscillating moisture conditions. It reached prominence between 17 000 and 12 000 BP. Progressive desiccation during the last 2 000 years saw the disappearance of these lakes. Shaw argues that Lake Palaeo-Makgadikgdadi at its maximum elevation (945 m) could only have been sustained by a large inflow, probably the Zambezi. Shaw and Thomas (1988) have reported on a large late-Quaternary palaeolake, Lake Caprivi, to the west of the Okavango depression. They argue that this palaeolake covered an area of 20 000 km^2 and was linked to the Zambezi via the Chobe River. Lake Caprivi was coeval with and an integral part of the 936 m Lake Thamalakane stage, which existed during the last glacial and again in the late Holocene. Changes appear to be linked to tectonic processes as well as climate change.

Vegetation change

The distribution of vegetation over the African continent can be explained firstly by macro-scale geological processes related to plate tectonics, and secondly by climatic changes, which are themselves in part a result of plate movement. Axelrod and Raven (1978) argue that Africa's historical biogeography from the Cretaceous period onwards can be explained by the separation of Africa from South America, Antarctica, and Australia, and subsequent topographic and climatic changes. Late-Cretaceous vegetation was related to Africa's latitudinal position. At that time Africa was still south of its present position and therefore only North Africa, which was located near the equator, experienced tropical conditions. By the mid Tertiary, Africa had almost reached its present position, and the tropical rainforests had therefore migrated to the present equatorial regions (*c.* 15° north and south of the equator). Post-Cretaceous drying resulted in the reduction of the area of rainforests, with savannas becoming more widespread. This was parallelled by an increase in the diversity of animals (Meadows 1996).

At the continental scale there is a clear relationship between the distribution of

Africa's biota and climate (Meadows 1996). The development of fynbos, for example, a unique floral kingdom in the south-western Cape, has been linked to the initiation of the winter rainfall region during the Tertiary. Given this strong link, any change in climate, past or future, will have a profound impact on the distribution of Africa's biota. In southern Africa, terrestrial ecosystems are sensitive to climatic change. This is because processes such as primary production are strongly related to precipitation, while vegetation structure can be linked to temperature and rainfall seasonality. Where rainfall and temperature gradients are gradual, as in the interior of South Africa, small changes in rainfall or precipitation can lead to large shifts in the geographical distribution of vegetation biomes.

Much of the evidence for vegetation change in southern Africa comes from the science of palynology, the study of fossil pollen. The pollen record in the region is limited, mainly because the relatively low acidity of soils precludes its long preservation. The longest pollen record comes from an organic spring deposit at Wonderkrater on the highveld (Scott 1982); it consists of a sequence of twelve radiocarbon dates extending back 35 000 years. Another long record was cored at Zoutpan, near Pretoria (Scott 1988). These data show that considerable vegetation and environmental changes have occurred in the last 30 000 years. Evidence from Boomplaas Cave in the Cango valley, southern Cape (Deacon and Lancaster 1988), has shown that species diversity decreased with the onset of the LGM, while the post-glacial climatic amelioration resulted in substantial increases in species diversity. While the LGM was dominated by *Asteraceae* (plants of the daisy family), the Holocene assemblage has been dominated by trees. Evidence from the Cango valley shows that around 18 000 BP the vegetation was semi-karroid shrubland, but with the post-glacial amelioration the vegetation changed to parkland, with *Oleas* (wild olives) prominent. By 10 000 BP, the *Oleas* had changed to an *Acacia Karoo* riparian woodland.

The **riparian zone** is the landscape zone which is in immediate contact with the river channel and whose physical features and plant and animal communities are determined by the flow of the river.

The post-glacial recovery was widely reflected by a wetter period, as evidenced by sites in the Eastern and Northern Cape, the Free State, and the highveld. The Holocene interglacial has generally been warmer. The pollen evidence for the last 6 500 years indicates that this period started with denser vegetation under moister, warmer conditions in most parts of southern Africa, but later changed to more open-type vegetation (Deacon and Lancaster 1988).

Climatic changes which have caused vegetation changes have at the same time created new habitats for fauna, as is well documented in southern Africa. Changes in climate have also had a significant impact on the distribution of human populations, particularly in relation to the availability of food resources. This was especially so prior to about 2 000 BP, when human populations were reliant on their habitat and not on agriculture. Howell (1979), using demographic studies of the !Kung San in Botswana, has shown that fertility is linked to environmental conditions. Lower birth rates during drought periods are primarily a response to a reduction in food resources.

Deacon and Lancaster summarize these vegetation changes as follows:

> climatic changes in the Late Quaternary have been the pervasive factor in determining the composition and nature of modern natural communities, and ... modern distributions were achieved only in the last 5 000–4 000 years. However, during the last 2 000 years and since the advent of European colonists, the natural communities have been considerably modified by land-use patterns. These patterns have lowered the diversity of ecosystems, and have ... produced conditions more similar to the late glacial than to the late Holocene. (Deacon and Lancaster 1988: 171)

Human-induced environmental change in southern Africa

Climatic changes

Much attention has recently been focused on global warming and modern climate changes. For Africa south of the equator, with the exception of 1941 the ten years with the highest average annual temperatures this century have occurred since 1983. Some of South Africa's severest droughts have occurred since 1982, while the most intensive three- to five-day rains ever recorded were associated with Cyclone Demoina in Kwazulu-Natal in 1984 (Schulze 1997). The heaviest snowfalls recorded for eighty years covered thousands of square kilometres in the Drakensberg in July 1996.

Global warming due to the enhanced greenhouse effect is a global problem, with increased $C0_2$ production over Western Europe, for example, causing climatic perturbations worldwide. The industrialized countries of the North have largely been blamed for the observed increases in greenhouse gases, but they alone are not to blame. Atmospheric emissions in the eastern highveld of South Africa (around Middelburg), for example, are estimated to be five to ten times higher than in the United States and Germany (Du Pisani and Partridge 1990). Acid rain is a serious problem to the east of industrialized areas in and around Gauteng. There has also been speculation as to the effect of methane production by large herds of game and domesticated cattle, and the effect of widespread veld burning on $C0_2$ releases and aerosol concentrations. The increased thinning of the ozone layer over South Africa is probably the result of worldwide increases in CFCs, but the impact is certainly felt at the local level. It is estimated that, since 1978, ozone has declined by 1 per cent per annum in the zone 30° to 40° S (Diab 1990).

South African climatologists have been active in the field of climatic modelling. All predictions point to significant climatic impacts in the region due to increased levels of $C0_2$. According to Tyson (1990), summer and winter temperatures may increase by as much as 4–6°C over the next thirty to fifty years. This is not an impossible scenario, as evidence from the southern Cape reveals temperature depressions of between 5 and 6°C during the LGM. Summer rainfall is likely to increase over the tropical parts of the southern hemisphere and decrease over subtropical regions, while winter rainfall regions and the Karoo are likely to suffer a diminution of precipitation.

These climatic and associated vegetation changes will have a significant impact on agriculture, as discussed by Du Pisani and Partridge (1990). From the agricultural perspective, the amount and seasonality of rainfall and the distribution of drought cycles are critical to agricultural production. The occurrence of the quasi-eighteen-year oscillation of rainfall has been stressed earlier; this is related to the El Niño Southern Oscillation (ENSO), which will be discussed in Chapter 12. General circulation models predict that extreme events generated by ENSO will occur more frequently in the mid-latitudes, mainly due to increased temperature contrasts between land and sea. Associated with changes in temperature and precipitation are possible increases in the magnitude and frequency of both droughts and floods. The impacts of these changes can be both negative and positive. Positive impacts include an increase in the geographical distribution of the frost-free growing season. This could benefit irrigation farmers, as it may be possible to grow two crops per annum. Negative impacts include a reduction in the potential of dryland maize production due to moisture stress, a reduction of wheat yields of up to 30 per cent due to increased temperatures, and a dramatic impact on the deciduous fruit industry in the Western Cape.

Changes are also likely to occur in the hydrology and water resources of the country (see Chapter 16). Schulze (1997) has modelled the possible impact of climate change on South Africa's water resources. South Africa is a dry country, with a mean annual rainfall of 483 mm, compared to a world average of 860 mm. Furthermore, the lithology of the country provides poor groundwater resources, which in effect means that over 80 per cent of its water resource requirements are met by rainfall and surface runoff. Of the 483 mm that falls per annum, only 9 per cent eventually contributes to streamflow, due to losses by evapotranspiration. Consequently, small changes in evapotranspiration rates, which are strongly controlled by temperature, can lead to significant changes in water availability. Higher evapotranspiration rates may be partly offset in some areas of the country by increases in the volume and intensity of precipitation associated with increased temperatures. As a result of higher temperatures, summer rains may start earlier in spring and last longer into autumn.

Increases in stormflow would result in flow volume increases of between 10 and 20 per cent in Johannesburg and 20 to 30 per cent in Cape Town. Flood magnitudes would also be increased: the 100-year flood in Cape Town may, for example, become the twenty- to thirty-year flood (Schulze 1990). Associated with increased evaporation rates would be changes in water quality. Warmer temperatures would result in the increased mobility of toxic substances such as copper, aluminium, mercury, cadmium, and lead, increases in water-borne diseases that are related to water temperature, possible increases in algal blooms, and eutrophication-related problems.

Eutrophication is the accumulation of nutrients in a water body which occurs naturally due to the breakdown of organic matter. The process is greatly accelerated by human-induced additions of nitrates and phosphates from fertilizer runoff and sewage treatment processes. The rate of organic breakdown will increase under a warmer temperature regime.

Vegetation changes

There is a need to establish the relative importance of climatic and anthropogenic factors in the process of vegetation change. Human impact on vegetation in southern African has a long history. Evidence exists for human-induced vegetation burning from around 100 000 BP and the introduction of domestic sheep into the Western Cape around 2 000 BP, followed by cattle around 1 300 BP (Avery 1992). On the east coast of southern Africa (near the present Lake St Lucia), slash-and-burn agriculture has been in existence since the third century. Evidence has also been forthcoming that the pre-colonial occupants of the Karoo, the Bushmen and the Khoikhoi, may have had a marked impact on the vegetation (Sugden 1989). It was, however, only at the beginning of the nineteenth century that permanent farm settlements were established in southern Africa. Vegetation in the region has thus been affected by both major climatic fluctuations during the Pleistocene and Holocene and agricultural activities over a considerable time period. A balanced perspective is necessary if we are to distinguish change resulting from farming and related anthropogenic activities from that resulting from natural environmental change.

These issues are encapsulated in the 'expanding Karoo' argument, an issue that has sparked considerable debate in southern Africa since the nineteenth century. This debate has been reviewed by Hoffman (1997) and is examined again in Chapter 15. Acocks (1953) produced a series of maps showing a rapidly expanding Karoo due to injudicious land-use practices and overstocking following European pastoralism (see Figure 2.6). It was suggested that the eastern Karoo had been altered from a perennial grassland to dwarf shrubland, with a concomitant reduction in productivity and a decline in carrying capacity. The idea of a 'pristine Karoo' dominated by perennial grasses has been questioned, however. Acocks's (1953) concept of a pre-colonial grassland in the eastern Karoo is not well supported by the historical evidence and, despite 100 years of scientific research, the published data are inadequate to support the theory. If anything, the data support the opposite trend (see Figure 2.7). Palmer et al. (1990), for example, have shown that the condition of vegetation has actually improved between 1950 and 1988 in the semi-arid succulent dwarf shrubland in the Eastern Cape. Hoffman (1997) suggests that it is more important to acknowledge that the dynamic seasonal response of grasses and shrubs is a vital component of any model seeking to explain long-term vegetation changes in the Karoo.

The concept of desertification, the diminution or destruction of the biological potential of the land, has come under severe criticism in the international literature. Natural cycles are often ignored when changes in vegetation are assessed,

FIGURE 2.6 Desertification of the Karoo

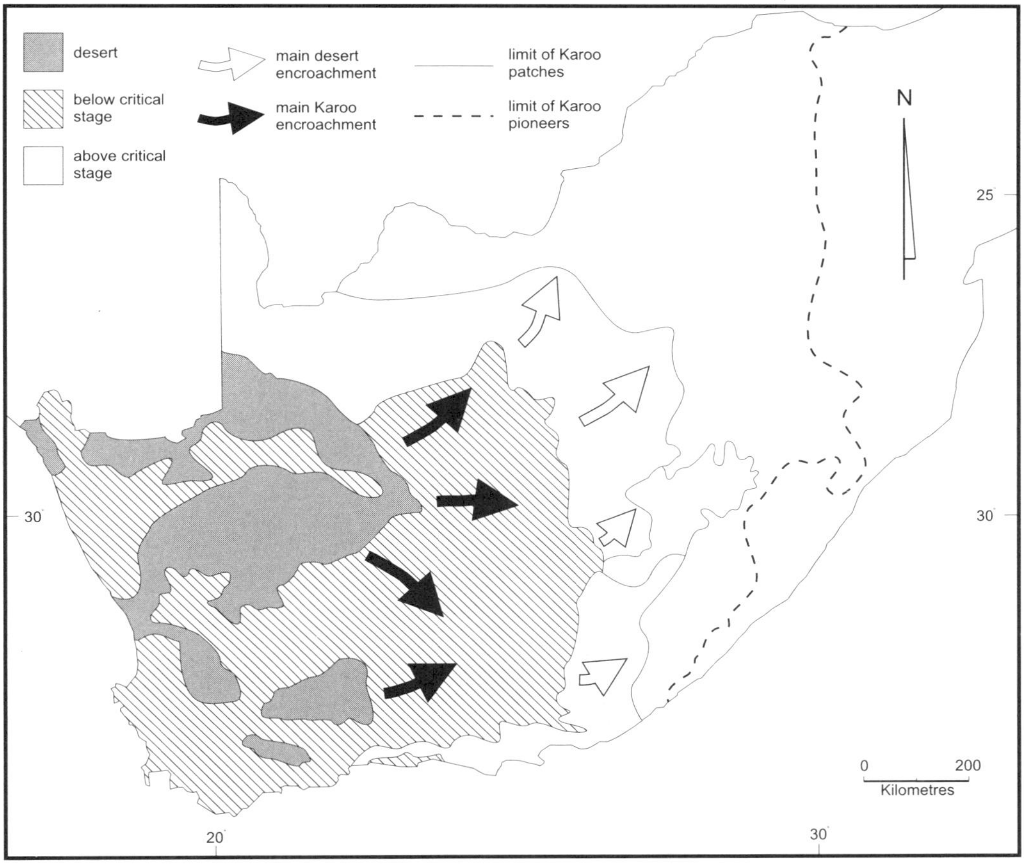

SOURCE: Acocks 1953.

giving rise to what has become known as the 'desertification myth' (Binns 1990). This myth seems to have been mirrored in the case of the Karoo.

Soil erosion and soil loss

Human-induced soil erosion has long been known to be a problem in southern Africa. General Jan Smuts once said that 'erosion is the biggest problem confronting the country, bigger than any politics' (Garland 1990). In the past, soil erosion was seen as the 'scourge of Africa', a problem created by African mismanagement of land. Recent studies, however, have refuted this assumption (Tiffen et al. 1993). While it is clear that human-induced soil erosion is a cause for concern, socio-economic and political factors are often the root cause of the problem. In South Africa, apartheid policies such as 'betterment schemes' and

FIGURE 2.7 Vegetation change in the semi-arid eastern Karoo

a b c

Vegetation change is influenced greatly by the response of grasses to grazing and to summer rainfall patterns, as these three photographs near Hofmeyr show (31° 27' S 25° 45' E). In September 1925 (a), after several years of heavy stocking and low rainfall, Pole Evans found large bare patches and little grass cover. Decades later, in August 1989 (b), Hoffman and Cowling (1990) recorded greatly increased cover of especially annual, biennial, and short-lived perennial grasses following several years of low stocking and favourable recent seasonal rains. By September 1992 (c), however, successive drought years had reduced the amount of visible grass cover to about half its 1989 estimates (see Hoffman et al. 1995)

PHOTO: Tim Hoffman.

the creation of 'homelands' have exerted pressure on the carrying capacity of the land, resulting in environmental degradation and accelerated erosion (see Chapter 9). The highly erodible nature of many South African soils (see Chapter 14) aggravates the problem. Unless it is seen in its proper context, there is a danger that the soil erosion question may fall into the same category as the desertification debate. Desertification has been well described by Stocking (1996: 326) as 'a notable example of a crisis constructed by institutions empowered to study it'. Exaggerated claims of desertification and massive soil loss based on limited data are overreactions to a very complex issue.

Betterment schemes were an apartheid policy of the 1940s applied to those areas then classified as Native reserves. The aim was to reorganize rural land-use practices and to centralize formerly scattered residences into nucleated settlements. The homelands policy is discussed in the next chapter.

The loss of soil fertility due to accelerated erosion can have serious consequences. Stocking (1996) has shown that by the mid 1980s Zimbabwe was losing up to Z\$700 million of phosphorus and Z\$900 million of nitrogen per year (R775 million and R1124 million respectively in 1985 South African currency). Commercial and arable lands were losing up to Z\$20 per hectare per annum, and communal arable lands up to Z\$80 per hectare per annum. The loss of nutrients was up to three times greater than their application by means of fertilizers. Loss of soil and soil fertility has negative social impacts: in its most extreme form it can lead to the collapse of agriculture and major migrations of people. If one adds up the costs of crop losses due to a decline in soil fertility, costs resulting from the silting of dams, where sediment accumulation takes up expensive storage space designed for water, as well as other ancillary costs, then the losses in monetary terms are enormous. There are clearly cases where soil erosion is a serious issue, though the extent of the problem may have been exaggerated. Stocking (1996) sums up the situation well, arguing that the problem of soil erosion should be seen for what it is, a problem for some people in some places.

Accelerated erosion is the human-induced loss of soil that exceeds the geological erosion occurring as part of natural long-term landscape development. Geological erosion rates, however, are by no means constant (Martin 1987). Rates for the last three million years, for example, are twice as high as those for the previous era of rapid sedimentation. Martin (1987) has compared modern sediment yields from KwaZulu-Natal rivers to palaeoyields from the Natal Valley, now in the Indian Ocean. He determines that the average modern rate of sediment yield (322.5 tonnes per square kilometre per year, or t km^{-2} $yr.^{-1}$) is twelve to twenty-two times higher than the geological average (14.7–26.9 t km^{-2} $yr.^{-1}$). It is tempting to explain this higher modern yield as anthropogenically induced, but Davies et al. (1977) have pointed out that sediment yields during interglacial periods have been as much as fifteen times higher than those for glacial periods. A number of authors have argued that present rates of erosion in South Africa are some two to three times the rate of soil replenishment through weathering. The question is: to what extent are the large increases in modern fluvial sediment yields anthropogenically forced, or do high modern yields form part of a natural fluctuation?

Lakes and fluvial systems

Human-induced impacts on lakes and fluvial systems have become significant forces for change. The construction of large dams, such as the Aswan on the Nile, Kariba and Cahora Bassa on the Zambezi, and the Gariep on the Orange, have had considerable environmental impacts on the ecology as well as on the physical functioning of the rivers. Environmental changes to modern rivers are also brought about by indirect means, which include land-use changes, agriculture, mining, and urbanization. The effects of these changes on the ecology of South African rivers are discussed by Davies and Day in their book *Vanishing Waters* (1998). The hydrological and geomorphological effects are outlined in Chapter 16.

Mining as an agent of environmental change

Mining in South Africa, as in many other southern African countries, plays a key role in the economy. Apart from the loss of aesthetic value, the major environmental impacts of mining include the generation of increased volumes of organic mineral, toxic waste, and groundwater pollution from contaminated recharge (Brink et al. 1990). In large mines, the pumping of groundwater leads to a lowering of the water-table which, if excessive, may lead to subsidence and the development of sinkholes. Serious water-quality problems are associated with mining. In the mining process, minerals are brought into contact with oxygen, water, and bacteria. Sulphides, in particular pyrite, oxidize to form acidic sulphate solutions of which the low pH and high salt load have negative impacts on the environment. Furthermore, the low pH enables other heavy minerals to be taken into solution, exacerbating the ecological problem. Discharge of these solutions from the mining operation is known as acid mine drainage (AMD).

Environmental change, social issues, and development

The issues of poverty, development, and environmental change in Africa are inextricably interlinked. While environmental change in developed countries has been caused by the adverse consequences of economic growth, the causes of environmental changes in developing countries are quite different. Cock (1991: 1) argues that 'South Africa, with its mix of First World environmental problems such as acid rain, and Third World environmental problems such as soil erosion, is a microcosm of the environmental problems facing the planet'. Acid rain, global warming, deforestation, and atmospheric and water pollution are caused most often by wealthy industrialized sectors of society, not the rural poor who are most often blamed for environmental degradation.

In developing countries, environmental degradation arises from the poverty of rural people and the fact that the environment often provides the only resource to which the poor have access. Unless the question of poverty is dealt with, therefore, little can be done about environmental degradation. Much of the blame for the environmental degradation in the former homelands of South Africa can be laid squarely at the door of apartheid policies. The homelands policy aggravated conditions in rural areas by means of forced removals and the crowding of people into small isolated pieces of land unable to sustain the population. It seems clear that as long as there are such vast inequalities in wealth, standards of living, quality of life, and access to resources in southern Africa, little can be done about environmental degradation.

While it is naive to consider Africa as some form of pre-colonial utopia, it is clear that 'protected areas will not survive in Africa without substantially increased participation of local people and, furthermore, that this participation will have to be far more thorough and extensive than [it] has conventionally been in the past' (Adams 1996: 379). The only ways to stop large-scale human-induced environmental change are to make local communities part of the decision-making process at every level where they interact with their environment, and to develop the economy of southern Africa as a whole in an environmentally sustainable manner. Likewise, it is important to assess and apply appropriate indigenous techniques for the management of environmental changes such as soil erosion. Economic development needs to reconcile the short- and long-term needs of the community with the sustainable conservation of natural resources.

Conclusions

It has been shown in this chapter that environmental change is a natural process that has existed since the origin of the earth over four billion years ago. Although environmental change is a natural process, human impact has accelerated the rate and nature of that change. The issue of global environmental change has received considerable attention in the media. The 'greening' of global politics and the emergence of programmes such as the International Geosphere and Biosphere Programme (IGBP) are evidence of the proportions of the environmental change debate. Research into environmental change has been led by physical and biological scientists; it is only recently that social scientists have begun to make an input into the debate. Greater attention must be paid to the socio-economic and political causes of environmental change if these problems are to be tackled successfully. Koch (1991: 21) puts this succinctly: 'we need to solve our socio-economic and political problems before people will be receptive to environmental issues'.

One of the great challenges for physical geographers and other environmental scientists is to identify the true nature and direction of environmental changes, including the evidence for long-term land-cover modification. Understanding past environmental change is critical in interpreting and understanding present and future change.

3 Historical geography

ALAN LESTER

Until recently, historical geography was concerned mainly with describing certain geographical patterns of the past such as settlement shapes, transport networks, and land-use arrangements. In this chapter, a different approach is taken – one informed by developments both within and outside the discipline of geography. Two of these, which were introduced in Chapter 1, have been crucial over the last three decades. Karl Marx's insights have been widely used to link social and spatial changes firmly to economic exploitation. Poststructuralism, which draws on the ideas of Michel Foucault, identifies the diverse ways in which power operates at a variety of scales within society, and this approach has also contributed to the unravelling of South Africa's historical geography that follows.

Rather than simply describing former geographical patterns, this chapter will therefore establish how core features of South Africa's geography have been shaped by political, economic, and social forces since the colonial period. In this way the region's geographies can be understood as the outcome of colonization and capitalism. Particular attention will be paid to those social groups which were the main beneficiaries of these processes, since they had the most power to shape spatial patterns and thus affect the geography of both southern and South Africa. Power of various kinds has been fought over and distributed between social groups through time. Furthermore, group identity has not been fixed: groups have continually adjusted their boundaries, including some people and excluding others, so as to maintain or create privilege. Racial appearance has been the most significant, but not the only, marker of group identity since the beginning of the colonial period.

The attempt to preserve power in white hands has shaped modern patterns of segregation at a variety of scales in South Africa. In the wider region, European colonialism has left further significant legacies in the ordering of space, including the boundaries of present nation states. The significance of racial segregation does not mean, however, that all whites were equally powerful or all blacks equally powerless. Nor does it mean that the groups wielding and resisting power were always defined solely by skin colour. Class divisions within the capitalist system remained critical and did not necessarily coincide with perceived racial differences. Furthermore, political and spatial relations were never simply created by colonialism. All colonial powers were forced to make accommodations with pre-existing structures of authority and to be mindful of provoking costly resistance.

The following sections examine the characteristic features which evolved in a sequence of historical periods:

- the mercantile period from the mid seventeenth to the mid nineteenth century;
- the industrial period from the middle to the end of the nineteenth century;
- the segregationist order, which lasted from the beginning of the twentieth century to the mid 1980s; and
- the new world order, which began in the late 1960s.

The period of mercantile capitalism

When the Dutch East India Company (the Vereenigde Oost-Indische Compagnie, or VOC) built a supply station at the Cape in 1652, this was intended to be a small, self-contained European settlement servicing the company's trading ships. The VOC was the largest of the great European companies creating enormous profits through extensive trade, or mercantile capitalism, and its ships connected Europe to the East Indies, where the trade in spices and silks was concentrated. Despite the VOC's limited intentions for its station at Cape Town, over the next 150 years a colonial society was established within the much larger western Cape surrounding it. The most evident characteristic of this society was its racial stratification, allowing white colonists to dominate slaves and indigenous blacks. The consolidation and spread of racial stratification during the period of VOC rule contributed greatly to the region's later geography, and it was the outcome of three main developments:

- the expansion of a colonial farming frontier;
- the associated dispossession and domination of the Khoisan; and
- the institution of slavery.

The creation and extension of racial stratification

Within twenty years of the settlement's establishment, Dutch and French farmers, brought to the Cape in order to produce more supplies for passing ships, were spreading into the interior. Although these farmers concentrated primarily on meeting their subsistence needs, most were still tied into the VOC's commercial economy, and this necessitated their acquisition of large areas of land. Surplus produce generated on their farms was either sold to the company or traded for items such as ammunition and tobacco, which were imported to Cape Town. The demand for land, reinforced by each generation's expectation of acquiring its own extensive farms, drove the colonial frontier forward into the interior.

By the early eighteenth century, farmers who could not afford land around

Cape Town were already located about eighty kilometres away, and during that century the white farming frontier pushed a further 800 kilometres inland (Christopher 1984). By the 1770s white farmers were already impinging on longer-established Xhosa settlements to the west of the Fish River (see Figure 3.1). The colonists' occupation of land in the interior came at the expense of the indigenous Khoikhoi and Bushmen, or San (collectively known as the Khoisan). The possession of firearms and horses gave colonial farmers a military advantage which was used first to secure those points in the semi-arid interior where water was available. Their organization into mounted commandos subsequently allowed for the domination of much wider spaces. Despite bouts of resistance, Khoisan political units were generally too small and too thinly spread to prevent white settlement, and resistance in the west was further undermined by a smallpox epidemic, brought by a European ship in 1713 (Guelke 1989).

Most of the colonists pushing furthest inland, unlike those settled in the west, lacked the capital to buy slaves. Those Khoisan not killed or dispersed in struggles over land were therefore incorporated on colonial farms as labourers and herders. However, the nature of colonial domination varied according to the location and the social standing of the colonist. During the early stages of their dispossession, frontier Khoisan were often able to nego-

FIGURE 3.1 The Cape Colony, *c.* 1780

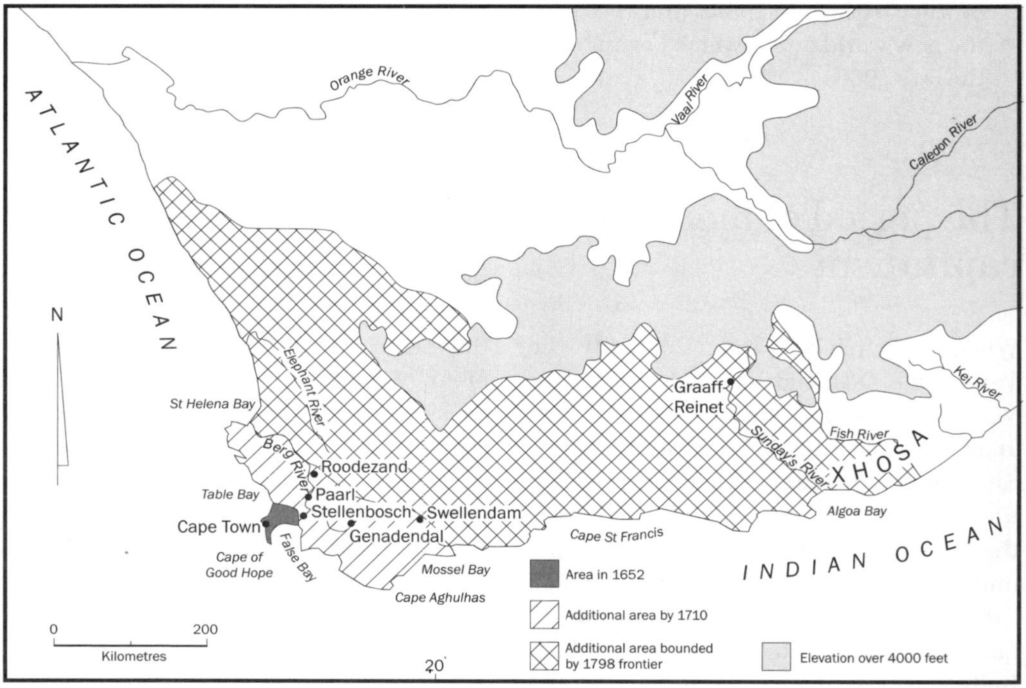

SOURCE: Thompson 1990.

tiate relatively favourable terms with their white 'employer'. Within these 'open' frontier areas, many male colonists, lacking the wealth to marry within colonial society, developed sexual relationships with Khoisan women, contributing to a large 'mixed-race' population known initially as '*bastaards*' and by the mid nineteenth century as 'coloured'.

The fairly fluid interaction between colonists and Khoisan occurred largely on the margins of the expanding colony, however. Once the balance of power in each district had shifted with the arrival of more white farmers, colonists were able to adopt more coercive methods. Khoisan labourers were retained by white farmers less by means of negotiation and more by withholding the labourer's stock, capturing his or her family, and frequent and often severe beatings. Thus a progressive shift in power relations allowed racial stratification to be extended gradually to the east and north during the eighteenth century. Thereafter, colonial domination remained harshest where resources were scarce and Khoisan resistance was most effective. In late-eighteenth-century Graaff-Reinet, for instance, farmers attempted either to exterminate raiding Bushmen or to take them captive and terrorize them into providing labour (Newton-King 1994).

If the domination of the Khoisan laid one foundation for racial stratification, the institution of slavery laid another. For fear of resistance, the VOC had forbidden the use of Khoisan as slaves, but allowed the importation of slaves from its other colonies in the East Indies to alleviate labour shortages. With a mercantile capitalist network extending across much of the world, it was possible for European slave traders from the fifteenth to the nineteenth centuries to use slaves from Asia and Africa, people whose different physical characteristics were used to justify their capture and subjection. Although richer white colonists could afford far more slaves than the poorer whites, this racial distinction between black slaves and white masters consolidated a general colonial conception that blacks were obliged to work for whites. Thus, Shell (1994: 415) writes, 'the slave period, 1652–1838, was the true gestation period of South African culture'.

We have so far examined the ways in which land was taken, and how slave and Khoisan labour was subjected to white authority within the early Cape Colony. But the increasing white control of black land and labour, which was occurring in all the European colonies, was not simply about strategies for material gain. It was also closely connected with European ideologies of race, religion, and gender, and Europeans' perceptions that they were superior to Africans.

Despite some shipwrecked sailors' positive representations of Africans, and male travellers' sexual attraction for indigenous women, from the seventeenth century the most common European reaction to encounters with Africans was a feeling of their own superiority. European technological and military advantage convinced travellers and colonists of this, but they were also persuaded by the illiteracy and the perceived lack of formal religion among sub-Saharan Africans. Impressions of the Khoisan were particularly derogatory. Sailors visiting the Cape found their lifestyles disgusting, and colonists involved in the most vicious frontier struggles for land adopted the word '*schepsel*' (sub-human) for them. Despite official tolerance of religious conversion, most colonists used their Christianity to mark a boundary between themselves and their slaves and Khoisan, thus preserving their own social superiority. By the early nineteenth century in Europe, the Dutch word for Khoikhoi – '*Hottentot*' – had come to represent the lowest form of human life (Lester 1996).

Racial domination went hand in hand with prevailing attitudes towards gender and the family. The control of black labour operated largely within each family household, rather than being imposed 'from above' by the colonial government. Isolated slaves and Khoisan labourers, like female and child members of the colonial family, were usually subject to the authority of the male head of household – a system known as patriarchy. As Scully (1994: 208) puts it, 'Boer masters represented all members of the farm as children ruled over by the senior male'. The colonial farmer, then, did not turn to the VOC government to enforce racial and gender stratification. Unless he pushed things too far, his authority over household women and black labourers was considered to be 'natural' within colonial society. Such thinking was just as much a part of the cement holding a racially stratified society together as was the economic domination of black labour.

Settler capitalism, colonial segregation, and expansion onto the highveld

In 1806 the Cape Colony was captured and subsequently administered on a permanent basis by the British government. Although they too began with limited intentions, by the late 1820s British officials and settlers were introducing profound changes to the Dutch colonial system. They brought with them a new kind of capitalism, regulated far more actively by the government. Their new system produced distinctive geographies of racial segregation in the eastern Cape and Natal and also helped to stimulate the extension of a different kind of colonial domination deep into the interior of southern Africa.

British reforms in the Cape were directed at replacing forced labour with waged labour, paid according to contracts. This was the practice which lay behind Britain's own rapid industrialization, and it was supported on both economic and humanitarian grounds. The restructuring of labour relations in the Cape first became apparent when further slave 'imports' were banned in 1807. Thereafter, slavery itself was abolished in 1834 and the Khoisan were accorded equal legal rights to the colonists. Despite attempts to soften the impact of these developments, for instance by allowing slaves to be kept on as 'apprentices' for another four years after emancipation, the reforms alienated many colonists. The official challenge to colonial farmers' control of labour was combined with other grievances in the frontier districts. These included growing landlessness as the Xhosa blocked further colonial advance to the east, and favouritism towards British settlers in the allocation of the land that was available. Led by men who had commercial interests and who saw the potential to grab land in the interior, some fifteen thousand Dutch-speaking colonists harbouring these grievances left the eastern frontier zone between 1834 and 1838, taking their black labourers with them. It was this emigration, along well-charted routes, which subsequently became known as the 'Great Trek' (Keegan 1996).

Ironically, as the trekkers were leaving the colony to extend white settlement and labour domination into the interior, the government reforms which had partially alienated them were succumbing to a more aggressive and racist British influence. This was settler capitalism. Not only did British settlers in liaison with officials dispossess the Xhosa of land in the eastern Cape, but they were also responsible for the first segregated spatial frameworks in the region (Lester 1997). Some four thousand British settlers were located in the Zuurveld on the eastern Cape frontier in 1820 (see Figure 3.2). They had been requested by the Cape authorities in order to secure the frontier between the colony and the Xhosa and to begin the Anglicization of the colony. Many of the settlers were poor when they arrived, but by the 1830s they saw the prospect of rapid economic gain through wool production. In order to acquire sheep, they could draw upon capital generated locally from trade with the Xhosa and money channelled from western Cape merchants and bankers. Although the capital for expanded wool production was available, however, land and labour were lacking. To many settlers the conquest and dispossession of the Xhosa seemed the natural solution.

Economic demands for the Xhosas' dispossession were complemented by the settlers' racial ideology, especially after 1834. At the end of that year, in response to land losses and colonial harassment, frontier Xhosa chiefdoms had launched a full-scale attack on adjacent colonial settlements. Genuinely shocked by the Xhosa assault as well as motivated by their desire for expansion into Xhosa territory, British settlers now set about portraying the Xhosa in the worst possible light. The settlers had arrived in the region with a negative image of local Africans (see Colour Plate 3, a well-known British cartoonist's view of the settlers' likely fate), and after their arrival many had suffered from Xhosa cattle raids. But in the wake of the war, leading settlers set about representing the Xhosa far more systematically as inherently savage and incapable of 'improvement'. The first local 'scientific' attempts to prove Xhosa inferiority through phrenology (the study of skulls) developed in this context (Lester 1998).

The Cape colonial government attempted to provide settlers with Xhosa land immediately after the war, but this plan failed, partly because humanitarians persuaded the British government not to believe settler representations of the Xhosa. However, the restraining influence of the British government was to be short-lived. By the end of the 1840s, humanitarians in Britain had lost much of their influence, and in the 1860s, after revolts in other colonies, most British politicians were coming to favour more aggressive and coercive colonial policies. When the Xhosa fought and lost another frontier war in 1846–47, there were no influential British voices raised against the annexation of their land and the establishment of a segregationist pattern upon it.

Within 'British Kaffraria' a model for later spatial arrangements was developed which, with various regulatory devices, became the forerunner of a process later adopted widely throughout the British

FIGURE 3.2 The Eastern Cape frontier

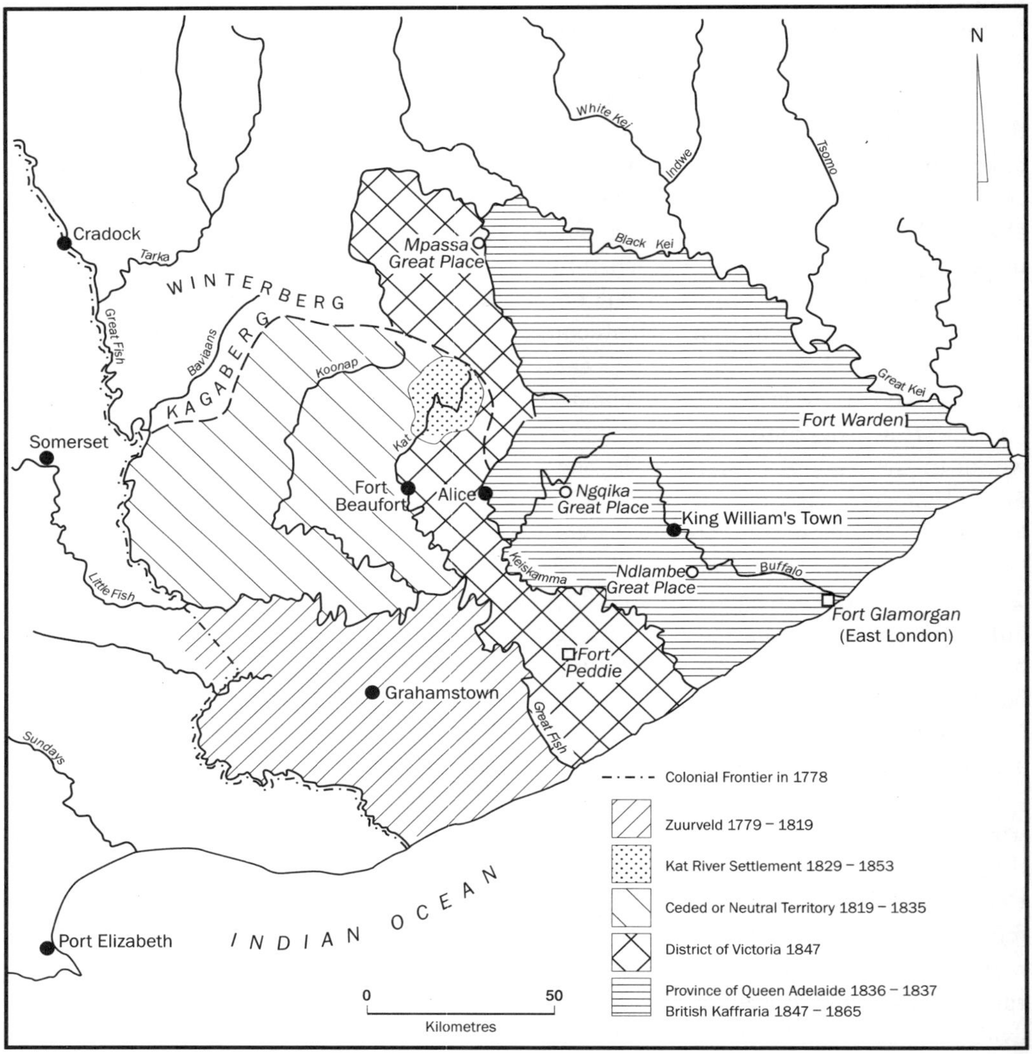

SOURCE: Lester 1996.

colonies of southern Africa. By allocating Xhosa land to white sheep farmers, the colonial government created a patchwork of white-owned farms and black 'locations'. The latter existed because the government could not afford to provoke the resistance which would ensue from total dispossession. Nevertheless, the locations served as useful sources of labour for the white farmers. A pass system was implemented to regulate the movement of these workers from their locations to white farms and back, and taxes were levied to ensure that Xhosa homesteads had to provide labour. Other black peasants remained on land outside the locations, but they were now considered to 'belong' to a white farmer and so had to pay 'rent' if they were not to be evicted. Further Xhosa and Khoisan resistance in 1850–52 and the frontier Xhosas' last desperate resort to spiritual assistance during the cattle-killing episode of 1856–57 could not reverse this spatial order. Indeed, the colonial government seized upon the weakness of the Xhosa in the wake of the cattle-killing to strengthen its own position (Lester 1997).

A similar framework was being consolidated in Natal. British emigration to Natal and large-scale land speculation had followed swiftly after its annexation in 1842. Overseen by Theophilus Shepstone, Natal's system was built on compromise with African authorities, as in the eastern Cape. African locations were recognized where chiefdoms remained intact, while other Zulu-speakers continued to cultivate government or 'white-owned' land. Taxation of Africans financed the colonial government and stimulated labour flows to white farmers, while African headmen and chiefs were allowed to administer their own hazily defined 'Native law' within the locations (Etherington 1989). Natal's system did not satisfy all white employers, though. The difficulty of prising African workers from the land led coastal sugar producers to 'import' indentured Indian labour from the 1860s onwards, establishing the basis for a large Asian population.[1]

White domination created a social layer which was placed on top of African structures in both the eastern Cape and Natal. The spatial result in each case was a patchwork of white-owned farms and African locations, with African peasants living on both. While neither the Cape nor Natal constitutions officially discriminated on the grounds of colour, in the Cape the property qualification required to vote was met by only a few Xhosa-speakers. They tended to act in liaison with white liberals and their presence allowed for colonial claims that the Africans were being 'civilized' – showing that Africans could aspire to reach the top social layer. In Natal, very few Africans ever qualified to receive colonial status. Instead, they were considered to be subject to 'Native law'. In both territories, struggles were fought to prise African men from the locations to provide labour, imposing extra strains on the female agricultural and domestic workforce, which was already hard-pressed by African patriarchy.

As these systems of spatial and legal segregation were being constructed in the British colonies, the Dutch-speaking (Afrikaner) colonists who had left the Cape during the late 1830s were working out their own compromises with African authorities on the highveld. Most of the emigrant Afrikaners wished to reinstitute a patriarchal form of labour domination within their own republics. In order to do so, the colonists had not only to defeat the Ndebele and Zulu armies resisting their occupation of land, but also to establish relationships of trade and tribute, and to secure military alliances, with other powerful chiefdoms. The main trekker republics – the Transvaal (the South African Republic) and the Orange Free State – were built upon both captive labour and negotiated labour tenancy and sharecropping.[2]

By the end of the 1860s, then, the region had a complex political geography. In the two Afrikaner republics the white landowners practised semi-subsistence farming. Surrounding African polities retained their political independence and forced compromises on those farmers, but their social structures were significantly affected by white missionaries and traders (Lester 1996). Within the eastern Cape and Natal, British colonial governments were regulating official systems of spatial segregation and labour procurement, also partly the result of compromise with African authorities. While these systems were being developed, in the western Cape, too, former slave, Khoisan, and mixed-race labour, coming to be known collectively as 'coloured', were brought more firmly under government regulation by the Masters and Servants laws.

The industrial period

The discovery and exploitation of diamonds in Griqualand West from 1868 and gold in the Transvaal from 1886 transformed the complex political geography of the region. Furthermore, together with a European scramble for territory, the 'mineral revolution' established the basis of southern Africa's modern economic, social, and urban geography.

The mineral revolution, confederation, and conquest

During the 1870s, the amassing of mining capital to extract the diamonds at Kimberley was responsible for the first extensive rail network in southern Africa, and this began the transformation of the region's economic geography. Kimberley was soon incorporated into the Cape Colony and linked by rail to Cape Town, Port Elizabeth, and East London using capital secured in London. Flows of capital and supplies further connected the diamond mines to merchants in the Cape and to black and white farmers in the eastern Cape, Natal, and the Afrikaner republics.

Kimberley's great mining companies, most notably Cecil Rhodes's De Beers, became so powerful that they effectively controlled the region's industrial economy. Their concern for a wide network of cheap labour recruitment overlapped during the 1870s with the British government's own plans for the region's political geography. These plans centred on the idea of a British-led confederation of territories. British officials reasoned that confederation of the Afrikaner republics and the two British colonies would facilitate not only more effective labour recruitment, but also a better transport system to service the mines, more co-ordinated financial planning, and the more effective crushing of black resistance. In line with this policy, the bankrupt Transvaal Republic was annexed in 1877 after the Southern Tswana around the diamond diggings had been confined to 'locations' and 'coloured' Griqua resistance had been crushed. In rapid succession, other African authorities in the region were conquered by British troops, bringing them under the control of a confederated state and providing its industrial economy with a cheap labour force: the Gcaleka Xhosa in 1878, and the Zulu and the Pedi in 1879. In 1881 Pondoland was taken under British authority and in 1884, after Cape forces had tried unsuccessfully to disarm them, the Sotho were given relative autonomy with British overlordship in Basutoland (modern-day Lesotho) – the first British High Commission territory in the subcontinent (Marks 1985a). Figure 3.3 shows the location of many of these late-nineteenth-century additions to the British Empire.

Confederation was dealt a severe blow by the regaining of Afrikaner independence in the Transvaal in 1881, but British imperial intervention in the region was aroused again by the gold discoveries of 1886. In a second major round of investment, the gold seams attracted even more international capital than had the diamond diggings. Johannesburg grew at a phenomenal rate as hundreds of mining companies established themselves on the richest gold deposits in the world. Part of their enormous investment went towards developing a further layer of transport infrastructure linking Johannesburg to the major ports of Durban, Port Elizabeth, and Cape Town. A process of amalgama-

FIGURE 3.3 Colonies and republics of Southern Africa, late nineteenth century

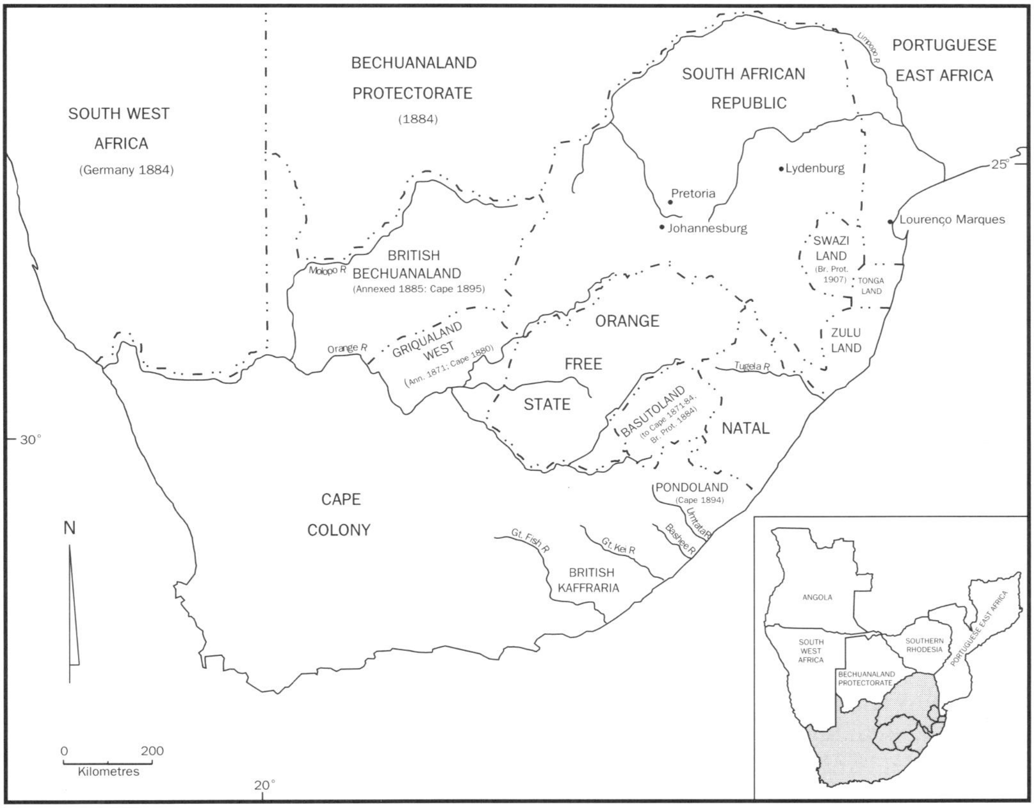

SOURCE: Oliver and Sanderson 1985.

FIGURE 3.4 The changing space-economy of South Africa

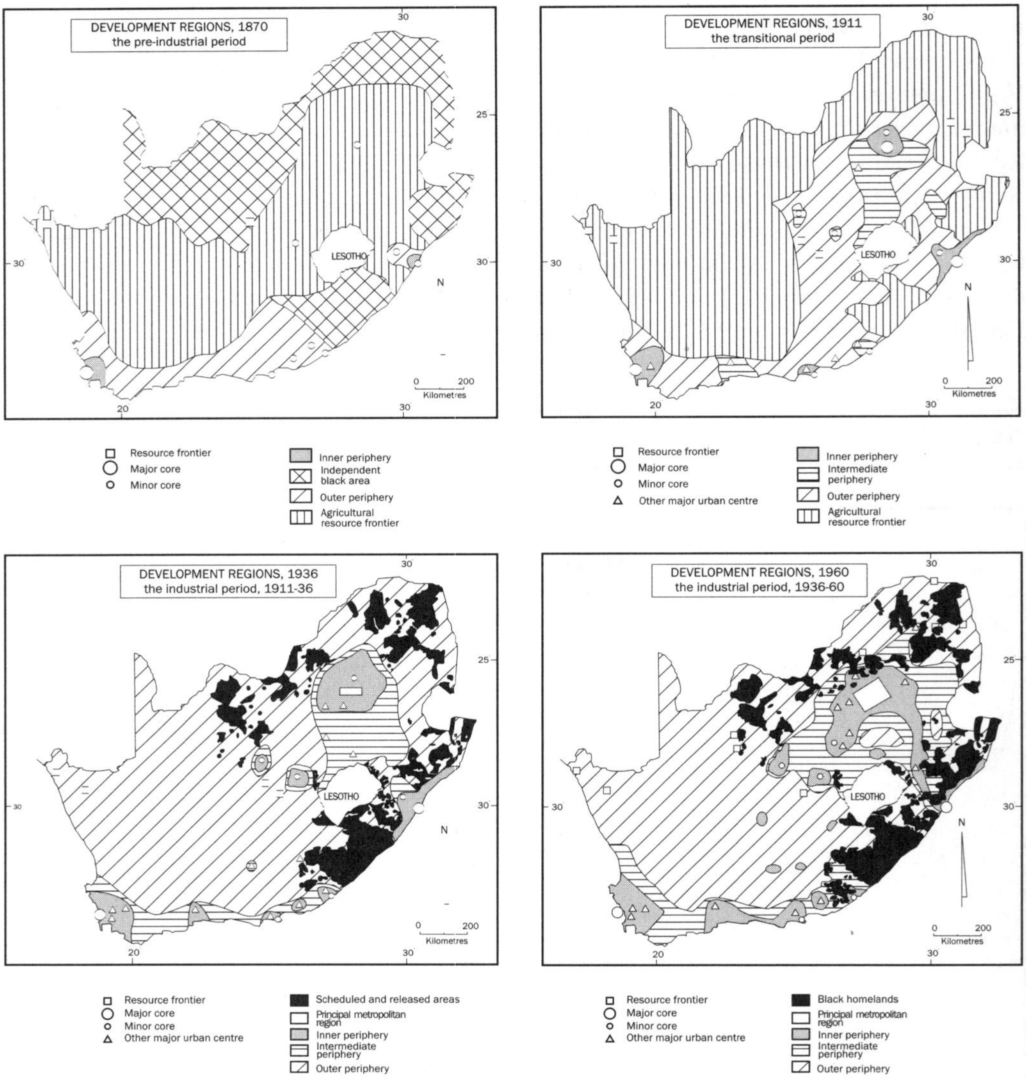

SOURCE: Lester 1996.

tions reduced the number of mining companies in Johannesburg, and the large consolidated firms grew to dominate the region's economy. Once an economic backwater in South African, let alone global, terms, the region was now a major feature in the international economy. Figure 3.4 reveals just how marked was the transformation of the South African space-economy between 1870 and 1911. The minor cores such as Port Elizabeth and East London had also expanded as a result of handling greater volumes of imports and exports.

The South African War and a new political geography

By the mid 1890s, a complex of key capitalist concerns had been established, based in minerals, but extending into all other sectors of the economy. Their concentration in Johannesburg, however, had created a contradiction which caused their owners concern. In the Cape Colony, British officials had been able to comply with many of the demands of the large diamond mine owners of Kimberley. But Johannesburg, in the adjacent Transvaal, was located within a different political system. The republic's Afrikaner government was primarily concerned with supporting a farming community lacking in capital, rather than ensuring maximum profits for large mining companies. A small group of major mine owners felt that they needed a different Transvaal government, more attuned to the needs of industrial capitalism (Van Onselen 1982). Their demands included comprehensive labour recruitment and political control for the mine owners rather than the farmers, a more efficient industrial transport system, and a sympathetic bureaucracy protecting their financial interests. During the late 1890s these demands dovetailed with the British government's own desire to maintain supremacy in the region, and they ultimately brought about the South African War of 1899–1902. After a protracted and costly guerrilla struggle, both the Afrikaner republics were defeated and brought under British control, thus ensuring that the mine owners' needs were to be met.

The Union of South Africa, created in 1910, was the final achievement of the earlier British goal of confederation. Although the Afrikaners of the two conquered republics were able to vote for a South African government led by Afrikaners, they were now part of a new political and economic geography, centred on the inland core of Johannesburg and established by English-speaking interests. Foremost among those interests were those of the big mining companies. These companies' plans for effective labour recruitment and for a capitalist restructuring of southern Africa were reflected in post-war government planning, and had already been largely realized by 1910 (Marks and Trapido 1979).

The scramble for southern Africa

As South Africa's economic and political geography was being restructured, so was the wider political geography of the region. The pattern of nation-state boundaries was being set by external intervention, not just by the British, but also by the German and Portuguese governments, which were engaged in global geopolitical competition. With the rise of other European industrial powers, Britain's former dominance of free trade in Africa and elsewhere was increasingly threatened during the late nineteenth century. When, in the 1880s, the British government was no longer able to deter French and German encroachment, the 'scramble for Africa' began, as each European power with a foothold on the African coast was drawn into a struggle to gain control of more territory. In southern Africa, Germany established its colony in South West Africa (modern-day Namibia), and Portugal further up the Atlantic seaboard in Angola and across the continent in Portuguese East Africa (Mozambique).

The British government incorporated the southern part of Bechuanaland within the Cape Colony in 1885 and deprived the Southern Tswana of much of their land. Like Basutoland, the Bechuanaland Protectorate north of the Molopo River was made a British High Commission territory (corresponding to modern-day Botswana). African authorities in both Basutoland and the Bechuanaland Protectorate made alliances with the British authorities, preventing the dispossession of most of their followers, but in a third High Commission territory – Swaziland – British and Afrikaner settlers with concessions from the Swazi king were able to secure two-thirds of the land. With British government support, all three of the High Commission territories evaded incorporation into the Union of South Africa in 1910, but they remained subject to coercive labour recruitment for the region's white-owned mines and farms (Crush 1987).

In the meantime, Rhodes had been exploiting the close connections between capitalists and the British government to secure official backing for the extension of the control of his British South Africa Company (BSAC) to the north of the Transvaal. Having produced a concession supposedly signed by the Ndebele king, Lobengula, giving him the right to exploit the mineral resources of the kingdom, Rhodes provoked a war so that the king's authority could be broken. BSAC control over Southern Rhodesia, as the territory was named, enabled influential investors to be reassured, labour to be channelled more effectively to the mines of the region, and profits to be made from settling the alienated land with white commercial farms once the anticipated gold reserves had not materialized. Despite some compromise after a Ndebele–Shona uprising in 1895, the BSAC confined Africans within the territory to locations and sold their best land to white farmers and land companies (Marks 1985b). This had become a familiar pattern in southern Africa.

To the west of Bechuanaland, and even before Rhodes had expanded his influence northwards, German control was being extended from 1884 out of the small coastal station of Lüderitz into the interior of South West Africa. Despite Nama and subsequent Herero resistance, mining companies took control of a third of the land, and, after a 1904 uprising, the German authorities directed a systematic policy of extermination against the resistors. To the east and west of BSAC territory, Portuguese imperial intervention had been proceeding in response to British and German expansion. From coastal trading enclaves, the Portuguese colonial government advanced its occupation inland, for example through southern Mozambique in 1895. The process of conquering northern Mozambique and the other area of Portuguese involvement, Angola, proved more difficult, but by 1908 a tenuous control had been established. Thereafter, with Portuguese assistance, Johannesburg mining companies were able to tap southern Mozambique more intensively for labour supplies.

The late-nineteenth-century pattern of imperialism in southern Africa involved more than just the forceful penetration of European capitalism and its labour demands, however. It was based upon an ideology of racial difference far more coherent even than that held by the mid-nineteenth-century British settlers described earlier. Persuaded by colonial revolts and by the defence of slavery in the American South, many European and North American academics had begun to argue that blacks were inherently inferior because of basic biological differences. From the 1860s they made this assumption central to the full range of social disciplines, from history and geography to anthropology and psychology. The clear implication was that Africans would not, and could not, benefit from humanitarian tutelage or policies of assimilation into ruling white colonial groups. Instead they had to be governed, for their own benefit, by Europeans. By the early twentieth century, such 'scientific' analysis of difference was being used by the dominant classes of Europe to legitimate not only the oppression of blacks in distant colonies, but also that of other Europeans such as the Irish, and the domestic subordination of women, the working classes, the insane, and criminals. 'Scientific' ideas were invoked to control any group which threatened political stability and economic growth. The mineral revolution, the growth of industrial capitalism, and the scramble for southern Africa thus did not occur within an ideological vacuum. They were set within a deep-rooted pattern of European thought which was reflected in social class structures and which manifested itself in the way colonies were organized at a variety of spatial scales.

The segregationist order

By the beginning of the twentieth century, industrial capitalism had transformed much within South Africa, but it had also left certain practices relatively untouched. Across large swathes of the highveld, partially commercial forms of labour tenancy and sharecropping continued to suit both poorer farmers requiring labour and African peasants needing access to land. During the first half of the century, however, continuing industrialization and the associated commercialization of white agriculture progressively ironed out such anomalies. Segregationism (the enforced designation of spatial areas for the exclusive use of particular social groups) proceeded hand in hand with the process and became an all-encompassing system by means of which which the modern capitalist economy would be managed in the region as a whole. It can perhaps best be thought of as a system which protected the boundaries around the main white beneficiaries of industrial capitalism, insulating them from challenges posed by those outside these boundaries.

White space, compounds, and townships in South Africa

In the industrializing cities a coherent form of segregation evolved. Africans had been policed and segregated in the eastern Cape towns of King William's Town and East London since the 1840s, but it was Kimberley that set the precedent for large-scale urban segregation within industrial South Africa. There the mining

FIGURE 3.5 De Beers mining compound at Kimberley

SOURCE: Parsons 1993.

companies introduced the first compounds for African migrant workers, ostensibly to prevent diamond thefts (see Figure 3.5). The compounds limited workers' independence and kept wage costs down, and this made them attractive to employers on the later gold mines on the Witwatersrand and to major industrialists across the region (Crush 1994; Mabin 1986). While African workers were housed as far as possible in Kimberley's compounds, white mineworkers were housed instead in the suburb of Kenilworth and coloureds were free to join other blacks in the poorer dwellings and shacks of the town. Even there, though, white fears of racial contact, stoked by scientific racism, demanded clearer spatial separation (Pirie 1991).

Subsequently, demands for urban segregation continued to be pressed in response to new challenges. White fears were compounded by the 'sanitation syndrome' between 1901 and 1904, when bubonic plague spread through some of the poorest housing of South Africa's cities. Associating the disease with blacks rather than with urban squalor, municipalities in Cape Town, Port Elizabeth, and Johannesburg reacted by clearing mixed-race areas close to the city centres and removing the black inhabitants to peripheral townships well away from the white suburbs (Bickford-Smith 1995; Parnell and Pirie 1991; Swanson 1995). Within these townships, whites who were thought of as 'experts' on the cultural traits of urban blacks attempted to regulate and control them (Robinson 1996).

During the 1920s the economy was diversifying through investments in secondary industry. This created a pressing demand for a larger urban African workforce and posed further challenges to the white urban sense of security. In response, the government's Stallard Commission reiterated the doctrine of urban segregation and the government passed the 1923 Natives (Urban Areas) Act. As well as reinforcing the principle that urban Africans were to be a temporary presence housed in segregated townships, the Act established that African housing in these townships was to be paid for by Africans themselves. The same principle had already been established in Durban, where separate African 'locations' were funded from a municipal monopoly on beer sales (Maylam 1990). Nevertheless, some urban Africans were still able to evade segregation, finding accommodation alongside lower-class whites, coloureds, and Indians in more squalid parts of cities, such as District Six in Cape Town and Sophiatown in Johannesburg (Lester 1996; Parnell 1988).

Increasingly coherent urban segregation was an early and critical component of segregationism as a whole, but it was not the only one. The system was also based on 'Native reserves' in the rural areas. These were the successors of the 'locations' developed by British colonial governments in nineteenth-century British Kaffraria and Natal. When not employed in white space, Africans were to live in such designated rural areas, where African chiefs had clung tenaciously to control over land, and where European supervision had been only tenuously superimposed. The government defined the reserves more clearly in the 1913 Natives Land Act, recognizing the bulk of South Africa as being for white ownership only. The Act was intended to encourage white commercial farming by excluding black sharecroppers as far as possible from white-owned land, causing them to be replaced by waged labourers. In principle, Africans were to farm autonomously only in the reserves. These amounted initially to 7 per cent, and later 13 per cent, of South Africa's land area (see Figure 3.6). Their environmental and economic impoverishment ensured that migrant Africans would continue to issue forth, providing cheap labour for the major employers of the cities and for the commercial farmers of the countryside. White administrators hoped that the system of reserves would continue to protect 'white space' and maintain black political and social exclusion, while still ensuring critical labour supplies.

During the first half of the twentieth century, then, there was broad consensus among whites that such a system of segregation was the most appropriate policy. Even a number of blacks could find some value in it. By preserving African access to land in the reserves it protected the authority of chiefs who were willing to compromise. And by discriminating particularly forcefully against Africans in 'white space', it allowed many coloureds and Indians to defend their intermediate status (Marks and Trapido 1987). Overall, however, the social boundaries of ruling group identity maintained by segregationism still excluded the vast majority of Indians and coloureds, prompting some to mobilize against discrimination. These boundaries, though, did not include all whites. Many Afrikaners, particularly the farmers who practised sharecropping or had labour tenants, lacked the capital to prosper within the competitive commercial economy. Rather than being the beneficiaries of South Africa's industrial capitalism, they were among its victims.

FIGURE 3.6 The 1913 and 1936 Land Acts

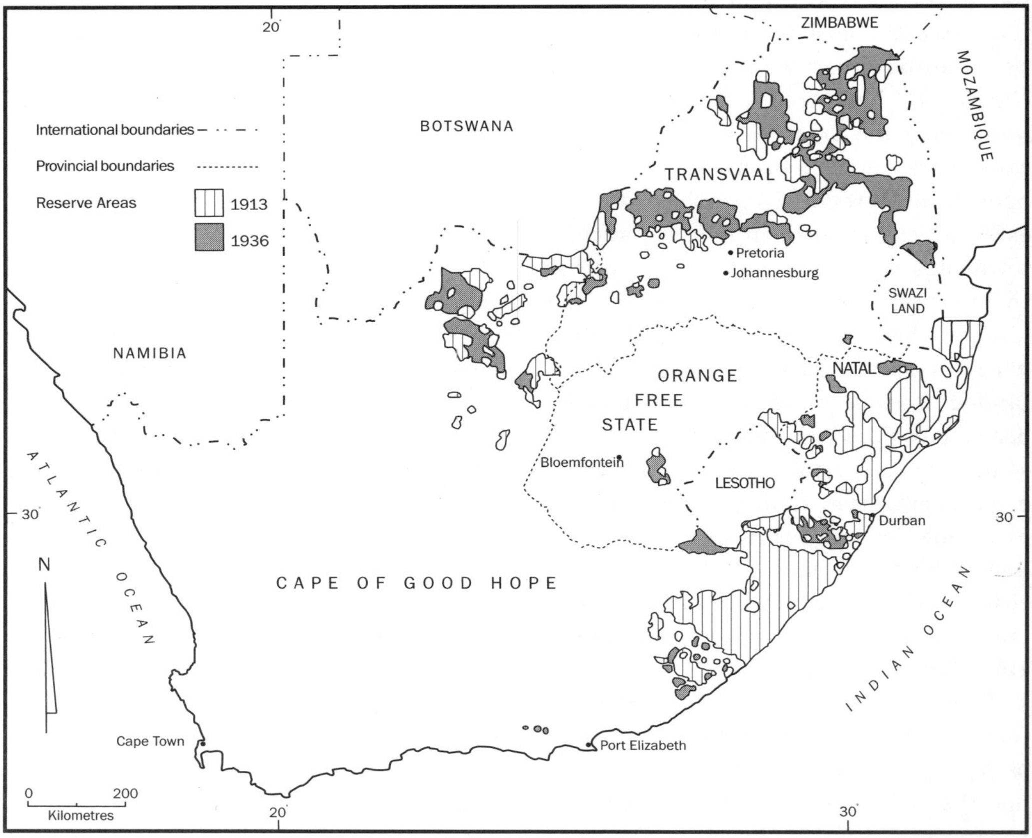

SOURCE: Christopher 1994.

Segregation in the wider region

Beyond South Africa, white settlers in each of the British colonies except Basutoland exerted considerable influence over local government during the early twentieth century. Although chiefly authority was maintained in Basutoland and Bechuanaland, discriminatory legislation in favour of whites was generally being entrenched. White settler land-grabbing was particularly extensive in Southern Rhodesia, and by the 1940s the government had installed a segregationist system similar in many respects to that of industrial South Africa, although not so coherently planned or so deeply ingrained.

In German South West Africa the colonial government instituted a harsh regime excluding Africans from extensive areas of mining production unless they were registered workers. After the First World War, having been captured by South African troops, the territory was entrusted by the League of Nations to South Africa's 'care'. African land dispossession then proceeded further as South African–based and British-financed companies bought out the German mining concerns. The Portuguese government, in the meantime, was treating Angola and Mozambique as its own provinces and using them for the direct benefit of the Portuguese economy. Its rhetoric of African 'improvement' and 'education' (similar to that employed in other colonial territories) obscured the continuation of forced labour practices on behalf of Portuguese capital. But in the 1920s and 1930s, despite Portuguese economic designs, South Africa's economic power was being consolidated in the subcontinent as a whole. With Portuguese co-operation in Mozambique, South African mining companies again stepped up their labour recruitment, and the South African government extended its rail and air network to incorporate surrounding colonies (Parsons 1993).

The apartheid order

In 1948, the Afrikaner National Party (NP) came to power in South Africa. It introduced a programme of apartheid that was intended to enhance the country's segregationist geography. In addition, apartheid was geared to redrawing the social boundaries of the dominant classes, so as to include more Afrikaners, and some attention needs to be paid to this aspect of the programme before its geographical effects can be fully appreciated.

Attempts to mobilize Cape Colony Afrikaners as an ethnic group had begun in the late nineteenth century, when Afrikaans was first threatened as a language of power and Afrikaner economic interests had to be defended from British competition. But a sharp ethnic divide within white politics developed only when the Afrikaner republics were threatened by British aggression. During the South African War, the death of 28 000 Afrikaner women and children in British concentration camps, through disease and neglect, left many Afrikaners embittered. After the war, within a system of industrial capitalism forged by English-speakers, it appeared that significant numbers of Afrikaners, particularly the poorer farmers, driven from the land by the decline of sharecropping and black labour tenancy, were sliding out of the privileged classes altogether. Some, having relocated to the cities, co-operated with English-speaking workers against industrial employers, but nationalist politicians attempted to guarantee their reincorporation in the ruling group by other means. These politicians endeavoured to unite as many Afrikaners as possible, regardless of their class, in a mass political movement – a movement directed against both British and black 'others'. Together with a reaction against the government's wartime reformism, the success of this movement allowed the NP to assume government in 1948 (Giliomee 1979; Lester 1996).

The NP's programme of apartheid was aimed at benefiting the various Afrikaner classes and resolving the contradictions between them (Posel 1991). In order to protect Afrikaner cultural identity, to uplift Afrikaners economically, and to shore up white supremacy as a whole, apartheid, like earlier segregationism, involved intrinsically spatial policies:

- Urban segregation was pursued even further than before to maintain and reinforce the existing pattern of white privilege. It was also intended to protect the imagined racial 'purity' of Afrikaners in the face of 'race-mixing' in unsegregated urban areas (Marks and Trapido 1987). By means of geographical separation, the social boundary between Afrikaners, as members of the dominant white class, and blacks, as their subordinates, would be reinforced.

- As a result of tighter 'influx control', the migrant labour system was bolstered as an alternative to African urbanization (Posel 1991). Arrests under the pass laws increased, and African women attempting to hold their families together bore the brunt of this initiative.
- Influx control in turn necessitated the more coherent planning of the reserves. The NP designated these, rather than the cities, as the proper geographical arenas for African life.

Each of these spatial intentions was realized by means of a barrage of legislation passed during the 1950s and 1960s. More rigid discrimination at all scales was founded in the first instance upon the Population Registration Act of 1950. This Act classified every South African as a member of a racial group, the four main groups being White, Asiatic (Indian), Coloured, and Native (Bantu, African, or Black). With all members of the population classified into groups, their geographical separation could then proceed at a variety of scales.

The Reservation of Separate Amenities Act of 1953 legalized segregated and unequal beaches, libraries, post office counters, and swimming pools. The Immorality and Mixed Marriages Acts sought to preserve white 'racial purity' more directly by preventing sex between members of different 'racial groups'. The education system was manipulated to the same ends of cultural 'purity' through the philososphy of Afrikaner 'Christian National Education', while black education was extended, but tied only to a labouring function, under the Bantu Education Act of 1953.

The 'essence of the apartheid policy', as Prime Minister D. F. Malan put it, was expressed in the Group Areas Act of 1950. The government insisted that separate zones eventually be demarcated for each race within every urban area (see the ideal form of the apartheid city, Figure 8.2, p. 203). Although African townships were designed more coherently, with security considerations in mind, the brunt of the Act was felt by coloureds and Indians who were not already formally segregated. Despite the municipality's reluctance, the 54 per cent of Cape Town's population classified as 'coloured' had to be contained within newly designated townships on the Cape Flats (see Figure 3.7). The breaking up of Cape Town's District Six is perhaps the most notorious case of the painful sundering of people and place which the Act inflicted (Western 1996). Durban's municipal government, with the backing of white residents and traders, was even more keen to remove Indians from its designated white group areas (Maharaj 1995).

Where the Group Areas Act was not used directly to enforce urban racial separation, other laws were invoked instead (Parnell 1993). The Johannesburg municipality used slum-clearance legislation to destroy mixed-race freehold areas such as Sophiatown. Its African residents were

FIGURE 3.7 Cape Town's group areas

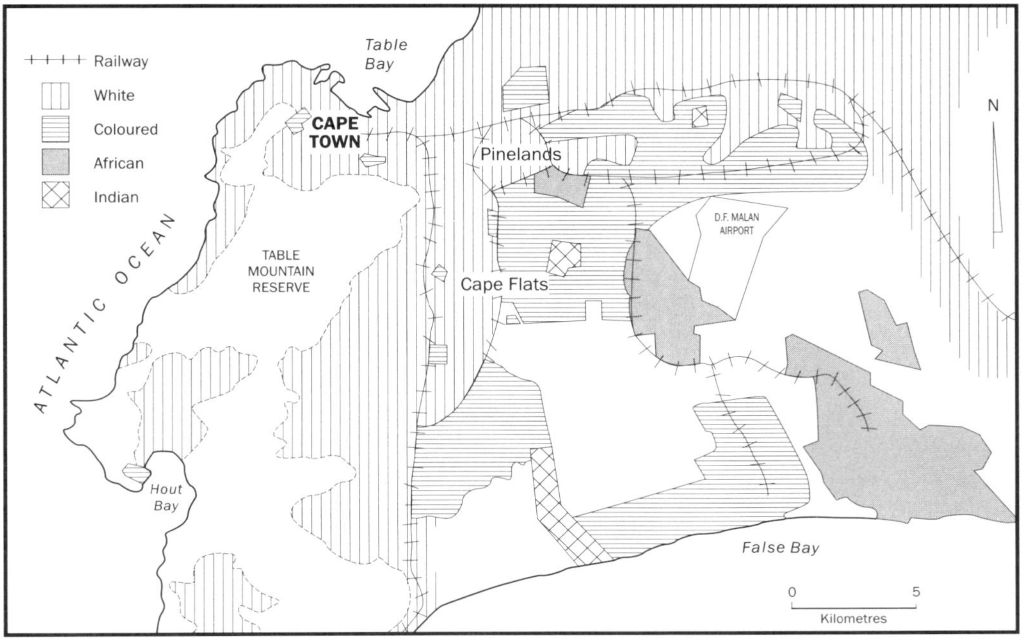

SOURCE: Lester 1996.

removed to the budding townships of Soweto, and it was zoned instead for working-class whites (Hart and Pirie 1984). The government's and employers' reluctance to shoulder the costs of housing evicted residents in the new townships led to most of the expense being passed on to those residents themselves. This in turn resulted in an exacerbated black housing shortage, the proliferation of shack settlements, and an issue around which black urban protest could be continuously mobilized (Maylam 1990).

From the 1960s, Africans were officially considered to belong to one of the ten ethnic 'homelands', created from the earlier reserves (see Figure 3.8), and hundreds of thousands of families were forcibly moved there from urban areas and white farms across the country (see Table 4.1, p. 105). In response to black resistance and the changing international environment, in which official racial discrimination was being discredited, the government intended the homelands to give the illusion of black self-determination (Lester 1996). Having removed many chiefs who had demonstrated their opposition in the 1950s, elections were arranged for black governments within the homelands, although the most popular politi-cal organizations were banned. In order to reinforce urban influx control, the South African government also encouraged industry

FIGURE 3.8 South Africa's self-governing territories and independent states

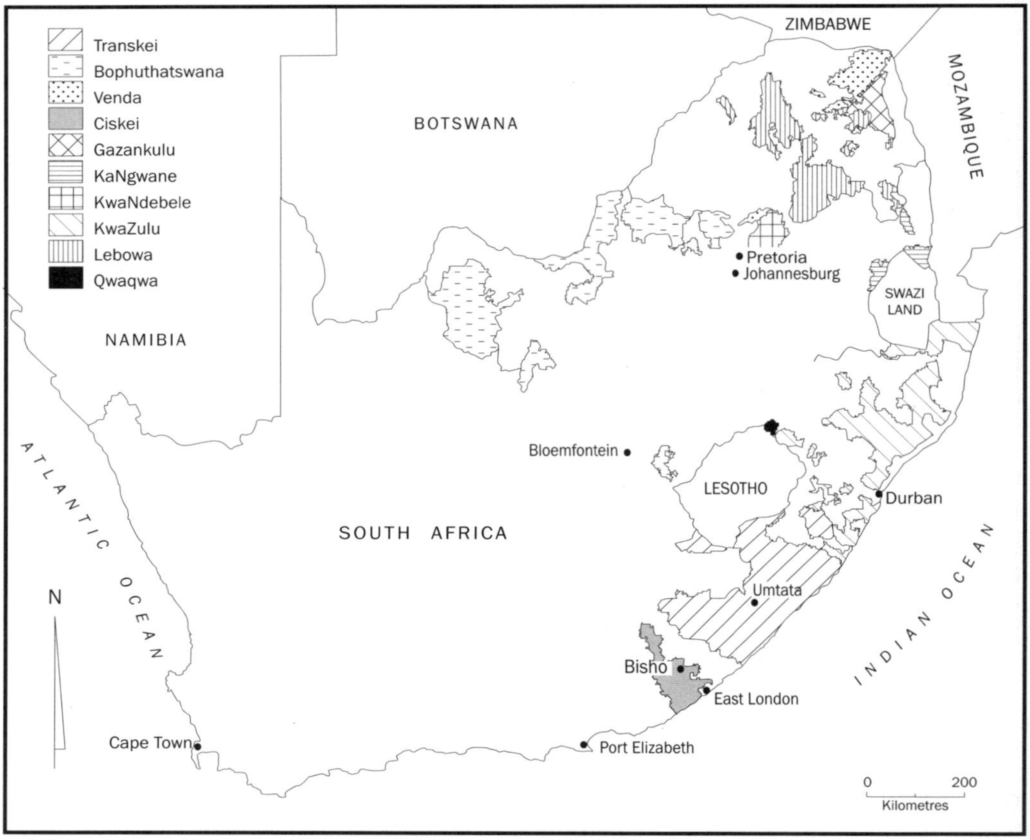

SOURCE: Lester 1996.

to relocate to areas adjacent to the homelands and to employ black labour from them during the 1960s, but this policy met with minimal success. Ultimately, the South African government recognized four homeland governments – Transkei, Bophuthatswana, Venda, and Ciskei (the 'TBVC states') – as fully independent, but few other governments gave them official recognition due to their lack of political legitimacy, low capital, poor infrastructure, and high unemployment, all of which ensured reliance on the South African economy.

By the early 1960s the government had reinforced earlier patterns of segregation in order to protect the social and economic privilege of an expanded white ruling group. However, the NP's forcible attempt to bolster racial privilege, unlike that of British imperialism some eighty years before, was swimming against the tide of international developments.

Southern Africa within a new world order

During the 1960s and 1970s, shifts in international capitalism, international politics, and black resistance were to pose irresolvable challenges to the apartheid system. In the face of these challenges the nature of apartheid changed considerably. Far from the initial, ambitious effort to redraw the boundaries of the dominant group and to maintain a segregated geography, by the early 1980s apartheid had become a reactive attempt to ensure the mere survival of white supremacy.

The economic challenges were of three major kinds. Firstly, racial segregation and domination, having for so long been generally compatible with capitalist production methods, were being outpaced by newer methods. Technological innovation had brought about the replacement of large numbers of cheap unskilled workers with more productive machinery, demanding fewer but better-skilled workers. Factory employers had been dissatisfied about the costs of retraining African migrant workers since the 1940s, and by the late 1970s all the major capitalists, including mining companies and commercial farmers, faced problems securing adequate skilled labour. Even with the growth of a small, skilled black middle class, generally poor black education provision still resulted in a shortage of skilled workers on the one hand, and mass unemployment among unskilled urban blacks on the other. Many urban blacks not only undermined apartheid's geography by filling in the spatial buffers between group areas with their shacks, they also continued to threaten revolt.

Secondly, under apartheid, only whites and the restricted black middle class were providing a significant internal market for certain products. For example, by the end of the 1970s car sales were only half the industry's production capacity (Price 1991). As international boycotts and sanctions spread in protest against apartheid's overt racism, other producers who relied on export markets were affected similarly.

Thirdly, international investors' assessments of the economic risks caused by internal instability led many of them not to invest further in South African companies. By the early 1980s even the Afrikaner business elite (itself having been created largely by apartheid policies) was expressing unease about apartheid's economic constraints (Lester 1996).

From the mid 1970s, the renewed costs of suppressing black resistance were added to the government's economic problems. When student demonstrators were shot by police in Soweto in 1976, the township erupted in rebellion. Urban blacks rapidly extended the revolt nationwide, and it became the most violent and costly challenge yet to the apartheid state. The future instability which it heralded prompted a particularly dramatic withdrawal of international investments. This had happened before, when investors were deterred by instability after the 1960 Sharpeville shootings, but then investment had resumed soon after government repression had secured surface calm. By 1978, South African capital was cut off, seemingly permanently, from many overseas sources of long-term loans. Foreign-owned firms were also repatriating more of the revenue that they earned within the country. And in the meantime, the internal security budget was soaring.

Regional decolonization

Further challenges to the structures of white minority rule in South Africa came from the rest of the subcontinent. While South West Africa continued to be occupied by the South African government and subjected to its apartheid laws, each of the British High Commission territories, followed by the Portuguese colonies, attained independence during the 1960s and 1970s. These newly independent countries subsequently became targets for South African–inspired destabilization, especially where left-wing governments harboured the military operatives of South Africa's banned liberation movements.

After the Second World War, African nationalist movements were being forged in all the British territories, and economic control was passing from British companies to a range of multinational corporations. Both tendencies persuaded the British government that it had more to gain from granting independence to these territories than it had to lose. Racial discrimination was officially eliminated and African nationalists in the High Commission territories prepared for independence. Basutoland and Bechuanaland became independent in 1966, as Lesotho and Botswana respectively, as did Swaziland in 1968.

White minority rule remained in Southern Rhodesia, however. In the 1960s the settler government adopted security legislation similar to South Africa's in order to suppress internal resistance. Defending itself against the trend of decolonization, Ian Smith's white government declared its own independence in 1965. In the face of sanctions levied by the British government, it was able to rely officially on the South African and Portuguese governments, and unofficially on multinational companies. In 1969 the white Rhodesian government declared itself a republic with policies similar to those of apartheid South Africa. By then, though, African resistance was developing into an extensive guerrilla war which ended only in 1979 when Rhodesia moved towards internationally recognized independence as Zimbabwe.

African resistance movements in Angola and Mozambique had also succeeded in gaining independence. Because the Portuguese government had showed no sign of granting independence willingly, guerrilla warfare had begun in Angola in 1961 and in Mozambique by 1964. By the mid 1970s, large areas of each colony were in the hands of guerrillas linking with other independence campaigns. Finally, in 1974, the fascist Portuguese government itself collapsed due to the global economic recession and the increased costs of defending its colonial rule. With independence conferred in 1975, 80 per cent of the white Portuguese settlers left Mozambique and 90 per cent left Angola, depriving each country of much of its skilled workforce (Parsons 1993). As a result of the decolonization of these neighbouring countries, the morale of blacks resisting apartheid in South Africa had been significantly boosted.

The abandonment of apartheid

In the aftermath of the 1976 Soweto revolt and in the light of the economic and security contradictions which now faced it, the

South African government made its first coherent attempts at the economic and geographical modernization of apartheid. Under Prime Minister B. J. Vorster, commissions advised the consolidation of a black urban 'insider' group which could provide both skilled labour and a political buffer against radical black demands. Outright racial discrimination was now to be replaced to a certain extent by economic discrimination, as a privileged strata of urban blacks were allowed to lease their township houses and to engage in registered union activity (Lester 1996).

Vorster's successor, P. W. Botha, inherited these reformist plans and co-ordinated them as part of a wider 'total strategy'. This strategy was intended to deal with internal black resistance, the independence of the surrounding states, and the economic constraints of apartheid. While the economically crucial urban blacks were to be co-opted by further reforms and material improvements, hostile neighbouring states were to be destabilized by military raids and support for destructive internal movements. Industry was again encouraged to decentralize to the homeland areas. This would both direct black employees away from the cities without the use of racist influx control and relieve water shortages on the Witwatersrand. In order to co-opt Indians and coloureds, Botha's government implemented a new constitution in 1984, which allowed them to elect representatives to their own, relatively powerless, legislatures, within the tricameral Parliament. Perhaps the most telling reform, however, was that seeking economic viability by means of new planning boundaries. During the late 1980s, 'development regions' were identified at the national scale, and Regional Services Councils at the local scale, which cut across homeland and township boundaries, ignoring racial divisions.

A combination of economic recession just when expensive improvements had been promised to urban blacks and the attempt to pass the costs on to those recipients themselves made Botha's 'total strategy' unworkable. Particularly severe rent and service levy increases in the industrial Transvaal's townships were the final straw for urban blacks who were experiencing reduced living standards and continuing exclusion from meaningful political representation. From 1983 onwards the insurrection begun by residents in these townships spread across South Africa, and by 1985 the government was involved in the bloody suppression of open rebellion. As a result, the townships became more politicized and more violent than they had ever been (Lester 1996; Price 1991). By the end of the 1980s many of the homeland governments had passed into the hands of leaders who claimed to support the African National Congress (ANC), such as Bantu Holomisa in the Transkei, and it was clear that no white minority government would ever be able to govern urban blacks without further bouts of expensive revolt. The boundaries of ruling-group identity would have to be redefined once more.

While F. W. de Klerk's subsequent government could have held out longer, the decision to negotiate with the ANC leadership was taken in 1990. This was partly because, as far as the NP was concerned, the timing was propitious. In the face of mounting costs and a lack of military success, reformists within the party had recently been able to secure a South African withdrawal from Namibia, allowing the South West African People's Organization (SWAPO) to gain power. The positive international reaction seemed to promise support for the NP's position in any subsequent negotiations with the ANC. Furthermore, the end of the Cold War had removed obvious white fears about Soviet backing for the ANC and its potentially socialist programme. Finally, within South Africa, secret negotiations with Nelson Mandela had already revealed that the ANC would rule out radical economic redistribution, in order not to deter international investment. It therefore seemed to the NP during 1990 that essential white minority and capitalist interests could still be preserved through negotiations (Lester 1996).

Conclusion

This chapter has examined the evolution of southern Africa's geography by assessing changing forms of power within a developing capitalist system. It was not cultural difference alone that first generated racial stratification in the region. Where there was intense conflict over scarce resources between Dutch newcomers and established Khoisan on the early frontier, colonists did develop aggressive and contemptuous racial attitudes. But coherent racial stratification tended instead to be forged behind the advancing frontier, where a semi-commercial economy was being tied into global mercantile capitalism. This economy was founded on patriarchy and forced labour. It involved both slavery and the domination of the Khoisan and it was sustained by a colonial culture based on the perceived inferiority of the indigenes.

British settlers were able to benefit from the racial hierarchy installed during the Dutch colonial period. But their lasting contribution was to gain the co-operation of the colonial government in reforming that system in line with new developments in global capitalism. Their commercial enterprise demanded the intervention of the government to subjugate the Xhosa and to appropriate their land in the eastern Cape. Necessary compromises with African power, both there and in Natal, though, resulted instead in a geography of segregated landholding and government regulation of black labour movement. Once this new pattern had been installed, largely through brute force, it too was sustained partly by more subtle forms of power. Even British liberals, whether consciously or not, legitimated the colonial domination of Africans by representing it as being in their own best interests.

Further developments in global capitalism lay behind the restructuring of the highveld. During the late nineteenth and early twentieth centuries, on the one hand, patriarchal power relations established by the trekkers were transformed, and, on the other, independent African authorities were subjugated. Although industrial capitalism and imperialism were not synonymous, in this period they were very closely identified. In combination they extended the British colonial pattern of reserve segregation across South Africa, they shaped modern forms of urban segregation, they created an economic core centred on the Witwatersrand and a migrant labour system extending from it across the region, and they established a complex of capitalist companies which, in various guises, continues to dominate the region's economy. They were also behind the extension of European domination within the subcontinent as a whole – a system which not only drew national boundaries, but constructed a vision of Africans largely as labourers, justifying that construction by means of 'racial science'.

The industrial system that was forged during the late nineteenth century disproportionately benefited South Africa's English-speaking, capitalist whites. During the twentieth century, an Afrikaner nationalist struggle was waged to extend the boundaries of ruling group identity, so that Afrikaans-speakers, too, could be included. Apartheid was intended to maintain the system as a whole, and to make it work in favour of relatively disadvantaged Afrikaners in particular. With apartheid, Afrikaner nationalist leaders further restructured South Africa's geography, hoping that they could remould and preserve general white privilege in the face of further shifts in global capitalism and international and regional politics. Today's legacies of urban inequality, violence, black marginalization, and the social, economic, and environmental deprivation of the former homelands cannot be divorced from this policy.

Notes

1 Between 1860 and 1911, 152 000 Indians were taken to Natal to work for a contracted five-year period. Following that, they could re-indenture for a further five years' work, return to India, or stay in Natal and attempt to start a new life.

2 Labour tenancy was the practice whereby African families lived on what was usually white-owned land, grazing and cultivating a portion of it and in return working some three to nine months each year for the landowner. Sharecropping was an arrangement through which landowners (usually white) who lacked their own oxen and ploughs invited Africans to live on, and cultivate, their land in return for a portion (often a half) of their crop.

SECTION II

Human geography

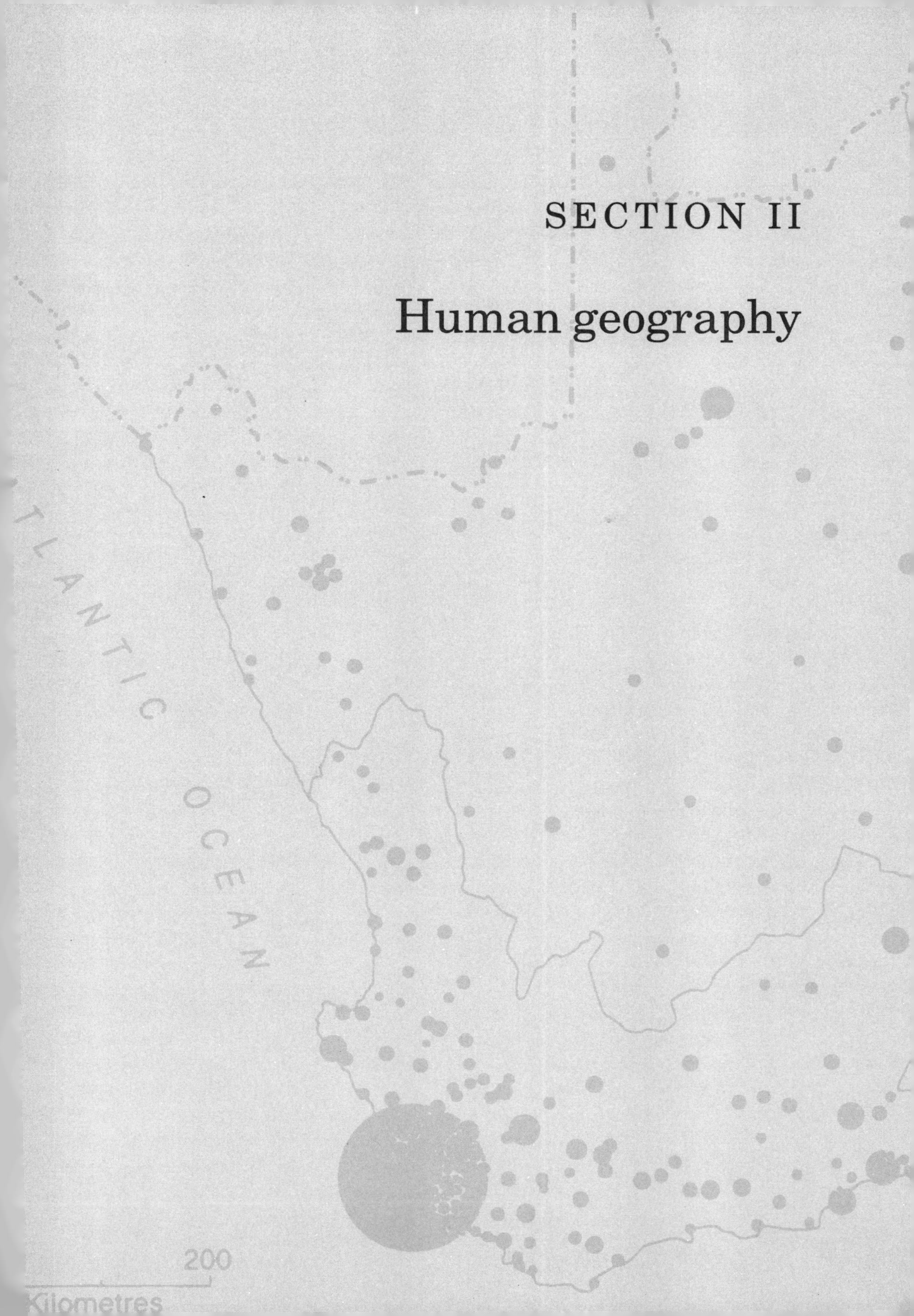

4 Political geographies of change in southern Africa

DAVID SIMON AND
MAANO RAMUTSINDELA

It is tempting to see the rapid pace and profound nature of geopolitical change in southern Africa since the late 1980s as a reflection of struggles and dynamics peculiar to the region. While it is certainly true that apartheid – and all that it stood for – represented a distinctively oppressive and discriminatory form of social engineering, it neither evolved from nor was finally eclipsed by purely local forces. It is more accurate and helpful to interpret these processes as changing interactions between broader global dynamics and those within the region and its individual countries.

Thus, for example, apartheid did not arise in a vacuum when the National Party took power in 1948. Many of the key measures, such as the Land Acts, the pass laws, and urban segregation measures, were already in existence. Ultimately, as Chapter 3 has shown, the origins of apartheid can in part be traced to earlier forms of social and economic segregation during Dutch and British colonial rule. Such practices were common in much of the colonial world; indeed, particular colonial powers frequently learned from one another, adopting and adapting experiences from other subject territories. For example, German 'native policy' during that country's brief but intense colonial occupation of South West Africa (1884–1915) borrowed much from existing British law and practice in the Cape Colony. Colonial authorities in Rhodesia, too, modelled key legislation and urban segregation policies on South African antecedents at various points during the twentieth century. Many of the most influential British and French military leaders, colonial officials, architects, and planners themselves served in several colonies, accumulating and translating experience from posting to posting.

Conversely, the Namibian independence process, the signing of peace agreements ending the bloody conflicts in Angola and Mozambique, and ultimately the demise of apartheid itself owed as much to the end of the Cold War in 1988/89 and associated agreements between the USA and former Soviet Union to cease regional conflicts as it did to struggles within the region.

In this chapter we shall survey selected global, regional, and national/local themes in order to highlight their interconnections and to counter the all-too-common tendency by South Africans to view their country as unique and as having more in common with Europe and North America than with the rest of Africa. It is clearly impossible to cover all significant issues and processes within the scope of an overview such as this. The choice of subject material reflects importance and topicality as well as our own research and expertise. The focus is contemporary, so our principal concern is with the period since the early 1980s, when the National Party under President P. W. Botha began a process of incremental, piecemeal reforms designed to modernize rather than phase out apartheid. These changes, including the legalization of African trade unions, the removal of 'petty apartheid' measures, and the introduction of the tricameral Constitution, were accompanied by tough repression of protest, and set up both internal and external contradictions which eventually contributed to the end of white minority rule.

Global geopolitical transitions

During the Cold War, Western European government leaders often proclaimed that the existence of the North Atlantic Treaty Organization (NATO) and stockpiles of nuclear weapons had helped ensure that there had been no wars in Europe since the end of World War II in 1945. Such justifications conveniently overlooked two closely related facts, namely that the post-war period witnessed record numbers of armed conflicts elsewhere in the world – principally in poorer regions in the South – and that these conflicts were generally exploited and exacerbated by the great powers in the cause of their Cold War rivalries. In other words, the Cold War was fought only verbally and through a highly militarized stand-off in Europe; the actual fighting and weapons testing under combat conditions took place in distant places, where local populations and societies bore the brunt. Examples include Central America, the Middle East, the Horn of Africa, and South-East Asia.

Southern Africa was no exception in this regard: the Portuguese sought to legitimize their colonial rule in Angola and Mozambique as resistance to the 'communist onslaught' in Africa on behalf of Western capitalism. This terminology was deployed by Ian Smith's illegal regime in Rhodesia during the late 1960s and 1970s, and by apartheid South Africa to justify giving military assistance to its allies at the time and, from 1975 onwards, its own military incursions into neighbouring countries. Such pro-capitalist, anti-communist rhetoric struck a chord with conservative governments in the USA, the UK, and Germany, in particular, and contributed to the generally cordial relations which South Africa enjoyed with those countries throughout the apartheid years. Along with their narrow commercial self-interest, this explains those countries' reluctance to implement rigid military and subsequently economic sanctions. One particularly striking example of covert collaboration was between South African military forces and the US Central Intelligence Agency (CIA) in seeking to prevent the then Marxist

MPLA (Movimento Popular de Libertaçâo de Angola) from taking power in Angola when the Portuguese eventually withdrew in 1975. Already at that stage, overt joint action would have been unthinkable, but the USA clearly perceived its Cold War strategic interests to outweigh public pressure to implement sanctions.

Similarly, the original United Nations–sponsored independence plan for Namibia, negotiated by the so-called Western Contact Group (USA, Canada, Britain, France, and Germany) during the late 1970s and embodied in UN Security Council Resolution 435 of 1978, faltered and then became embroiled in Cold War geopolitics and apartheid South Africa's 'total strategy'. The basis of the logjam throughout the 1980s was the issue of 'linkage', in terms of which US President Ronald Reagan's conservative Republican administration and President P. W. Botha's Cabinet tied the granting of Namibian independence to the withdrawal of the 25–30 000 Cuban troops and advisers bolstering President José Eduardo dos Santos' MPLA government in Angola.

By 1988, events within South Africa were beginning to suggest that change might occur, and the South African military reversal at the major battle at Cuito Cuanavale in southern Angola early that year had a significant impact on morale and the myth of invincibility. These circumstances, however, coincided with the negotiations to end the Cold War, conducted between the outgoing Reagan administration and Mikhail Gorbachev's debilitated Soviet Union. These talks provided the momentum and vehicle for an agreement known as the Namibia–Angola Accords to be accepted and signed by Angola, South Africa, and Cuba at UN Headquarters in New York in December 1988. The accords set out a timetable for the Cuban withdrawal, South African military disengagement, and the commencement of the Resolution 435 process on 1 April 1989. Namibia finally gained independence on 21 March 1990, in one of the final acts of African decolonization.

This event held substantial significance for South Africa too, and illustrates well the contingency and interrelation of geopolitics. Not only had the South African government rid itself of a millstone round its neck and boosted the morale of black South Africans, but, by so doing, President F. W. de Klerk gained a tangible measure of international respect and recognition. Occurring a mere six weeks after the unbanning of the African National Congress (ANC) and release from prison of Nelson Mandela – and although Mandela was a star guest at Namibia's independence ceremony and overshadowed De Klerk – the event confirmed the beginning of De Klerk's rise to statesmanhood in international eyes and gave new momentum to preparations for multi-party talks on South Africa's future.

Despite the importance of Cold War geopolitics in fanning and exploiting local conflicts, it is important to point out some of the other sources of schism in Africa. One very durable division which is a direct colonial legacy is that between anglophone and francophone (and to a lesser extent also lusophone, or Portuguese-speaking) countries. Almost invariably, independent African states have retained their former colonial language as the official language, either for reasons of Europhilia among the elite or out of political expediency in seeking to avoid conflict over which indigenous language would replace it. There are a few exceptions, such as some North African states, in which French or English has joint official status with Arabic. In Kenya, Kiswahili, the local lingua franca, enjoys joint status with English. Tanzania went furthest, making Kiswahili the sole official language.

The linguistic barrier between anglophone, francophone, and lusophone countries is symbolic of broader differences in terms of administrative structures and cultures, the orientation of trade – usually still disproportionately focused on the former colonial power – and military, security, and economic alliances. Although such divisions have not directly spawned conflicts, the barriers – at least at an official level – have remained surprisingly pervasive until recently. Gradually, however, the situation is changing as Cold War considerations, the end of decolonization and apartheid, and the rapid deregulation of world trade foster greater mutuality of interest. Perhaps the most dramatic recent example is provided by Mozambique (a former Portuguese colony) and Cameroon (a former German, French, and British colony) joining the Commonwealth, hitherto exclusively an association linking Britain and its former colonial possessions. Furthermore, the new ruling elite in Rwanda includes many former exiles educated in North America and with strong ethnic ties with eastern Uganda. There now seems to be a marked shift towards closer links with anglophone Africa, a tendency reinforced in the Great Lakes region by the admission of the Democratic Republic of Congo (DRC – formerly Zaire) to membership of the Southern African Development Community (SADC) in late 1997. (SADC is discussed in the following section.)

Ethnic and religious differences have, directly or indirectly, spawned many violent conflicts in Africa. Perhaps the clearest example is that of Sudan, where attempts by successive Muslim-dominated governments to impose Islam and Islamic law on the mainly Christian and animist populations in the south of the country triggered a long-running armed conflict which has wreaked widespread destruction, impoverishment, starvation, and appalling abuses of human rights. By contrast, in neighbouring Somalia, a country with remarkable ethnic homogeneity, bitter rivalries between competing clans and warlords all but destroyed Somalia as a state in the early 1990s. One consequence of this conflict has been secession and the declaration of the separate state of Somaliland by the clans living in the north-east (the area once colonized by Britain as British Somaliland) in an effort to insulate themselves from the conflict further south.

In the two sections below, we shall outline some of the most important political geographical changes in the southern African region and in South Africa.

Southern Africa in flux

The impacts of geopolitical and social transitions within southern Africa will be discussed here under three broad categories, namely contested national boundaries, changing territorial organization within individual states, and multilateral initiatives to establish regional institutions. Naturally, many other important regional and cross-border initiatives and processes – which fall beyond the immediate scope of this chapter – are occurring in southern Africa at present (Simon 1998).

One of the apparent ironies of post-colonial experience in southern Africa is that hardly any of the national borders inherited from colonial rule have been changed. These had been drawn up, often with little knowledge of prevailing local conditions, to suit imperial ambitions and rivalries and for administrative convenience (Katzenellenbogen 1996). It might therefore have been expected that the new rulers of independent African states would seek to redraw their borders to accord better with ethnic and cultural boundaries and physiographic features. However, African leaders quickly realized the potential for catastrophe if territorial claims against neighbours turned into conflict, so one of the first resolutions of the Organization of African Unity (OAU) after its establishment in 1963 was to accept existing national boundaries. This principle has guided OAU policy ever since.

Across the continent there have been several attempts to secede from existing states, for example Katanga from Zaire, Biafra from Nigeria, and Eritrea from Ethiopia, which have led to protracted and bitter conflicts. Another form of conflict has been military action to wrest territory from neighbouring states, such as Morocco's invasion of part of the former Spanish Sahara, preventing the establishment of the independent Sahrawi Arab Republic; Libya's invasion of the Aouzou Strip in northern Chad; and military efforts to seize pockets of land in the oil-rich Nigeria–Cameroon border zone. Eritrea ultimately (re)gained its independence in 1991, becoming the first 'new' state on the continent since decolonization. As mentioned above, it was soon followed by Somaliland. These two have gained international recognition (only partial in the case of Somaliland) by virtue of their previous separate status.

Unsurprisingly, one consequence of the preservation of inherited boundaries has been the frequency of conflicts across them. In other words, many African boundaries remain geopolitically problematic. Sometimes this reflects the artificiality of national boundaries dividing particular ethnic, clan, or other groups, resulting in conflicts over resources claimed by one side or the other. Sometimes, too, rivalries have been cynically exploited by political leaders to bolster their domestic positions. Most political borders in Africa and other regions of the South are porous, in that crossing them between the few, widely spaced official border posts is easy. Sometimes these boundaries are felt by local populations to have little if any legitimacy, or simply to be a nuisance. The Angola–Namibia border, for instance, divides many Ovambo kin groups and even separates some villages from their traditional grazing land and water sources. The villagers therefore cross over on a daily basis, often at great risk to life and limb during the conflict that enveloped the area from 1974 until 1989. The dangers have again increased, as a result of the Namibian government's concerted efforts to prevent cross-border movements, ostensibly in order to keep out UNITA (Uniâo Nacional para a Independencia Total de Angola) forces and to prevent them from obtaining supplies and harassing Namibians. When the Namibians first erected a security fence along the border in 1992, local people in the Omusati and Ohangwena regions (see Figure 4.1) simply cut the wire in order to carry out their normal activities.

Borders and borderlands represent zones of opportunity as well as constraints, however, providing great incentives for smugglers to cross them with goods when commodity shortages occur in one country or when wide price differentials for particular commodities exist between neighbouring states. Likewise, when the state attempts to monopolize trade in particular goods or when trade in certain materials such as ivory becomes illegal, smugglers can make money by taking goods across borders, as they can when conflicts envelop key producer areas. Permeable borders make policing

FIGURE 4.1 Namibia's new regional councils and local authorities

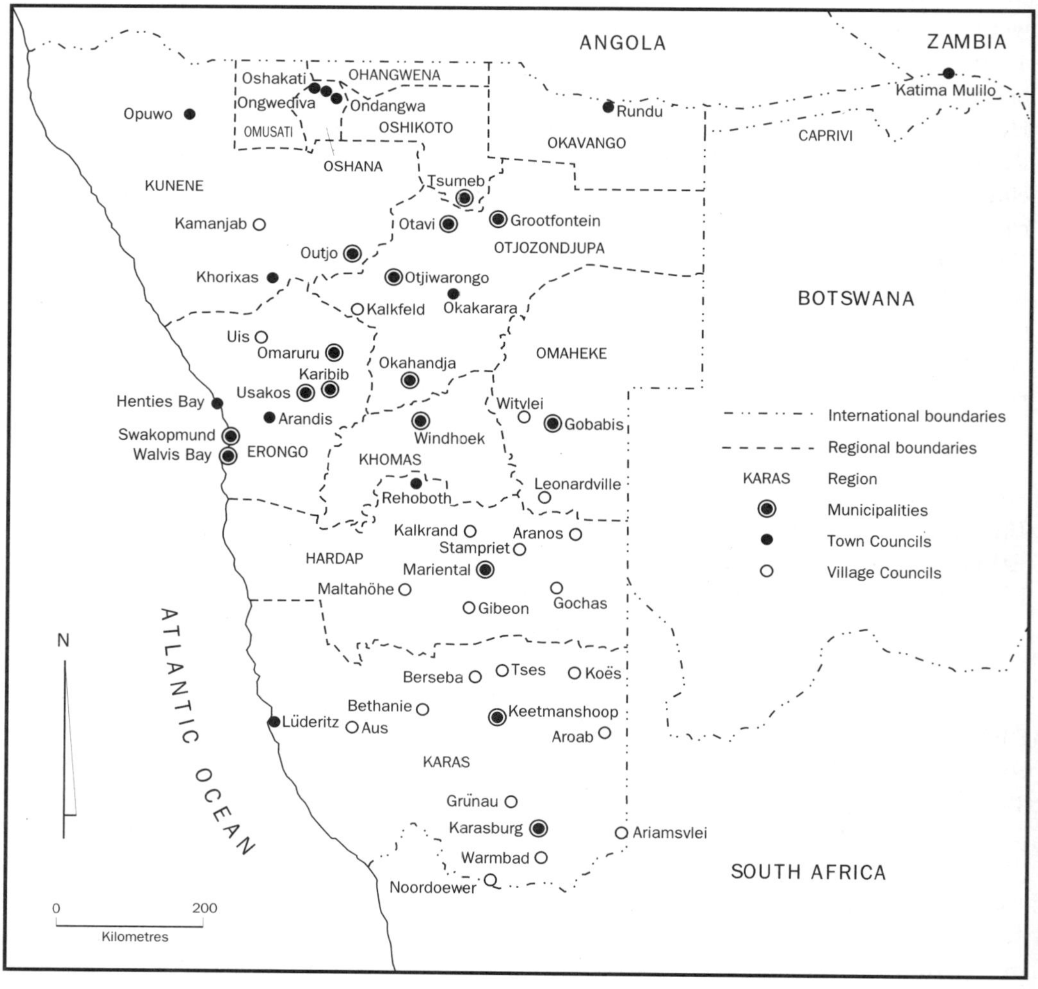

difficult, and a substantial proportion of specific crops or valuable commodities such as gold or diamonds may 'leak' out by means of smuggling and other types of informal trade, with resultant losses of potential tax and duty revenue to the state. In southern and central Africa, Angola and Zaire (now the Democratic Republic of Congo) have recently lost major diamond revenues in this way, and it has now been documented that elements within the South African security forces colluded with the UNITA and RENAMO (Resistência Nacional de Moçambique) guerrilla movements to smuggle large quantities of poached ivory and hardwood timber out of Angola and Mozambique during the wars in those countries in the 1980s.

For all these reasons, recent thinking in political science and political geography has increasingly emphasized the need to move away from earlier formalistic studies of borders as lines on the ground and on maps, and instead to analyse them as integral elements in border zones, which are the arenas within which interactions, opportunities, and constraints occur (Nugent and Asiwaju 1996).

Contested national territories

In southern Africa, territorial changes have occurred only where the alignment of boundaries has been imprecise or anomalous and has therefore become contested. One major and rather distinctive exception, which ultimately failed to materialize, was an abortive secret land deal between South Africa and the late King Sobhuza II of Swaziland in 1982, in terms of which South Africa planned to cede Ngwavuma, the northernmost district of KwaZulu, and the KaNgwane 'homeland' to Swaziland as part of a covert treaty in exchange for Swaziland accepting ethnic Swazis from South Africa and guaranteeing not to permit ANC activities on its soil (Griffiths and Funnell 1991).

For the Swazi king, this deal would have resulted in a long-held dream of recreating a greater Swazi nation and national state, lost in nineteenth-century conflicts with Boer and British invaders. Ngwavuma would also have provided direct access to the sea, thereby eliminating Swaziland's landlocked and dependent status. Conversely, P. W. Botha's government calculated that this territory was a small price to pay for the riddance of up to one million black Africans and a recently signed mutual 'non-aggression' pact which would greatly impede the ANC's ability to infiltrate guerrillas into South Africa. Indeed, the deal would have considerably advanced the National Party's dream of creating a white majority state by depriving black Africans of their citizenship, since both Chief Mangosuthu Buthelezi of KwaZulu and Enos Mabuza, Chief Minister of KaNgwane, were opposed to taking 'independence'.

The deal was to be pushed through without public discussion or referendum, but publication of the enabling proclamations in the South African *Government Gazette* in June 1982 unleashed a storm of protest from those affected in both countries. The KwaZulu authorities duly won their challenge to the legality of the moves in the courts, and, when the Appeal Court upheld the verdict, the South African government appointed a commission to investigate. This commission conducted its inquiry for two years without reporting publicly, and the government then announced that Swaziland would have to negotiate with the homeland authorities in order to finalize the plans. For obvious reasons, this never happened, and the deal was effectively dead (Griffiths and Funnell 1991).

The best examples of border changes actually implemented in southern Africa are provided by the Orange River boundary between South Africa and Namibia, and the enclave of Walvis Bay. Although clearly of great political and economic significance, these issues had been largely academic before Namibian independence, as that territory was occupied and controlled by South Africa. Following independence, the matters were resolved peacefully by means of bilateral negotiations. Although South Africa had previously claimed that the border ran along the northern bank of the Orange River, the river's midpoint was quickly agreed upon, in accordance with international law. South Africa eventually relinquished control of Walvis Bay to Namibia on 1 March 1994, following a remarkable process of bilateral negotiation and a period of joint administration. The critical turning point came in the 1993 Multi-Party Negotiating Process, in which Namibian diplomats were able to exploit the anomaly over Walvis Bay's envisaged position within South Africa's planned new provinces to Namibia's advantage (Simon 1996a).

In view of the fact that South Africa had been an oppressive colonial power in Namibia, the peaceful resolution of the long-standing boundary disputes was a major achievement. It is ironic, therefore, that in the late 1990s there has been a simmering dispute – in the course of which shots have been exchanged – between Namibia and its supposedly fraternal neighbour Botswana, over an ultimately insignificant and seasonally inundated island in the Chobe River. Known as Kasikili in Namibia and Sedudu in Botswana, the sandy island will now have its fate decided, at great cost to the litigants, by the International Court of Justice, as a bilateral joint technical committee failed to resolve the matter.

One further possible and far more dramatic future change to the political map of southern Africa should be mentioned. Lesotho is a classic enclave state, entirely surrounded by South Africa, on which it is highly dependent for wage labour remittances and the vast majority of its imports and exports. During apartheid, foreign aid formed one of the largest sources of government income, but recently this has started drying up, and the numbers of migrant workers are falling (to about 80 000 in 1996), with the result that the remittances sent home from South Africa are also in decline. Development options are extremely limited in the mountain kingdom, which has been wracked by political uncertainty in the late 1990s. A fundamental debate has been under way in recent years over the country's future viability as an independent state, and some groups advocate seeking incorporation into a democratic South Africa. Although relatively unlikely, not least because the advantages to South Africa and many Basotho would be outweighed by the costs, the equation could change (Lemon 1996a).

Internal territorial organization

Where the governments of newly independent states espouse radically different ideologies from those of their colonial or minority predecessors, it might be expected that they would seek to reorder the national territory into spatial units which accord more closely with changing realities and new policies. Surprisingly, however, colonial provincial boundaries have often been retained. This could be for reasons of political expediency (such as an unwillingness to stir ethno-regional sensitivities by opening up possibilities for expansion) or because of a lack of resources to undertake the major investigations that such changes would require.

Angola and Mozambique provide the most obvious examples of unchanged colonial boundaries in southern Africa. In both countries, overtly Marxist–Leninist liberation movements, the MPLA and Frelimo (Frente de Libertaçâo de Moçambique) respectively, came to power at independence in 1975, at a time when many countries in the South were experimenting with variants of socialism. Both Angola and Mozambique embarked on radical projects of socialist transformation, with far more rapid nationalization of commercial farms and properties than intended, because of the sudden exodus of virtually all the Portuguese settlers. Angola was embroiled in a destructive civil war and suffered successive South African invasions from the outset, but Mozambique made impressive gains in social service provision and meeting basic human needs before civil war destroyed much of the infrastructure and caused mass population displacement. Neither country changed its inherited Portuguese provincial boundaries, however, apart from Angola's division of Lunda Province into two, Lunda Norte and Lunda Sul, in 1978 (Sidaway and Simon 1993).

Zimbabwe, which gained internationally recognized independence in 1980 under Robert Mugabe's ZANU (Zimbabwe African National Union) government, also retained existing provincial boundaries. Although ZANU espoused Marxist rhetoric during the 1980s, the reality of its policies was generally far from radical.

By contrast, Namibia (like South Africa – see below) gained independence at a time when geopolitical options were far more limited, the 'space' for socialism having seemingly evaporated with the collapse of the USSR and the increasing international hegemony of Western liberal democratic ideologies. SWAPO (South West African People's Organization), which was destined to form the new government, effectively abandoned its radical socialist rhetoric ahead of the pre-independence elections in 1989, committing itself to multi-party democracy and a mixed and open economy with a liberal foreign investment policy, while striving to overcome the divisive apartheid legacy. These objectives have indeed characterized policy since 1990, although some critics have increasingly questioned the level of political commitment to redressing the impact of apartheid.

Nevertheless, Namibia instituted a full and careful process of replacing the regional and local boundaries that had been enforced by the South African apartheid state. The recommendations of the First Delimitation Commission were accepted and implemented, establishing a new system of thirteen regions (see Figure 4.1), each comprising between six and twelve constituencies. Boundaries were drawn so as to bridge the historically institutionalized division between the homelands, with their communal tenure and farming systems, and commercial lands in former white areas (see the discussion of South Africa below), and to facilitate accessibility, senses of community, balanced political representation, and regional development. Similarly, the boundaries of urban local authorities and their constituent wards were now explicitly non-racial and designed to promote urban integration. Although the new regional councils and smaller local authorities – especially those established from scratch in the former homelands – faced major imbalances in powers, resources, and demands upon them, their early performance has generally been encouraging (Sidaway and Simon 1993; Simon 1996b, 1996c; Swilling 1997).

Regional institutions

There is currently an unprecedented degree of flux and debate regarding the nature and role of regional institutions within southern Africa. This reflects their origins and purposes during the era of white minority rule in South Africa, and the changing needs and expectations as a result of political transitions within the region as well as the profound systemic changes in the world economy and its management (Sidaway and Gibb 1998). Two potentially conflicting trends are currently evident. On the one hand, regional trading blocs are emerging in various parts of the world, in an effort to promote South–South trade and especially to enhance the South's collective bargaining power with the major economies of the North. On the other hand, the new World Trade Organization (WTO) is implementing a major initiative towards 'free trade' which involves a phased removal of protectionist measures imposed by individual countries or trading blocs. Many current arrangements will therefore require modification, as the WTO has strong powers of enforcement.

The Southern African Customs Union

By far the oldest regional institution is the Southern African Customs Union (SACU), established in 1910 to protect the three weak former High Commission territories – now Botswana, Lesotho, and Swaziland (the 'BLS' countries) – when the Union of South Africa was created (see Figure 4.2). A customs union is an agreement between countries in terms of which trade barriers among the members are removed and common tariffs or quota restrictions are imposed on trade with non-members. SACU has always been run by South Africa on terms which, after their independence, the BLS countries deemed highly favourable to itself. Successive renegotiations led to revision of the treaty in 1970 and a more equitable share-out of customs revenues, but dissatisfaction has remained and each state has at one time or another actively considered withdrawal. Ultimately, though, the high proportion of government revenues accruing from SACU has served to retain their membership.

Following Namibia's independence and its membership of SACU in its own right, the size of its revenue share proved difficult to calculate and was contested. By 1991, South Africa claimed that it was

FIGURE 4.2 Membership of SACU

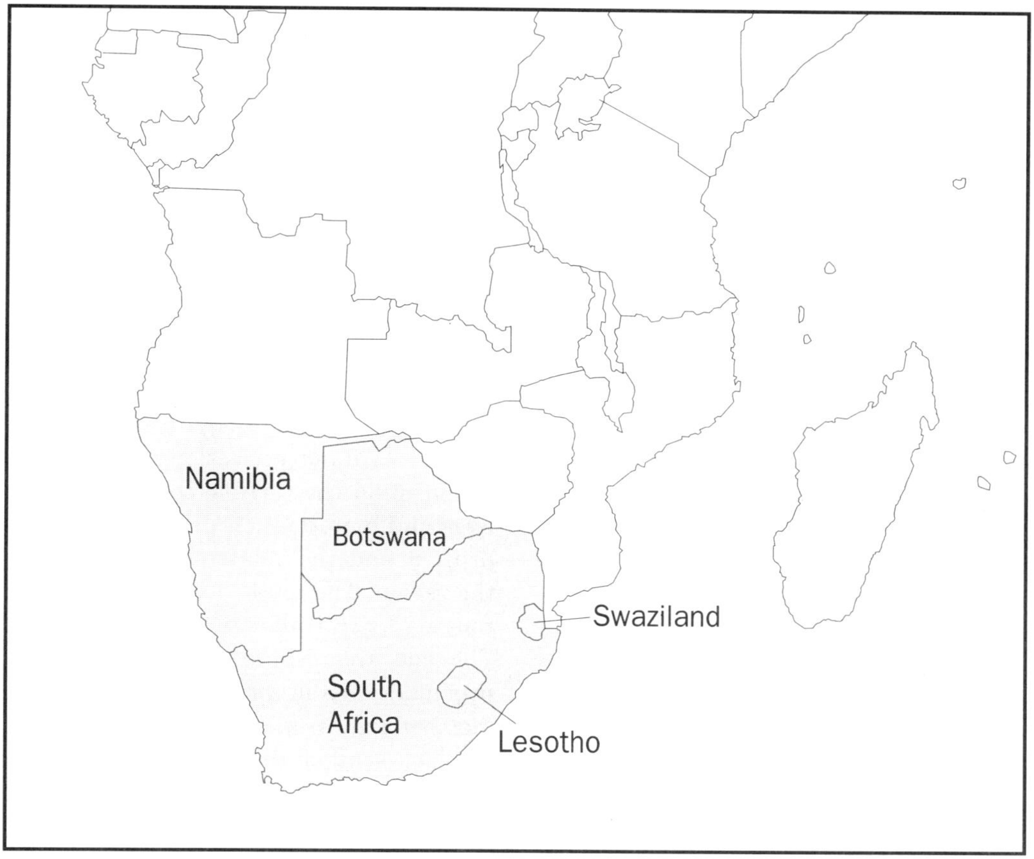

incurring losses from SACU and sought to change the nature of the organization. Protracted negotiations among the member states have continued, with some delays during South Africa's own political transition. To date, however, no firm conclusions have been reached, either to continue as before, to change the terms of operation, to broaden the membership, to merge with other institutions, or to disband SACU entirely. SACU remains one of the most asymmetrical customs unions in the world, in terms of the vast difference in economic size and strength between South Africa and the other members.

The Southern African Development Community

The Southern African Development Community (SADC) is a rather different type of organization. It was originally initiated in 1980 as the Southern African Development Co-ordination Conference (SADCC) by the Frontline States (Angola, Botswana, Zambia, Mozambique, and Tanzania) opposed to apartheid South Africa, along with Lesotho, Malawi, Swaziland, and the newly independent Zimbabwe. In 1990, Namibia joined as the tenth member. As its name suggests, SADCC was not a customs union but a rather looser grouping designed to reduce the members' individual and collective dependence upon South Africa, to promote economic co-operation and regional integration for mutual benefit, and to obtain additional development resources and international co-operation for their 'economic liberation'.

SADCC consciously sought to avoid the errors of many previous attempts at regional institution-building in various parts of the South, which had been undermined by unrealistically high aspirations as well as rivalry among members. The objectives set were accordingly modest and only a small secretariat was established. The bulk of work was undertaken by civil servants of member states on the basis of portfolios distributed among members according to existing expertise or strategic involvements. Heads of state met annually at a summit to take stock and prepare for the donors' conference thereafter. This conference was the main method of raising funds, in a format in which SADCC priorities were offered to various bilateral donor governments and multilateral donor organizations, each of which in turn had its own policies and priorities. SADCC projects were regional in scale, involving more than one member state, in order to ensure that the organization's efforts were strategic and complementary to the development efforts of individual member countries.

During the 1980s, SADCC gained importance as a vehicle for mobilizing funds and international support in the face of increased South African destabilization of its neighbours. Approximately 60 per cent of development funds were committed to the transport and communications sector, especially the rehabilitation and upgrading of railway and harbour corridors in order to diversify efficient options and provide alternative routes to those through South Africa. This was rightly seen as the principal way in which dependence on South Africa could be reduced. Despite criticism by some observers, SADCC was generally quite successful in relation to its objectives, especially bearing in mind the terrible cost in terms of lives and material destruction of destabilization and the related conflicts within the region. The total cost has been put at roughly US$60 billion for the 1980s alone. The one area where negligible progress was made was in increasing trade among member states; this remained at a mere 4–5 per cent of the members' combined total trade throughout SADCC's existence.

Changing international and regional circumstances, especially the constitutional negotiations to end apartheid in South Africa, forced a re-evaluation of SADCC's role and operations. In 1992, it was relaunched as the Southern African Development Community, with new objectives, including the establishment of common institutions to promote more formal economic integration. Although the new South Africa joined as the eleventh member following the 1994 elections, and Mauritius (the first francophone and island member) as the twelfth in 1995 (see Figure 4.3), progress towards a new level of activity has been slow. This reflects

FIGURE 4.3 Membership of SADC

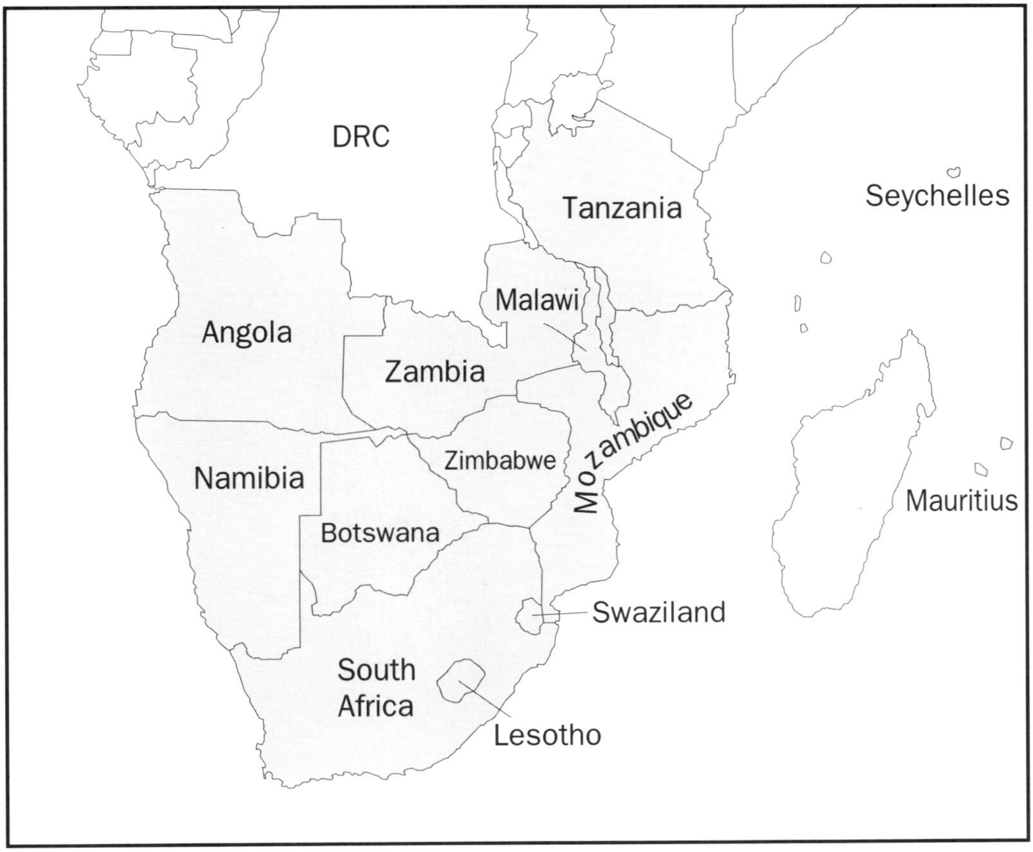

the ambivalence of members. SADC has hitherto not actively sought expansion of its membership, but the organization has been responsive to changing geopolitical and geo-economic circumstances; in late 1997, Laurent Kabila's Democratic Republic of Congo and the Seychelles applied for, and were promptly granted, membership. (Under Mobutu Sese Seko, Zaire had deliberately been kept out.) Hence, SADC now has fourteen members, including three francophone countries, two of these also being offshore island states. Consequently, some changes to SADC's basis of operation may be expected as the regional coherence is arguably diluted and the degree of overlapping membership with COMESA increases.

The Common Market for Eastern and Southern Africa

The Common Market for Eastern and Southern Africa (COMESA) has by far the largest geographical scope of the three regional institutions, with twenty-four members stretching from the Limpopo River to the Red Sea (see Figure 4.4). The objectives of this organization are explicit in its name. It was launched in 1994 as a successor to the Preferential Trade Area for Eastern and Southern Africa (PTA) in a manner and for reasons similar to those underpinning the metamorphosis of SADCC into SADC. A preferential trade area is a less complete economic arrangement than a customs union: trade barriers between member states are reduced but not abolished, and each member is still free to charge its own tariffs or impose its own quotas on trade with non-members.

The PTA was originally established in

FIGURE 4.4 Membership of COMESA

1981, in pursuance of the OAU's Lagos Plan of Action to promote regional economic co-operation and integration on the continent as a way of enhancing trade between peripheral nations and promoting Africa's collective bargaining position relative to other powerful states and trading blocs. It achieved significant success in some respects, especially the promotion of trade among members by means of the PTA clearing house. This was a mechanism designed to remove the constraint of having to pay for imports and exports within the PTA using hard currency – which was inevitably in short supply. Instead, trade was valued in the UAPTA (unit of account of the PTA), with balances adjusted at set intervals. PTA travellers' cheques were produced, so that bank charges and profits on currency exchange remained within the region. The principal beneficiaries of the PTA were the economically strongest members, especially Zimbabwe in the south and Kenya in the north. This asymmetrical distribution of costs and benefits has been a persistent problem with regional institutions.

Again, since 1994, COMESA has made little progress towards its new goals. Recent allegations of financial mismanagement by the Secretary-General and his subsequent suspension in mid 1997 have not helped, but the principal reason for delay has been the ongoing stand-off with SADC over ways to rationalize the now overlapping aims, objectives, and methods of the three regional organizations in the post-apartheid era. This problem has been exacerbated by the increasing overlap of membership. Neither SADC nor COMESA has been willing to disband, and the basis of any merger has also been disputed. SADC has suggested that it should be the sole body in southern Africa and that COMESA should concentrate on central and eastern Africa – in other words that the two bodies should cover distinct regions and not have overlapping membership. COMESA rejected this proposal, however. During early 1997, SADC called on its members to withdraw from COMESA, and by the end of June Mozambique and Lesotho had indicated their intentions to do so. Namibia was rumoured to be considering this step too. COMESA responded with alarm and called for a joint summit between the two organizations. This now seems likely in an effort to reach some workable accommodation.

South Africa: the imprint of politics

Socio-political context

The political geography of South Africa reflects the political ideologies and systems pursued by its successive governments. Their dominant political view was that the population groups found in South Africa could not live together as one nation in one country; hence the division of land between whites and blacks by means of the Natives Land Act of 1913, the Group Areas Act of 1950, the Population Registration Act of 1950, and so forth. When apartheid was firmly entrenched, blacks were further divided along ethnic lines, a process which led to the establishment of homelands by the Black Authorities Act of 1951 and the Promotion of Black Self-Government Act of 1959 (with Transkei, Bophuthatswana, Venda, and

Ciskei – the 'TBVC states' – becoming nominally independent) (see Figure 3.8, p. 80). The policy of dividing blacks into smaller groups was aimed at protecting white minority rule by preventing blacks (as a majority) from forming a consolidated front. The underlying assumption was that blacks were not a 'nation' but instead comprised scattered, distinct, and mutually antagonistic clans and tribes – a typical European view of blacks as savage and uncivilized, an attitude that is discussed in more detail in Chapter 3.

Non-whites who were designated as 'coloureds' and Indians were not allowed to live in white areas but did not have their own homelands. Coloureds were considered to have no desire to be a nation with an own independent state (February 1991), and Indians were seen to be an unassimilable immigrant community not suitable to be part of South Africa. These two groups were, instead, confined to their segregated group areas scattered in many parts of the country. But a map of South Africa showing distinct designated areas occupied by different racial groups could not have been drawn without removing blacks, coloureds, and Indians from one area to another. The forced removals which ensued were not only aimed at separating racial groups, but also included the separation of one black ethnic group from another (a process discussed more fully in Chapter 6) and the consolidation of people scattered throughout the reserves into villages (Wilson and Ramphele 1989) (see Table 4.1). As a result of the practice of swopping people and places, an estimated 3.5 million blacks were forcibly removed between 1960 and 1983 (Platzky and Walker 1985). The policy of homeland formation and the forced removal of millions of black people were denounced as an international crime against humanity by the United Nations in 1981 (Resolution 36/172A).

The rejection of apartheid by its victims (represented mainly by the liberation movements) in South Africa and by the international community forced President P. W. Botha's government to introduce constitutional reforms in the 1980s. As a result of these reforms, which centred on the creation of the tricameral Constitution, separate Houses of Parliament were established for coloureds and Indians in the Parliament of South Africa without essentially changing the principles and practice of the Group Areas Act by which population groups were required or restricted by law to reside in designated areas. This was an attempt by the apartheid government to institutionalize and enforce the principle of self-determination for coloureds and Indians while maintaining racial divides. It is not surprising that, three years after the adoption of the 1983 Constitution, P. W. Botha still maintained that the Group Areas Act was one law which would never be revoked in his lifetime.

Another development during the 1980s was the establishment of nine 'development regions' which transcended the existing four provinces of the Cape, Natal, the Orange Free State, and the Transvaal, and the homelands. Although development regions (see Figure 5.3, p. 131) were created in order to address socio-economic development in the homelands after the failure of industrial development plans launched in the 1960s, the underlying motive was to provide the basis for the future political map of South Africa, whether defined in federal or confederal

TABLE 4.1 Estimated numbers of people forcibly removed in South Africa, 1960–83

	Eastern Cape	Western Cape	Northern Cape	Orange Free State	Natal	Transvaal	Total
Farms	139 000		40 000	250 000	300 000	400 000	1 129 000
Black spots and homeland consolidation	19 000		40 000	40 000	115 000	400 000	614 000
Urban	151 000	32 000	20 000	160 000	17 000	350 000	730 000
Informal settlements	12 000		50 000	50 000			112 000
Group areas		409 000		14 000	295 000	142 400	860 400
Infrastructural and strategic	80 000				18 500	5 000	103 500
Grand total							3 548 900

SOURCE: Platzky and Walker 1985.

terms (Lemon 1987). It was an attempt by Botha's government to 'search for means of incorporating blacks within central institutions, and the retention of race or ethnic identity as the basis for political representation' (Cobbett et al. 1987: 18). The regions were founded on apartheid ideology and were intertwined with support for the homelands. Nevertheless, they benefited black people more than the previous regional development initiatives which had all but excluded them.

Under internal and external pressure, the apartheid state failed to sustain its segregated multinational project. Internal pressure was exerted by structures such as the United Democratic Front (UDF) and labour unions. Organized white business and intellectuals were becoming aware of the necessity of cutting back on apartheid spending – a move which led to ideological schism within the ruling National Party. Although the cost of apartheid cannot be accurately measured in monetary terms, critics have estimated (conservatively) that 10 to 21 per cent of the 1985/86 budget was devoted to financing its machinery. The official demise of the system was sealed in 1990 when liberation movements such as the African National Congress, the Pan Africanist Congress (PAC), the South African Communist Party (SACP), and thirty-three other organizations restricted in terms of the emergency regulations of the mid 1980s were unbanned following F. W. de Klerk's speech of 2 February 1990.

Post-apartheid territorial reorganization: provinces

When formal apartheid rule came to an end in the 1990s, it left behind a divided South Africa with gross inequalities and a population distribution which still strongly reflected apartheid planning. The first inclusive population census of 1996, for instance, shows that post-apartheid provinces such as KwaZulu-Natal, Eastern Cape, Northern Province, North West, and Mpumalanga, which contain large tracts of former homelands, had a high non-urban population (see Table 4.2). Furthermore, these provinces, excluding the Northern Cape, had the highest unemployment levels (43.3 per cent in the Eastern Cape, 32.2 per cent in KwaZulu-Natal, 36.4 per cent in Mpumalanga, 36.6 per cent in the North West, and 47 per cent in the Northern Province), compared to the Free State (24.4 per cent), Gauteng (28.7 per cent), and the Western Cape (17.3 per cent) (South African Institute of Race Relations 1996). The coincidence of the non-urban population and low levels of development in most of the homeland-dominated provinces is a living testimony to the role homelands played as dumping grounds for the 'surplus' black population.

TABLE 4.2 Preliminary 1996 census estimates of urban dwellers, South Africa

	Urban	Percentage	Non-urban	Percentage
Northern Province	490 000	11.9	3 638 000	88.1
Mpumalanga	1 014 000	38.3	1 632 000	61.7
Gauteng	6 911 000	96.4	260 000	3.6
North West	1 060 000	34.8	1 983 000	65.2
Free State	1 718 000	69.6	752 000	30.4
KwaZulu-Natal	3 341 000	43.5	4 331 000	56.5
Eastern Cape	2 188 000	37.3	3 677 000	62.7
Western Cape	3 703 000	89.9	415 000	10.1
Northern Cape	535 000	71.7	211 000	28.3
National totals	20 960 000	55.4	16 898 000	44.6

SOURCE: Central Statistical Services 1997.

The major task of the ANC-led Government of National Unity (GNU), which came to power after the April 1994 national elections, was to address the effects of decades of apartheid rule and to build a democratic, non-racial, and non-sexist future (African National Congress 1994). The vision for a democratic South Africa could not be carried out without changing the political map which represented the balkanization of the country. It was therefore necessary to do away with homeland boundaries and to reincorporate the TBVC 'states' into South Africa. Their reincorporation was carried out by the Convention for a Democratic South Africa (CODESA), later followed by the Multi-Party Negotiating Process, and was entrenched into the 1993 Interim Constitution of the Republic of South Africa. Resolution 31 adopted by the Negotiating Council of the Multi-Party Negotiating Process on 5 October 1993 ensured the restoration of South African citizenship to residents of the TBVC 'states'.

Reincorporation was followed by the delimitation and demarcation of regions (Fox 1995) – a process carried out by the Commission on the Delimitation/Demarcation of Regions/States/Provinces (CDDR). A large number of submissions by political parties took the development regions as their point of departure, although the PAC favoured the use of the four provinces of the Cape, Natal, Orange Free State, and Transvaal (Muthien and Khosa 1995; Khosa and Muthien 1997). After receiving proposals from political parties, interested organizations, and communities, the commission proposed nine provinces for post-apartheid South Africa. Hence the emergence of new provinces which transcend racial/ethnic and tribal divides (see Figure 4.5). The renaming of regions as provinces was suggested in a technical report of September 1992, and their boundaries were defined according to Schedule 1 of the 1993 Interim Constitution and formed the basis on which the 1994 national elections were held. As the new provinces are a modification of the 1982 development regions and were based on, *inter alia*, socio-cultural considerations, some of them have retained ethnic compositions established during apartheid. For example, the Eastern Cape has a Xhosa majority, the majority of the Zulu live in KwaZulu-Natal, the Southern Sotho in the Free State, and the Tswana in the North West. As South Africa has been a state without a single nation, the government is involved in the project of building a 'rainbow nation' which seeks to represent a common national identity for all citizens of South Africa irrespective of race, colour, creed, or sex. The 'rainbow nation' concept represents a complete reversal of the apartheid nation-building project, which sought to use racial and cultural differences to promote the development of distinct 'nations' within South Africa. Minority political parties such as the Freedom Front, the Conservative Party, and the Volksfront do not support the idea of one nation in South Africa, and wish to maintain a divided country.

The establishment of new provinces has led to boundary disputes, however. While these are sometimes seen as a 'bad omen' of things to come, affected communities have raised various concerns which are worth noting. The low population of the Northern Cape (which, according to the 1996 census, comprises only 2 per cent of the country's population) almost

FIGURE 4.5 South Africa's nine new provinces in relation to former homeland areas

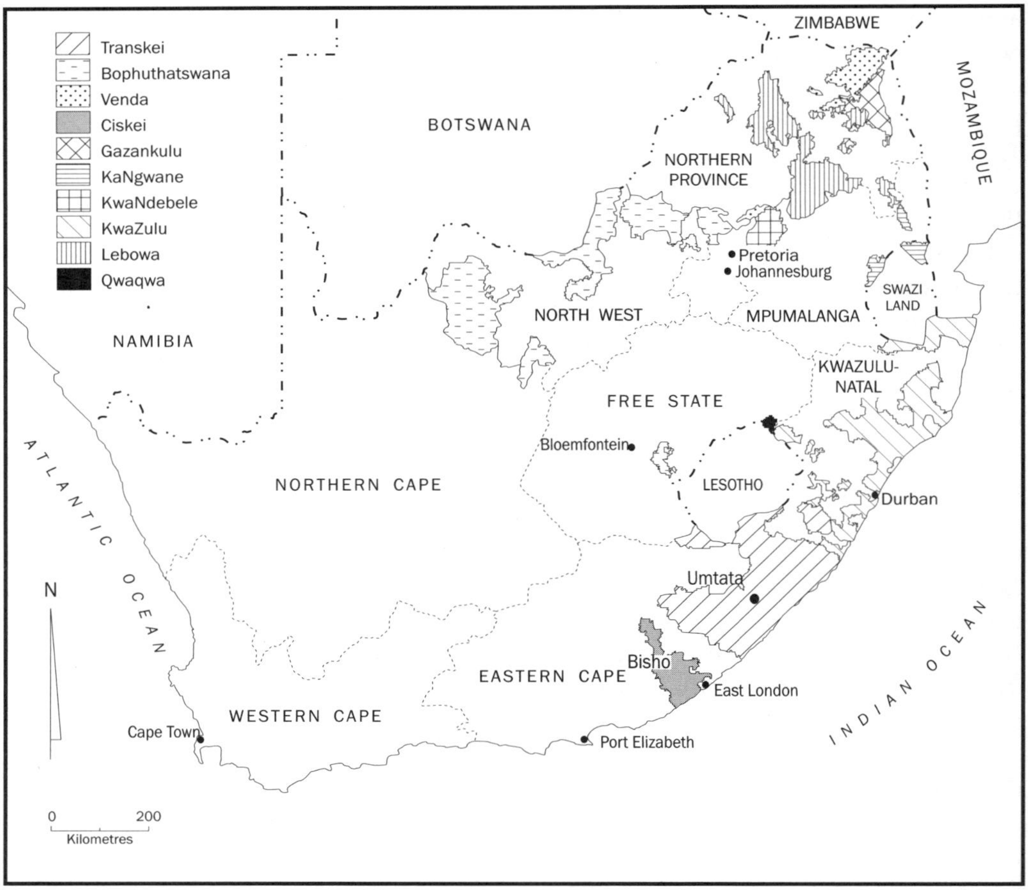

resulted in the whole province being eliminated and absorbed by neighbouring provinces. A number of areas around the country, such as Namaqualand, Clanwilliam, Pondoland, East Griqualand, Umzimkulu, Sasolburg, Kuruman, Brits, Moretele, Moutse, Groblersdal, and Bushbuckridge experienced provincial boundary disputes. The last-named area is used here to illustrate how the complexities of apartheid division and negotiated change have created such disputes.

The Bushbuckridge boundary problem stems from the process of delimitation/demarcation of regions/provinces which was set in motion with the establishment of the CDDR by the Multi-Party Negotiating Process on 28 May 1993 (see Figure 4.6). Many of the political parties and other organizations, including the Northern Transvaal Regional Political Discussion Forum, proposed the inclusion of Bushbuckridge in the Northern Province. The CDDR accepted the proposals

FIGURE 4.6 The disputed Bushbuckridge area

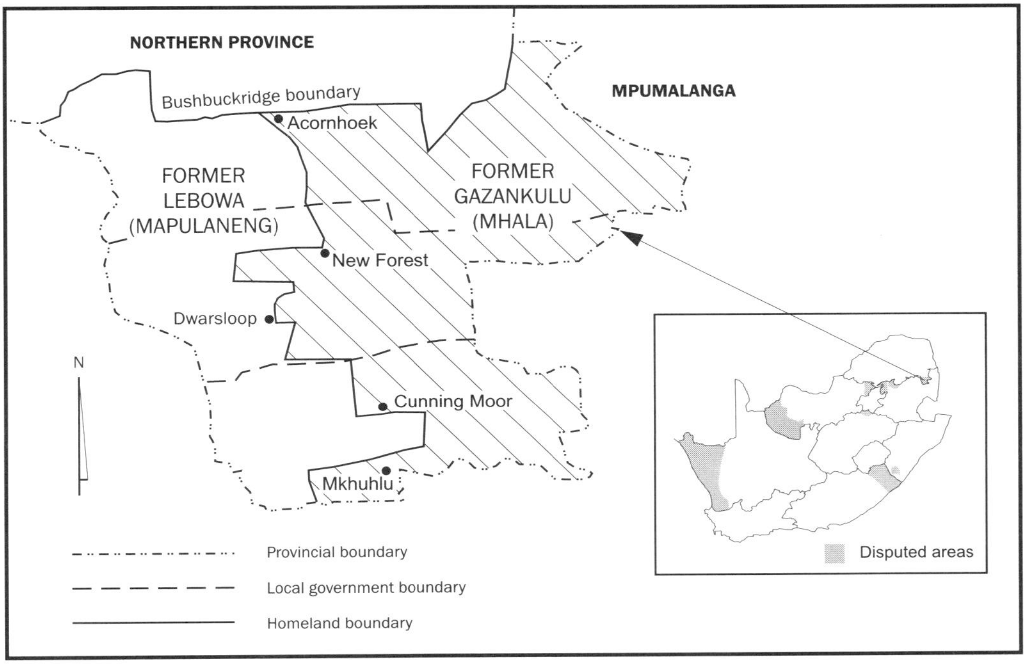

and on 31 July 1993 recommended its inclusion. Two reasons can be given for this. First, the two sections of Bushbuckridge (Mhala and Mapulaneng) were administered as parts of the former homelands of Gazankulu and Lebowa respectively, both of which now form a major part of the Northern Province. Second, Bushbuckridge was also part of the 1982 (apartheid) Development Region G – a region which has become the Northern Province in post-apartheid South Africa. Thus Bushbuckridge was allocated to the Northern Province mainly because the CDDR followed homeland boundaries.

The majority of the residents of Bushbuckridge, however, wanted their area to be transferred from the Northern Province to Mpumalanga because of their geographical location, their daily connections with Mpumalanga towns, and the perception of better development prospects there. The 1993 Interim Constitution made provision for solving this and similar disputes. Residents in disputed areas are entitled to petition the Secretary of Parliament for a referendum, and the change must be approved by the provincial legislatures and a two-thirds majority in the National Council of Provinces (NCOP) and a joint sitting of both Houses of Parliament. For example, in the new Eastern Cape province an attempt was made to call a referendum as a way to solve the disputes on splitting the province in half, but the required number of signatures to instigate a referendum could not be obtained.

Besides applying constitutional measures to the letter, disputes can also be solved by negotiation between political parties and/or provincial officials, and by commissions of inquiry. In the case of Bushbuckridge, the dispute was placed squarely in the hands of the then provincial premiers, Mathews Phosa (Mpumalanga) and Ngoako Ramatlhodi (Northern Province). The governments of the two provinces refused to hold a referendum, but instead sought to settle the disputes through ANC regional structures. Although both premiers agreed on the transfer of Bushbuckridge in 1995, the demand by the Northern Province for the incorporation of Groblersdal from Mpumalanga, in compensation for the loss of territory, complicated the transfer. In an attempt to resolve the dispute, the Minister of Provincial Affairs and Constitutional Development, Mohamed Valli Moosa, submitted an enabling bill to Cabinet in 1996. The bill was later withdrawn, because the ANC parliamentary caucus did not believe that it (the bill) would obtain the two-thirds majority required, as the two provinces differed on their terms of transfer. This and other disputes were not even on the agenda when the final Constitution was negotiated in Cape Town from 1994 to 1996. Failure to resolve the dispute resulted in intensive protests by residents (especially from May to June 1997), threats to make the area ungovernable (terminology used during anti-apartheid campaigns) and to boycott the national polls in 1999, plans for a sit-in at the Union Buildings in Pretoria, demands for intervention by President Mandela, and threats to take the government to the Constitutional Court. A proposal has been tabled for the joint management of Bushbuckridge by the two provinces, by which the area would be administered by the Northern Province, while Mpumalanga could provide services on an agency basis. This proposal amounts to creating a special administrative unit for Bushbuckridge, but does not solve the problem.

Restructuring the local state

The process of knitting together the state after the divisions created by apartheid involves the changing of local governments as well, because these are crucial in promoting non-racialism and the distribution of resources. The ANC's Reconstruction and Development Programme (RDP) sought to amalgamate an estimated 800 segregated local authorities into fewer than 300 authorities. The demarcation of their boundaries was aimed at ensuring that informal settlements on the outskirts of towns and cities, and urban settlements displaced behind former homeland boundaries, were incorporated into the jurisdiction of new local authorities (African National Congress 1994). According to the 1996 Constitution, local governments are, among other things, to provide democratic and accountable government for local communities, and to promote social and economic development. In black rural areas local governments have even brought together communities previously divided along tribal lines. The new local government boundaries of Bushbuckridge, for example, ignore the boundary between the former, ethnically based Lebowa and Gazankulu homelands in the Northern Province.

The restructuring of local governments, too, has not been without problems. Conservative whites, for instance, have seen the process as an expensive political hoax, a disruption of established communities, and a way of killing the goose that lays the golden egg; thus a typical *swart gevaar* (black peril) mentality prevailed and continues to do so. Local government elections held on 1 November 1995 were postponed in parts of the Western Cape, the whole of KwaZulu-Natal, and in Louis Trichardt and the Groblersdal/Motetema area in the Northern Province. The roots of the problems are many and varied. Gerrymandering was widely suspected in the Western Cape, while violence threatened democratic elections in KwaZulu-Natal. The Greater Louis Trichardt Transitional Local Council refused to amalgamate with the tribal areas of Sinthumule and Kutama, and obtained a court interdict to prevent the elections in 1995 (they were ultimately held on 29 May 1996). Other problems in the restructuring of local governments are related to the maintenance of racial divides. The white towns of Kuruman and Taung, for instance, are part of the Northern Cape, while their surrounding townships are in the North West. A similar situation holds for Groblersdal, where the white towns of Groblersdal and Marble Hall form an exclusive local government (in Mpumalanga) while surrounding areas for blacks belong to the Northern Province. Divisions of this nature are bound to affect the use of resources, which remains skewed in accordance with apartheid planning and administration.

Some of the major challenges facing post-apartheid local governments are the delivery of and payment for services, and the need to change structures of white power. It is worth noting that during P. W. Botha's 'reform' era a unified system of local government was avoided because it would have undermined the structures of white power (Lemon 1995). Since apartheid laws which guaranteed such structures have been scrapped, measures under new banners (such as community interest/identity) are likely to underpin post-apartheid local politics. Indeed, such notions affected the debates in the early 1990s over the integrated 'one city one tax base' approach versus more disaggregated two-tier models for the government of the major metropoles. Here the choice was between a strong or a weak metropolitan authority controlling the smaller substructures. One such variant was eventually adopted, with a relatively weak metropolitan authority seeking to integrate and co-ordinate the several metropolitan substructures (Lupton and Mather 1997; Swilling 1997; Lemon 1996b; also see Chapter 8). Figure 4.7 shows the four substructures delimited in Johannesburg after a hotly contested process centred on how Soweto and the other large, impoverished townships would be included and their upgrading funded.

The Masakhane Campaign is being used as a strategy to encourage residents (blacks in particular) to pay for municipal services rendered to them. The campaign is important, especially because township residents developed a strong culture of non-payment for services as a protest against apartheid. The payment for services within a particular local government has its own complexities, because new local governments (especially in urban areas) are a fusion of 'First' and 'Third World' conditions. These discrepancies, in

FIGURE 4.7 The new boundaries of the Johannesburg metropolitan region, with four substructures

SOURCE: Lupton and Mather 1997.

turn, make it difficult to manage the payment of the same rate for the same service (say, electricity) by residents from areas with profoundly different levels of basic infrastructure, let alone to strike a measure of understanding between the victims and beneficiaries of apartheid. Ratepayers' associations therefore mushroomed within specific areas during the 1995 local government elections. It is evident that some of the activist traditions built up by civic associations during the anti-apartheid struggles of the 1980s and early 1990s have begun to dissipate since

1994. This reflects the general legitimacy of the new local authorities, the loss of key personnel to other sectors, and other factors. Some civics are becoming more akin to conventional ratepayers' associations, but other communities are realizing that the struggle for resource redistribution will need to be resumed in a more activist way even under a legitimate government.

Conclusion

The interplay of global, regional, and local politics which informs our chapter is a useful context for understanding the political geographies of southern and South Africa. Although the break-up of the Soviet Union has affected the axis of global power politics which was contested across Africa and southern Africa in particular, the pendulum of the same politics still swings over the continent. In the late 1990s the US made a surprise 'discovery' of Africa as a market for investment – an apparent change of mentality from the long-standing 'African crisis' syndrome – while Africa's former colonial masters in Europe are still determined to influence, and even to control, events in the continent. The playing of politics by Europeans on African soil is likely to remain for some time to come. The appointment of Kofi Annan (of Ghana) as the seventh United Nations Secretary-General in 1997 was, for instance, preceded by strong-arm political tactics by the 'leading' nations whose vested colonial interests in Africa were visible.

The liberation of southern Africa and, ultimately, South Africa has created a new spirit of 'African renaissance', with South Africa taking its appropriate position as an important role-player in African affairs. South Africa took up the challenge by offering its skills on negotiation politics to the troubled Great Lakes region in early 1997. In southern Africa, the Mandela government has shown the political will to pay for some of the damage caused by apartheid (for example, it wrote off Namibia's colonial debts) and to contribute to the economic recovery of the region (through SADC and other bilateral programmes). These initiatives have been interpreted variously as 'bullying' or a handshake with former comrades, depending on the motives attributed and the perspective of the observer.

This chapter has highlighted domestic challenges facing the new democracy in South Africa. While the nation-building efforts driven by RDP projects and those of its successor, the Growth, Employment, and Redistribution (GEAR) strategy, are necessary and politically correct, the demand for redistribution is too high to be met within a short time. Some communities (such as the people of Bushbuckridge) are even resorting to seeking a change of place and thereby hoping to improve their fortunes. On the other hand, most privileged white communities remain determined to preserve the economic status quo within the new political dispensation. The neo-liberal, market-orientated nature of GEAR certainly restricts the scope for redistribution at national level. Political dynamics of this nature provide a useful framework for analysing the changing political geographies of South Africa, the region, and the continent.

5 Economies and economic development

ETIENNE NEL

The world has been subjected to profound economic changes since World War II which have impacted, to varying degrees, on all nations and communities. The popular notion of the 'global village' epitomizes the transition to a new world order which is increasing marked by:

- the reduction of barriers to trade;
- the rise in the significance of multinational corporations (MNCs);
- a relative reduction in the power of the nation state; and
- the universalization of consumerism and most forms of economic activity (Berry et al. 1997).

These trends have been facilitated by rapid technological advances in communications and transport and encouraged by international bodies such as the United Nations, the World Bank, and the International Monetary Fund (IMF). The impact of economic globalism has varied considerably from place to place, resulting in the emergence of 'winning' and 'losing' nations which are tied together by unequal patterns of production and trade. In an attempt to correct social and economic disparities in the world, a succession of development strategies has been pursued by nations, global institutions, and, more recently, by individual localities. The nature of the development theories and policies adopted has varied considerably.

This chapter will explore the key economic changes which have occurred since World War II and the mixed fortunes of applied development strategies. As in other chapters the scale of analysis will have a global, regional, and national focus. It must be emphasized, however, that scale distinctions have become increasingly artificial in a world moving towards ever-greater degrees of integration.

The world economy since 1945

The collapse of the socialist and communist political economies in the late 1980s and early 1990s has promoted capitalism to the position of undisputed global economic paradigm.[1] Capitalism has exerted a tremendous impact, shaping space, influencing labour and resource utilization, and articulating demand. Global interdependence has been actively promoted under capitalism, resulting in a situation marked, according to R. J. Johnston, by the reality that 'there is but one world economy, to which all places are linked ... [This] provide[s] the framework within which regional differences have evolved and are evolving' (quoted in Knox and Agnew 1989: 5).

The most obvious regional difference is that between the *North* and the *South* (see Figure 5.1). The North includes the economically advanced countries of North America, Europe, Russia and its former satellites, Japan, Australia, and New Zealand. These are largely wealthy states; for example, in 1994 it was estimated that Switzerland's Gross Domestic Product (GDP) per capita was US$27 500 (World Bank 1994). The South is the rest of the world, which is broadly similar to the popular notion of the Third World. The South has 75 per cent of the world's population but only 20 per cent of global income and 10 per cent of global industry. It consists of poor nations, one of the poorest of which is in southern Africa – Mozambique, with a GDP per capita of only US$80. Such industry which does exist in Southern countries is normally owned by foreign-controlled multinational corporations.

The world is divided into the North and the South by the Brandt Line, which is based on Gross National Product (GNP): the cut-off point in 1984 was a per capita GNP of $7 000. There are some exceptions to the general poverty of the South and the wealth of the North, for example the Organization of Petroleum Exporting Countries (OPEC) and the newly industri-

FIGURE 5.1 The world economy: major divides

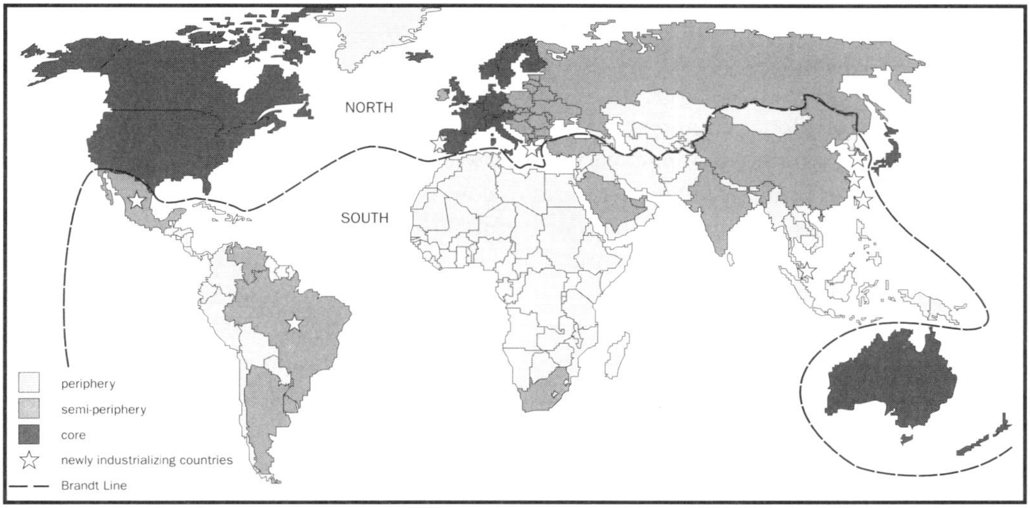

alizing countries (NICs), which have relatively high GNPs and are often characterized by advanced, technology-based economies.

Another useful categorization of the world is the division between the *core*, *periphery*, and *semi-periphery*, following the 'World System Theory' of the socialist thinker Immanuel Wallerstein. Core countries are characterized by high wages, advanced technology, and a diversified economy, while peripheral countries tend to have low wages, rudimentary technology, and a simple production mix. The semi-periphery, which broadly parallels the NICs, is seen as having a mix of the other two sectors. Wallerstein's theory maintains that the core dominates and exploits the rest of the world, and that this is reflected in trade and investment patterns, the flow of wealth, and the location of economic control (Knox and Agnew 1994).

The contemporary global economic system owes its origins to the Bretton Woods settlement concluded between the victorious Allied powers at the end of World War II. This system laid the basis for a world order based on stable economic relations anchored by a strong US dollar, fixed currency exchange rates, and the establishment of global financial organizations to co-ordinate economic activity and to encourage sound management and free trade. This was reinforced by the military might of the USA, targeted support for anti-communist states,[2] and general economic recovery as a result of vast direct US investment in many Western European countries through the Marshall Aid Plan. The Bretton Woods system permitted the economies of Northern countries to flourish in an era of stability, support, and security, and contributed to the full reconstruction of the economies of Japan and Germany (Scammell 1980; Brett 1985). The era of economic stability was based largely on what is known as Keynesian economic theory (after the great economist J. M. Keynes), which views state intervention and management of the economy as essential.

The major global organizations which owe their origin to the Bretton Woods agreement are:

- the International Bank for Reconstruction and Development, more commonly known as the World Bank, which was established as a bank for countries experiencing financial problems;
- the International Monetary Fund, which focuses on economic policy mechanisms;
- the World Trade Organization (WTO), which pursues global free trade. The WTO was set up in 1995 as the successor to the General Agreement on Tariffs and Trade (GATT), which had been established in 1948. Both organizations have sought to reduce tariffs and other trade restrictions.

International economic stability and support for trade by the above organizations helped MNCs emerge as powers in their own right, providing a basis for what is now referred to as the new international division of labour (NIDL) (this is discussed later in the chapter).

The post-war era of economic stability ended abruptly in the 1970s. Economic stringency dominated in the USA as a result of the cost of the Vietnam War, the cost of supporting anti-communist states and paying for goods imported as a result of liberal trade policies, and other outflows of capital. In 1973 and again in 1979 there were huge rises in the price of oil. Oil prices quadrupled in 1973 when the OPEC countries sought to end what they saw as the exploitation of their wealth, and the petroleum-dependent world was plunged deep into recession. As a consequence the global economy was struck by the following problems:

- the abandonment of fixed exchange rates and the fixed value of gold;
- global economic uncertainty and recession; and
- currency devaluations around the world, as countries had become too closely linked to the now vulnerable US dollar.

Reduced global dependence on the US dollar and the emergence of floating exchange rates led to investment insecurity and tremendous flows of currency around the world in search of safe investment opportunities (Brett 1985).

The increased price of fuel imports and the restriction of markets caused by global recession led to the closure of many ageing, cost-ineffective industries in the North and their frequent replacement by MNCs using cheaper labour in the South. The traditional areas which had dominated global economic activity in the immediate post-war period, for example the Ruhr in West Germany, northern England, and the north-eastern USA were characterized by textile, coal, and iron and steel manufacturing industries. These industries could no longer compete effectively because they were inefficient, using dated technology which was costly to run and maintain and used fuel excessively. In addition, rising salaries and the strength of trade unionism had squeezed many firms out of existence. These factors, along with the search for profits, encouraged the relocation of firms to areas of lower labour cost and resulted in the decline of these once-dominant regions to the status of 'rust belts' (Healey and Ilbery 1990).

The rapid growth in service-sector employment and the emergence of high-technology industry in certain parts of Northern countries have partially offset the effects of the crisis. The 'Sunbelt' cities of the southern USA and high-technology corridors such as Silicon Valley in California indicate that there have been some regions of resurgence. The switch in manufacturing techniques from 'Fordism', based on the mass-production techniques championed by Henry Ford, to the 'flexible production' of small quantities of high-value goods by skilled workers has become a key feature in the North. The rise in the importance of high-technology industries (often associated with scientific universities) focusing on aerospace, communications, defence, and biotechnology has been another distinguishing feature of Northern economies.

The economies of the South were severely handicapped by the 1970s' recession, and they had to seek additional funds to balance their economies now that they were paying so much more to import oil. Ironically, they were frequently granted loans by European banks with money that originated from the oil revenues (paid in dollars) to Arab countries. Their failure to repay these debts in the 1980s has bankrupted many states and forced them to restructure their economies based on the dictates of the World Bank and the IMF. Commencing with Mexico in 1982, most countries in the South have been forced to accept structural adjustment programmes

(SAPs) which restrict state economic intervention, promote balanced economies, and rely on free-market principles. These packages derive from the economic experiments in Britain and the USA in the 1980s. Critics vary over the implications and effects of these policies, which often seem to have aggravated economic conditions (Todaro 1994).

By the early 1980s many Northern countries were renouncing their Keynesian policies of state economic management in favour of development driven by market forces. Known variously as Thatcherism and Reaganomics (after the British and American leaders of the time), these policies overturned the principles of Keynesianism in favour of:

- reduced state economic intervention;
- encouraging privatization of state-run industries;
- reducing social welfare expenditure; and
- moving to facilitate and encourage the private sector.

The connection between these policies and the dictates of structural adjustment programmes should be obvious, and they have been adopted by many countries throughout the world. This chapter will later discuss the adoption of such policies by South Africa's Growth, Employment, and Redistribution (GEAR) strategy. Elsewhere, following the collapse of communism, many ex-communist states have attempted to emulate the economic success of Europe and North America. China has undoubtedly been the most successful and, even though not having fully abandoned its previous political ideology, has become the world's fastest-growing economy, with growth rates of up to 20 per cent per annum being recorded in its special economic zones.

The internationalization of manufacturing

Linked to the changes detailed above has been the internationalization of manufacturing, production, and control, a process marked by increasing dependence between places on a vastly increased spatial scale. Production, trade, and consumption have moved from a national to a global focus, and a key causal feature of this integration has been the new international division of labour (NIDL). International integration in terms of the production system of a single MNC can now be dispersed to multiple locations around the world. The following paragraphs introduce this increasingly acronym-ridden branch of geography, examining NIDL, MNCs, NICs, and EPZs.

As a result of the new international division of labour, a firm's labour and other activities are spread between multiple countries to reduce costs, maximize profits, and ensure market share. This is frequently achieved by exploiting cheap, non-unionized labour in Southern countries. In complex industries, such as car manufacturing, the globalization of production involves regions and countries specializing in the production or design of specific components in which they have a competitive advantage. Figure 5.2 shows the countries which manufactured components of the 1989 Pontiac Le Mans. In this way the regional specialization typical of Fordism has been replaced by regional dispersal.

FIGURE 5.2 The new international division of labour: production of the Pontiac Le Mans

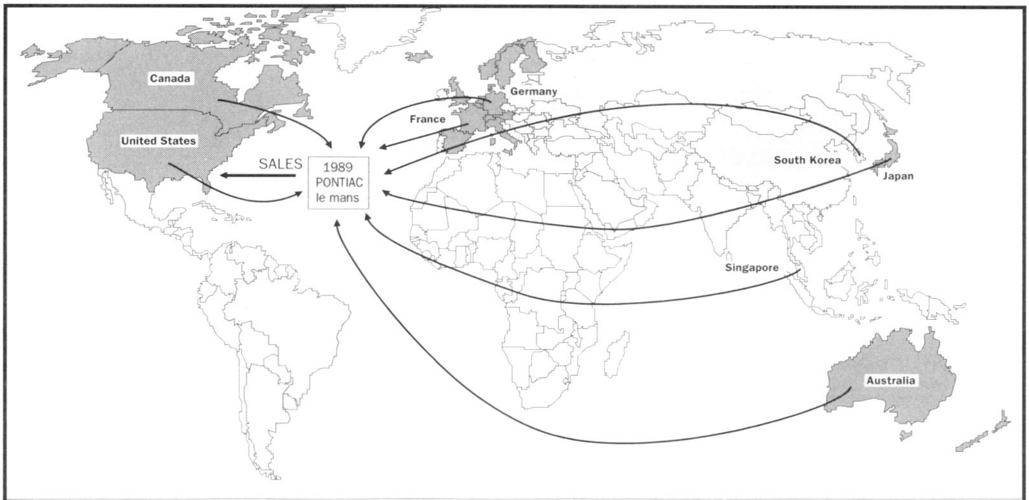

Canada
transmission and automatic transaxle

United States
transmission and automatic transaxle
manual transaxle
fuel pump
fuel injection system
rear axle components
steering components

France
manual transaxle

Germany
design

South Korea
1.6 litre engine
electrical (wiring) harness
brake components
manual transaxle
tyres
stamping of exterior body parts
rear axle components
windshield glass
battery
final assembly

Australia
2.0 litre engine

Japan
sheet metal

Singapore
radio

The NIDL is driven by large corporations which have a global reach – the multinational corporations. By 1995 there were 37 000 MNCs in the world, a fivefold increase since 1970. In 1969, United States MNCs produced $140 billion worth of goods, which was then more than that produced by any national economy, with the exception of the USA and the Soviet Union. MNCs now employ 65 million people and account for 10 per cent of total employment in the North. The world's largest MNCs, such as General Motors, Exxon, Ford, and IBM, are each economically stronger than many of the world's nations. General Motors, the world's largest corporation, employed 756 000 people in 1995 and had an annual turnover of nearly US$125 billion. In the same year South Africa's total Gross National Product was US$80 billion, equivalent to that of the Ford Motor Corporation (Knox and Agnew 1994).

As part of the NIDL a number of Southern nations have tended to monopolize new industrial activity, to attract MNCs, and to emerge as distinctive economic powers in their own right. Known as the newly industrializing countries, the most prominent are the South-East Asian 'Tiger Economies' – Singapore, Hong Kong, South Korea, and Taiwan – as well as Brazil, Mexico, Portugal, and Greece (see Figure 5.1). These countries are characterized by rapid economic growth, the operation of MNCs within them, and an increasing share of the global market. Despite the 1997–98 recession in South-East Asia, and in the light of the recent rise to power of China as an economic giant, the region is set to become the locus of half of the world's economic activity in the early years of the twenty-first century.

Key growth factors in these countries include the export-orientated nature of an economy based on industrial activity and the support rendered by national governments to enhance productivity and efficiency. From a planning perspective, the establishment of Export Processing Zones (EPZs) – export-orientated industrial zones adjacent to international ports or airports – offering incentives such as zero tax ratings to firms, has lured in MNCs and helped drive the economic booms which have been experienced. These zones can also be found in southern Africa's strongest economies, as will be seen later.

Development strategies

One of the dominant themes in recent global economics has been the search for appropriate development strategies which can bring about improved economic and social conditions in the South. National development strategies have been pursued by all countries in the South in the post-independence era. Initially these strategies were strongly influenced by the development experiences of both the capitalist and communist North. In recent decades, however, they have focused more on small-scale, localized interventions using indigenous resources and skills rather than seeking to develop on Northern lines. In this section earlier development theories are briefly discussed as a prelude to an examination of a range of recent development strategies.

Early development theories

1 The *diffusionist strategy*. From the 1960s many countries in the South adopted this strategy, which provided a 'recipe' for development based on the Northern growth experience. 'Stage models' which allegedly showed how national economies evolve from a 'backward' traditional stage to advanced capitalism and high mass consumption were developed by theorists such as Milton Friedmann and W. W. Rostow. In practice, however, the economic growth which was achieved did not spread, or diffuse, as it was supposed to, from the limited enclaves of growth to the rest of the country (Fair 1982).
2 The *dependency strategy*. During the 1970s the limited success of the diffusionist model was contrasted with what appeared to be successful experiences in communist China and South America. Underdevelopment in Southern countries was blamed on colonialism, which had stripped resources from them and created dependence on the North. It was argued that development could be achieved only by means of independent action and delinking from the world economy. Despite the moral legitimacy of such arguments, the ability of small countries to delink themselves from what has become a highly integrated world economy is limited, as the experience of Tanzania vividly illustrated (Fair 1982).

3 The *basic-needs approach*. This approach sprang from the failure of many Southern countries to achieve high rates of economic growth and from the parallel revival of neoclassical (free-market) ideas. The strategy was also influenced by the belief in the 1970s that China's achievement of mass welfare through simple, non-technical methods might well provide a model for addressing the basic needs of the very poor; this was an alternative to the fallacy that South could attain the development levels of the North (De Souza 1990). The basic-needs approach promotes the removal of legal and financial barriers to individual improvement. Emphasis is placed on improving the conditions of the poorest and facilitating their access to resources. Finally, service extension and the use of small-scale, labour-intensive modes of production are deemed essential.

The weaknesses of this approach include the high cost of attempting to apply it on a large scale, especially in the light of the poor transport and infrastructure systems in the South. The revelation of problems with the Chinese experience and the difficulties of winning support from the elites in the South, together with problems of providing back-up services, have limited the strategy's success. Nevertheless, it remains a contemporary development alternative.

The neoclassical revival

The revival of free-market, neoclassical economics in development thinking (sometimes referred to as a 'counter-revolution') has drawn great attention to price structures, reduced government intervention, and the freeing of the market. The adoption of such principles by international bodies such as the World Bank and the IMF has had a fundamental influence on world development. A significant proportion of world aid, especially to the South, is now conditional on the recipient country adopting policies which are approved by the IMF. This usually leads to the introduction of SAPs, as mentioned earlier. SAPs have been criticized for the following reasons:

- They simplify the problems of the South and prescribe a single, universal cure – structural adjustment.
- The North appears to benefit more than the South, since weaker Southern currencies result in cheaper imports of raw materials.
- Debt repayment is having a disastrous effect on Southern budgets, with employment, health, and education suffering as a result.

Development from within: local self-reliance

The failure of foreign development strategies, as well as government bankruptcy, mounting unemployment, and the economic and debt crises, has forced many communities in the most marginal areas to adopt an inward-looking focus and launch independent development endeavours. Strategies which are variously referred to as survivalist, local coping, self-help, and self-reliance are becoming prominent in many developing countries (Taylor and Mackenzie 1992). Such strategies focus on using locally available resources and skills in an appropriate fashion. Often the aim is survival, although there are many instances of the successful adaptation of available (and often degraded) natural resources to compete on the open market, for example by peasant farmers in the Machakos district of Kenya (Binns 1995).

Advantages of this approach include the fact that local people are able to control the process and are not dependent on external support and skills. On the negative side, there are limits to what communities can do without outside support. In addition, local self-reliance is not an alternative to national development, but rather should be seen as an adjunct to it. Tentative support for such initiatives is being given by the World Bank, and South Africa's Reconstruction and Development Programme (RDP) was influenced by this approach.

Regional development and the rise of the locality

Cuts in regional development expenditure by the state have occurred as part of a counter-revolution in development thinking. Interestingly, the intervention which now occurs often tends to be driven by a desire to unlock or work in conjunction with the private sector. A case in point, described later, is South Africa's policy of Spatial Development Initiatives (SDIs). The encouragement of urban redevelopment corporations, dockland renewal schemes, zones of high-technology industry, and EPZs is typical of the reduced, market-driven approach to development intervention which now prevails (Healey and Ilbery 1990). Although globalization has inevitably diminished the importance of the nation state, the same is not true of the locality. Individual cities or areas now compete in the global market as centres of investment and comparative advantage. Such action takes the form of what has come to be referred to as 'urban entrepreneurialism' or 'local economic development' (LED). Leading cities such as New York, London, and Tokyo posture as world capitals, while other centres compete for a share in the global market, as centres of supply, production, control, or consumption (Judd and Parkinson 1990). Political liberalization and reduced central state control have accelerated these trends. There are over forty-seven identified urban economic strategies, ranging from investing on the stock market to tourism promotion, marketing, infrastructure provision, and support for new firms.

Southern Africa: economy and development

From an economic perspective, defining southern Africa is relatively simple. The self-identification of bonds of economic and political compatibility in the region facilitated the establishment of the Southern African Development Co-ordinating Conference (SADCC) in 1980, which was replaced by the Southern African Development Community (SADC) in 1992 (discussed in Chapter 4). While the former began its life as a regional grouping to oppose apartheid South Africa and lessen dependence on that country, the latter body represents the new entity formed after the collapse of apartheid and the reacceptance of South Africa into the region. The countries constituting SADC are Angola, Namibia, South Africa, Botswana, Zimbabwe, Mozambique, Lesotho, Swaziland, Zambia, Tanzania, Malawi, Mauritius, the Seychelles, and the Democratic Republic of Congo (see Figure 4.3, p. 101). One of the region's defining characteristics was opposition to former white rule in Rhodesia, Angola, Mozambique, and apartheid South Africa; in return for this SADCC countries experienced destabilization and civil wars which exacted a heavy economic price (Libby 1987).

The economies of southern Africa

Southern Africa's regional economy is small, being valued at US$170 billion in 1995 and only contributing a humble 0.6 per cent of the world's GDP (Louw 1992). Considerable inequality prevails in the region, with South Africa dwarfing the economies of its neighbours. Most countries are heavily dependent on primary economic activity such as the production and export of minerals and agricultural products. Secondary industry, or manufacturing, is not a major element in the region, although exceptions are found in the cases of South Africa, Zimbabwe, and Mauritius. Agricultural activity is the dominant form of employment and a key facet of the economies of most countries, but the region periodically has to import food. The most severe case was the 1992 drought, which resulted in emergency food imports from North America.

Table 5.1 clearly shows the economic dominance of South Africa in terms of a range of key indicators. The depressingly low economic scores at the national and per capita levels in Mozambique, Malawi, and Tanzania reveal the existence of vast development backlogs in the region, as do the generally low levels of life expectancy. The small population size of countries such as Lesotho and Swaziland impacts negatively on their investment attraction and government revenues.

An issue of concern is the narrow base of many of the region's economies, most of which tend to rely on the production and export of only one or two primary products. The near-collapse of the Zambian economy in the 1970s after the fall in the world price of copper, the export of which generates 91 per cent of the country's foreign earnings, exemplifies the dangers of an economy which is not adequately diversified. Examples of other countries' dependence on a narrow economic base include Angola's heavy reliance on oil export from its Cabinda region, Botswana's dependence on diamonds, Namibia's on diamonds and uranium, and Tanzania's on coffee (Hodd 1991).

The economies of Zimbabwe, Mauritius, and South Africa are more diversified, but have been subject to various extraneous forces. Mauritius relies on export-orientated industry producing mainly clothing and textile products for the world market, a situation which results in vulnerability to global price fluctuations. In Zimbabwe, policy shifts between proclamations of socialist transformation and nationalization of property on the one hand, and measures of support for the free market on the other, have deterred investment. This flirtation with what Sylvester (1991) has termed 'contradictory development' has not allowed the country to reach its development potential. South Africa's

TABLE 5.1 Key economic indicators in the SADC region

Country	GDP (US$ billion) 1995	GDP per capita (US$) 1995	Exports (US$ million) 1994	Agriculture (% GDP)	Industry (% GDP)	Services (% GDP)	Life expectancy (years) 1994	Population (million) 1995
Angola	12.1	1 050	2 995	–	–	–	–	10.5
Botswana	4.3	2 750	1 845	5	52	43	68	1.4
Lesotho	0.9	500	134	11	45	45	60	1.4
Malawi	1.4	200	325	28	22	50	44	9.8
Mauritius	3.8	3 250	1 347	11	33	56	70	1.1
Mozambique	1.3	90	13	64	15	21	44	18.1
Namibia	3.2	2 000	1 321	12	26	62	59	1.6
South Africa	126.3	3 300	25 000	4	42	54	63	45.1
Swaziland	1.0	1 200	343	–	–	–	–	1.0
Tanzania	4.2	200	519	61	12	26	51	28.7
Zambia	4.1	400	1 075	16	47	37	48	9.4
Zimbabwe	6.3	550	1 775	22	35	43	60	11.1
Total	168.9	–	36 692				–	139.4

SOURCES: Ahwireng-Obeng and McGowan 1998; Economist Intelligence Unit 1998b; World Bank 1994.

economy stagnated during the last years of apartheid, there was large-scale disinvestment, and up to the mid 1990s economic growth rates were consistently negative (Libby 1987).

Economic policies in the region have taken a diversity of forms. Zambia and Tanzania pursued policies of African socialism based on a blending of African traditional value systems with ideas adopted from European socialism. The pursuit of villagization schemes designed to create foci of rural development, at the expense of industrial development, impacted severely on Tanzania's economy. Other countries, such as Mauritius and Botswana, followed free-market policies, encouraging investment and export-orientated industrialization. Botswana's economy was able to grow at levels of up to 13 per cent per annum in the 1980s, while neighbouring countries such as Zambia were declining at 3.7 per cent per annum. Although the high growth phase has slowed somewhat, Botswana's economy is still growing at 7 per cent annually. While the 'magic of the market' has definitely counted in favour of the two countries, their small populations and initially limited economic bases have contributed to their high growth rates (Harvey and Lewis 1990).

Angola's and Mozambique's initial courting of socialism led to destabilization by South Africa, periods of civil war, and economic stagnation. Both have since sought to participate more directly in the world economy, and in the case of Mozambique much rests on the ability of South Africa to revive the country's economy by means of ambitious projects such as the Maputo Development Corridor project, which provides an economic and transport link from Gauteng province in South Africa to southern Mozambique. Zimbabwe's pursuit of 'contradictory development' and threats of land nationalization have taken a heavy toll on the economy, including the suspension of international loans, the deterrence of international investment, threats to agricultural self-sufficiency, and the arousal of the ire of the world's leading economic powers.

The inability of southern Africa to develop independently of the global system is a vivid reality. The region's role in the NIDL is that of a supplier of raw materials and a limited number of manufactured products. With the exception of larger South African firms, the region does not participate as an investor or a supplier of services and sophisticated goods to the global market. The situation of global dependence has been accentuated in recent years by the devastating impacts of the 1980s' debt crises and the resultant imposition of SAPs on most of the countries in the region. Although there has subsequently been some evidence of better economic management and freer trade, there is also increasing impoverishment which, in part, precipitated the food riots in Zambia in the 1980s and the more recent civil unrest in Zimbabwe in 1997 (Mwanza 1992).

At a broader level, the future of southern Africa depends on its ability to attract foreign investment and to generate greater levels of internal economic activity. Even though links with the global market encourage dependence, the region's economic history and trade imply that if there is to be any realistic hope of expanding employment opportunities and overall levels of wealth, further international integration is inevitable.

Within the region, high levels of dependency exist. Tens of thousands of migrant workers still seek employment in South Africa, although numbers have declined from the peak of 250 000 people in the late 1980s (Libby 1987). In the case of Lesotho, earnings from migrants in South Africa are one of the key sources of national income. The country has become even more dependent now that revenues accrue for the sale of water to South Africa from the vast Lesotho Highlands Water Project. In direct terms, this project led to a 10 per cent real GDP growth in Lesotho in 1997 (Economist Intelligence Unit 1998a). At a broader level, the operation of the Southern African Customs Union and the Common Monetary Area ensures high degrees of dependence on South Africa by the other partners, namely Lesotho, Swaziland, Botswana, and Namibia.

Development strategies

Development policies pursued in the region have been influenced by international trends. The imitation of Western models and diffusionist strategies influenced many countries in the post-independence period. The Zimbabwean growth-centre policy is a good example of continued belief in diffusionist principles. Despite the ideal of achieving balanced growth through the designation of dozens of development points, it is only in Harare's neighbouring city of Chitungwiza that significant growth appears to have occurred.

In the 1970s, dependency notions of delinking from the world economy, nationalization of wealth, land redistribution, and socialist transformation were applied to varying degrees in Mozambique, Angola, and Tanzania. Zambia and Zimbabwe have pursued the principles more in policy terms than in practical applications. The poor economic performance of the first three countries and their subsequent attempts to seek greater degrees of participation in the capitalist world market are testimony to the limited success of their initial programmes.

From the 1980s, as has been detailed above, structural adjustment programmes have led to the grudging liberalization of the economies of all countries in the region and the pursuit of more market-driven strategies. The general move to neo-liberal, export-orientated strategies is most evident in Mauritius, Botswana, and Namibia. Generous industrial support measures and tax rebates apply to international firms setting up plants in these countries. All three have Export Processing Zones, with Mauritius being an acknowledged world leader in this regard. Botswana's sole EPZ, at Selibwe Phikwe, has experienced far less success than the other industrial zones in the country, where under the state-administered Financial Assistance Package firms have received incentives which are almost as attractive. In Namibia the EPZ in Walvis Bay has shown signs of attracting investment in the late 1990s (Economist Intelligence Unit 1998a).

Although there is naturally considerable evidence of low-scale, basic-needs intervention in areas such as food supply, farming, and water provision by charitable and non-governmental organizations, it is only in South Africa that policy has evolved to support local development initiatives through the decentralization of control to local governments. This will be described later.

Towards regional integration

Southern Africa has a long history of integration initiatives leading to what are now overlapping systems. As shown in Chapter 4, a number of systems currently exist: the Southern African Customs Union (SACU), the Common Market for Eastern and Southern Africa (COMESA – formerly the Preferential Trade Area), and, most recently, SADC. In August 1992, following political change in South Africa, the country was admitted to the recently renamed SADC. According to the Windhoek Treaty, the focus of SADC has shifted to an emphasis on development integration and 'enhancing the capacity of the regional economy to produce and exchange goods and services' (Louw 1992: 11) The goals of SADC are:

- to promote deeper economic co-operation and integration and the principles of balance, equity, and mutual benefit;
- to promote common economic, political, and social values; and
- to strengthen regional solidarity, peace, and security.

Longer-term goals include the co-ordination of monetary and fiscal policies, the elimination of internal tariffs, freer movement of capital and labour, and ultimately a regional parliament.

Since 1992, although there have been moves to promote closer regional integration and there is evidence of closer cooperation between individual countries, there is little to suggest that significant regional-level change has occurred. The most noteworthy international development projects which have been embarked on include:

- the Lesotho Highlands Water Project, involving South Africa's construction of mega-dams in Lesotho and its purchase of water from that country;
- the reconstruction of the Cahora Bassa Dam's electricity transmission lines and, more generally, the sharing of electricity resources;
- the construction of the Trans-Kalahari highway linking the Namibian coast to Botswana and Gauteng; and
- the construction of the Maputo Corridor between Gauteng and Maputo in southern Mozambique.

Earlier, SADCC had attempted to revive the Beira Corridor from Mozambique to Harare, but this was unsuccessful. Major obstacles appear to be the trade rivalry between South Africa and Zimbabwe and the sheer dominance of the South African economy (which is approximately three times the size of the other ten countries combined), which impede prospects for integration on an equal basis. According to Carim (1993: 3), the country 'stands out as an island in a sea of chronic stagnation and abject poverty'. Southern Africa's trade with the rest of the world is only 30 per cent that of South Africa's. It is therefore inevitable that South Africa will dominate, if not overwhelm, its neighbours if integration is further pursued. In addition, the bias in favour of external trade and fact that the countries of the region produce similar products militate against opportunities for closer integration.

South Africa: economy and development

South Africa's space-economy is marked by tremendous contrasts and inequalities. Even though the country actively participates in the global economy, co-operates with the global financial institutions, receives considerable international investment, and is itself a base for key multinational corporations, it is regarded as one of the most unequal societies in the world. Any understanding of the country's economy must take cognizance of the tremendous differences which exist between the sophisticated modern sector marked by world-class facilities and industries, and the abject poverty of millions subsisting in the shadow of the nation's wealth, living in squatter camps and isolated rural villages, eking out an existence from the informal sector or sub-subsistence agriculture and state pensions. This is a set of circumstances produced by the interplay of capitalist development and apartheid. Currently the political will clearly exists to attempt to correct these economic imbalances, and the government has devoted considerable effort to identify and instigate development programmes to this end. The country's democratic transformation might be threatened if the benefits of recent economic growth are not spread to the entire population. In the words of the Stutterheim Development Foundation (1997: 2), 'South Africa witnessed a political miracle and is now faced with the need for an economic miracle'.

This section will consider recent changes in the economy and their unequal impact, before proceeding to an examination of applied development policies.

The South African economy: from apartheid to the 1990s

South Africa's economy has always relied heavily on primary products. The production of minerals, particularly diamonds and gold, helped to fund the establishment of the country's modern corporations and made it the biggest and most sophisticated economy in sub-Saharan Africa. Without any doubt, profits accruing to local and foreign investors were enhanced by the payment of low wages to black labourers and the subservient position to which they were confined for most of the twentieth century (Wolpe 1972). The country relied heavily on a system of extreme exploitation of labour, referred to as 'racial Fordism' (Pickles 1991), using a police state to ensure a low-waged, servile working class, while ensuring whites positions of authority and privilege. According to Rogerson and Pirie (1979: 325), the 'expansion of capitalism in South Africa resulted in the institutionalization of massive structural imbalances' in racial and spatial terms.

The Second World War provided a major boost for the economy: the number of manufacturing firms in the country increased by 8 per cent and output by a staggering 115 per cent. High growth in the global economy after the war impacted positively on South Africa, and by the 1960s its economy had developed a large and expanding service sector and manufacturing was increasingly characterized by capital goods production (Nattrass and Ardington 1990). As in all aspects of social life, there was strong state control of the economy, channelling development, controlling trade, protecting local firms from imported goods through the use of high tariffs, and creating monolithic state-supported enterprises such as ISCOR (the Iron and Steel Corporation). Despite the intensification of opposition to apartheid in the 1960s, the economy achieved its highest-ever growth levels of up to 6 per cent per annum. By the 1970s South Africa had become largely self-sufficient in consumer and many capital goods; the cost, however, was dependence on industries sheltered by excessive tariffs and a cheap labour policy (Lipton and Simkins 1993). In trade terms, the country relied on the export of agricultural and mineral raw materials to fund manufactured imports, as it continues to do. This is the classic role of a peripheral nation in the world economy.

The 1970s' oil crisis, South Africa's economic isolation in the 1980s and 1990s, and associated disinvestment seriously affected the economy, causing many years of low and even negative growth. Following the Reynders Commission of 1972, the country gradually began to actively encourage export-orientated industry rather than the traditionally supported import-substituting industrialization. In the 1990s, export incentives were provided to boost export-orientated manufacturing, and protective duties were radically lowered after the Government of National Unity signed the GATT treaty (Cassim 1988). The reductions of tariffs in accordance with GATT have hit the clothing, textile, and car industries particularly badly. Thousands of workers have lost their jobs because of cheaper foreign imports. The development of a modern, externally competitive industrial base capable of surviving on the global market has been achieved only by a very small number of firms, particularly in the defence sector. South Africa continues to enjoy international renown for the quality of its fruit and drink exports, while mining, though slipping drastically in relative importance, remains a critical element. The country is one of the world's leading producers, if not the top producer, of gold, coal, uranium, chrome, platinum and vanadium, iron ore, zinc, and manganese. Because these local minerals are not adequately refined, however, the products exported have low added value and further diversification of the economy is limited.

The 1990s have witnessed a return to positive economic growth. Foreign investment has clearly returned, albeit it at a significantly lower rate than was initially hoped for. It has been limited by investor uncertainty and the fact that international capital is highly mobile, generally preferring the more guaranteed returns from NICs and Northern economies. More concerning, though, is the fact that growth in the 1990s has been jobless. Even though total economic output is increasing, the number of jobs is actually decreasing. Instead of the economy generating an envisaged 200 000 jobs in the first year and a half of the existence of the government's Growth, Employment, and Redistribution strategy (South Africa 1996), 80 000 jobs were in fact lost. The worst-affected sector has been that of mining, where some 250 000 employees have been retrenched since 1987. This parallels trends in the North in the 1970s, when firms introduced cost-cutting measures such as the greater use of machinery while reducing employment levels and closing older plants.

Economic policy in the 1990s swung dramatically from the inwardly focused development of the isolation era to more active participation in the global market. Despite the ANC's initial support for nationalization and the redistribution of wealth, as spelt out in the early policy documents and the Reconstruction and Development Programme, external pressure and internal debate have resulted in subsequent policy evolving along neo-liberal lines. The GEAR strategy was initiated in 1996, supplanting the RDP and aiming to expand the private sector, improve output, reform fiscal matters, and encourage trade and investment. These objectives are clearly in line with the world's economic orthodoxy, but the strategy, while encouraging economic growth, has not shown signs of alleviating the poverty of the majority, and job creation remains an area of major concern (Nattrass and Seekings 1996). Anticipated growth rates of 6 per cent per annum have not been attained, and in 1997 the economy grew by only 1.7 per cent, down from the 1995 peak of 3.4 per cent.

There are clear areas of recession and resurgence in South Africa's space-economy, which correspond to the national core and periphery (see Figure 3.4, p. 70). The areas of recession, the former homelands and the isolated rural peripheries, are characterized by failed development strategies, high levels of poverty, and spatial dislocation as a result of apartheid policies. Resurgent are the core metropolitan areas and the regions around them: Gauteng, the Cape Peninsula, Durban–Pinetown, and Port Elizabeth–Uitenhage. It is in these areas that approximately 81 per cent of the industrial, commercial, and business wealth of the country is concentrated. What transition there has been to high-technology industry is a feature mainly of defence-related industry in Gauteng and the Cape area, for example, and the development of the country's first major scientific, high-technology plant at Capricorn on the Cape Flats. A certain degree of flexible production, along with the outsourcing of manufacturing contracts, is discernible in the South Africa's cores, particularly with respect to the vehicle industry; this is another process experienced earlier in the cores of Northern countries (Rogerson 1991).

The economic sophistication of parts of the core can obscure the existence of the more marginal sectors of the economy within them and in the rest of the country. In stark contrast to the wealth produced in the cores, many millions subsist on low wages and through dependence on informal-sector income or state welfare grants. In the case of Durban, it is estimated that fully one-third of the city's households are involved in some form of informal activity (May and Stavrou 1989). Due to its hidden and frequently illegal nature, estimates on the economic contribution of the informal sector are inevitably imprecise, varying from 16 to 41 per cent of total national production.

Of great concern is the scale of unemployment in the country; while official sources place it at approximately 30 per cent of the potentially economically active population, in many rural areas unemployment figures in excess of 70 per cent have been estimated (Development Bank of Southern Africa 1991; Economist Intelligence Unit 1998b). In such cases, remittances from migrant family members in the core areas and the all-important monthly or bimonthly pensions are all that keep many of these areas functioning economically. In the light of these considerations and the overriding importance of generating employment, discussion now shifts to an examination of past and present development interventions in the South African space-economy.

Spatial development interventions and job creation

South Africa has a long history of state intervention in the space-economy, much of which is discussed by Tomlinson and Addelson (1987). Regional development interventions commenced in the 1940s, with the state offering targeted support to firms settling in peripheral areas. From the 1960s, the policy was linked to homeland development, initially subsidizing firms setting up on the borders of homelands and later within them. In the 1980s subsidies were increased, as was the number of designated growth points, and an attempt was made to promote development in non-homeland areas. However, as Figure 5.3 indicates, the homeland bias prevailed. Simultaneously, the country was subdivided into a number of development regions based on perceived economic

FIGURE 5.3 South Africa's industrial development points and development regions, 1981–91

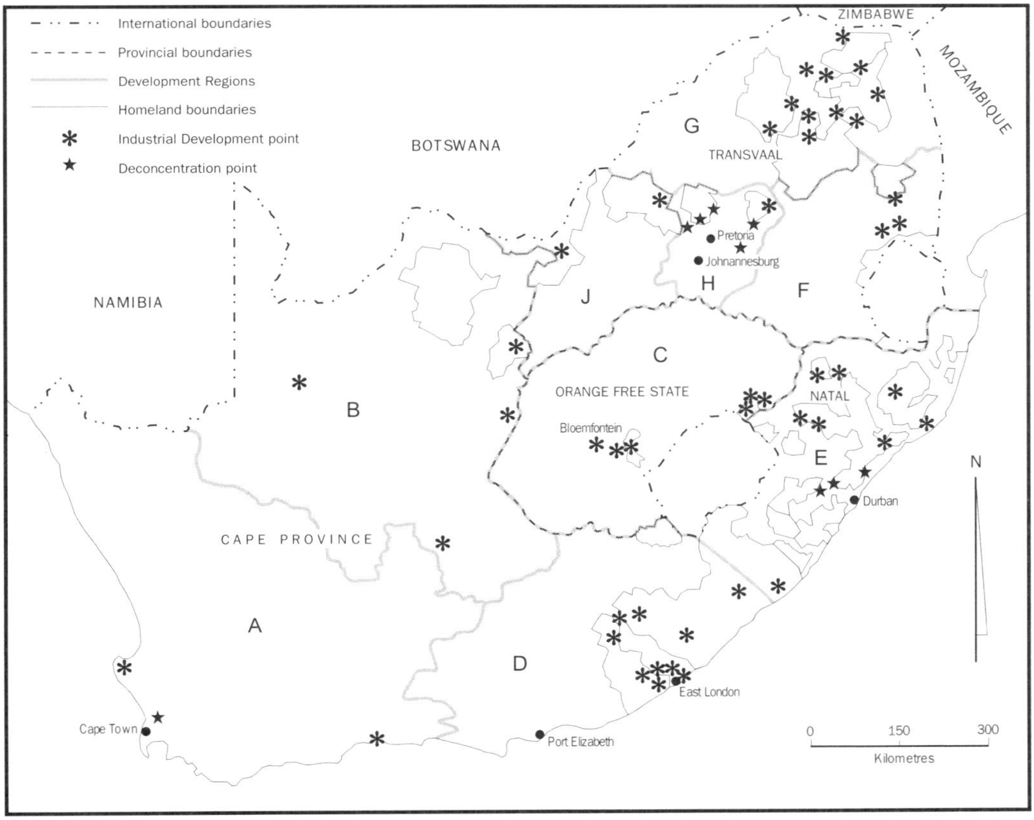

similarities in selected areas. The overlap with the country's present provinces is obvious.

The state's development policies had a tremendous impact: by 1992, 30 per cent of the nation's firms, employing some 400 000 workers, had received state support. However, the policy was economically irrational. Surveys indicate that over half the firms set up in development points simply because of the incentives and not because of the inherent advantages of the site. In many cases, firms appeared to use the state subsidies, considered to be the best available in the world (see Table 5.2), to ensure profitability, and many firms claimed that the suspension of subsidies would jeopardize their operations (Tomlinson and Addelson 1987). This reality came to pass after 1991 when all previous incentives were gradually phased out. The effects on many of the previous development points has been appalling. In the Transkei the number of firms in the four industrial development points fell from 107 to 77 and the number of jobs from over 17 000 to fewer than 9 000 in under two years, and in some places such as Sada in nearby Ciskei seven of the eight assisted firms had closed by 1998. The curtailment of the policy has clearly caused problems for the residents of the former homelands. In a sense they are twice-over the victims of apartheid: first they were forcibly moved to the homelands, where many found employment in state-supported firms, and then the demise of apartheid deprived most of those who had jobs from that luxury (Pickles 1991).

Two key development programmes

TABLE 5.2 Example of incentives available to firms in Development Region D under apartheid's regional development strategy, 1981–91

INCENTIVE	GRANT
Rail rebate	40–60% of actual cost
Harbour rebate (from East London)	50% of actual cost
Electricity	Subsidized to the Johannesburg rate
Wage subsidy (for 7 years)	R70–100 per month
Training grant	125% of actual cost
Interest rebate (on building and housing loans)	40–60%
Relocation allowance (in 1981)	R500 000
Price preference on government tenders	10% above other quotes

SOURCE: Tomlinson and Addelson 1987.

were launched prior to the post-1994 development initiatives of the Government of National Unity. The first was the revised and scaled-down Regional Industrial Development Programme, which gave tax relief to approved manufacturing firms. The incentive was available anywhere in the country, but only to a limited degree in the core areas (Rogerson 1991). By 1997, when the programme was finally phased out, 2 184 applications had been approved, with an employment-generating capacity of 28 472 jobs, but most firms were located, or intended to locate, in the immediate vicinity of the Cape Town, Durban, and Port Elizabeth core areas. Secondly, a well-intentioned public-works programme based on the provision of minimal skills training and the offering of temporary employment was attempted in the 1990s. Overseen by the state and the Independent Development Trust (IDT), it appeared to have enjoyed only limited success. Though training and the payment of part-time salaries to improve infrastructure, such as roads and agricultural facilities, were a valuable contribution, the failure to create many permanent jobs was a cause for concern (Nel 1996).

In 1994 the Government of National Unity came into power and the RDP was adopted as the de facto national development programme (African National Congress 1994). Though noble in ideal, the programme proved difficult to apply and in 1996 the RDP ministry was closed. While the RDP's principles are still used as policy guidelines, the GEAR strategy has become the basis of the state's economic policy. The development ideals of the RDP – such as the provision of housing, safe drinking water, and employment opportunities – remain key development objectives of numerous government ministries. The Department of Water Affairs has been outstanding in this respect.

The Rural Water Supply Programme is overseen by the department and is designed to use labour-intensive construction and maintenance methods to bring domestic water supply to rural villages which previously had to rely on the manual collection of water. By 1997, 1.7 million people had gained access to domestic water supply under this programme. In addition, the Working for Water Programme, also run by the Department of Water Affairs, has created over 8 300 jobs in 38 project areas around the country. The programme has been designed to create jobs and increase the amount of water available to the country by clearing alien vegetation which is currently clogging water courses.

Housing is a further example where RDP principles have eventually been implemented. The government's initial plan was to build one million houses in six years, but it has been slow to achieve its initial targets. By the end of 1997, 400 000 houses had been or were being built and 673 000 home loans had been approved. Such massive cash injections have an important impact in low-income areas across the country.

Development programmes in the late 1990s

There were at least half a dozen development programmes running by the late 1990s aimed at stimulating big and small businesses throughout the country and promoting development within particular regions through the Spatial Development Initiatives (SDIs).

Manufacturing Development Programme, tax holiday scheme

This programme was launched in 1997 to promote new investment by larger firms with assets exceeding R3 million. Firms were offered a tax holiday of up to six years depending on their ability to meet certain criteria: that they be labour intensive (two years' exemption), locate in one of 57 targeted areas – which have a high degree of overlap with the SDIs mentioned below – (a further two years), and produce certain approved goods (two years). The selected centres are shown in Figure 5.4. Although there are numerous development points in the former homelands, a clear break with the previous, apartheid-based system has been made. By the end of 1997, 42 applications from firms had been approved, largely focusing

FIGURE 5.4 South Africa's tax holiday scheme: spatial component

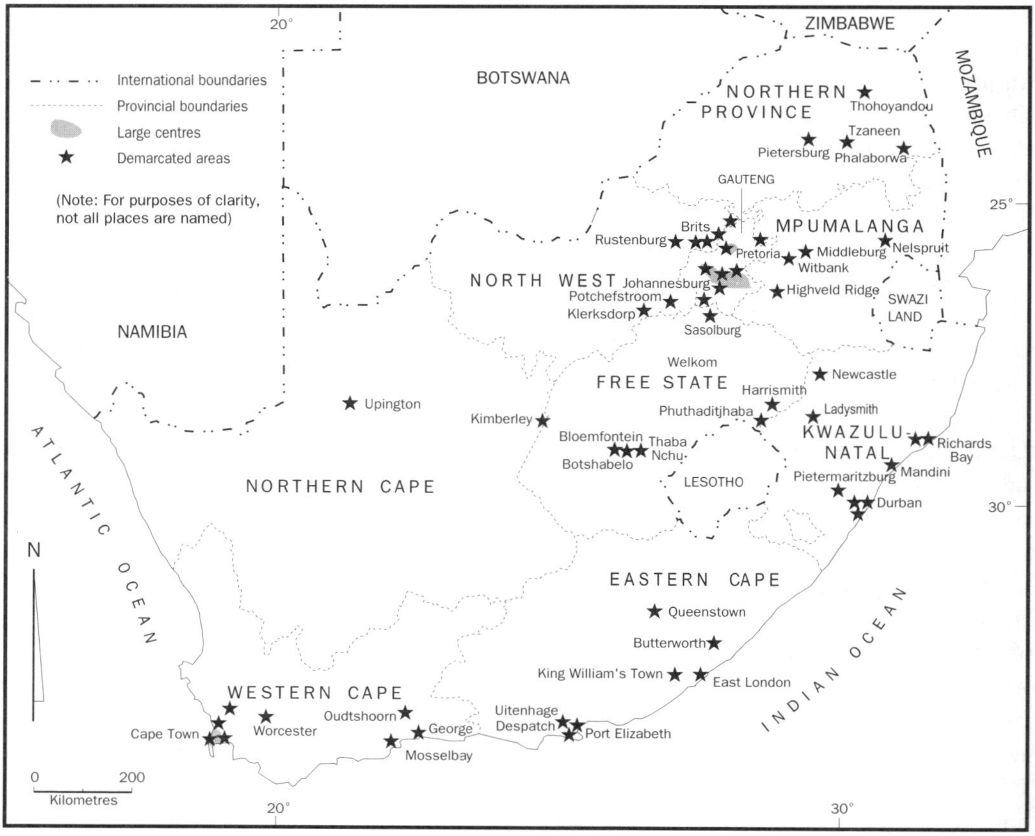

on the Gauteng, Durban, and Port Elizabeth areas and offering some 4200 employment opportunities (South Africa 1998a).

The Small–Medium Manufacturing Grant

This grant, which also became effective in 1997, is for firms smaller than those eligible for the tax holiday scheme, and allows for three years of financial incentives and up to three years of reduced tax. By the end of 1997, 218 firms, aiming to employ nearly 7000 people, had been approved, locating mainly in the Western and Eastern Cape, KwaZulu-Natal, and Gauteng.

Spatial Development Initiatives (SDIs)

This is one of the country's most ambitious programmes, jointly run by the Departments of Trade and Industry and Transport. It seeks to promote a series of development corridors in the country and, in some cases, to neighbouring countries as well (see Figure 5.5). The concept is based on the development and improvement of existing transport infrastructure and the support or initiation of a range of economic activities, including industry, tourism, and agriculture, in the vicinity of the transport corridors. The strategy is seen as a mechanism which can improve economic conditions in marginal areas and which will boost the nation's economic development. The most important SDI is the Maputo Corridor project, which promises to transform parts of Mpumalanga province and southern Mozambique as well as providing Gauteng with a reliable and necessary sea link.

Industrial Development Zones (IDZs)

Related to the notion of SDIs are Industrial Development Zones, which are planned to be state-of-the-art industrial zones, with infrastructure and services designed to facilitate international competitiveness and trade. Support for firms will take the form of innovation and research support measures, investment incentives, and the tax holiday programme. Management will be undertaken by joint public–private corporations. Eight IDZs have been designated, all at major ports or in the vicinity of airports (see Figure 5.5). The most advanced is that of Coega, near Port Elizabeth, where there are plans for a dedicated deep-water port, associated with zinc refining and cement and fertilizer production. In many ways, therefore, the IDZs are the South African equivalent of the Export Processing Zones found elsewhere in the world, although without the provision of non-unionized labour.

Small-business promotion

The promotion of small, medium, and micro-enterprises (SMMEs) is a mechanism to draw traditionally marginalized entrepreneurs more fully into the economy. The small-business sector currently absorbs 44 per cent of new entrants to the formal-sector job market. The chief forms of support rendered at present emanate from the policy initiatives of the Department of Trade and Industry, which established the Ntsika Enterprise Promotion Agency to actively promote SMMEs and to offer training and targeted support. In addition, a policy of approving and financially assisting small-business support

FIGURE 5.5 South Africa's Spatial Development Initiatives and Industrial Development Zones

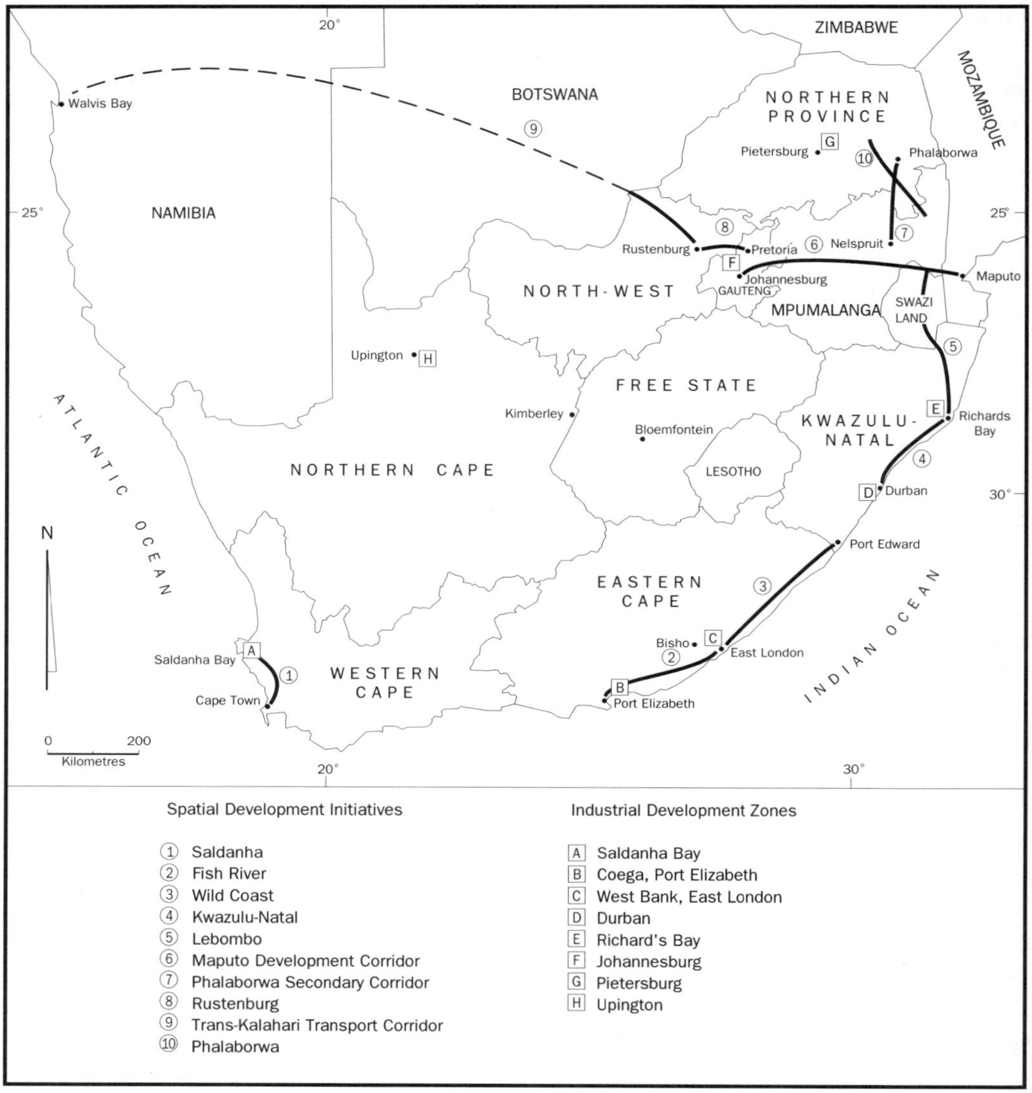

and advice centres, known as Local Business Service Centres, operates. Such centres can be controlled by either a local authority or a locally approved interest group. By 1997 there were 35 centres of this type in the country. Major efforts are being made to encourage state organizations and local authorities to procure inputs from SMMEs so as to assist this category of business.

Local economic development (LED)

The principle of 'developmental local government' has been enshrined in the White Paper on Local Government (South Africa 1998b). This policy obligates local governments to take a greater role in the economic development of their local areas, in job creation, and in the improvement of socio-economic conditions. Such an approach is in line with international moves to urban entrepreneurialism mentioned earlier in this chapter. However, because of the absence of targeted state support or guidance, the lack of trained staff at local government level, and the frequent bankruptcy of local authorities, it is only the larger authorities which are active in this area. In Cape Town, Pretoria, Durban, and Johannesburg the metropolitan authorities have established dedicated economic development units which are pursuing policies of small-business support, the provision of facilities for informal traders, tourism promotion, and the encouragement of investment. Some of these issues are being pursued, at a much reduced level, by some of the smaller authorities. One of the interesting alternatives, employed in the town of Stutterheim, is that of having an independent development corporation pursuing LED in the town in collaboration with the local authority (Stutterheim Development Foundation 1997).

Conclusion

As the preceding discussion has vividly illustrated, economic patterns and processes around the world reveal striking similarities. Any understanding of South Africa's economy and the development initiatives within it cannot ignore the broader regional and global context of which it is a part. Global crises and economic forces impact directly on the country, influencing policy selection, trade, and investment. The use of various development mechanisms, such as EPZs, IDZs, and LED, and a universal shift to neo-liberal economic policies indicate the degree of unity which exists in the world. Increasingly in the future, connectivity will increase and the attainment of the 'global village' will become even more of a reality than it is at present.

Notes

1 Capitalism involves the private or individual ownership of the means of production and property. Behaviour is seen to be economically rational and motivated by a desire for profit.

2 Examples of anti-communist states were South Africa under National Party rule and Rhodesia following the Unilateral Declaration of Independence. Their communist 'enemies' were post-colonial states such as Angola and Mozambique.

6 Dismantling dystopia:[1] new cultural geography for a new South Africa

BELINDA DODSON

'Cultural geography is a subfield of human geography that focuses on the impact of human culture, both material and non-material, upon the natural environment and the human organization of space'. Thus reads the definition in the 1986 edition of the *Dictionary of Human Geography* (Johnston et al. 1986: 63). By the time the 1994 edition was published, the entry on cultural geography had changed significantly:

> Traditionally a subfield of human geography that focused upon the patterns and interactions of human culture, both material and non-material, in relation to the natural environment. Today the term 'cultural geography' covers a diverse range of studies with only tenuous links to a common tradition. (Johnston et al. 1994)

As this semantic shift suggests, cultural geography today is a complex and contested enterprise, one that does not lend itself to either ready definition or tidy demarcation. Part of the difficulty lies in the fact that culture itself is so hard to define – drawing renowned cultural theorist Raymond Williams (1976: 76) to the conclusion that it 'is one of the two or three most complicated words in the English language'. In 1973, American geographer Wilbur Zelinsky offered the following definition:

> culture can be regarded as the structured, traditional set of patterns for behavior, a code or template for ideas and acts ... In its ultimate, most essential sense, culture is an image of the world, of oneself and one's community. (Zelinsky 1973: 70)

But scanning the 1994 *Dictionary of Human Geography* for an up-to-date version would be in vain. Instead,

> [t]he concept of culture traditionally used by cultural geographers has ... been replaced by a theory of cultural politics, in which different cultural discourses engage each other in a

constant struggle for power.
(Johnston et al. 1994: 112)

Applying these concepts to the South African context brings further complexity. No one needs to be reminded that South Africa is a multicultural society. Terms such as 'ethnic mosaic' and 'rainbow nation' are attempts to capture that diversity in a way that is celebratory rather than denigratory, and thus to overcome the legacy of apartheid. South Africans, in all sorts of ways, are renegotiating their cultural identities and affiliations as these become matters of personal preference instead of official prescription. Given this state of flux in the discipline of cultural geography and the actual cultural geography of contemporary South Africa, this chapter cannot hope to be comprehensive.[2] Nevertheless, it aims to share with the reader some ideas on how theoretical developments in cultural geography might contribute to a better understanding of the 'new South Africa' – and, indeed, vice versa. Long neglected in favour of the political and economic, arguably in reaction to the apartheid state's own geographical project of ethnic 'divide and rule', cultural geography in South Africa is surely due for a renaissance.

Culture, geography, and cultural geography

The major tradition in cultural geography has been American, associated with Carl Sauer and the Berkeley 'school' (one of the longest established 'modern' branches of geographical enquiry shown in Figure 1.1, p. 12).[3] This tradition has focused on historical rather than contemporary landscapes, on pre-modern ('folk') rather than modern cultures, on rural rather than urban places, and on material rather than non-material culture (Jackson 1992). Its methods are field-based observation and archival reconstruction, empirical rather than theoretical. The products of such research are often maps and accounts of the origins and diffusion of particular material artefacts (houses, churches, field patterns, and so on) or cultural practices (language, religion, diet, and the like), particularly those of various immigrant groups in the Americas. The key concepts in such enquiry are culture group, culture area, culture hearth, cultural diffusion, and cultural landscape. The implicit – and sometimes explicit – ideology of such traditional cultural geography is conservative. Rubinstein's 1996 fifth edition of *The Cultural Landscape: An Introduction to Human Geography* and De Blij's 1993 fourth edition of *Human Geography: Culture, Society, and Space*, for example, are richly illustrated with such material and well worth examination to see how American undergraduate students are introduced to the world through this cultural tradition.

A new cultural geography has emerged in the 1980s and 1990s, partly in reaction to such conservatism; also in response to intellectual developments in human geography (discussed in Chapter 1); and finally from geographers' attempts to understand the messy real-world cultural geography of the late twentieth century. The essence of this new cultural geography is captured by Peter Jackson in the preface to his 1989 book *Maps of Meaning*:

> It looks at the cultures of socially marginal groups as well as at the dominant, national cultures of the elite. It is interested in popular culture as well as in vernacular or folk styles; in the urban as well as the rural; in contemporary landscapes as well as relict features of the past. Drawing on the literature of cultural studies and social theory, it introduces a variety of new perspectives on the geographical study of culture besides the landscape approach to which cultural geographers (particularly in North America) remain so devoted. (Jackson 1989: ix)

New cultural geography sees culture not as an abstract realm somehow above other social relations, but rather as something 'socially constructed, actively maintained by social actors and supple in its engagement with other "spheres" of human life and activity' (Mitchell 1995: 102). Mitchell goes on to argue that cultural geographers should 'begin focusing instead on how the very idea of culture has been developed and deployed as a means of attempting to order, control and define "others" in the name of power or profit' (1995: 104).

New cultural geographers such as Mitchell have accused traditional cultural geography of being object-obsessed, naive, and atheoretical; the traditionalists have made counter-accusations of shoddy scholarship and intellectual posturing. Cultural geography has thus re-emerged as one of the most exciting areas in contemporary geographical scholarship.

The wider context of these debates in contemporary human geography is the complex phenomenon represented by the deceptively simple term 'globalization'. We live in a world which 'has witnessed not only radical economic restructuring at the hands of globalized markets and flexible accumulation, but also a globalization of culture' (Johnston et al. 1995: 228). The hyper-mobility of people, capital, and ideas has created global systems of cultural production, consumption, and exchange. Thankfully, globalization has not meant simply the 'coca-colonization' of the world by Western (American) culture. Instead, the outcome has been 'new patterns of cultural complexity, diffusion, innovation, and polycentralism ... a destabilization of sites and places, in which hybridity and the refashioning of new identities seems to proliferate' (Johnston et al. 1995: 228–9). This is what many commentators describe as the condition of postmodernity (Harvey 1989). 'New cultural geography' is thus a response to a new geography of culture, to new patterns and mechanisms for which the old models of culture hearths and cultural diffusion provide inadequate and incomplete explanation.

New approaches to cultural geography have not altogether displaced the traditional paradigm. Indeed, there is much that such traditional geographical methodologies can still contribute to an understanding of the cultural geography of South Africa before, during, and after apartheid. It is, however, an unpalatable truth that the tools of traditional cultural geography were also the very tools by which the colonial and apartheid states were created and maintained, and that the map can be a powerful weapon indeed. In what follows, old and new approaches are combined in an attempt to point out some of the directions that an academic re-engagement with South Africa's cultural geography might take.

Modernity and postmodernity
Whereas modernism and postmodernism refer to modes of understanding and enquiry, modernity and postmodernity are the terms given to eras in world history. Modernity refers to a 'particular constellation of power, knowledge and social practices which first emerged in Europe in the sixteenth and seventeenth centuries, and the forms and structures of which changed over time and extended themselves over space until, by the middle of the twentieth century, they constituted the dominant social order on the planet' (Johnston et al. 1994). Several commentators now argue that we have moved into a postmodern era, and that there has been 'a distinctive shift in the temporal and (more pertinently) spatial organisation of economic, social, political and cultural processes' (Cloke et al. 1991: 170).

Re-mapping the colonial encounter

The whole project of cultural classification, naming, and mapping had its origins in the era of European colonial expansion, forming an essential tool of conquest and control.[4] The first 'cultural geographies' of southern Africa are to be found in the maps and journal entries of early colonial explorers, settlers, and administrators (see Figure 6.1). Yet, even as it was classifying and naming culture groups and mapping their distribution, colonialism was simultaneously disrupting the very patterns it was setting down on paper. Colonizing cultures imposed themselves, their lifestyles, their values and their landscapes on new territories. In the process, certain colonized cultures and peoples were obliterated, whether by genocide or cultural assimilation. Others were dispossessed and displaced, sometimes moved halfway around the world. Pre-colonial patterns of population movement, and thus cultural interaction, were altered by the imposition of new political boundaries and other social controls. Everywhere, both colonized and colonizer were subject to cultural change and adaptation, with the process of cultural exchange being a two-way process even though grossly unequal in favour of the colonizer. Throughout the colonial world, each new wave of conquest and settlement consciously or unconsciously inscribed something of itself onto the landscape, erasing the traces left by earlier groups, while itself evolving in response to new natural and cultural encounters.

Both 'old' and 'new' cultural geography have something to contribute to an understanding of southern Africa's 'colonial encounter'. Had South African geographers ever developed a scholarly tradition to match that of the Berkeley school, there might exist a larger body of literature tracing the arrival and spread of various pre-colonial culture groups and their impact on the landscape through settlement, fire, land use, and property relations. Using a variety of methods and drawing on a number of other disciplines, a diligent cultural geographer would be able to reconstruct the landscapes and livelihoods of pre-colonial cultures – their seasonal migration patterns, their settlement forms, their agricultural and pastoral practices, their sacred sites and burial grounds. Such work has largely been done

FIGURE 6.1 Colonial map, from *Pioneer Travellers in South Africa*

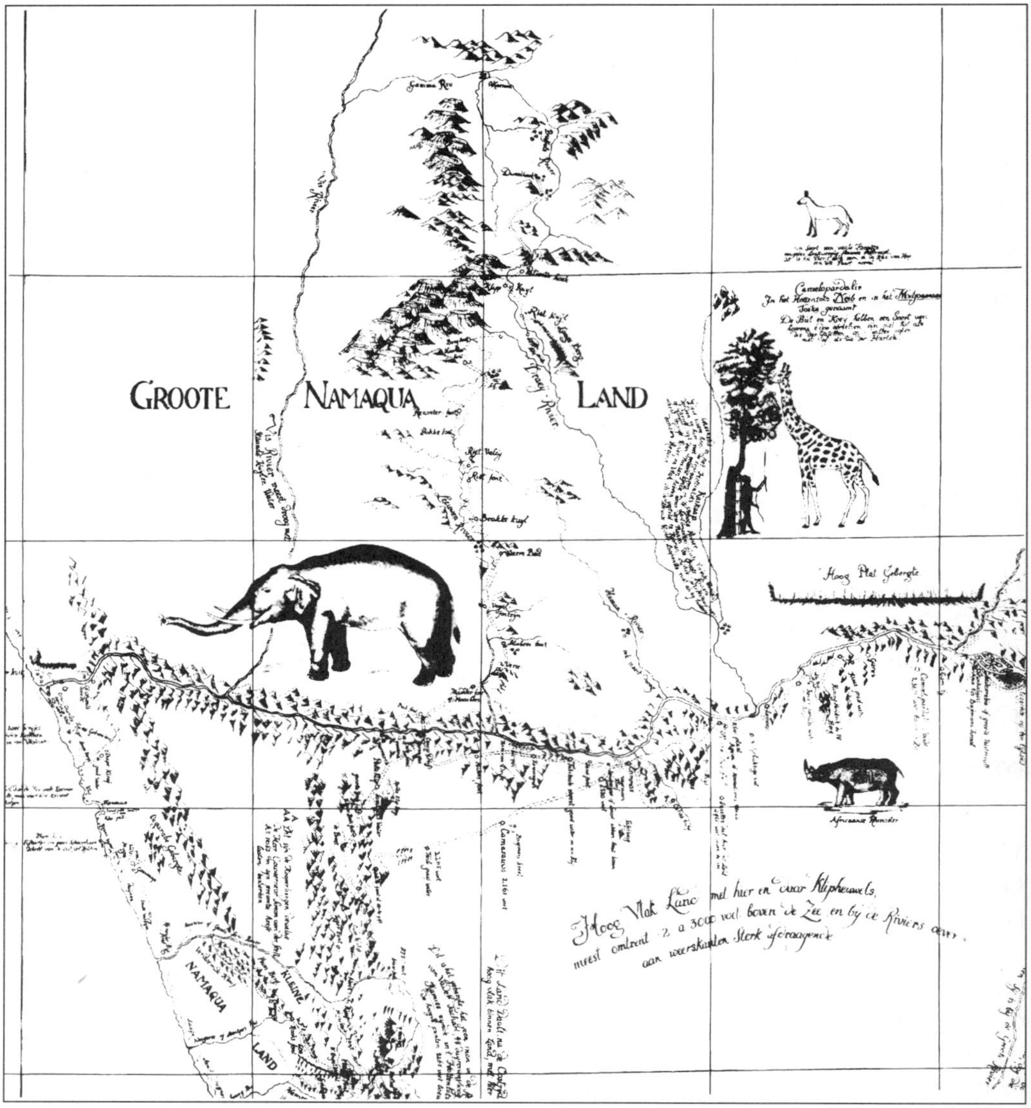

SOURCE: part of Map 3 of the Gordon Collection, 1777–87, redrawn in 1952 by Marie Kenyon for reproduction in Forbes, V. S. 1965. *Pioneer Travellers in South Africa*. Cape Town: Balkema by archaeologists and anthropologists, with only secondary interest in the spatial patterns and relationships that a geographer would consider central, and there remains considerable scope for further geographical research. Even for the colonial period, it has been archaeologists, historians, and architects who have contributed most to the study of the cultural landscapes that European settlement created.[5] A traditional cultural geography of southern Africa – the identification of culture hearths, the mapping of cultural diffusion, the classification and location of material artefacts – is thus a task that remains incomplete. New cultural geography would bring additional insights, being more concerned with reading the landscape as a text to uncover the ideas, beliefs, and symbols of different culture groups; with reconstructing the social relations and deconstructing the discourses through which new geographies and landscapes were produced. Here again, anthropologists, archaeologists, and art historians are leading the way and geographers have been slow to follow.[6]

If academic cultural geography in South Africa remains underdeveloped, this is certainly not due to any lack in the 'real' cultural geography of the country. Our cultural landscape is the material product of a wealth of diverse cultural influences that have created unique juxtapositions and hybrids. Most obvious in material terms are the settlement patterns, building styles, town plans and agricultural practices of the various European settler groups. Thus one can trace particular housing styles to reconstruct the spread of Dutch settlement from the Cape into the interior (see Plate 4, colour section), see the legacy of Dutch and French settlers in the vineyards and farmsteads of Stellenbosch and Franschhoek (see Plate 5), or compare British colonial edifices in Durban or Cape Town with the city halls, railway stations, churches, and post offices of other parts of the empire such as Canada, India, and Australia (see Plates 6 and 7). Street plans and the relative centrality of religious or adminis-

Discourse is the shared mental constructs, language, and rules by which particular social groups produce and communicate knowledge. In simple terms, discourse is the way people think and talk about the world differently, depending on who and where they are. For example, one can distinguish:

- academic from popular discourse;
- apartheid-era from post-apartheid-era discourse;
- scientific from religious discourse.

Social and cultural change is reflected in shifts in discourse and in struggles over the control of discourse.

Deconstruction is 'an attempt to undo claims to truth and coherence by uncovering the incoherences within texts' (Johnston et al. 1994). Geographers apply deconstruction both to landscapes themselves ('reading' the landscape) and to representations of landscape in writing, art, and film.

'Deconstructing the discourses' therefore means unravelling how different culture groups construct identities for themselves and others and subsequently represent those identities in the texts and landscapes that they produce.

trative buildings in a town's layout both provide important clues to its history of settlement by one or other culture group. Pietermaritzburg, for example, is an interesting hybrid of Dutch and British settler elements (Haswell 1979): large blocks, irrigable erven, wide streets, and its very name show its Voortrekker origins, on which have been superimposed buildings that are archetypes of high-Victorian British imperial style.[7]

Toponymy – the naming of places – is another cultural practice which can be read on any map of South Africa. Names such as East London and Newcastle are direct transplants from England; Port Elizabeth, Durban, and Harrismith all commemorate one or other British colonial worthy; while Pretoria and Pietermaritzburg were named for Voortrekker heroes. Of course, these and other places also had African names, many of which are today being elevated to official status. In more physical terms, statues and monuments are cultural signifiers that were enthusiastically erected by colonial settler groups, and, like every other former British colony, South Africa has an abundance of statues of Queen Victoria with her regal stance and formidable imperial gaze (see Plate 8).

These were landscapes of power – imperialism inscribed in material form. As countless examples show, their imposition dislodged pre-existing landscapes and land uses. At the Cape, a traditional Khoikhoi meeting place became a Dutch maritime victualling post and eventually a British colonial port city. What had been seasonal grazing land for the pastoral Nguni people of the eastern seaboard became the Voortrekker town of Pietermaritzburg, later the capital of the British colony of Natal. Xhosa territory on the eastern Cape frontier was taken over by British settlers from 1820 onwards, the landscape 'Europeanized' into farms, villages, and towns. Changes in landscape reflected changes in control; changes in control were usually effected by force, with the British, Dutch (or 'Boers'), and Africans fighting a number of wars over competing claims to territory.

Far from being simple or monolithic, the colonial settlement of southern Africa was further complicated by a deep cultural cleavage between the colonizing nations themselves.[8] One of the legacies of apartheid, indeed one of its deliberate constructions, is the idea of a common white South African cultural identity, playing down the linguistic and other cultural differences between English and Afrikaners. It is worth reminding ourselves that South Africa was formed in 1910 as the union of two British settler colonies, the Cape and Natal, and two Boer republics, the Transvaal and the Orange Free State, after a series of bitter wars culminating in the South African or 'Boer' War of 1899–1902 (described in Chapter 3). In 1910, as now, a new political dispensation had to set about deliberately constructing a new, national (albeit all-white) cultural identity. European settlers both groups may have been, with many common cultural attributes, but their cultural differences were and are at least as significant.[9]

Colonization brought to South Africa culture groups from places other than just the colonizing nations. Perhaps the most significant single group in numerical and landscape terms has been the Indians, brought originally from one British colony to work as indentured labour on the sugar plantations of another.[10] Indians brought with them their religions, languages, architecture, dress, food, kinship, and social systems – all of which have proved remarkably enduring in their new geographical context, some even more so than in their countries of origin. The Indian influence is most striking in KwaZulu-Natal, where mosques and temples are integral parts of the cultural landscape and few towns are without a significant Indian-owned retail sector, typically selling Indian-style clothes, foods, and spices. This example is an illustrative case of the material base of cultural geography: it was the economic need for cheap labour in the colonial agricultural economy that transplanted a culture group from one place to another, thus creating a new cultural geography and transforming the cultural landscape.

The particular position of the Indian immigrants in the colonial capitalist economy served to keep them marginalized and isolated from contact with other culture groups, thus promoting the retention of cultural practices and attributes by hindering processes that might have led to greater cultural exchange and assimilation. Similarly, in the former Cape Colony, slavery had both a material base in the colonial political economy and significant cultural impact. Its legacy can be read in the cultural landscape, with slave quarters and slave bells being common features on many of the older Cape farms. The Muslim religion of many of the slaves, most of whom came from various Dutch and British colonies in 'the East', survives both in practice and in material form at the mosques, shrines, and burial sites that dot the Cape Peninsula and its surrounds. Plate 9 is one example, showing the Muslim shrine on Robben Island.

Religion can be understood as both a cultural practice and a material presence in the cultural landscape. It provides a rich field of enquiry for the cultural geographer, although again this has been under-researched in South Africa. The ritual of church services and the reproduction of traditional ecclesiastical architecture and decoration provided British settlers with both a sense of group identity and a reminder of 'home'. Temples played a similar role for the Hindus, and mosques for the Muslims. For the Dutch settlers who penetrated the country's interior, the establishment of a *kerk* (church) and the monthly *nagmaal* (Communion service) were important territorial and cultural markers, a claim to space as well as an affirmation of identity and belonging.

The function of the church went far beyond white settler society's cultural continuity, however. Christianity played a central part in the whole process of colonization, with religious conversion of the 'natives' being tied to wider processes of social control. Missions, and the Christian church more broadly, were also instrumental in the spread of formal, European-style education. Many mission stations included schools for 'native' children, who were subjected not just to a curriculum of dubious relevance to their needs and circumstances but also to religious conversion and 'civilization', being encouraged to abandon their own traditions for European ways and mores. British settlers' sons and daughters, meanwhile, were taught at educational institutions modelled on the public schools and universities of England and Scotland. These have been described as 'Little England on the Veld', with all the cultural and class baggage the expression conveys.

Together, churches, missions, and schools provided both the institutional structures for the establishment of European cultural hegemony and the physical structures that displayed that hegemony in material form in the landscape. Another part of the legacy of what has been described as 'muscular Christianity' is the centrality of sport in South African culture. Ironically, particularly given their elitist class associations, the colonial sports of rugby and cricket have been mobilized as unifying forces in post-apartheid nation-building. The geography and history of sport provide a rich field of research which, although well established in other countries, has been barely explored in South Africa. A similar point could be made about the cultural geography of music, food, and dress, each of which acts as a repository of cultural influence and meaning and might well provide evidence of greater interaction between culture groups than a superficial reading of South Africa's past would suggest.

Even more fundamental to cultural identity than religion or sport, diet or dress, is language, the medium through which culture is conveyed and reproduced. Linguistic analysis thus provides a further tool for the cultural geographer. Afrikaans, for example, while based on Dutch, has a strong 'Malay' influence, and it is now believed that the role of the Cape Muslim community in the development of Afrikaans was far more important than previously acknowledged. The Bo-Kaap's 'Malay Quarter' thus becomes as much a part of the historical geography of Afrikaans as the town of Paarl. The very development of Afrikaans as a language distinct from its parent tongue of Dutch is indicative of the processes of cultural change that emigrant groups undergo as cultural and other ties to their countries of origin weaken over time. In another example, while linguists continue to debate the evolution of the various Bantu languages, linguistic analysis has already provided clues to the origins and spread of African peoples and thus of the early settlement history of southern Africa. This too has been distorted in white South African nationalist myth-making, with the country's cultural history and geography being deliberately misrepresented in an attempt to privilege European over African claims to political and cultural hegemony. Accent, dialect, and slang are other aspects of language that cultural geographers study; the *Dictionary of South African English on Historical Principles* (Silva 1996) is therefore one more place where geographical research could begin.

Also warranting further research by cultural geographers are the interactions between South Africa's various culture groups, whether indigenous or immigrant, and the region's natural environment (see, for example, Daniel 1973). From the earliest times, through the use of fire and hunting weapons, human culture has had an impact on nature. Agriculture and pastoralism increased this impact further, modifying plant and animal species composition and allowing more permanent human settlement. It is nevertheless true to say that the environmental impact of southern Africa's pre-colonial cultures was slight. Here, as elsewhere, dramatic environmental change came with the colonial encounter, making much of the literature on the environmental impact of colonialism in North America, South Asia, and Australia equally relevant to South Africa, and locating the region's environmental history in a web of inter-colonial cultural and ecological interactions.[11] Reflecting these connections, many of our introduced animal and plant species did not originate in Europe but were brought to southern Africa, deliberately or accidentally, from other parts of the 'New World'. Other non-native species were brought to the region from Europe. Typically, settlers wanted to make their new environment look as much like 'home' as possible, leading the British to build houses and plant gardens that might have been transplanted directly from the shires of England. Many introduced species, plant and animal, are now integral to what South Africans regard as their own national or regional landscape – jacarandas in Pretoria, oaks in Stellenbosch, Indian mynahs in Durban. However, while the impact of some species introductions has been relatively benign, others have proved highly invasive, with irreversible and often highly damaging environmental consequences. Another negative environmental impact of colonial culture was that of hunting. Although pre-colonial cultures had certainly hunted the indigenous fauna, colonial settlers with their guns and their lust for trophies magnified the animal mortality rate several times over. Combined with these powerful cultural forces was the economic drive to extract profits from commercial crop, timber, or livestock production – to clear, to plough, to plant, to fence. Through this combination of mutually reinforcing cultural and economic pressures, the natural environment has been markedly changed.

In ecology, language, religion, and architecture – indeed every aspect of culture, whether material or non-material – it is the admixture of African, Asian, and European cultures that has produced the South African cultural mosaic and its associated landscapes. The 'rainbow nation' is an appropriate if controversial metaphor, for it captures the way in which cultures have been juxtaposed rather than integrated, with full cultural interaction and exchange being prevented by the state's segregationist and later apartheid policies. South Africa may yet become a 'melting pot', but the human-modified environment will always bear testimony to the legacy of colonialism and the enforced separation of one culture group from another over centuries of the country's history.

Apartheid as cultural geography

Engineering society and space

Apartheid itself can be interpreted as an applied exercise in traditional cultural geography, for what were group areas and homelands if not the all-too-rigid delineation of culture areas for different cultural groups? Arising in the colonial era, further entrenched under the Union's policy of 'segregation', and finally reified by the apartheid state, cultural distinctions on the basis of race and ethnicity were for centuries the fundamental organizing principle of the South African social formation and political economy. The result was that practically any discussion of culture and ethnicity was denounced by the political and intellectual left as promoting the interests of the apartheid state.

Only in the last few years has the whole question of cultural identity regained some credibility in scholarly debate, although it remains politically highly charged.[12] Once again, this leaves the field wide open for the development of a new, critical cultural geography, even as it makes this chapter's task quite difficult owing to the paucity of sources outside the ideologically tainted official record.

Another difficulty is that much of the cultural geography created by apartheid remains so familiar and taken-for-granted. Nevertheless, there are at least two reasons why apartheid's cultural geography warrants revisiting. Firstly, in this early post-apartheid era, it is all too easy either to ignore or tacitly accept the geography that apartheid left behind. Imprinted in both material and non-material terms, this is a geography that continues to pervade our everyday lives, constraining and frustrating the creation of post-apartheid cultural forms, practices, places, and spaces. Secondly, it is important, if cultural studies in South Africa are to advance, that the role of a particular conception of culture in underpinning apartheid is acknowledged and confronted.

Apartheid was fundamentally and profoundly cultural and geographical. Space was appropriated and organized to serve the state's agenda of white cultural 'purity' and political hegemony. Vibrant, culturally heterogeneous communities and places were destroyed and replaced with more culturally homogeneous townships and homelands. In constructing this geography, the apartheid state concealed crude discrimination in a cloak of cultural diversity, while simultaneously reducing culture to the narrower concept of 'ethnic group'. In addition to being designated as white (European) or non-white (non-European), people were in turn classified as black, Asian, or 'coloured', before being distinguished again as, for example, Zulu or Xhosa, Indian, Chinese, or Malay. Thus culture was debased by the application of the Population Registration Act. Social interaction between the races was strictly proscribed and interracial sexual relations and marriage were forbidden altogether. Under the Group Areas Act, every town-dwelling South African had to live in a racially defined, state-declared group area. Africans were further assigned to a particular ethnic 'homeland' – KwaZulu for the Zulu, Transkei and Ciskei for the Xhosa, Bophuthatswana for the Tswana, and so on – in a classic example of 'divide and rule'. When not providing labour for the white-controlled economy, it was to such a homeland that most black South Africans were obliged to return. Four of these homelands were even given the status of 'independence', with apartheid's architects citing Europe as a supposedly equivalent grouping of culturally distinctive nation states.

Figures 3.8 and 8.2 (pp. 80 and 203) show how these classifications were defined at the regional and urban scales. At the local level, schools, universities, sports clubs, restaurants, bars, hotels, cinemas, museums, art galleries, and theatres – all the sites at which cultural interaction, display, and consumption takes place – were racially segregated using a range of legislative and spatial means. Wherever possible, the different races were kept geographically apart by the construction of completely separate facilities under the Separate Amenities Act. In places where people of different races were permitted to come together, there were usually separate entrances, seats, toilets, canteens, queues, and counters for whites and non-whites. Among the many artefacts vying for the status of apartheid icon must surely be the 'whites only' sign on park bench, post office entrance, or public convenience. Most of the institutions of 'high culture' – the art galleries, museums, and theatres – were restricted to whites only, meaning that all other South Africans were denied a range of cultural experience and expression.[13]

Through such geographical and legal strategies, the usual forces of cultural interaction, exchange, and development were thwarted. The government's stance was that the maintenance of cultural differences was essential to the social order of the country since they represented fundamental, immutable attributes of individual and group identity. Intercultural (and in particular interracial) interaction was portrayed not as a positive force for cultural enrichment but as an erosive influence leading to the diminution and weakening of culture, with the maintenance of 'European' culture and values being the boldly claimed justification for the entire apartheid system. Such 'tidying up' of the country's cultural geography could be achieved only by a combination of Orwellian bureaucratic machination and Machiavellian state violence. The state was not the sole agent in constructing and policing the cultural geography of apartheid, however. Both capital (for example the mining companies) and the public aided and abetted the process, whether through passive acceptance or active collaboration.

Official records show that the state was assiduous in recording and mapping culture. At the national scale, the decennial census has left a detailed archive of the geography of racial and ethnic distribution as well as of cultural attributes such as language and religious affiliation (see Figure 7.2, p. 166). Although tainted by its associations with the racist past and its role as an 'archive of power', this record warrants serious revisiting by geographers. Of course, the accuracy of census data is open to question, as information may well have been inaccurately reported, misrecorded, or even deliberately distorted for political ends. Nevertheless, there are patterns to be observed, and analysing those patterns over time provides an important insight into the geographical processes involved in apartheid's creation and maintenance. The most important of these were:

- forced removals to ethnically defined black homelands;
- the prevention of movement of blacks into urban ('European') areas under influx control policy;[14]
- the designation of the western Cape as a coloured labour preference area; and
- the destruction through physical demolition of communities such as Johannesburg's Sophiatown and Cape Town's District Six. Figure 3.7 (p. 79) shows the impact of these policies around Cape Town.

Another official source that provides fascinating insight into both the 'real' cultural geography of the country and the apartheid government's vision of that geography is to be found in the 1955 report of the Tomlinson Commission (see Figure 6.2). Set up to establish a spatial framework for the 'separate development' of black South Africans, this commission devoted much of its effort to the detailed

FIGURE 6.2 Possible consolidation of 'Bantu areas' from the Tomlinson Commission Report

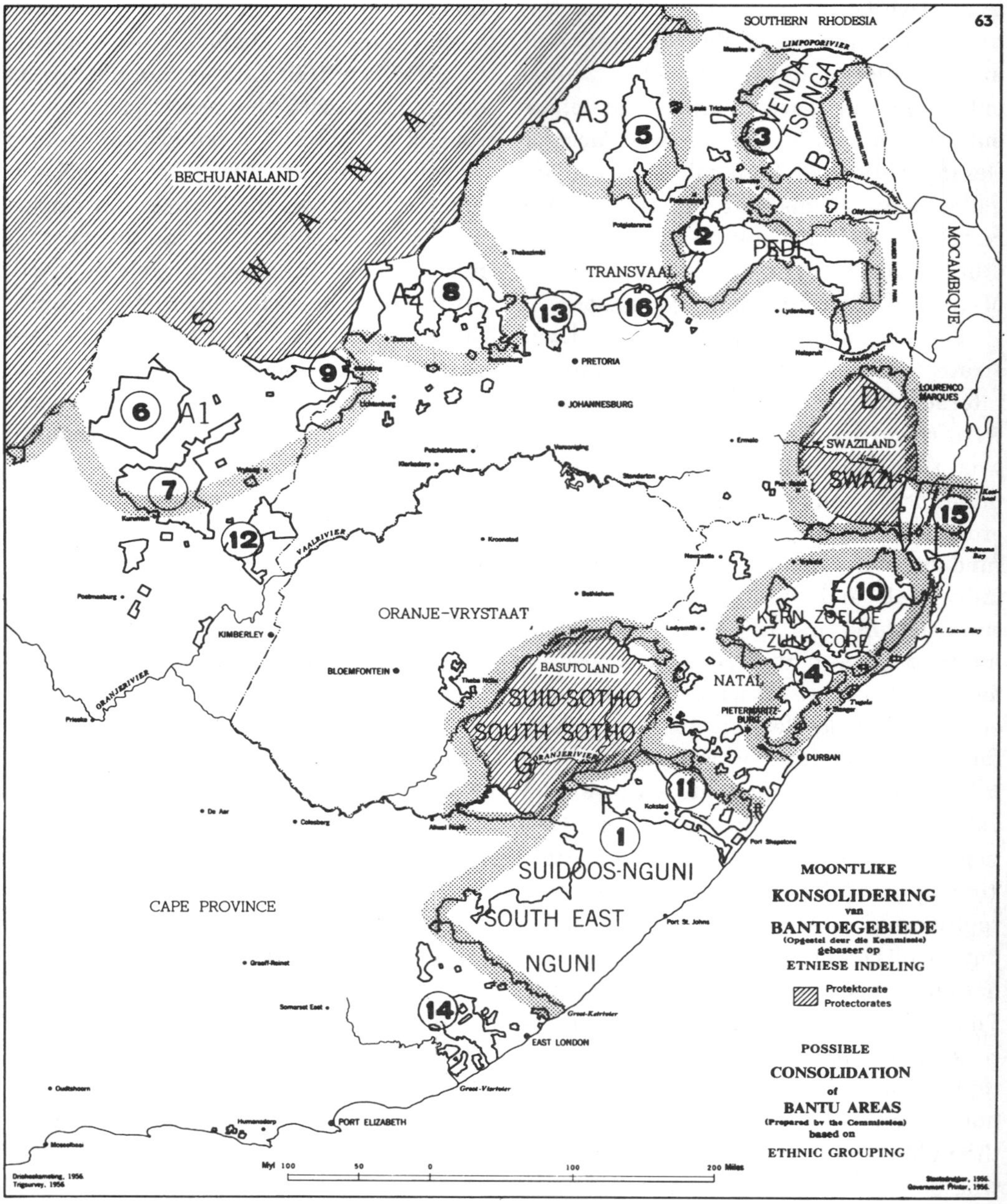

SOURCE: Tomlinson Commission Report 1955: Figure 63.

ethnic mapping that divided Africans into the linguistically based 'nations' on which grand apartheid was to be based. The Tomlinson Report (South Africa 1955) provides a useful baseline for studying the subsequent cementing of the country's cultural geography into rigid, ethnically defined zones over the four decades of apartheid rule.

While the cultural geography of ethnicity and language have, in one sense, been over-documented, there is nevertheless scope for further geographical research. Work such as the language atlas compiled by Van der Merwe and Van Niekerk (1994) at the University of Stellenbosch is an important start, but it remains very much within the traditional paradigm of cultural geography. A new cultural geography of language in South Africa would include analysis of the political agendas and discourses involved in classifying and mapping language, both during and since apartheid. Such research is important not least because language or 'ethnic' group was one of the primary criteria in the designation and delineation of the country's nine new provinces in 1994. Language and ethnicity thus remain the basis of political organization in the post-apartheid era; they are as much as ever tools for territorial claims and political gain (see Christopher 1994).

Apartheid's landscapes

Also deserving of academic attention are the cultural landscapes that developed under apartheid. Here too there is scope for a new cultural geography of the country to be written and mapped. The 'old' cultural geography, with its emphasis on material landscape forms, provides an inadequate framework, although it does provide a starting point. What are (or were) the distinctive cultural landscape features of apartheid? In identifying such landscape features, it becomes difficult to separate the cultural from the political, although it is surely true to say that in South Africa the political was cultural (and vice versa). The buffer strip between urban group areas might be one, migrant workers' hostels another, the endlessly replicated 'matchbox house' of the black townships a third; there are many more. Today, in an effort to cleanse the country of its racist past, as well as in response to economic forces such as demand for land and housing, some of these landscape elements are being erased forever, and should perhaps be declared national monuments to ensure their preservation. Some landscape forms, such as the tent camps and 'toilet towns' that were the destination of victims of forced removals, were by their very nature temporary, if not as temporary as their occupants might have wished. Others remain to haunt us, and for many South Africans the township house remains either a depressing daily reality or even an elusive dream (see Plate 14).

At the same time as it was building a monotonous, modernist landscape of cost-effective, mass-produced housing in official township areas, in other places the state was destroying buildings that did reflect the diverse cultural heritage of South Africans, demolishing houses and other structures that ended up in the 'wrong' group area or in so-called black spots. Cape Town's District Six, perhaps the most famous example, is recorded in photograph, painting, literature, and song,

Modernism is the philosophical assumption that there is a degree of order in the world, that it can be understood rationally, and that it is fundamentally structured in ways that we can observe and comprehend. Apartheid's planners were thus modernists in that they imposed spatial order on the landscape by enforcing cultural separation.

as well as in the memories of those who lived there, thus constituting a powerful, even iconographic cultural artefact despite – or indeed because of – its physical destruction. The site is still bare, but for the now-stranded churches and mosques that were saved from demolition and the few new buildings, such as the Cape Technikon, that have been allowed to encroach. Its future remains contested and uncertain, although perhaps there is something to be said for the suggestion that it should remain undeveloped and be declared a 'peace park'.

The apartheid state was well aware of the power of the built environment as an agent of repression and control. The country is full of examples of prisons, police stations, magistrate's courts, and one-time 'pass offices' begging analysis of the use of space and design in the exercise of political power. Such structures were the visible, symbolic presence of state authority, buildings from which apartheid's functionaries carried out their daily activities of surveillance and suppression. The plans and designs of such structures attest to both their immediate function and their broader purpose in creating and maintaining a culture of fear and domination: many have high-windowed or windowless exterior walls concealing the activity within; officials worked behind locked doors and metal grilles; the entrance would usually have a guardpost where an armed soldier or policeman would control entry and exit; the perimeter wall or fence would often be topped with razor wire or broken glass as an added deterrent. At a larger scale, the street layout of townships such as Soweto was deliberately planned to allow rapid sealing-off in the event of political unrest, with few entrance and exit points and strategically located transport nodes and security posts.

Within such a constrained physical, social, and legal environment, opportunities for cultural expression in the built environment were few for 'non-white' (especially African) South Africans. Nevertheless, among the many forms of resistance to apartheid was cultural resistance, and some of that cultural resistance did take material form in the landscape. In African townships, where drinking of alcoholic beverages was meant to be restricted to state-controlled beer halls, there were hundreds of illicit shebeens (taverns, usually in private homes), often run by feisty black 'mamas', and with something of the flavour of a prohibition-era speakeasy in the USA, whose style urban black South Africans often consciously imitated. Robbed of the freedom to shape the wider landscape, people (especially black women) used domestic interiors as sites for cultural expression and resistance, of a sort, to the drab uniformity of township architecture and planning. Graffiti was another means of cultural expression, notably for black youth, and included both political slogans such as 'Free Nelson Mandela' and territorial markers related to gang activity (another element of American culture that has a tragic imitation in South Africa). A particularly interesting landscape form was the 'people's parks' that sprang up in townships during the 1980s – open patches of land that were taken over as sites for recreation and public meeting, decorated with bits of scrap and rubber tyres and named after heroes of the struggle such as Steve Biko, Nelson Mandela, and Oliver Tambo. Also included among the cultural landscape forms of apartheid-era township life should be the patrolling army and police vehicles; the bullets, dogs, and teargas used to disperse political gatherings; and the burning barricades, stones, and petrol bombs used in counter-attack.

So-called coloured and Indian South Africans experienced a similarly alienating landscape of repression and violence, and also suffered restrictions on their choice of where and in what sort of dwelling they could live. Relative to Africans, however, their greater economic power, urban property rights, and marginally higher political status allowed them slightly greater control over their material environment, and there are elements of architectural design and decoration that might be considered distinctive to these culture groups. An example would be the large, often lavish houses accommodating wealthy multi-generational families in 'Indian' group areas of cities such as Durban and Pietermaritzburg, given a particular South African twist by being located in an ethnically rather than income-defined residential area. With the end of apartheid, some of the occupants of such dwellings have moved into previously white group areas, so that the landscapes of South African residential areas are likely to become increasingly shaped by income and class rather than ethnic group.

White South Africans had far greater power to impose and inscribe their own cultural forms and values on the landscape. Yet there is little that constitutes a distinctive white South African cultural landscape, beyond that created by colonial settlers and discussed in the earlier part of this chapter. For the most part, the places where they have lived and worked since 1948 resemble urban areas anywhere in the 'New World' capitalist West: the bland placelessness of suburbia, shopping malls, freeways, and high-rise office blocks. North Americans or Australians would notice perhaps two differences from their own suburban cultural landscape: the evident domestic security infrastructure of burglar bars, burglar alarms, walls, fences, and signs warning of guard dogs or armed response services; and, in older suburban houses, the servant's quarters where until a few years ago almost every white South African household accommodated a black or coloured female domestic worker.[15] If there is a distinctive white South African cultural landscape, it is probably to be found in rural rather than urban areas – the Karoo or Free State *plaas* (farm), perhaps, although even here there would be similarities with, for example, a sheep station on the Australian outback. Again, what makes the South African cultural landscape South African is not its individual features but its juxtapositions: not the suburban house and garden but the poky servant's quarters tucked behind the two-car garage; not the sprawling, *stoep*-enclosed, tin-roofed stone farmhouse but the rudimentary and unkempt farm labourers' cottages a kilometre or two away.

One might try to seek a more distinctively South African cultural landscape in statues, monuments, and memorials. Here, however, the imprint of apartheid was surprisingly slight, with the colonial era and the period between 1910 and 1948 each leaving a far more 'monumental' legacy.[16] Perhaps because it had such a hold on power, the apartheid state had little need to employ statues, monuments, and triumphal arches to display that power, which was all too real, immediate, and obvious to all over whom it was exerted. Perhaps it was a more deliberate and cunning strategy, a refusal to construct specific sites that might be targeted for occupation or defacement by opponents of apartheid. What this also shows, of course, is the continuity between those earlier political and cultural systems and the one that controlled the country from 1948 to 1994. A similar conclusion can be drawn from looking at place names. These remained largely unchanged after 1948, except for a few instances such as the cynically named Triomf (triumph), the white suburb that was built where Sophiatown had been razed.[17]

One apartheid-era monument of particular and explicit cultural significance is the Taalmonument (Afrikaans language monument) near Paarl, but it was erected with money raised from public subscription, not state funds. Nearby Franschhoek houses the Huguenot Memorial Museum commemorating the contribution of French Huguenot settlers to South African culture; Grahamstown has the English equivalent in the 1820 Settler Monument. Both were also erected by private foundations. While the Paarl and Franschhoek monuments now seem culturally chauvinistic, the Grahamstown monument provides an interesting example of post-apartheid cultural redefinition. It comprises a multi-purpose building that provides a centre for an annual festival of the arts, originally intended to celebrate English language and culture in South Africa but increasingly inclusive and multiculturally South African in nature. Its bland modernist architecture is part of the reason why it could thus be recast in a new role. The same might be said of many other apartheid-era public buildings, whose very functionalism and lack of ornamentation provide a flexibility of purpose and meaning even as they allow a convenient negation of an often sorry past. Simply change the name and remove the bust of Vorster or Verwoerd from the foyer.

More difficult to erase or deny is apartheid's impact on the landscape in environmental terms, in many ways the product of cultural as much as political–economic forces. Most drastic is the widespread land degradation in what were the black homelands, where fuelwood extraction, overgrazing, and poor farming practices denuded vegetation, eroded soil, and depleted water resources. While African culture, for example the practice of paying *lobola* (bride price) in cattle, has often been blamed, the more fundamental causes were the artificial overpopulation and economic stagnation caused by inequitable land distribution, ethnic segregation, and the promotion of migrant labour. Ironically, both the colonial and apartheid governments were great proponents of 'nature conservation'. Yet this was almost always achieved at the expense of African people, who were forced off their land to make way for nature reserves, and allowed back in only as game guards or workers in whites-only tourist camps. The result is that, while white South African culture attaches virtual iconic status to game reserves and other wilderness areas, to most black South Africans such preserved nature represents profound and fundamental alienation. Chapter 18 provides case studies which show that a major effort is needed to re-establish a 'conservation culture' among black South Africans. Certainly the natural environment is likely to remain an area of contested cultural meaning.

Studies of the material landscape, built or 'natural', still provide only part of the picture. A true understanding of the cultural geography of apartheid demands an interpretation of what Jackson (1989) calls 'maps of meaning': not the material forms themselves but what places meant to people and what people did in them; who was included and who excluded; who had ownership and control and who did not. As those meanings, practices, and power relations change, it is important that the old ones are recalled and recorded. Just as there are discontinuities, so there are continuities, and these too need to be considered. Cultural change will eventually and inevitably be reflected in the landscape, but for now the geographical legacy of apartheid remains an impediment to the creation of a truly multicultural society, demanding imagination and innovation in dismantling the dystopias of the past.

Creating a new cultural geography

South Africa's cultural geography has to be understood in terms of power relations and social constructions, rather than simply maps and material artefacts. Such an understanding provides insight into the struggles over place, space, and meaning, as well as over language, religion, and education, that are part of the country's ongoing post-apartheid cultural transformation.

In addition to these internal dynamics, the end of apartheid has also brought about new international cultural linkages. Cross-border migration is just one of the ways in which South Africa is developing new links with the rest of Africa. The country is now a member of the Southern African Development Community (SADC) and has been readmitted to the Organization of African Unity (OAU), as well as to African professional, cultural, and sporting organizations. Affiliations with neighbouring African states in the Commonwealth has also brought about new cultural interactions and opportunities. Such cultural re-linking has brought South Africa into global networks of information, entertainment, and consumption. Some see a threat of 'coca-colonization' of South African culture, but international experience shows that globalization everywhere meets countervailing forces, so that global cultural influences are reworked into a particularized local form and meaning.

In addition to signalling the likely directions of cultural change in post-apartheid South Africa, new 'postmodernist' theories (discussed in Chapter 1) and geographies of culture have much to tell us about the failure of apartheid. The entire apartheid system was based on a flawed, modernist conception of culture. Contrary to apartheid's vision, culture is not fixed and immutable but fluid and negotiated, not begotten but made. This does not mean that cultural identity can be imposed from above, but that it is continually reworked by individuals in the patterns and practices of their everyday lives. Thus South Africans who once rejected state-imposed ethnic classification are now actively reclaiming and reasserting their ethnicity. Others seek to construct a new nationally rather than ethnically based cultural identity, describing their cultural identity as 'South African' rather than English or Afrikaans, Zulu or Xhosa, Indian or coloured. Geographical strategies, such as choosing the community in which one lives, are an important element in such cultural reorientation. Certainly the forces of fragmentation, such as Zulu nationalism, Muslim fundamentalism, and the call for an Afrikaner *volkstaat*, are at least as strong as the forces of unification, and the future cultural geography of the country will be determined by the ways in which these opposing forces play themselves out.[18]

If South Africa does succeed in celebrating diversity while avoiding cultural discrimination, chauvinism, or conflict, it will be an achievement indeed. And while the state can certainly be involved in a facilitatory capacity, surely a lesson from apartheid is that culture is beyond, if not above, politics. Whatever the outcome, space, place, and nature will all be involved in the process, and geographers have a constructive role to play.

Notes

1 *Dystopia*: a bad or evil place, according to the *Penguin Dictionary of Human Geography* (Goodall 1987).

2 Among its omissions are art, music, theatre, film, and literature, all of which have been studied by cultural geographers elsewhere but remain lamentably under-researched by South African geographers (although see Badenhorst 1992; Lammas 1992, 1993). The chapter also omits detailed discussion of other southern African countries, although South Africa shares with them similar cultural attributes. I should here acknowledge my own ignorance of their cultural geography, partly a function of the academic and cultural isolation of apartheid South Africa.

3 The University of Aberystwyth in Wales was another 'hearth' of cultural geography through the work of people like C. D. Forde.

4 See, for example, Matthew Edney's *Mapping an empire: the geographical construction of British India 1765–1843* (1997).

5 Although see Christopher 1976.

6 This was evident at an interdisciplinary conference, 'New approaches to interpreting past landscapes', held at the University of Cape Town in January 1997. Geographers were conspicuous by their absence.

7 Geographers Ron Davies and Rob Haswell have explored the legacy of British and Dutch colonial settlement in the cultural landscapes of the old Cape and Natal provinces, although regrettably neither has yet developed his insights into a full-length published study.

8 There is a parallel with Canada, with its 'two solitudes' of English and French Canadians.

9 This is another topic worthy of further geographical research, in particular the interrogation of the cultural landscape for different 'maps of meaning'. Monuments, memorials, and significant sites related to the Anglo-Boer (or South African) War certainly warrant further attention.

10 Indentured labourers had their passages paid from India in return for work over a contractual period. Many indentured labourers stayed in South Africa at the end of their contracts instead of returning 'home' to India.

11 See, for example, Crosby 1986. Grove (1995)

locates South Africa's environmental history firmly in a broader colonial context.

12 See, for example, the special 1993 edition of the *Journal of Southern African Studies* (vol. 20 no. 3), part of the proceedings of a conference held at Rhodes University earlier that year and entitled 'Ethnicity, Identity and Nationalism in South Africa: Past, Present and Future'. The introduction by Wilmsen, Dubow, and Sharp provides a useful summary of the status of research into ethnicity and culture in South Africa at the time.

13 Of course other sorts of art, music, and theatre flourished elsewhere, constituting an important part of the struggle against apartheid.

14 For example, the paper by Fox and Tipler (1996) assesses the first two of these processes.

15 'Fortress suburbia' has become even more fortress-like in the post-apartheid era, while the tradition of live-in domestic workers has declined dramatically, with servants' quarters now commonly converted into flats for grandparents or teenage children.

16 The famous Voortrekker Monument near Pretoria, commemorating the Great Trek, predates formal apartheid.

17 The name Sophiatown has recently been reinstated, and claims for the restoration of land rights are pending.

18 International comparisons are also instructive, and geographical work such as that by Audrey Kobayashi (1993) on Canadian multiculturalism has a number of parallels in the South African context. In Canada, just as in South Africa, there is a tension between the construction of a new, unified national identity, on the one hand, and allowing diverse cultures – aboriginal, settler, and immigrant – to flourish on the other.

7 Population: structure and dynamics in a crowded world

HANNES AND IZAK VAN DER MERWE

Why population studies?

The systematic study of a country or geographical region usually focuses first on the physical characteristics of landscape and natural resource endowment. Yet geography remains fundamentally anthropocentric, placing humanity at the centre of life on earth. Human inhabitancy of the physical environment, human interaction with the environment, and the characteristics of the human population of any region dominate the systematic understanding of the complexities of human existence in a finite world. Aside from scientific necessity, human planning considerations and business practice require the study of population. Plane and Rogerson (1994) point out that few subjects have been influenced by as many disciplines: demography, geography, planning, economics, sociology, mathematics, and ecology. The reasons for the interest in population include, firstly, the need to understand the world around us, and, more practically, the need for planning in both the public and private sectors to enhance the very existence and survival of people. Planning applications in the public sector for which knowledge of population distribution, growth, and movement is essential include efficiency in facility provision and location (schools, hospitals, sanitation services, transportation, and many others). Private-sector applications include targeting niche markets and planning the optimum location and size of business establishments. Both fields require knowledge and understanding of population trends and distributions at global, regional, national, and especially local levels – which are the domain of geography as a discipline.

This chapter provides some insights into the dynamics which govern the survival and growth of human populations. Key concepts and theories that explain the growth, distribution, and structure of population will be introduced at three scales: global, African, and South African.

Population structure and composition

Humanity is not simply one great homogeneous mass, but consists of wonderfully diverse groupings of individuals with distinctive characteristics. Table 7.1 presents an overview of a range of such criteria for the countries of the Southern African Development Community (SADC), Africa, and the world.

Demographic characteristics

The physiological attributes age, gender, and ethnicity are the most obvious demographic characteristics of populations. The age and gender structure of a country's population is typically depicted in a population pyramid diagram. Figure 7.1 (p. 162) gives three examples: South Africa, Malawi, and the former West Germany. The pyramid bars show the number (frequency) or proportion of the total population (expressed as a percentage) in each five- or ten-year age cohort. The shape of the pyramid reveals much about the growth dynamic in that population. A broad-based pyramid indicates large numbers of children in the lower age groups and thus a young and fast-growing population. The converse – a narrow base and a bulge towards the top – means relatively large numbers in older cohorts (an ageing population) and stable population growth.

The German pyramid shows the mature profile of a stable population with low proportions in the youthful categories, while Malawi displays exactly the opposite – a fast-growing population with large proportions of children in the younger cohorts. South Africa's population shows encouraging signs of slower growth, with lower proportions evident in the two youngest age cohorts.

The concept of a **cohort** is often used in demographic studies. It designates a group of people sharing a vital event, such as being born in the same year or period (an age cohort).

Another significant aspect of the age structure of any population is the proportional distribution of children, adults, and the elderly. The age thresholds between these groups are normally taken as 15 and 65. The relative size of the three groups indicates the extent to which the productively active adult group has to support the other two dependent groups. The *dependency ratio* expresses this relationship. Taking the 1995 world average dependency ratio of 61 per cent, which shows that every 100 productive adults must support 61 dependent people, the precarious situation on the African continent (89 per cent – see Table 7.1) is evident. The adult producers in Africa support a large dependent group, mostly children, for whom services such as schooling must be provided. South Africa (71 per cent) is in a slightly better position than its subcontinental neighbours (averaging 86 per cent), but still carries a heavy burden.

Dependency ratio: $D = (y + e)/a \times 100$, where y = number of youths, e = number of elderly, and a = adults. Multiplication by 100 provides either a percentage or an index which indicates how many people every 100 productive adults have to support.

TABLE 7.1 Comparative population indicators for the Southern African Development Community

	ANGOLA	BOTSWANA	LESOTHO	MALAWI	MAURITIUS	MOZAMBIQUE
Total population 1996 (million)	11.5	1.5	2.1	11.4	1.1	16.5
Population density (per km^2)	9	3	70	97	550	21
Urbanization level (%)	31	27	23	13	41	33
Population growth rate 1990–2000 (%)	3.6	3.0	2.7	2.6	1.1	3.0
Fertility 1995–2000 (births per woman)	6.7	4.5	4.9	6.7	2.3	6.1
Crude birth rate (per 1 000 population)	48.4	34.6	34.9	47.4	19.3	42.6
Crude death rate (per 1 000 population)	16.9	5.7	8.6	19.6	6.5	17.6
Gender ratio (males/100 females)	97.5	95.9	96.3	97.4	99.8	97.5
Dependency ratio (0–15/100 adults)	100	84	86	98	50	92
Migrants 1990 (% of total population)	0.3	1.8	1.4	12.1	0.8	0.1
Refugees 1995 (thousands)	10.7	0.5	0.05	90.2	0	0.3
Per capita GNP 1994 (US$)	620	2 800	700	140	3 180	80
Human Development Index	0.283	0.741	0.464	0.321	0.825	0.261

Where years are not stated: 1995 or 1995–2000

SOURCES: Haldenwang 1997; Haub and Yanagishita 1996; United Nations Population Fund 1996.

(SADC)

NAMIBIA	SOUTH AFRICA	SWAZILAND	TANZANIA	ZAMBIA	ZIMBABWE	SADC	AFRICA	WORLD
1.6	38	1.0	30.5	9.7	11.5	141	748	5 804.1
2	31	59	34	13	29	20	25	44
36	55	27	24	37	35	35	34	45
2.7	2.0	2.8	2.9	2.8	2.4	2.7	2.8	1.5
4.9	3.3	4.5	5.5	5.5	4.5	5.1	5.4	3.0
35.1	26.6	36.4	41.1	41.9	36.0	38.2	39.4	23.7
9.2	7.9	9.2	13.5	16.2	13.2	13.0	12.7	8.8
99.0	98.7	92.0	97.9	97.8	98.4	98.0	99.5	101.5
84	71	84	94	99	88	86	89	61
0.6	3.1	9.4	2.3	4.1	8.0	3.4	2.5	2.3
1.1	91.9	0.6	883.3	141.1	2.2	1 222	6 775	14 674
2 030	3 010	1 160	90	350	490	550	660	4 740
0.573	0.649	0.586	0.364	0.411	0.534	?	?	?

FIGURE 7.1 Three age–sex pyramids

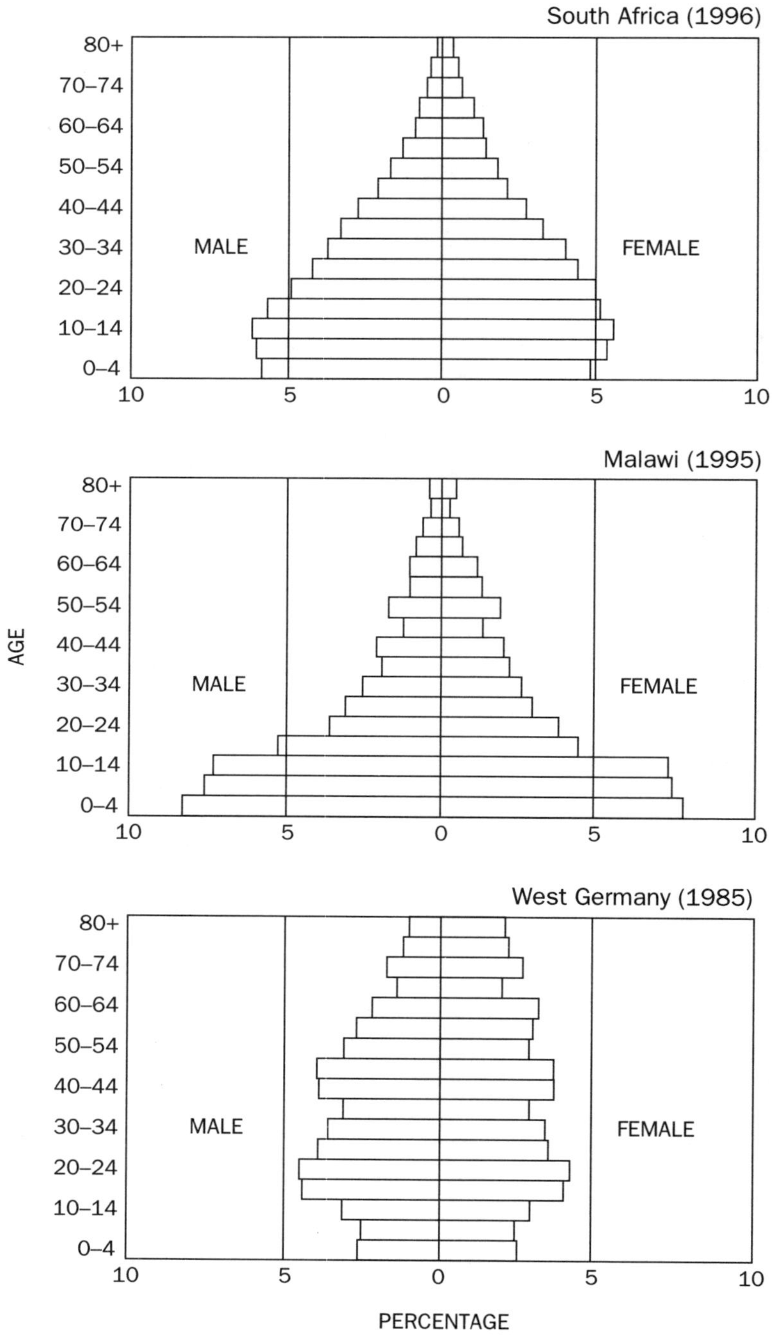

The ratio between the sexes is normally roughly similar; that is, there are equal numbers of males and females in the population. The fact that more males than females are born is probably an evolutionary defence against the fact that the survival rate of males in all mammal species is lower – men die earlier and women therefore have a higher life expectancy at birth. The sex ratio is expressed as the number (or percentage) of males per hundred females. In geographical terms this ratio is significant at international and national levels, since it reflects migration patterns. In Namibia, for example, the Wambo population of the north of the country have a relatively low male proportion because many men are employed on the diamond mines in the south. In South Africa many peripheral regions, particularly the former homelands, also reveal this pattern, since male labourers have migrated to metropolitan areas in search of employment opportunities. One reason for the disagreements over the preliminary 1996 census results for South Africa was that the male–female ratio was a disproportional 48/52 per cent.

Ethnicity and language

Ethnicity has always been a significant element in the South African context because of the country's history of segregation and apartheid. While the new Constitution's founding principles unequivocally put the country on the path to 'non-racialism and non-sexism', the practical application of the enshrined principles of redress and restitution, affected through measures such as 'affirmative action' to 'advance persons or groups of persons' (South Africa 1996a), means that ethnicity will remain part of everyday reality. Table 7.2 provides some insight into the differential dynamics driving the structural

TABLE 7.2 South Africa's population, basic demographic characteristics

Group	% of total	Birth rate (TFR)	Growth rate (%)	Doubling time (years)
Asian	2.5	2.3	1.3	52
Black	76.8	4.2	2.4	29
Coloured	8.4	3.7	1.5	50
White	12.3	1.7	0.7	100
Total South African population (1996)#: 37.9 million				

Controversy surrounds this total declared by Central Statistical Services (South Africa 1997). Various other estimates (Calitz 1996; Haldenwang and Boshoff 1996; Steenkamp 1997), based on earlier demographic models and census results, range between 42 and 43 million.

characteristics of these groups. It is clear that South Africa's white population has a very slow growth rate and would have a population pyramid like Germany's. In contrast, the black population has a much higher total fertility rate and growth rate. The Asian and 'coloured' populations lie between these two extremes.

The **total fertility rate** (TFR) indicates the number of children that would be born to 1 000 women passing through the childbearing age.

Cultural diversity finds expression in the rich variety of languages spoken in South Africa. The country's language policy is defined in the Constitution: 'the official languages of the Republic are Sepedi, Sesotho, Setswana, siSwati, Tshivenda, Xitsonga, Afrikaans, English, isiNdebele, isiXhosa and isiZulu'; furthermore, 'recognising the historically diminished use and status of the indigenous languages of our people, the state must take practical and positive measures to elevate the status and advance the use of these languages' (South Africa 1996a: 4). In 1991 isiZulu, isiXhosa, and Afrikaans dominated the linguistic scene (see Table 7.3): approximately 8.3 million people, or 22 per cent of South Africa's total population, regarded isiZulu as their first or home language in 1991. Almost 18 per cent were isiXhosa speakers, comprising 6.6 million, while Afrikaans speakers totalled 5.7 million (15 per cent). The other prominent languages were Sesotho sa Leboa (formerly called Northern Sotho), Setswana, and English. The speakers of the nine African languages jointly constituted approximately three-quarters of the total population of the country.

Figure 7.2 depicts where each of the eleven official languages is numerically dominant and simultaneously serves to illustrate the linguistic diversity of the country. African languages are concentrated in segregated single-language rural regions, the former ethnic homelands discussed in earlier chapters. The Nguni language group (isiXhosa, isiZulu, and siSwati) occurs in the east and along the south coast, the Sotho languages (Sesotho, Sesotho sa Leboa, and Setswana) in the west and on the inland plateau, Tshivenda in the far north, Xitsonga interspersed in the east of the Northern Province, and isiNdebele in the west of Mpumalanga province. The core area of Afrikaans is the sparsely populated south-western part of the country, adjacent to its historic point of origin. English, on the other hand, is dominant in the few small but densely populated metropolitan areas. Metropoles generally display a cosmopolitan character, with a mixture of all the languages, since they were the destination areas of migrants from the ethnically exclusive rural homelands.

On a regional level, the selection of a suitable language medium for effective communication in public management, health delivery, commerce, marketing, newspaper, television, and radio coverage, and as a teaching medium in school, is usually dependent on the dominant language pattern.

Socio-economic structure

This set of attributes embraces a wide range of structural characteristics that will differ according to the development

TABLE 7.3 Home languages spoken in South Africa, 1991

Language	Number of speakers	Speakers as % of total population
isiZulu	8 343 590	22.1
isiXhosa	6 646 568	17.6
Afrikaans	5 702 535	15.1
Sesotho sa Leboa	3 530 616	9.4
Setswana	3 482 657	9.2
English	3 414 900	9.0
Sesotho	2 420 889	6.4
Xitsonga	1 439 809	3.8
siSwati	952 478	2.5
Tshivenda	673 540	1.8
isiNdebele	477 895	1.3
European immigrant	109 825	0.3
Oriental	25 505	0.1
Other	495 597	1.4
Total	37 716 404	100

SOURCE: Van der Merwe and Van Niekerk 1994.

level of the community and, particularly in South Africa, its ethnic characteristics. *Family structure* usually reflects the value systems prevalent in communities. Marital status is pivotal in many respects, because marriage has become institutionalized as a means to enhance group survival and growth. Census data usually report whether people are single, married, divorced, or widowed, since these states greatly affect fertility levels and hence the rates of population growth.

FIGURE 7.2 Home language distribution

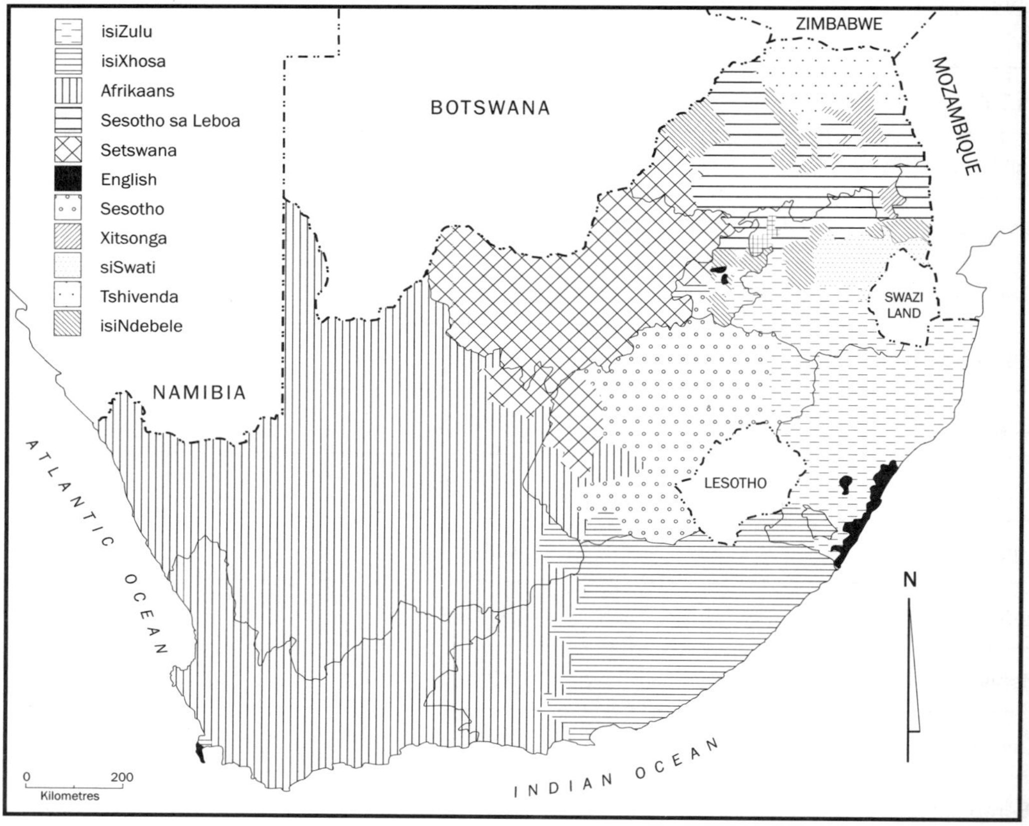

Family size is naturally linked to marital status and must be distinguished from *household size*. While marriage statistics are available and indicate that South Africa had 5.6 million married couples in 1995 (South Africa 1996b), the household denotes a group of people living together and sharing resources (South Africa 1996c). Household size may well be equated with the 'extended family', prevalent among black South Africans, which encompasses an original 'nuclear family', consisting of a married couple and their children, and several generations of relations to the nuclear family all residing under the same roof. Table 7.4 indicates that black rural households are the country's largest, while white households are the smallest, regardless of their location. Interestingly, coloured households are smaller in rural than in urban areas.

The next crucial set of characteristics is economic in nature, indicating the popu-

TABLE 7.4 Socio-economic indicators by race group in South Africa (1995)

	Asian	Black	Coloured	White	Total
Persons/household	4.2	4.7	4.6	2.9	4.3
Persons/household: urban	4.2	4.3	4.8	2.9	3.9
Persons/household: rural	4.6	5.1	3.9	2.9	4.9
Married couples (%) [Total = number]	3.9	65.1	9.0	22.0	5 640 830
White-collar occupations (% of total)	6	42	8	44	33
Blue-collar occupations (% of total)	3	71	14	12	67
White-collar females (% of total)	40	48	54	50	49
Blue-collar females (% of total)	25	35	40	20	33
Informal sector (% of total)	2	80	10	8	
HDI* (1991)	0.836	0.500	0.663	0.897	0.677

* Calculation of the Human Development Index (HDI) is based on various income and education standards as discussed in this section

SOURCES: South Africa 1995, 1996b, 1996c.

lation's status and level of well-being. Usually the level of education, occupational status, and income are so closely associated that they are used as interchangeable indicators of well-being. All these variables are measured and reported by the census and related surveys and can be combined to calculate a comprehensive Human Development Index (HDI) (Thiessen 1997; South Africa 1995) from standardized measures of longevity (life expectancy), ability to communicate and participate in society (years of schooling and adult literacy), and ability to live a decent life (per capita Gross Domestic Product). The higher the index, the better the development of that group of people. Table 7.4 shows that the whites were most developed according to this index, followed by Asians, coloureds, and blacks. South

Africans were on average better-off than all the other SADC countries with the exception of Botswana (see Table 7.1).

Occupation is broadly subdivided into white-collar (professional, administrative, and sales) and blue-collar (production and trades) groupings. People in the former group require longer training at tertiary level, earn more, marry later, have children later, and therefore have smaller families. At the national level South Africa is somewhat unique in that people's characteristics are determined largely by ethnicity, and so it is not unreasonable to equate whites with white-collar occupations and blacks with blue-collar occupations. A comparison of black and white characteristics in Tables 7.2 and 7.5 reveals, as suggested above, that white people on average earn more and have much lower birth and growth rates than do black people.

TABLE 7.5 Income and expenditure of South African households (1995)

	Income (R)	Expenditure (R)
RSA total	40 771	35 688
Asian	70 992	65 438
Black	23 288	19 820
Coloured	31 835	29 121
White	102 857	91 021

SOURCE: South Africa 1996c.

Well-being displays uneven patterns in national space: whereas 18 per cent of the people in the Western Cape live in poverty, 70 per cent in the Northern Province do so (Editors Inc 1997). The estimated 34 per cent of the country's formally unemployed are largely concentrated in the poorest provinces – the Northern Province and Eastern Cape. Furthermore, compared to the national average household income of R40 771, households in the richest province (Gauteng) earn R71 000 and those in the poorest (Eastern Cape) R24 000. Urban (R55 000) and rural (R23 000) average incomes are likewise glaringly divergent. People in traditional (R14 000) and informal (R15 000) surroundings do relatively badly compared to those in formal settlements (R52 000).

Population distribution

Population distribution in space is determined by a number of interacting factors, such as physical conditions, climatic extremities (for example low temperature), geology (energy and mineral resources), presence of disease, availability of water, and quality of soil. Earlier chapters have shown that the impact of colonial and apartheid policies must be added to these criteria when evaluating the population patterns in our subcontinent. Plate 10 (colour section) is a good example of these points: it shows the distribution of population across southern Africa in 1990, and even a cursory glance reveals the vast areas of the western arid interior which are very sparsely populated, particularly in Namibia and Botswana. Also apparent are the shapes of South Africa's homelands as areas of high population density, especially along the south-eastern seaboard and in the Northern Province. Other high-density areas are the better-watered areas of Zimbabwe and Malawi. Finally, the map shows that the areas of highest density are the large urban centres: Johannesburg is clearly visible, as are Cape Town, Beira, and Harare.

Population distribution in South Africa

The South African population distribution has been greatly affected along ethnic lines by apartheid policies. Apart from the general trend of increasing urbanization, overall distribution patterns are unlikely to show major changes in the short or medium terms. Figures 7.3 and 7.4 show the patterns of population in South Africa which have persisted for decades without any major changes. The urban population (Figure 7.3) is concentrated in five major and several smaller metropolitan areas. Cape Town, Port Elizabeth, East London, and Durban have coastal locations, while the behemoth of South African urbanization, the Witwatersrand (encompassing Johannesburg and its surrounding satellites), and Pretoria are located on the interior highveld. Secondary centres such as Potchefstroom/Klerksdorp, Free State Goldfields, Bloemfontein, Kimberley, Pietersburg, and Pietermaritzburg form urban clusters of secondary stature, while the multitude of some 700 towns of various sizes accommodate the rest of the urban population spread around the country. Within a national urbanization average of 55.4 per cent, the 96 per cent and 90 per cent urbanization levels in Gauteng and the Western Cape contrast sharply to the 12 per cent of the Northern Province (South Africa 1997), clearly indicating the differential rates of population concentration between the provinces. With an estimated urbanization rate of 4.9 per cent during the 1990–95 period, the process of urban concentration of people in South Africa has clearly outstripped the total population growth rate of 2.0 per cent per annum during the 1990s. Chapter 8 deals with these issues in more detail.

Generally speaking, climatic determinants (discussed in Chapter 12) originally encouraged the larger concentrations of rural settlements (see Figure 7.4) in the north-eastern interior and along the eastern and southern coastal belt. This relationship can be seen to this day, but now the homeland areas stand out as islands of high black rural population density in a sea of considerably lower white rural density as a result of the policies of past governments.

Population density and overpopulation

The previous section showed that populations are unevenly distributed in southern and South Africa. This is a recurrent phenomenon elsewhere in international and national space. Globally, the greatest densities of population are found in the Netherlands (at 354 people per km^2), India, Japan, and China. But national figures may be misleading, since the total national space is used in calculations (see the equation below). In Egypt, for ex-

FIGURE 7.3 Urban population distribution, 1991

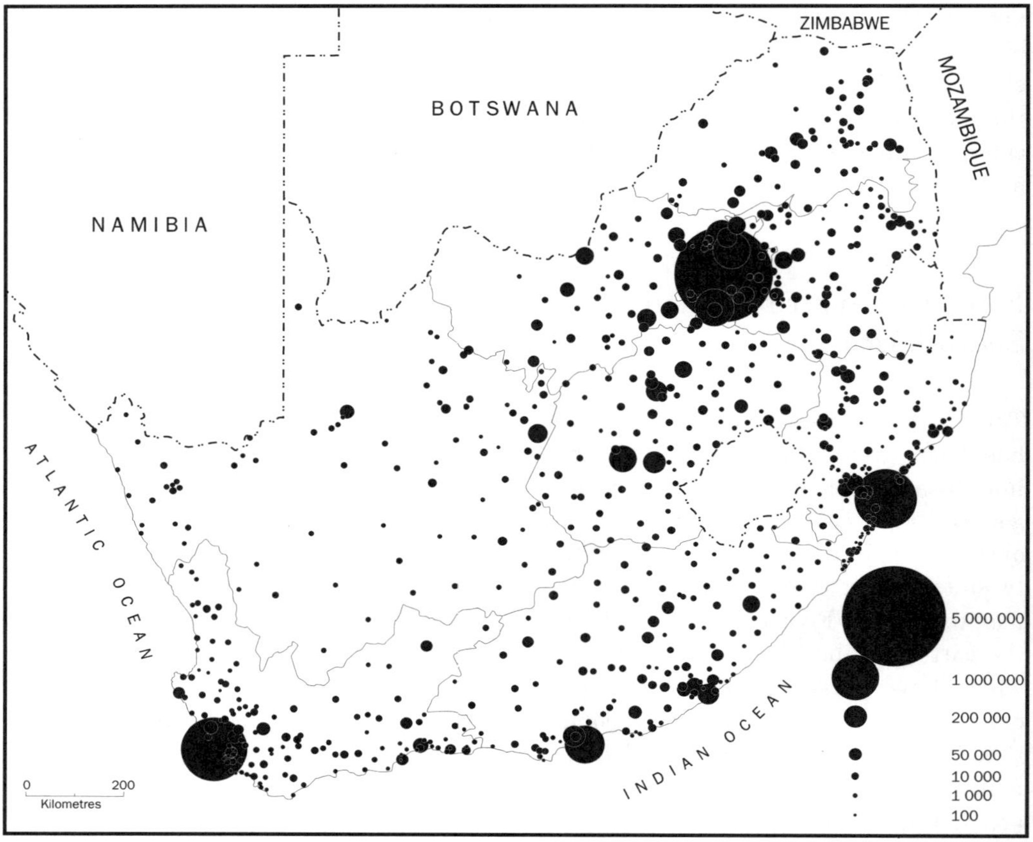

ample, only 4 per cent of the country is habitable, largely in the Nile Valley, which contains 95 per cent of the country's 63 million people. Using national space (995 450 km^2) as the norm, rather than the Nile Valley (40 000 km^2), a density of 63 people per km^2 can be calculated, compared to 1 575 per km^2 – the latter obviously being the more accurate measure.

Population density = population number ÷ area (in square kilometres, or km^2)

A central concept in the study of population distribution is that of *overpopulation*. Semantically the term means 'too many people' – presumably for the space (land) or resources available. When is a country or region overpopulated? Japan and the Netherlands have extremely high population densities and relatively few natural resources, yet are among the most advanced and prosperous nations of the world. Witherick (1990: 134) suggests the following more accurate and sensible definition: 'Overpopulation is said to occur when and where the number of people

FIGURE 7.4 Rural population distribution, 1991

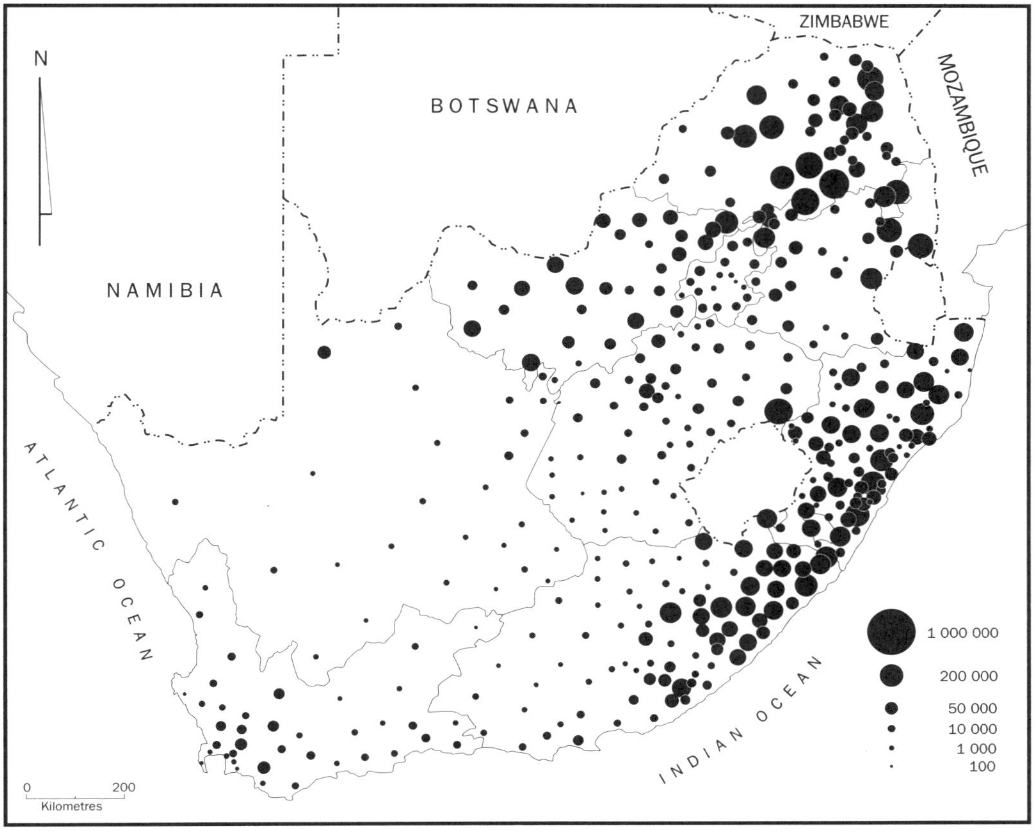

exceeds the supporting capacities of current resource use and the prevailing mode of production'. This would explain why Africa, with a total population density of about 18/km^2 is considered overpopulated, while Europe, with 155/km^2, is not. This argument could be extended to some of the previous homeland areas in South Africa. Table 7.1 shows how population density varies among countries in the subcontinent, the densest population distribution often coinciding with the highest standards of living, as in Mauritius.

Views on the interaction between population, resources, and the environment became formalized after Thomas Malthus argued in 1798 that unchecked population growth increases geometrically (1, 2, 4, 8, 16, etc.), while food supply grows arithmetically (1, 2, 3, 4, 5, etc). According to this view, population growth inevitably outstrips food production, unless growth is checked by external measures. These predictions had at first been disproved by the opening up and exploitation of new lands as colonies which could absorb people and increase food production. Furthermore, the harnessing of fossil fuels and new technology have also increased agricultural production. Esther Boserup presented the counter-argument to the Malthusians in 1965 when she proposed that increasing population growth results in technological innovation and social evolution so that food production can keep pace with population growth.

The debate continued with the influential Club of Rome report entitled *The limits to growth* (Meadows et al. 1972), which used computer simulations to predict that the world system would collapse well before 2100. This would occur due to the exponential growth in food, industrial output, and population, which would exhaust the resource base, causing a growing population to be faced with worsening pollution and rising death rates due to deteriorating food and medical services. While its arguments are not all valid, the report expanded the Malthusian approach to the debate. Alarmist views were further consolidated by estimates in the early 1990s that the global economy was in some areas beginning to 'overshoot' its sustainable limits (Thiessen 1997).

The debate about overpopulation is encapsulated in the concept of *carrying capacity*. This concept has been used by natural scientists to mean the maximum number of organisms that can be sustained by a food-producing system without impairing that system's ability to continue producing. Excessive numbers slowly destroy the system and hence the organisms themselves. This concept has been transferred and applied to human populations (Barrett 1992). It has been argued, along Boserupian lines, that carrying capacities are not fixed through time, but can be raised or lowered by development and technology.

Population growth

According to the latest estimates, world population is expected to increase from its current 5.8 billion to 7.9 billion in 2020 and 10 billion by 2050 (see Figure 7.5), before reaching a peak of 10.8 billion by 2085 and then levelling out. This means that the production system currently has to contend with feeding and caring for an additional 86 million people each year. The 1997 Conference on Population Growth in Addis Ababa agreed that in many African countries the population grows at 3 per cent per year, adding some 20 million additional job seekers and people to feed annually. At this rate the African population will reach 850 million by 2000 and 1 600 million by 2050. What is alarming about these figures is the fact that income in sub-Saharan Africa, already the poorest region in the world, is falling. To improve well-being, real economic growth must exceed the population growth rate; thus an economic growth rate of 3 per cent per annum, sustained over the next decades, is required merely to maintain current levels of living; higher rates are required to improve conditions.

Although the South African growth rate of 2.0 per cent compares favourably with those of its neighbours and indeed with the continent as a whole (see Table 7.1), Calitz (1996) has warned that it is considerably higher than countries similarly categorized as upper-middle income on the global development scale, which have an average rate of 1.8 per cent. The differential growth rates for South African racial groups, reported in Table 7.2, show why population policy (as will become clear later) has in the past focused more

FIGURE 7.5 World population growth

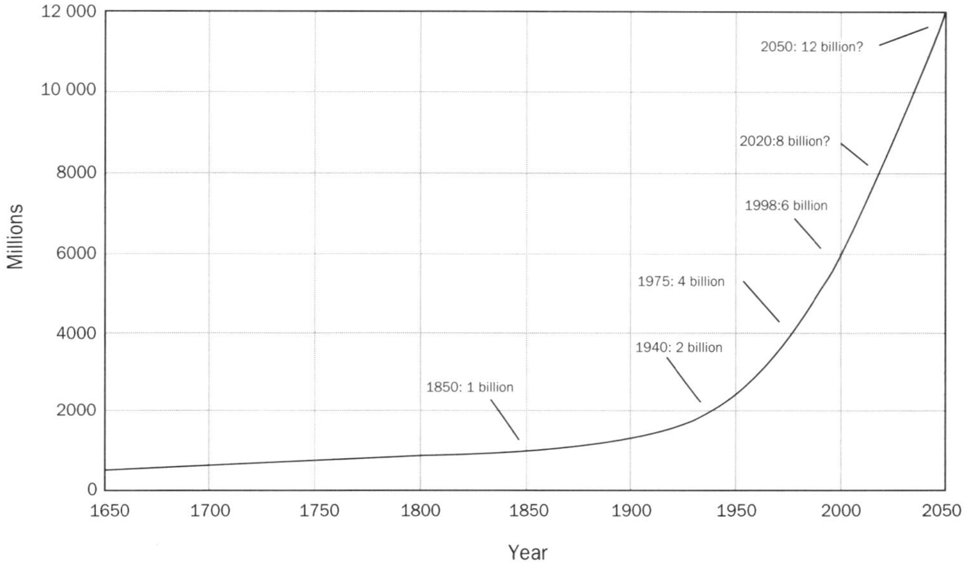

pertinently on, and continues to target, the development of the majority black population groups as part of a strategy to influence demographic behaviour. To do so effectively, the underlying demographic processes must be clearly understood.

Components of population change

Population change is the result of interaction between three processes: births, deaths, and migration. Interaction between these three can cause populations to *increase* (births exceed deaths, or in-migration exceeds out-migration), to *decrease* (deaths exceed births, out-migration exceeds in-migration), or to *remain stable* (births and deaths, and in- and out-migration, are in balance). Fertility and mortality are key concepts impacting on births and deaths. This section will examine these concepts, before addressing the impact of migration and then drawing all three together in terms of the theory of demographic transition.

Fertility as a concept

Fertility is defined as the measured capacity of women (or a nation) to generate births. The number of births is the most important determinant of total population growth, and the *crude birth rate* (CBR) is a simple and most common fertility measure, for which the formula is given below. Because this measure does not reflect the structure of the population (for example the number of women in the reproductive age groups), several standardized measures are commonly used to compare subsections of a population or a country. The crude birth rate shows annual additions to the population directly, but compares them to the whole population, a large proportion of which is not involved in the process. Therefore the *general* and *age-specific fertility rates* (GFR, ASFR) are used; these are standardized measures according to age and gender composition. Since both are fairly cumbersome to calculate, demographers have devised the *total fertility rate* (TFR) as a single, sensitive index that indicates the number of children that would be born to 1 000 women passing through the childbearing age – assuming no women die during the childbearing period and that age-specific births remain the same as in the year of measurement. The latter value at 2 100/1 000, or 2.1 births per woman, is the level of replacement, the point at which the parents are replaced by the children, indicating a stable population (zero growth). Table 7.1 reveals that the TFR for South Africa is 3.3, slightly above the global rate of 3.0 but half the rates of Malawi and Angola.

Crude birth rate (CBR) = (births in a specific year) ÷ (total mid-year population) x 1 000
General fertility rate (GFR) = (number of annual live births) ÷ (total number of women of childbearing age (15–44 or 49)) x 1 000
Age-specific fertility rate (ASFR) = (number of annual live births to women in a specific age cohort) ÷ (mid-year total women in that specific age cohort) x 1 000
Total fertility rate (TFR) per woman (replacement level) = (age-group interval) x (sum of age-specific birth rates over all childbearing age groups) x 1 000

Several factors are responsible for the levels and trends of fertility experienced by specific countries and individual groups. Biological factors such as the age of marriage and the duration of sexual abstinence and breastfeeding determine the length of time a woman is exposed to sexual relationships, conception, and hence childbearing. Induced abortion as a controlling measure is often still illegal in developing countries, but South Africa has recently become one of the few African countries to legalize this practice. Natural sterility, often caused by disease, is declining as a factor, but modern contraception has become the major fertility reducer. Acceptance levels in sub-Saharan Africa are still below 10 per cent, but in South Africa contraception is practiced by 82 per cent of whites, 72 per cent of Asians and coloureds, and 52 per cent of blacks (Calitz 1996). The problem is that in Africa men still prefer larger families than do women, although estimates for both are declining (Roudi and Ashford 1996).

Socio-economic factors, such as the economic value of children as labour in the household from a young age and for support to ageing parents, strongly influence the number of children parents will have. The concept of *wealth flow* underlies this mechanism. Where children are able to realize substantial earnings from a young age (wealth flows from children to the family), more children are an asset and thus desirable. If parents are forced to pay for education and other services for children (wealth flows from the family to children), they become real liabilities and smaller families result. Recent labour legislation in South Africa prohibits child labour below 15 years of age and therefore supports the drive for lower fertility in the country. The status of women in society, especially their educational attainment, is crucial to the lowering of fertility rates. It is no coincidence that women form 64 per cent of all illiterates in developing countries (Yanagishita and Yinger 1995). Literate mothers have greater access to information, and this increases the effectiveness of public campaigns on how to maintain and improve their children's health and thus their chances of survival. After all, 'the most important health care worker is the mother' (Barrett 1992: 124). Higher infant survival rates lengthen breastfeeding periods and remove the need for 'extra' children to compensate for those who may die, and natural contraception allows parents to plan families. Greater female participation in formal employment similarly lowers fertility rates (Roudi and Ashford 1996). Career orientation, less time for childrearing, and potential loss of earnings encourage women to plan for and have fewer children.

Mortality as a concept

Mortality is simply defined as the occurrence of death. If fertility adds numbers to the population through births, mortality provides the counterpoint where deaths reduce population size. Since the control of death, unlike that of birth, is a generally accepted aim of all societies, it has been exceptionally successful. Consequently, drastic reductions in death rates the world over have been achieved long before similar reductions in birth rates.

Indices developed to measure mortality are quite similar to those measuring fertility. The simplest is the *crude death rate* (CDR), which expresses the number of deaths in a given period per 1 000 population. The measure is distorted by the age structure, since populations with a larger proportion of younger people will experience relatively lower numbers of deaths; therefore the *age-specific mortality rate* (ASMR), instead, is calculated for the purpose of comparison. If, for the same purpose, a summary measure is required, the *standardized mortality rate* (SMR) may be more useful, since it indicates what the death rate would be if the target population had the same age and sex composition as some standard population. In addition, because the health and well-being of children is such an excellent indicator of social and environmental conditions in a community, the *infant mortality rate* (IMR) is commonly used. The most sophisticated and useful instrument emerging from the study of mortality is the *life table*, from which is derived *life expectancy*, or the average number of years a person may expect to live. Life expectancy measures the general health and durability of a population, since the higher life expectancy, the lower the mortality rate.

Mortality is influenced by environmental factors such as droughts, floods, storms, and earthquakes, irrespective of victims' development level. However, poorer sectors of society bear greater exposure to these factors. Socio-economic factors such as unsanitary, high-density living space lacking in services such as clean water, sewerage, refuse removal, and medical services promote disease and higher mortality levels. Probably the most frightening spectre affecting mortality in Africa and our subcontinent at present is the danger posed by AIDS. Roudi and Ashford (1996) indicate that, since its dis-

Crude death rate (CDR) = (deaths in a specific year) ÷ (total mid-year population) x 1 000
Age-specific mortality rate (ASMR) = (number of annual deaths of persons in a specific age cohort) ÷ (mid-year total persons in that specific age cohort) x 1 000
Infant mortality rate (IMR) = (number of annual deaths of infants under one year of age) – (total number of live births in that year) x 1000
Life expectancy at birth: calculation of life tables

covery, 4.5 million (75 per cent of the world's 6 million) cases have been identified in sub-Saharan Africa alone, and a further 8.5 million infected cases are suspected today. The World Bank (1996) identifies fifteen nations in central Africa which face declining life expectancies due to AIDS. SADC members include Malawi, Mozambique, Tanzania, Zambia, and Zimbabwe. The United Nations estimates that life expectancy from 2000–2005 in the above countries will decrease from an average of 57 years without AIDS to 49 because of its devastation. Some countries face vast decreases in life expectancy during that time, such as Uganda (10.3 years), Kenya (10.7 years), Zambia (13.2 years), and Zimbabwe (14.8 years). In South Africa 11 per cent of all adults and 6 per cent of the total population (2.4 million) were infected by 1997, and the number is increasing by about 90 000 annually (Steenkamp 1997). While the demographic effect of AIDS has been limited, because of its eight- to nine-year gestation period the pandemic will become significant by about 2005.

Since mortality (death) is the first factor that individuals and public policy endeavour to affect by ensuring longer, healthier lives, it is the first vital statistic of any population which will be so influenced. Measured against the current world CDR of 8.8/1 000 reported in Table 7.1, the African figure is substantially higher and the SADC figure higher still. South Africa's relatively higher level of development is reflected in a death rate lower than the world average. Close neighbours such as Mozambique, Malawi, Zambia, and Angola experience rates almost double this average.

In demographic terms, infant mortality is even more important, since parents will have more children to compensate for possible losses. Infants are most susceptible to the hardships associated with underdevelopment. Africa and the southern African subcontinent display the same negative deviation from the world's average of 57 infant deaths for every 1 000 live births. Botswana, South Africa, and Namibia have managed to reduce their levels to well below the world average – a sign of success in the fields of health services and generally better living conditions. While estimates for the South African situation vary (Editors Inc 1997; Calitz 1996), there seems to be consensus that the IMR for blacks (50/1 000) is substantially higher than those for coloureds (26/1 000) or for Asians and whites (both below 10/1 000). Similarly, the provincial figures vary according to development level, with more than 55/1 000 in the poverty-stricken Eastern Cape, compared to 25/1 000 in the much wealthier Western Cape. Ashford (1995) has shown how average world figures have improved from 156/1 000 in 1950 to their current levels.

She contrasts 1993 figures of 69/1 000 in the developing world with 9/1 000 in the developed world, and reports that the IMR in sub-Saharan Africa is above 100/1 000, compared to 53/1 000 for North Africa.

African and SADC life expectancy, while being generally higher for females than for males, is, as can be expected, much lower than the world average. Men and women in South Africa and Botswana can expect to live substantially longer than their peers in neighbouring countries, with Malawi, Zambia, and Mozambique worst off. World Bank (1996) figures indicate that, while the world average increased from 61 in 1980 to 65 in the mid 1990s, the gap between the least developed (43 years) and most developed (78 years) nations is expected to widen even further to a difference of 37 years by the year 2000.

The trends in fertility and mortality have implications for understanding the dynamics of population growth. *Natural change* occurs in populations because of variations in the balance between the natural life processes of births and deaths. Whether this balance is positive or negative depends on the relative dominance of one of the processes. *Natural increase* or *growth* is obtained by subtracting the number of deaths from the number of births in a given year. While this index is a rather rough measure of the population growth rate and potential, it does provide an indication of the population growth element which may be targeted for effective population development and control policies. Furthermore, it must be borne in mind that 'even if fertility levels drop below replacement level, population grows because of "population momentum" – i.e. each woman has fewer children, but many more women from previous cohorts are giving birth' (World Bank 1996).

Migration

The third and last element affecting population growth is migration – the permanent or semi-permanent movement of people between countries and between locations or regions within a country. Demographically, such movements affect both the area of origin (from which migrants originate) and the destination (to which they move). Migration between countries (by emigrants leaving or immigrants arriving) should not be confused with within-country movement (out-migrants leaving an area and in-migrants arriving in an area), nor should migration and circulation (short-term repetitive movement such as commuting) be confused, although both circulation and migration are embraced by the same term – *mobility*. Salient factors generally affecting migration are a mixture of economic, social, and environmental forces. The various factors act together to push migrants away from an area of origin and to pull them towards a destination – from areas of low to areas of high opportunity (Ashford 1995). Migration, more than any other process, is responsible for the kaleidoscope of peoples coexisting across the globe.

Economic factors are among the major reasons why people migrate. Migration is encouraged by opportunities for employment or the exploitation of natural resources, and facilitated by a good connecting infrastructure. In Europe, 'labour migrants', doing jobs shunned by local workers, generally form between 5 and 10 per cent of the population. In South Africa, migrant labour has always been regarded as temporary, and families were not allowed to accompany male workers. The large numbers of labourers from neighbouring countries and from various domestic origins are classified in the census as temporary workers employed in the country. Furthermore, since the advent of democratic transformation, there has been a growing illegal immigrant problem. Recent estimates of the current illegal population range from ministerial pronouncements of 2 million to alarmist estimates of between 8 and 12 million; the consensus figure is 4 million (about 10 per cent of the total population). In 1996 the South African state apprehended 180 173 illegal immigrants and repatriated them to their countries of origin (most often war-torn Mozambique).

Socio-political factors have created a widespread refugee problem as a result of political or other forms of oppression or persecution. In 1995 some 25 million refugees were registered worldwide (Kane 1995; Ashford 1995). High-profile examples abound in Africa; in Rwanda, for example, resource competition among the most densely settled population on the continent, combined with historical animosity between rival Hutu and Tutsi tribes, led to violent conflict resulting in hundreds of thousands of deaths and millions fleeing to Zaire, Tanzania, and other neighbouring countries. Even countries as far afield as South Africa received 8 166 applications for asylum from Rwandans and Angolans in 1996 (Steenkamp 1997).

Environmental events such as drought, desertification, resource depletion, and overpopulation often precipitate crises of survival, forcing the relocation of large numbers of people. Such moves often take place within countries, so that their extent is less obvious, yet they produce the largest class of displaced people in the world (Barrett 1992). Land degradation, usually attributable to bad management practices and over-exploitation, is causing the displacement of millions of people worldwide. While some environmental events such as floods or landslides cause limited spatial displacement that may be of a temporary nature only, many events have permanent effects. Of growing concern is the extent and frequency of technologically triggered environmental devastation – a form of threat not readily heeded by governments. The World Bank furthermore estimates that some 10 million people are displaced internally by sanctioned public-works programmes every year (Kane 1995).

Although not of critical concern, it is important to recognize that most migration decisions are made by individuals reacting to particular perceptions about conditions at origin and destination. This means that migration can be highly selective in nature. For instance, while the world has been astounded by the success story of the democratic transition in South Africa, some 3 000 people, largely professionals whose loss the country can ill afford, have officially elected to emigrate annually since 1994. Some sources put these figures substantially higher.

Demographic transition

The availability and thorough analysis of historical population data for some developed countries (especially Scandinavian countries such as Sweden and Denmark) have enabled researchers to gain insights from which a comprehensive model of population development has been formulated: the process termed *demographic transition*. While present world population growth seems to be attributable to the widening gap between birth and death rates, the theory of demographic transition suggests that all countries or societies undergo a differential process of demographic regulation through the change from high to low rates of fertility and mortality (see Figure 7.6). First suggested by Thompson in 1929 and later formalized by Notestein in 1945 (Barrett 1992; Witherick 1990), the theory states that societies try to keep births and deaths in balance to ensure continued growth within the limits set by the carrying capacity dictated by land, the economy, customs, and other factors.

Where mortality is high, fertility must be high too, to compensate for losses in population numbers (Stage 1 in Figure 7.6). Control of fertility is thus not important, but societies universally strive to minimize the rates of mortality, and that is the first vital process to be so affected. Death rates are lowered once disease control and higher levels of food supply are achieved. Thereafter, continuance of the previously high fertility produces high rates of natural increase, which eventually threaten carrying capacity and thus the community's well-being (Stage 2). Only then does the lowering of birth rates become a priority. A time lag follows before birth rates are similarly brought under control (Stage 3) – the end of the transition process. At both the beginning and the end of the transition process (Stages 1 and 4), natural growth is insig-

FIGURE 7.6 The demographic transition model

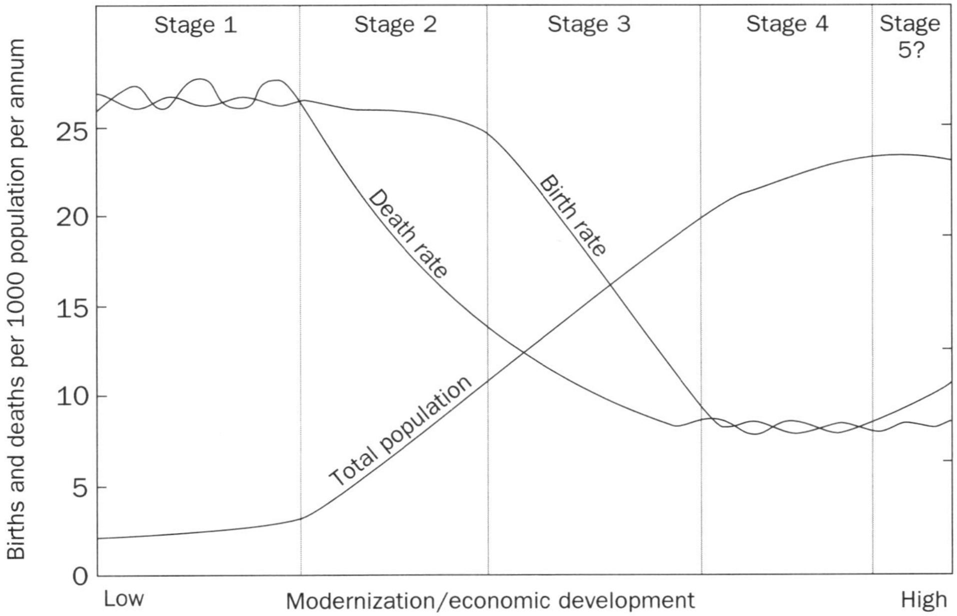

nificant – vital processes are in balance, resulting in a stable population. The critical factor for population growth is the duration of the response time of the birth rate to drop to meet the death rate, as a 'population explosion' takes place in this period.

Most of South Africa's SADC neighbours find themselves firmly in Stage 2 of the transition model; South Africa, however, has evidently just entered Stage 3. Mortality is already relatively low, and fertility has apparently started a steep decline over the last ten years. Demographic planning at government level throughout the subcontinent should aim at steering population growth to the stability of Stage 4 in as short a time as possible.

Population policy

Because the human population changes dynamically, it is understandable that various forms of policy have been devised to manipulate the processes discussed above. Policy takes various forms, but can be divided into those formulated at the international level and which have no binding force, and those at national level implemented by governments and their agencies.

International policy development

International population concerns have led to several organized initiatives since the middle of the twentieth century. United Nations–sponsored international meetings and conferences on population have been held in Rome (1954), Belgrade (1965), Bucharest (1974), and Mexico City (1984), and these were capped by the International Conference on Population and Development (ICPD) in Cairo in 1994. The latter has set the agenda and targets for population development and control at the international and national levels for the decade bridging the twentieth and twenty-first centuries. The conference was exceptional because, for the first time, 180 countries (bolstered by more than 1 200 NGO representatives) reached consensus on an approach to stabilize the world's population. It placed the causes and effects of rapid population growth in the context of human development and social progress (Ashford 1995). Even though the ICPD document is non-binding, it legitimizes a framework of thinking internationally – especially regarding the pivotal role of women in the process and the intertwined relationship between population and the environment. The strong emphasis on women's empowerment was also the main theme of the Fourth World Conference on Women, held in Beijing in 1995 (World Bank 1996).

Ashford (1995) considers most significant the following aims from the ICPD policy guidelines on reproductive health, poverty, and environmental protection:

- bringing about more equitable relationships between men and women, and empowering women to participate more fully in development;
- reducing and eliminating unsustainable patterns of production and consumption;
- formulating and implementing population policies as part of social and economic development;
- taking steps to eradicate poverty;

- providing universal access to family planning services and reproductive health care;
- improving the health of infants and children;
- increasing access to education, especially for girls;
- improving the status of women, and expanding opportunities for young women (education and job possibilities); and
- involving men in childrearing responsibilities and family planning.

The programme of action also sets specific goals for the target year 2015 in several areas:
- providing universal access to a full range of safe and reliable family planning methods and related reproductive health services;
- reducing infant mortality rates to below 35 infant deaths per 1 000 live births and mortality rates of children under five years of age to below 45 per 1 000 live births;
- closing the gap in maternal mortality between developing and developed countries and aiming to achieve a maternal mortality rate below 60 deaths per 100 000 live births;
- increasing life expectancy at birth to more than 75 years, and, in countries with the highest mortality, aiming to increase life expectancy at birth to more than 70 years; and
- achieving universal access to and completion of primary education, and ensuring the widest and earliest possible access by girls and women to secondary and higher levels of education.

National policies

National population policies have been increasingly adopted ever since the 1984 Mexico conference, but many developing nations, particularly in Africa, have done so only in the 1990s (Ashford 1995). Government policy may be used to increase or decrease population growth. Pro-natalist policies have been followed when population growth has fallen below replacement levels. Methods include the restriction of access to fertility control measures such as contraception or abortion and the encouragement of large families through financial incentives and low-cost service provision.

Anti-natalist policies aim to reduce growth by lowering birth rates. This is achieved by means of incentive schemes and by providing free and ready access to family planning services. Examples of such policies abound in developing countries such as India (since the 1950s), Pakistan, South Korea, and China. Aims for reduction include growth-rate limits or total number limits as targets. As yet, many African countries have no population policies, although Kenya is an exception, with policies initiated in 1967. Most officially view their birth rates as 'high', but a number view theirs as 'satisfactory', even when this has dire connotations for countries with low or negative economic growth rates. Zimbabwe is often credited with the best and most successful population policy in Africa (Spies, cited in Haldenwang 1993). The latest census results seem to indicate similar policy success in South Africa in the 1980s.

Government policy is most visibly employed to affect population distribution patterns. Some measures may be indirect and others coercive in nature. In the developed world, economic measures such as regional development strategies control where development takes place and where employment opportunities and housing are provided. Coercive strategies have been more typical in developing countries. Recent examples have been the forced relocation of people as a result of the building of large dams, such as Aswan in Egypt and Kainji in Nigeria.

South Africa has in the past been notorious for the application of coercive apartheid policies with the explicit spatial aim of concentrating people according to ethnic grouping in the nine homelands and in specific urban townships. In addition, industrial development was used as a spatial development policy, and this impacted on the distribution of economic activity and hence the distribution of the population. Current policies, such as the Spatial Development Initiatives described in Chapter 5, will also attract people to particular destinations in the country.

South Africa's population policy

The first comprehensive population policy in South Africa was the family planning programme established in 1974. This promoted access to contraceptive services, especially for the black population, in order to lower growth rates. Family planning clinics were established and contraceptives were provided free of charge. The programme functioned inadequately and operated independently of other health services. In 1984 the Population Development Programme (PDP) was established to implement the recommendations of the Report of the Science Committee of the President's Council on Demographic Trends in South Africa. It set the target of achieving an overall fertility rate of 2.1 by the year 2010 in order to stabilize the population at 80 million by the year 2100 – the maximum the resource base (especially water) could support. While focusing on family planning and fertility reduction to reach this goal, intervention in other influential aspects, such as education, primary health care, housing, and economic and human resource development, were targeted. Also, mortality, migration, urbanization, and rural development were indirect concerns.

The policy, although refreshingly different from the previous efforts, lacked political legitimacy. It was implemented by the Chief Directorate of Population Development (CDPD) in the Department of Health and Population Development and by population units in the various provinces. The policy suffered funding shortages and administrative shortcomings and lacked the support from an overarching socio-economic development plan for the whole country. From 1990 the CDPD's focus consequently shifted to the formulation and implementation of population information, education, and communication (IEC) programmes, which promoted small-family norms. The objective was to influence family size preferences and the reproductive behaviour of subgroups with high fertility.

In 1996 the Government of National Unity produced its new policy in the Draft White Paper for a Population Policy (South Africa 1996d). The preamble of this document states explicitly that the new South African policy is embedded in the guidelines formulated at the conclusion of the Cairo ICPD, which the country endorsed. The White Paper reflects a recognition of the fundamental interrelationship between population, development, and the environment. The policy aims to resolve a long list of population concerns, including:

- the poor population data and knowledge base, and inadequate analytical capacity;
- the problems of youth and the aged, the low status of women, and insecure family and community life;
- demographic problems, including high incidence and rates of unwanted fertility, teenage pregnancy, infant and maternal mortality, and international immigration; and
- problems of population growth in excess of economic development, social service backlogs, and population pressure on the environment.

The policy contains a comprehensive set of guiding principles encapsulated in the vision statement: 'to contribute towards the establishment of a society which provides a high and equitable quality of life for all South Africans and in which population trends are commensurate with sustainable socio-economic and environmental development' (South Africa 1996d: 24). Policy objectives have been designed to address the aforementioned concerns by means of a co-ordinated, multi-sectoral, interdisciplinary, and holistic approach. This places South Africa in the vanguard of current international policy orientation and initiatives. The policy is to be implemented by existing government agencies and restructured provincial population units, with the Office of the Vice-President responsible for monitoring the process.

Sources of population data

One fundamental requirement for the study of population is accurate information at all scales – population numbers, their geographical distribution, structure, and change and its causes. Unfortunately, such information is still lacking or at best of poor quality for many parts of the world, and especially so in under-resourced developing regions where such data are most in need. Since the 1950s the United Nations Organization (UNO) has done much to improve the availability and standardization of demographic information from developing countries. Similar population data are now collected in national population censuses across the globe. According to Witherick (1990), these aspects include:

- total population and its internal distribution across the country;
- sex, age, and marital status (vital for growth projection purposes);
- birthplace and nationality;
- language, literacy, and educational level;
- economic characteristics such as employment and occupation;
- rural/urban location and living conditions; and
- household and family structure and fertility levels.

South Africa has conducted twelve national censuses, all in the twentieth century (1904, 1911, 1921, 1931, 1946, 1951, 1960, 1970, 1980, 1985, 1991, and 1996), providing the country with a fairly substantial basis for the scientific study of the population. Not all these censuses were conducted at the same level of accuracy in all regions of the country – as the controversies surrounding the 1996 census results show. Normally countries carry out census surveys at ten-year intervals, although some (Japan, for example) do so at five-year intervals.

Censuses the world over follow similar procedures (South Africa 1997). To ensure that information on the spatial distribution of population is obtained, the process starts with the demarcation of a network of spatial units spanning the total national space. Enumeration areas (EAs) are delimited such that a fixed number of census officers are enabled to conduct the census questionnaire survey. For the 1996 South African census some 86 000 EAs were demarcated and 100 000 enumerators and supervisors were employed to conduct the survey. The size of EAs are determined by the norm of 100–250 households per EA. Their extent may range from single street blocks in densely populated urban areas to many thousands of square kilometres in the sparsely populated western interior. At the next level of spatial demarcation, EAs are grouped into 370 magisterial districts and nine provinces, and spatial studies can therefore analyse and map results at district, regional, and provincial level. Research can also take place for individual urban areas, analysing data at EA or neighbourhood/ward level. Rural data are limited to the EA level only, and this follows farm property boundaries where they exist.

The enormous effort required to conduct, code, and capture census information in computer-readable form results in substantial time lags between survey and result publication. The latest South African census, for example, was conducted in May 1996 and the final results were released in September 1998 – a full two-and-a-half years later. The spatial framework from one census to the next usually remains fairly consistent, enabling temporal comparisons of spatial units such as districts. Areas undergoing rapid development due to urbanization may be subdivided into smaller units, but it will always be possible to aggregate spatial information to the same previous areal units for the purpose of comparison. No census survey is absolutely accurate. A post-enumeration survey of the 1996 census exposed an average undercount of 6.8 per cent, with more accurate figures obtained from rural than densely settled urban areas. Accuracy checks are performed on most censuses and they take the following forms:

- post-enumeration surveys of approximately 1 per cent of the population;
- controlling age-group statistics against official lists (for example school enrolments); and
- demographic modelling.

A very important addition to census statistics are the *vital* (or *life*) *statistics* of a nation, consisting of data on marriages, births, adoptions, divorces, and deaths – the elements governing fertility, mortality, and thus population growth. In South Africa this collating task is also performed by Statistics South Africa (formerly the Central Statistical Services) from data registered in magisterial offices in each town and by appointed local residents, 'justices of the peace', in remote rural areas. Most governments require these events to be officially registered and reports on them are obtainable as statistics from Statistics South Africa and the Government Printer in South Africa.

Conclusion

This chapter has aimed to introduce some of the main concepts required for comprehending the dynamic processes that drive population growth, which is surely one of the most serious challenges of our time. It is hoped that the reader's interest has been sufficiently stimulated by the scientific challenge of population studies to encourage continued and deeper academic understanding of the population phenomenon. At the same time, the aim was to provide some factual basis to the population dilemma in southern and South Africa. From that angle, it is hoped that the reader will remain informed about, and practically contribute towards, population education, policy-making, and the sustainable development effort whereby a healthy and balanced society may be built.

8 Urbanization and urban forms

ANTHONY LEMON

In the next century, the most relevant unit of economic production, social organisation and knowledge generation will be the city.

George Yeo, Information Minister of Singapore, 1995

An urbanizing world

The twentieth century has witnessed an urban explosion. Only Britain, the cradle of the Industrial Revolution, had an urban majority at the beginning of this century; today this is true of all developed countries, and early in the twenty-first century it will be true of a majority of developing countries. United Nations figures for 1994 show that 45 per cent (2.5 billion people) of the world's population are urban dwellers. This is projected to rise to 50 per cent by 2005, and to over 60 per cent by 2025. The world's urban population is growing by 2.5 per cent annually, three times as fast as the rural population. This means that urban areas are absorbing 61 million more people each year.

This increase in the proportion of the world's population living in cities is known as *urbanization*. It is distinct from the process of *urban growth*, which refers both to the increasing population of individual cities and to expansion in their physical size; such growth may occur without urbanization, particularly where (as in many developed countries) rural population growth is equal to or greater than that of urban areas. A third trend, which is again common in many developed countries, is the spread of urban values and lifestyles beyond the confines of the city: this may be termed *urbanism*.

In 1800, only 3 per cent of the world's population was urbanized; some of today's largest cities, such as Tokyo, with 25 million people, have a population as large as the global total at that time. Nineteenth-century urbanization was closely associated with the Industrial Revolution in Western Europe and North America. Some countries urbanized very rapidly: in the USA, between 1820 and 1870, the estimated average annual growth rate for the urban population was 5.5 per cent. The urban population of Japan, which industrialized later, increased by around 6 per cent a year during the 1930s. Such growth rates actually exceed those of urban populations in most developing countries today; the urban population of Africa, for instance, grew by less than 5 per cent a year during the 1980s. What is unprecedented in the last three or four decades is the number of countries undergoing rapid urbanization, and the number of cities worldwide that are growing rapidly. The scale of change in the twentieth century is shown by the dramatic rise in the number of very large cities. According to the United Nations Centre for Human Settlements (HABITAT) (1996), cities with over one million people increased in number from 24 in 1921 to 41 in 1941, 113 in 1961, and 198 in the early 1980s. By the mid 1990s this number had increased to 281, of which 179 were in less developed countries, including 20 of the world's 30 largest urban agglomerations (see Figure 8.1). Most of the latter are now growing more slowly than the lesser 'million cities': only Dhaka (Bangladesh), Lagos (Nigeria), Karachi (Pakistan), Jakarta (Indonesia), and Bombay (India) exceeded 4 per cent growth per annum in the 1980s.

The actual populations of the world's largest cities depend on the definitions used: thus Mexico City, the largest city in the developing world, had a 1990 population of 15 million, but this figure would be several million larger if it measured the wider region where different metropolitan areas fuse or overlap in a polycentric manner. Tokyo, however, the biggest city in one of the core nations, remains the world's largest, with a 1990 population of 25 million, and 31.6 million in the Greater Tokyo Metropolitan Area.

Today it is developing countries which are experiencing the highest rates of urbanization. Three-quarters of the people in developed countries are already urban dwellers, but only half this proportion – 37 per cent – is found in developing countries. As this proportion increases, the developing world will come to have the lion's share of the world's urban dwellers: 71 per cent in 2000 and 80 per cent in 2025, according

FIGURE 8.1 Global distribution of large urban agglomerations, 1990

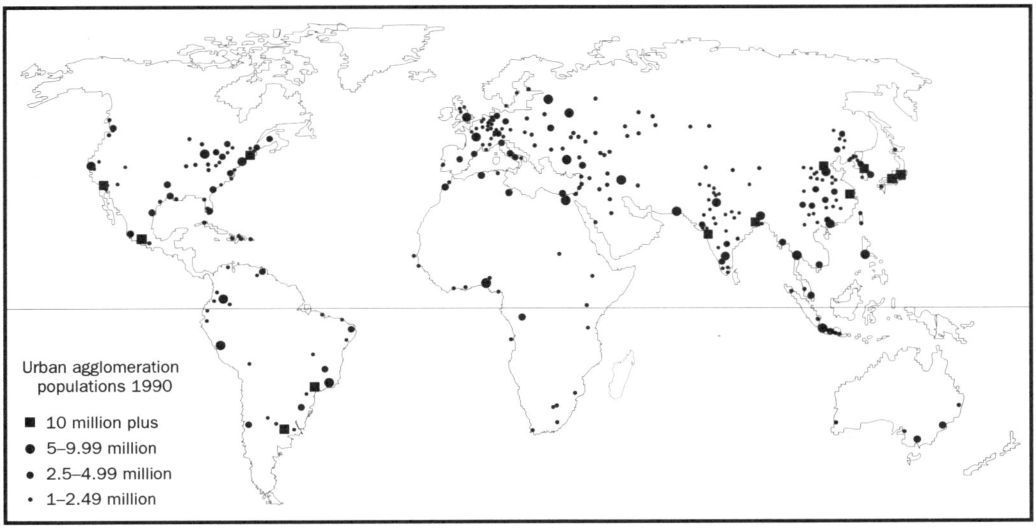

to United Nations predictions. Latin America is already highly urbanized, therefore the major projected increases in urbanization are in Africa and Asia (see Table 8.1). Since 1950, the pace of urbanization has been particularly high in sub-Saharan Africa. This reflects the small base from which it began, which was in part due to colonial restrictions on the right of Africans to live in urban centres. Urbanization in Africa and the developing world generally differs critically from that in the developed world, because it is not supported by widespread industrialization. This has led some to argue that in many developing countries levels of urbanization are excessive and detrimental to economic growth. We shall return to this theme, but first it is important to look at some of the changing characteristics and problems of cities in the developed world, in order to appreciate what is distinctive about the southern African and developing-world experience of urbanization and urban growth.

Urban trends in developed countries

As the major cities of the developed world have grown, most have experienced decentralization of both population and production. This process is reflected in the succession of terms coined to describe successive stages of the process – *conurbation* (introduced in 1915), *metropolis*, *megalopolis* (first used in 1961), and even Toynbee's vision of 'Ecumenopolis, the single globe-encompassing city of the not very distant future' (Toynbee 1970: 8). Such massive expansion has produced major problems of urban management and government in dynamic urban regions.

Residential patterns within capitalist cities have been strongly influenced by what is known as the 'bid–rent curve'. Business uses placing a high value (bid) on centrality have paid a premium (rent) for use of the most accessible central sites, leading to the formation of central business districts, or CBDs, and the development of high-rise buildings. Much early industrial development also occurred close to the centre of the city. The poorer inhabitants, lacking the financial means to commute long distances, needed to be close to their employment; this led them to live in inner areas of relatively expensive land, which they could afford only by living at very high densities. The middle classes lived in suburban areas at lower densities, while the rich, who enjoyed higher personal mobility and could afford longer commuting journeys, lived in spacious outer suburbs or in the rural fringe. Elements of this pattern are present in most Northern industrial cities, although in some Western European countries many wealthier people opt to live in flats in central areas, preferring accessibility to space; in central Paris, for example, some affluent residential districts are close to low-income areas.

The inner parts of many developed-world cities have become areas of multiple deprivation. Physical decay adds to the problem of poor-quality building. The economic base of these inner areas has declined as many small industrial firms have either closed down or moved to locations more accessible to the modern road systems, while many service functions such as docks and warehouses have also

TABLE 8.1 Global urbanization levels in 1920, 1994, and 2025 (%)

	1920	1994	2025
World total	19	45	61
LESS DEVELOPED REGIONS	10	37	57
Africa	7	34	54
East		21	41
Central		33	54
North		45	63
Southern		48	67
West		36	59
Asia	9	34	55
East		36	58
South-East		33	55
West		66	80
Latin America	22	74	85
MORE DEVELOPED REGIONS	40	75	84
Europe	46	73	83
North America	52	76	85
Oceania	47	70	75

SOURCE: Potter 1992; United Nations 1995.

closed or relocated. The more successful inner-city dwellers have moved out, leaving behind the poor, often characterized as an 'underclass', who lack the resources to move and who therefore remain trapped in deteriorating areas of high unemployment, deprivation, and often crime. A low tax base makes improvement difficult without external support. Such areas provide a milieu where the newcomers to the city can find refuge, as can drop-outs. In countries such as Britain and the USA inner cities are also home to large ethnic minority populations. These may display a high degree of segregation, as in American ghettos, where socio-economic mobility has not necessarily allowed blacks to escape the ghetto in the face of white racism. There may be elements of choice too, as ethnic minorities seek protection and solidarity in their own communities; the relative importance of choice and constraint is the subject of extensive debate in the work of social geographers on ethnic residential segregation.

Some inner-city areas have experienced redevelopment. This may take the form of a gradual process of *gentrification* – the transformation of dilapidated property into upmarket homes for 'yuppies' (young urban professionals) and higher-income groups who value accessibility. Gentrification impinges on the availability of accommodation for the less well off, and can be seen 'as part of that social polarisation which characterises the post-industrial, transactional city' (Carter 1995: 294). Some explain gentrification in terms of changing consumer demand, but Neil Smith (1979) sees it as part of the structural adaptation of capitalist cities to the post-industrial world, as capital seeks areas of maximum returns on development. More dramatic redevelopment occurs in wholesale urban renewal schemes, including dockland and waterfront schemes; the London Docklands provide one of the best-known European examples, and a South African instance of this international trend is Cape Town's Victoria and Alfred Waterfront. While spectacular in their visual and economic impact, these developments add to social polarization and do little to improve employment prospects for the underclass.

The outer areas of Western cities also have their problems. In some outer estates these derive from isolation, by cost if not time and distance, from the facilities and services of the central city. Social problems are accentuated where municipal housing allocation policies have created 'sink estates' by concentrating single-parent families and ethnic minorities. The result of such concentration, when combined with high unemployment and a lack of community facilities, is often vandalism and crime.

As rioting, crime, and violence become more common, sometimes spreading beyond inner cities and peripheral estates, the middle classes in some cities have begun to feel insecure. One response, especially in the USA, has been an increase in gun ownership and the willingness to use weapons in self-defence. In spatial terms the reaction has been to create walled and defended housing developments patrolled by security guards.

In recent decades many urban dwellers in the most developed countries have moved out beyond the suburbs, leading to a process known as *counter-urbanization*. This involves a shift to smaller places, not necessarily villages, and it seldom involves more than superficial adoption of a rural lifestyle or values. It seems to have reached its peak in the 1970s in the USA and Western Europe, but continues at a lower level of intensity. In some places it is paralleled by the out-movement of industry and offices to smaller 'freestanding' towns which are often the fastest-growing settlements in developed countries. A more recent trend, foreseen by the American geographer Brian Berry (1970), is the phenomenon of increased home-based work facilitated by sophisticated electronic forms of communication. As yet, however, the impact of this trend on settlement patterns falls far short of the reversal of spatial patterns of urban growth envisaged by Berry, just as the process of 'megalopolitanization' falls far short of Toynbee's ecumenopolis.

Attention so far has focused on capitalist cities characterized by a free land market (albeit modified by the planning process) and competition between individuals, commercial concerns, and indeed cities themselves as they increasingly engage in local economic development initiatives (discussed in Chapter 5). Some urban geographers have recognized distinctive *socialist cities* where private profit is outlawed and the 'community' or local state allocates resources. Much of the vast literature which is critical of capitalist cities implies a contrast with an unspecified ideal socialist city; few attempts have been made to build models of socialist cities, although visual differences in housing are obvious enough at a practical level in the former communist states of Eastern Europe, with their extensive landscapes of state-constructed flats. Although critical analyses were possible in Eastern Europe and the former USSR before 1989, the collapse of communist political structures has made them outmoded; in the Third World even Havana, the Cuban capital, is being influenced by market forces (Carter 1995: 326). Moreover, the lack of data in the public domain has hindered research. Finally, the relative brevity of communist rule means that the effects of previous eras remain important influences on the form of the cities concerned.

Colonial cities

Most Third World cities have colonial origins, although in Asia and Latin America colonial structures were often imposed on indigenous urban institutions. The colonial inheritance typically includes an eccentric location, skewed employment structures, and cultural pluralism. Locations are frequently coastal, reflecting an external orientation towards the colonial power; this inheritance is strongly reflected in the current distribution of large cities in the Third World (see Figure 8.1). As these cities were instigated as export and administrative centres, development of industrial employment was usually very limited: the colonial power sought to exploit any market for goods with its own products. Residential segregation of colonizers and colonized was characteristic of most colonial cities: it was imposed by the British from their earliest colonial settlements onwards (Christopher 1983) and was equally characteristic of francophone Africa (Winters 1982).

Most African cities did not exist before colonial rule. Thus Salisbury (modern Harare) was founded in 1890 when the

British Union flag was raised on the previously uninhabited site of Fort Salisbury, and Johannesburg in 1886 on the site of a diggers' village which was being developed on unclaimed land close to a major gold-mining complex. The development of a colonial cash economy, which made available goods and services in exchange for wages, established powerful currents of labour migration to mines, plantations, and urban employment; most migration was between rural areas, but perhaps a quarter of all migrants went to the towns (HABITAT 1996). Colonial control over African urban integration was strongest in the white settler colonies of eastern, central, and southern Africa, where influx controls were applied by British administrations to limit the growth of the African urban population and to relate it to labour demand.

Relaxation of such controls led to rapid urbanization in the early years after independence. Many African governments responded to the resultant urban problems by commissioning expatriate firms to draw up master plans similar to those produced in the colonial period. These plans influenced approaches to land-use planning in central areas of cities, but failed to guide or constrain the speed and direction of growth in peripheral areas. They were almost never supported by the level of capital expenditure necessary to implement their infrastructural projections. Most planners were still Europeans, and worked in a regulatory framework largely derived from the former colonial power. Housing policies, too, changed little in the early years of independence, with a focus on the redevelopment of decaying core areas, the removal of 'slums' or squatter areas, and the construction of large public housing estates. The latter made little impression on housing shortages, and many of the houses went to middle- or upper-income groups. Building standards more appropriate to developed countries made housing too expensive for most households, so an increasing proportion lived in informal or squatter settlements whose quality tended to reflect the degree of security enjoyed by the inhabitants as much as their material resources.

Third World cities – parasitic or generative?

In an influential paper, Hoselitz (1957) argued that cities can be either generative, stimulating growth in their wider regions, or parasitic, inhibiting such growth by extracting surpluses without contributing to rural development in their hinterland. In the colonial period they were primarily parasitic, but Hoselitz believed that they later came to be agents of modernization, spreading the benefits of development. Such concepts underlay much early post-colonial planning, and were exemplified in the first two Zambian National Development Plans, which envisaged a spreading or *trickle-down* of growth from urban cores to rural areas (Chapter 5 introduces this process as diffusionism). The reality has generally proved otherwise; as the Swedish development economist Myrdal argued, once economic success is established in a city or region, it tends to grow cumulatively through a process of *backwash*, at the expense of the surrounding area, which is drained of some of its most able and enterprising people through selective migration, and which suffers from unequal exchanges of goods and money with urban areas. Myrdal did recognize the existence of *spread* effects – benefits generated by urban areas in the surrounding regions – but saw these as limited and occurring spontaneously only at a late stage of economic development; if developing countries wanted to generate such effects, government intervention would be essential (Lemon 1977). Many African governments have accentuated backwash effects by excessive concentration of scarce resources in their capital cities, often intending them to be enclaves of development for prestige purposes. This has emphasized the *primacy* of these cities, enhancing them as magnets for migration. The American planner Friedmann, in his well-known *core–periphery* model of regional development, recognizes that transition from urban primacy to a more spread-out pattern of development including sub-cores rarely happens without government involvement. However, attempts by Third World governments to spread growth through the designation of 'growth poles' have frequently misfired too, relying on inappropriate notions of spread and trickle-down, and typically designating an unrealistic number of growth poles in relation to the size of the modern economy (Rogerson 1989).

Primacy is the domination of a country by a single large city, known as the **primate city**, which is much larger and more influential than any other cities.

Some have argued that many developing countries are 'over-urbanized' in relation to national income. Urbanization has occurred without economic growth, industrialization, and increased agricultural productivity to sustain it. In Latin America, average levels of urbanization in 1980 were similar to those of developed countries, but with less than one-third of the per capita Gross National Product. Michael Lipton's 1977 book *Why the poor stay poor: urban bias in world development* blames this situation on the systematic bias of the upper classes dominating the political economy, who invest in urban infrastructure and services at the expense of other productive sectors of the economy such as agriculture. This is a powerful argument, and it influenced the World Bank's structural adjustment lending programme from the late 1970s onwards, leading to a new emphasis on the rural sector in the hopes of reducing income disparities and levels of rural–urban migration (Harris 1992: xvi). It is now recognized, however, that the theory was based on inadequate empirical understanding of the reasons why people migrate to cities. Increasingly, the notion of over-urbanization is seen to rest on the questionable assumption that developing countries should necessarily follow similar paths to those of developed countries in the nineteenth century. Inadequacies of data in any case make such comparisons difficult, while part of the difference is explicable in terms of the failure of urbanization rates in some developing countries to decline in the face of a stagnant or declining economy (HABITAT 1996).

While the growth of cities in the developing world may not spread benefits to other regions, in much of the world there appears to be a relationship between urbanization and economic development.

Economies of scale and higher labour productivity in large cities lead to higher Gross National Product per capita in more urbanized countries. It has also been argued that rising levels of urbanization are strongly associated with growing and diversifying economies, and that the newly industrializing countries, whose economic performance in the 1970s and 1980s has been widely envied, have high rates of urbanization. Simon (1997), however, on the basis of African evidence, finds relationships between urbanization and prevailing socio-economic conditions complex and often unpredictable. In Zambia, economic decline as a result of collapsed copper prices did lead to declining rates of urbanization, and by the late 1970s there was an urban exodus from the Copperbelt. But in many other countries, including Mozambique, Tanzania, Ethiopia, Rwanda, Côte d'Ivoire, and Gabon, negative economic growth rates in the 1980s were accompanied by continuing rapid urbanization. In some cases, economic and political crises actually increase rates of urbanization, especially where they impact strongly on rural areas, as in the case of the civil wars in Mozambique and Angola in the 1980s; in others, such as Somalia and Rwanda during 1997, urban-based insurrections have led large numbers of people to flee the cities.

Migrants and marginality

The scale, direction, and nature of migration flows are fundamental to our understanding of urbanization, yet there is surprisingly little research on migration within developing countries. This has led to widespread assumptions which are not always accurate, arising in part simply from the high visibility of migrants to Third World cities. In most of them, however, natural increase rather than migration is now the main element of population growth. Rural–urban migration flows are not necessarily the largest, but are often exceeded by rural–rural flows; and by no means do all rural–urban migrants go to the largest cities. Even in rapidly urbanizing countries, there is invariably significant reverse movement. Migrants in different countries and regions vary in terms of age, gender, level of education, the extent to which the migration is considered permanent or temporary, and the extent to which the move is part of a complex household survival strategy. In each source region, the factors which influence migration are complex and often constantly changing. What is not generally true is the often-repeated assumption that migrants who go to the cities are influenced by false perceptions. In most cases there is sufficient contact with the city for them to know a good deal about urban conditions. In many cases their migration may be seen as putting a foot on the bottom rung of a ladder of opportunity which is non-existent in rural areas; they may be over-optimistic about the chances of climbing higher, but they are under few illusions about the nature of life at the bottom. Such a life will be one of crushing poverty, revolving around the insecurities of informal jobs, squatter or slum housing, and minimal access to basic services, but it may still represent an improvement on rural conditions.

In most African cities it was clear by the late 1960s or early 1970s that the rate of migration was greatly exceeding the rate of formal employment creation. For many years the residents of squatter settlements were seen as poor, mostly migrant households which contributed nothing to the urban economy but increased pressure on the provision of urban services. Oscar Lewis viewed them as living in a 'culture of poverty' on the margins of modern urban life. Janice Perlman (1976: 195) refuted this in her study of Rio de Janeiro, concluding that the poor 'are not economically and politically marginal, but are exploited and repressed... not socially and culturally marginal, but are stigmatized and excluded from a closed social system'. Informal activities provide cheap services such as gardening and car-washing, vehicle and other repairs, decorating, carpentry, and construction for the urban middle classes; in many cities they also include manufacturing activities which, because of their unregulated nature, operate cheaply and often provide inexpensive inputs for larger capitalist firms in the formal sector. Politically, the urban poor tend to be quite conservative (although South Africa is an exception to this): they have rarely participated in revolutionary changes of government. Socially and culturally, they often retain close links with rural areas, particularly where, as in Africa, many are first-generation urbanites. Many, especially women, also engage in urban agriculture where conditions permit, to supplement incomes which have failed to keep up with prices. Economic and cultural differences between town and country may thus be blurred, leading to the labelling of Third World urbanization as 'the ruralization of the city'.

In Africa, migration to cities has slowed down in the 1980s and 1990s, and natural increase is the dominant component of an overall urban population growth rate of 4.5 per cent per annum. The informal sector has nevertheless expanded in the face of economic stagnation or decline, with increasing numbers of hawkers and food-sellers in downtown areas, more and larger open-air markets, increasing numbers of private taxis and minibuses, and an explosion of small trades of all kinds. Even the educated are by no means guaranteed jobs, as structural adjustment programmes lead to the contraction of the public sector and the disbandment or privatization of parastatals. Many of those in the government service receive nominal salaries which necessitate supplementation through corruption or informal-sector jobs. The centre of gravity in many cities has shifted from the CBD, with people moving to the periphery where land is cheaper and more accessible and where harassment from police and formal planning restrictions are largely absent. Few services follow: in many cities, refuse is not collected, roads turn into quagmires in the wet season, water must be purchased expensively from street traders, there are few telephones and little electricity, basic drugs have disappeared from public clinics, and professional medical care is extremely difficult to obtain. While urban health indices are still generally higher than those in rural areas, the gap may be narrowing as both governments and individuals spend less on health and nutrition. As Drakakis-Smith observes,

> Real improvement in any one of these basic needs is heavily dependent on improving the access

> of the poor to employment with fair and adequate recompense. Not to recognize this basic fact is to condemn many Third World cities to an unstable and uncertain future. (Drakakis-Smith 1987: 109)

This reality, or something very similar, is as true for the 'underclass' in developed-world cities as for the poor in the cities of southern Africa. It is to the latter that we now turn.

The growth of towns in southern Africa

Prior to white settlement in southern Africa, the only notable concentrations of population were at the semi-desert edges of the Kalahari in present-day Botswana and in part of what is now Zimbabwe. These Tswana 'towns' were concentrations of herders and cultivators forced by aridity to concentrate in a few areas where springs were available. Impressive stone ruins are located over much of Zimbabwe in a wide arc from the Khami ruins near Bulawayo via Great Zimbabwe near Masvingo to Nyanga near Mutare, but such ruins are not necessarily evidence of urban settlement.

The first permanent European settlements were small coastal administrative and trading centres established by the Portuguese in what are now Angola and Mozambique from the sixteenth century onwards. Dutch settlement at the Cape began in 1652, but the population of Cape Town was only 16 428 at the time of the British census in 1806. The main impetus to urbanization came with the development of diamond mining around Kimberley from 1867, and more especially the rapid growth of goldfields on the Witwatersrand after 1886. This was the first significant urbanization of the interior. By 1896 Johannesburg had a population of 102 000 people, half of them Africans, and was served by railways from all the South African ports and Lourenço Marques (present-day Maputo), which had themselves become rapidly growing cities. Urban settlement in central Africa at this time had barely begun, with the founding of Salisbury in 1890 and Bulawayo in 1893. Mining was again the main impetus to large-scale urban development in the interior, reflecting the essentially exploitative nature of European colonialism. Extensive mining in the Copperbelt began in the 1930s, leading to the development of what is today Zambia's major urban agglomeration, with six distinct towns. (The geological origins of southern Africa's immense ore deposits are described in Chapter 11.)

Other towns grew up as administrative capitals of Boer republics (Bloemfontein, Pretoria, and initially Pietermaritzburg) and colonial territories, although some of the latter were too small and undeveloped to support even one large town – Bechuanaland (modern Botswana) was administered from Mafeking (now Mafikeng, part of Mmabatho in the North West province of South Africa) outside the territory. Elsewhere the network of smaller marketing and service centres reflected the density of population, and of white agricultural settlement in particular. In southern Africa there are fewer centres than might be expected at the lower levels of the central place hierarchy, reflecting the low intensity of rural development over much of the region, particularly in the underdeveloped homelands.

Central place theory is an explanation of the number, location, size, spacing, and functions of settlements within an urban system consisting of distinct levels.

Portuguese, Dutch, German, and British settlers all left their own stamp on the towns they founded (see Chapter 6). The brief period of German rule in what is now Namibia is recalled by distinctive buildings in Windhoek and the coastal towns of Swakopmund and Lüderitz. In Mozambique and Angola most Portuguese settled in the cities, living in the *cidade de cimento* (cement city) – so called because all the buildings were made of bricks or cement – from which most blacks were excluded; blacks lived in the *cidade de canico*, where buildings were of cane, corrugated iron, or other cheap materials. Public buildings reflected the anti-modernist notions of 'typical' Portuguese architecture held by Dr Antonio Salazar, the Portuguese dictator for three decades, who favoured neoclassical colonial architecture intended to reflect the 'glories' of lusotropical culture (Sidaway and Power 1995). As such ideas receded, modern blocks of reinforced concrete became the norm.

In South Africa, distinctive Cape Dutch architectural styles are found in Cape Town and smaller towns such as Stellenbosch. Most smaller Afrikaans *dorps* (towns) are dominated by Dutch Reformed churches in a central position, as in Graaff-Reinet, for example (see Plate 11, colour section) After the British occupation in 1806, Cape Dutch was slowly superseded by British influences including Georgian and, for public buildings and churches, neoclassical and Gothic revival. Many Victorian buildings survive in cities such as Pietermaritzburg, Port Elizabeth, Johannesburg, and Grahamstown (see Plate 7). The Union Buildings in Pretoria (see Plate 6) were designed by Sir Herbert Baker, whose school dominated public architecture in the first half of the twentieth century. Subsequently the influence of Louis Kahn and the American school tended to predominate.

Urbanization levels in southern Africa

Current levels of urbanization vary considerably among the countries of the region (see Table 8.2). Such data must be treated with caution, as much depends on official definitions. This is particularly true of South Africa, where large informal settlements within the former black homelands were classified as rural in official statistics, but in practice constituted part of the functional urban regions of cities such as Durban and Pretoria; the true level of South African urbanization is probably around 60 per cent, which is closer to the expected level for a country at its level of economic development. Zambia's relatively high level of urbanization reflects exceptionally high rates of rural–urban migration in the early post-independence period: the urban growth rate was 8.9 per cent in the years 1963–69 and 5.8 per cent in 1969–80. It slowed to only 3.7 per cent in the 1980s, when there was

TABLE 8.2 Urbanization in southern Africa, 1994 and 2025

Country	1994 urban pop. ('000)	2025 urban pop. ('000)	1994 rural pop. ('000)	2025 rural pop. ('000)	Per cent urban (1994)	Per cent urban (2025)
Angola	3 357	14 799	7 317	11 820	31.5	55.6
Botswana	391	1 651	1 053	1 328	27.1	55.4
Lesotho	446	1 973	1 550	2 199	22.3	47.3
Malawi	1 427	7 083	9 416	15 265	13.2	31.7
Mozambique	5 088	21 468	10 439	13 671	32.8	61.1
Namibia	545	1 939	955	1 110	36.3	63.6
South Africa	20 457	48 673	20 097	22 278	50.4	68.6
Swaziland	251	933	581	715	30.2	56.6
Zambia	3 936	11 467	5 260	7 664	42.8	59.9
Zimbabwe	3 454	10 878	7 548	8 758	31.4	55.4

SOURCE: United Nations 1995.

little net migration to the cities and actually net out-migration from the Copperbelt in the wake of the collapse in world copper prices (Potts 1994, cited by Simon 1997).

Zimbabwe, the most developed economy in the region apart from South Africa, has a similar urbanization level to three much less developed countries – Namibia, Mozambique, and Angola. This reflects, on the one hand, the effects of settler-colonial controls on black urbanization in Zimbabwe until 1980, and the continuing concern of the government to avoid the growth of a large urban squatter population. Wars have resulted in increased urbanization levels in Namibia, Angola, and Mozambique: the liberation war in Namibia led to major concentrations of essentially rural people close to the Angolan border, and the horrors of post-independence civil wars in Mozambique and Angola caused large numbers of people to flee to the cities, especially Maputo and Luanda, in search of food, relative security, and basic services which were non-existent in the countryside. Four smaller, landlocked countries are the least urbanized at present: Lesotho and Malawi, the least developed, are predictably at the bottom of the list. Botswana has enjoyed rapid economic growth, but this is very much dependent on its diamond mines, which have not become major urban magnets. Its urban population is, however, among the most rapidly growing in the region, but starting from a very low base (see Table 8.3).

The predictions of urbanization levels for 2025, shown in Table 8.2, are highly speculative. Their most notable feature is the degree of convergence which is envisaged, with all except Malawi and Lesotho reaching urbanization levels of between 55 and 69 per cent. Optimistically, such levels could reflect economic growth and a parallel increase in levels of urbanization to match those of existing semi-developed countries. Pessimistically, they could reflect a continuation of those negative factors such as rural poverty and insecurity which have contributed to the existing urbanization process. In the 1980s, certainly, urbanization has continued in the face of economic stagnation and decline in most southern African countries, as Table 8.3 shows.

Colonial, segregation, and apartheid cities in South Africa

An enduring contradiction has pervaded South African cities throughout the period of white minority rule. Whites regarded the cities as their own cultural domain, and were reluctant to accept black urbanization, but they needed black labour (Chapter 3 explores this theme further). The attempt 'to secure labour-power without labourers' (Maylam 1990) was thus a major feature of state policy, leading both to controls on rural–urban migration and to segregation of those whose presence in the city was unavoidable. Similar attitudes prevailed under settler-colonial rule until 1980 in Rhodesia, where the white government consciously imitated some aspects of apartheid in the 1970s, and under South African rule in South West Africa (now Namibia) where urban reforms from the late 1970s tested the ground for those which came later in South Africa itself.

Davies (1981) has identified three main phases in the development of South African cities. The *settler-colonial* period lasted from the beginnings of white settlement in 1652 until the early years of Union after 1910. The Natives (Urban Areas) Act of 1923 marked the beginning of the conscious nationwide pursuit of urban *segregation*, which was superseded by the more rigid and far-reaching policy of urban *apartheid* after the Group Areas Act of 1950. In practice there are strong threads of continuity between these phases, and elements of all three persist in South Africa's emerging *post-apartheid cities*. All three phases are characterized by forces of control and relations of minority dominance common to the social formation of all colonial cities (Lemon 1991). These depended crucially upon access to political power, but also included a range of other controls: over access to the means of production and higher levels of employment, and to the educational and training opportunities which facilitate socio-economic mobility; over land resources and access to services and amenities; and over spatial relations through segregation and influx control.

TABLE 8.3 Urbanization and GNP growth rates, 1980–90/91

COUNTRY	AVERAGE ANNUAL GROWTH RATE (%)		
	Urban population 1965–80	Urban population 1989–90	GNP per capita 1980–91
Angola	6.4	5.8	–
Botswana	12.6	9.9	5.6
Lesotho	7.5	7.0	–0.5
Malawi	7.4	6.2	0.2
Mozambique	10.2	10.4	–1.1
Namibia	4.6	5.3	–1.2
South Africa	3.2	3.7	0.7
Zambia	6.6	6.2	–
Zimbabwe	6.0	5.9	–0.2

No centralized authority was exercised over black urbanization before 1923, and even municipal controls were absent through most of the nineteenth century. This reflects the limited pace of black urbanization: even in the first two decades of the twentieth century the proportion of urbanized blacks was only 12–13 per cent. Employers developed their own forms of containment, as the compound system introduced in Kimberley in the 1880s (see Figure 3.5, p. 74) rapidly spread to mines throughout southern Africa, and was adopted by municipalities, factories, warehouses, and docks as well. Even in the mining towns, however, many blacks lived outside the compounds. Conscious segregationist policies in this period were directed mainly at Indians, who were seen as a more direct threat to whites. Some segregation was achieved in the private housing market through the inclusion of racial clauses in suburban property deeds. In the early twentieth century it was white fears of the spread of contagious disease which encouraged the establishment of 'native locations' by cities such as Cape Town, Port Elizabeth, and Johannesburg. Many such locations themselves became a health threat, owing to their siting near sewage farms and refuse dumps, and in Johannesburg it was a serious influenza epidemic in 1918 which aroused the civic conscience and led to the first black housing schemes.

State intervention to limit and manipulate black urbanization was strongly influenced by the ideas of the Stallard (Transvaal Local Government) Commission of 1922, which was concerned to restrict black numbers so as to minimize expenditure on locations. The 1923 Natives (Urban Areas) Act empowered, but did not compel, municipalities to establish locations and a rudimentary system of influx control. The Act embodied the central principles of segregation (and hence relocation) and influx control, together with the expectation that black urbanization would be self-financing and that blacks would be co-opted, initially in an advisory capacity, to help the system operate (Robinson 1996). All these principles were developed and intensified in subsequent legislation, including the 1945 Natives (Urban Areas) Consolidation Act, which allowed local authorities to strengthen influx controls.

Black urbanization accelerated markedly to meet labour demands during the Second World War and the early post-war years of rapid industrial growth. Housing provision lagged far behind, and squatter settlements mushroomed around major cities. The dramatic growth of the urban black population became a central issue in the 1948 election campaign, in which the National Party's use of *swart gevaar* (black peril) arguments contributed to its unexpected victory. The policies of urban apartheid which it implemented resisted the realities of social and economic change for four decades, until the strength of those realities finally began to break down apartheid controls in the 1980s.

Apartheid's designers viewed urban blacks in 'Stallardist' terms, as a labour force whose workers (and their dependants) 'belonged' in the homelands, and as a necessary evil whose numbers should be minimized. To achieve this, the central state assumed more power over the lives of urban blacks at the expense of municipalities. It even attempted to strengthen homeland links by segregating black ethnic groups within the townships, although with limited effectiveness. The most crucial piece of legislation was the Native Laws Amendment Act of 1952, which laid the basis for all state intervention controlling the distribution of black labour between town and country, and between towns. It strengthened the provisions of the 1945 Act, made them mandatory for all local authorities, and introduced rural labour bureaux to control and canalize the exodus of homeland blacks. Rigorous implementation of these measures led to well over half a million prosecutions annually for transgression of the pass laws in the late 1960s.

For blacks with legal rights to reside in urban areas, vast new townships were built as far as possible from white residential areas, and usually close to industrial areas. From 1967 onwards, such townships were built in the homelands wherever possible, and many blacks were removed from existing townships to new ones. Rural–urban migration characteristic of Third World cities was increasingly dammed up behind artificial homeland boundaries as large informal settlements began to grow in these areas, housing migrants anxious to live as close as possible to urban opportunities. A new apartheid concept of 'frontier commuters' was born, to describe those who were, in theory, crossing international frontiers (in the case of 'independent' homelands) on their way to work in industries that had been encouraged to locate 'on the borders' of South Africa. By 1979 over 700 000 people were commuting in this way, with Pretoria and Durban attracting by far the largest numbers (Lemon 1982).

Within the city, the Group Areas Acts of 1950 and 1966 led to a major re-ordering of the segregation cities which had preceded them. The ultimate goal was the establishment of areas for the exclusive occupation of each officially recognized racial group – with group areas for whites, 'coloureds', and Indians, and townships for blacks. Plate 12 shows how this resulted in a fragmented urban structure in the small Karoo town of Graaff-Reinet: the white area lies in the foreground in the loop of the Sundays River; the coloured group area and the black township can be seen in the background. Markedly different housing densities and environmental conditions were to be found between the group areas and townships (see Plates 13, 14, and 15).

Implementation was slow in the 1950s, but eventually well over 1.5 million people were moved: these included 126 000 families, mainly coloureds and Indians, and perhaps a million blacks who were moved to townships (later designated 'development areas') to make way for group area planning. By the end of 1987, more than 1 300 group areas had been proclaimed. Whites, who numbered 67 per cent of the non-black population in these areas, had been allocated 87 per cent of the land. Nevertheless, even this massive replanning exercise left some 10 per cent of the urban population (including blacks) living outside their designated areas. These included black workers living in white areas, mainly live-in servants but also migrant workers in mining compounds and in state institutions such as hospitals and prisons (Christopher 1989, 1994).

Pre-apartheid cities were already highly segregated, as detailed studies of Durban and Port Elizabeth have shown. In some cases black/white segregation would have been almost total if domestic servants had been excluded. Why, then, did the state not content itself with simply 'tidying up' existing segregation cities? The answer must be seen in the context of the overall aims of apartheid. Officially, the object was to minimize contact and so, it was argued, prevent friction between ethnic groups which were mutually incompatible. To this end, 'buffer strips' were often proclaimed to act as barriers between different groups, 'islands' of a different ethnic group were removed, and group areas were planned so as to minimize the need for members of one population group to travel through the area of another group. Most importantly, however, apartheid was seen as a lasting solution to the problems, as many whites perceived them, of South Africa's multiracial population. It was therefore essential to plan the cities in a way which would endure. This demanded group areas which could expand indefinitely, which in turn implied a broadly sectoral arrangement, modified as necessary according to local conditions.

The changes dictated by these objectives are shown, in idealized form, in Davies' (1981) models of the segregation and apartheid cities (Figure 8.2). Zones of mixing between whites and coloureds were eliminated, as were inner areas of Indian and coloured housing and enclaves of one group in areas generally occupied by another. Blacks were concentrated

FIGURE 8.2 Segregation and apartheid city models

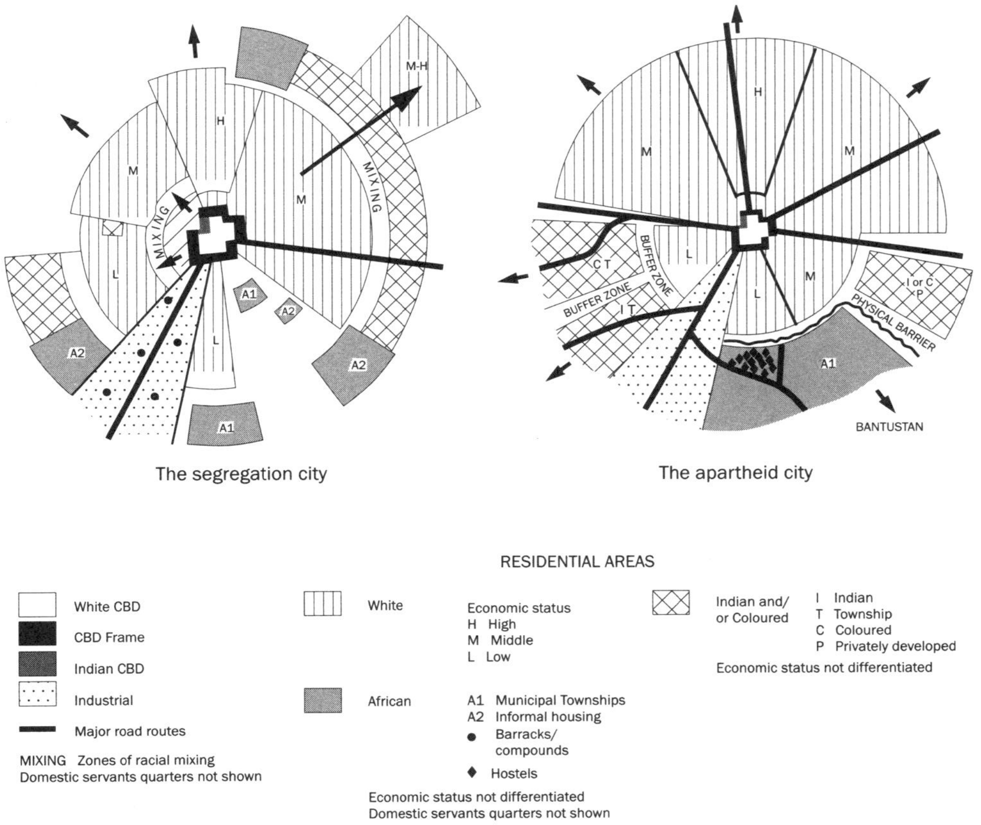

SOURCE: Davies 1981.

where possible in a single large township, with migrant labour hostels clustered together in one section of it, close to industrial areas. Buffer zones or physical barriers such as rivers separated group areas and townships from one another. Fully half the periphery of the city, normally that which was topographically or climatically most attractive, was available to middle- and higher-income whites for future residential expansion. Figures 3.7 (p. 79) and 8.3 illustrate these features for Cape Town and Johannesburg respectively.

Apartheid urban planning also condemned many people to long and expensive commuter journeys. Nowhere is this more starkly illustrated than in Pretoria, which has an extensive commuting hinterland that includes the Odi and Moretele magisterial districts of the former 'independent' homeland of Bophuthatswana as well as parts of the former KwaNdebele homeland (see Figure 8.4). Commuters from the most distant settlements in both these former homelands spend up to seven hours or more on

FIGURE 8.3 Johannesburg's group areas

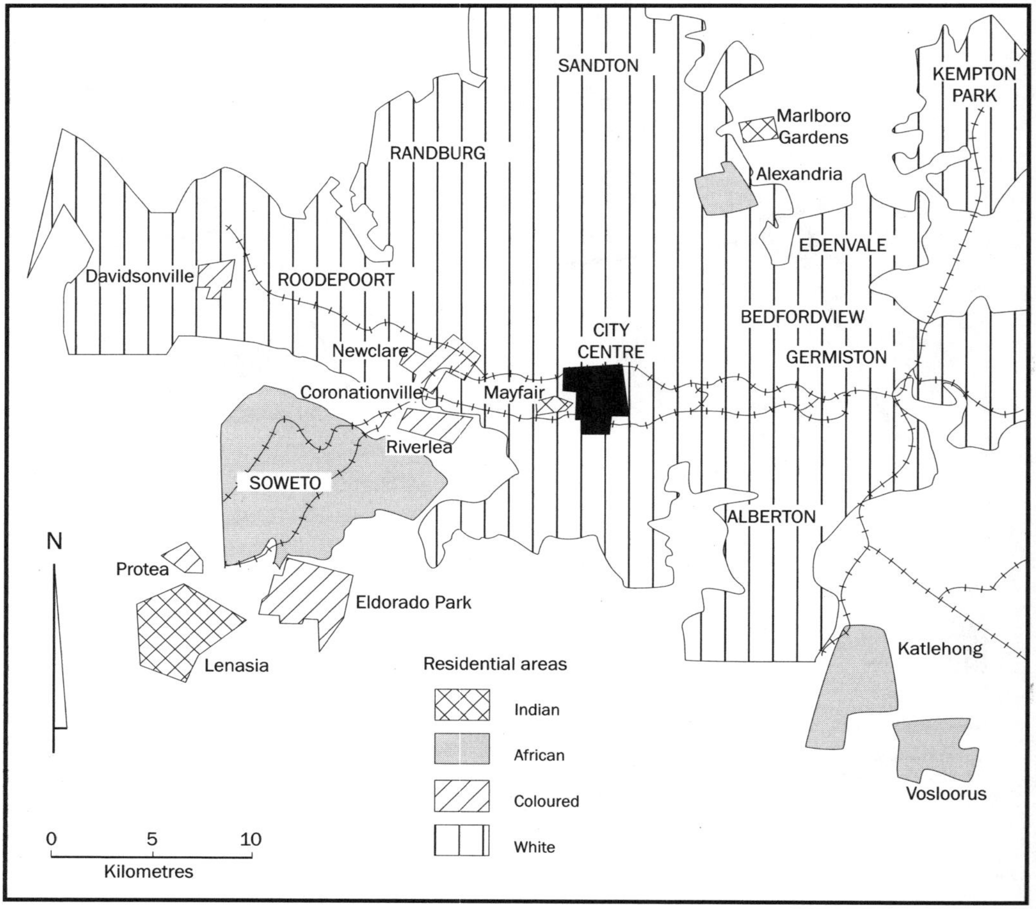

travelling (Hattingh and Horn 1991). Nearly half of Bophuthatswana's population already lived within fifty kilometres of Pretoria in 1970, but both homelands experienced substantial internal migration as their own citizens and blacks of other ethnic groups sought to position themselves as closely as legally possible to urban job opportunities.

In the early 1980s, nearly 60 per cent of Pretoria's 400 000 black commuters travelled from Bophuthatswana, and most of the rest from Pretoria's own black townships of Atteridgeville and Mamelodi. Bophuthatswana commuters came from both the squatter areas of the Winterveld and the formal townships of Mabopane, Ga-Rankuwa, and Temba. All these urban areas rely heavily on the Mabopane–Belle Omre commuter rail link, which transports more than 400 000 people daily. The complexities of apartheid urban planning

FIGURE 8.4 Pretoria and adjacent former homelands

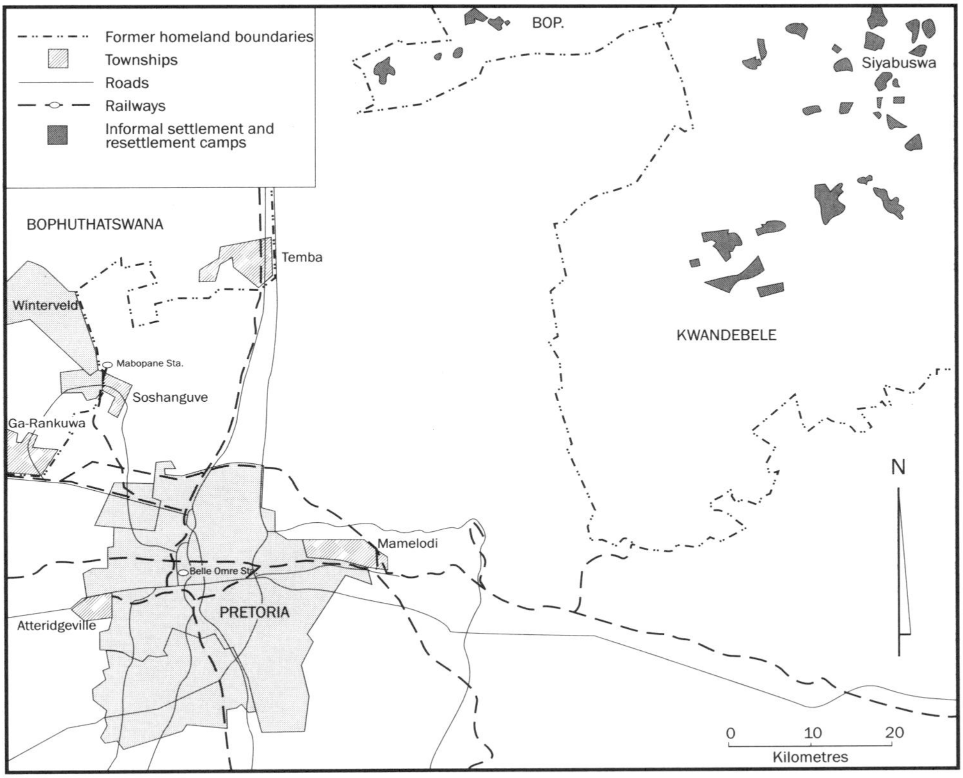

are symbolized by the creation of Soshanguve, formerly Mabopane East, which was excised from Bophuthatswana in order to house increasing numbers of non-Tswana squatters in the Winterveld who were unwelcome in the newly 'independent' homeland: its name reflects its origins as a home for *So*tho, *Sha*ngaan, *Ngu*ni, and *Ve*nda people. In practice this merely slowed down the pace of urbanization in the Winterveld and had even less impact on its ethnic diversity.

The number of commuters from distant KwaNdebele was much smaller, but the number of daily buses from its resettlement camps to Pretoria grew from 2 in 1979 to 105 in 1981 and 263 in 1984, by which time the South African government was paying the bus company a subsidy greater than the entire GDP of this impoverished homeland (Lelyveld 1987: 122). There is no more dramatic illustration of the lengths to which apartheid planning was prepared to go to achieve physical and social separation in the face of economic integration.

Readers with an interest in how apartheid planning affected urban structures are directed to Lemon (1995), which gives detailed examinations of group areas planning and the spatial displacement of black urban growth for a wide range of cities in South Africa and beyond: Cape Town, Port Elizabeth, East London, Durban, Pietermaritzburg, Bloemfontein, Kimberley, Johannesburg, Pretoria, Mafikeng–Mmabatho, Windhoek, and Harare. But perhaps only those who have experienced the miseries inflicted by both influx control and group areas planning can fully appreciate them. Many South African novels have done something to convey the grim realities to a wider audience: Alan Paton's *Cry, the Beloved Country* (1948) and Nadine Gordimer's *A World of Strangers* (1958) are two classics which will always be worth reading, the former illustrating the trauma of families split by the migrant labour system and the evils of urban life confronting migrants, and the latter revealing the completely separate worlds of different races in Johannesburg and the grim struggle of the poor for survival. Among academic geographers, John Western (1981) has provided the most penetrating study of the human impact of group areas. Probing the images of Cape Town held by its coloured people, he demonstrates the humiliating damage which removal from and destruction of homes had on self-esteem and security, and the way in which people had to learn again their place on the edges of a callous urban society.

'Reforming' apartheid: the beginnings of urban transition

The 1980s are now viewed as a decade in which attempts were made to reform apartheid in what was surely a doomed attempt to win the acquiescence of significant numbers of blacks (Lemon 1995). In 1986 the state recognized for the first time that 'orderly' black urbanization was not only inevitable but potentially a positive force for development. The ending of racially discriminatory influx control meant, in theory at least, that blacks needed no longer suffer insecurity arising from impermanence in the cities, and that migrant labour and the splitting of families were no longer unavoidable. Henceforth much more attention was to be focused on the provision of housing, infrastructure, and social services. New Regional Services Councils were given revenue from turnover and payroll taxes to improve infrastructure, and the Independent Development Trust was set up with state funding to develop housing and other upgrading projects in black areas. But the new strategy still sought to control and channel black urbanization; the Group Areas Act and the planning process left material constraints on land and housing unchanged, while existing industrial decentralization and deconcentration programmes continued. In the context of prolonged unrest and violence in the townships in 1984–86, the state was acutely conscious of the threat posed by large urban black concentrations; as Mabin (1989) observed, the new strategy sought to take advantage of informal settlements on the metropolitan peripheries and further afield to fragment the urban black population in ways which facilitated indirect political control.

Measures to control slums and illegal squatting in the late 1980s further signalled the government's intention of continuing to control black urbanization. The squatting legislation did, however, have positive aspects, giving farmers and local authorities power to establish informal housing areas, and providing more land for squatting by the establishment of 'informal towns'. These measures, together with the relaxation of influx control, resulted in accelerated urbanization which, with natural increase, led to annual population growth of 600 000 a year in South Africa's seven major metropolitan areas.

Legally enforced residential segregation came under increasing pressure in the 1980s, with growing numbers of middle-class Indians, coloureds, and blacks able to afford houses in white areas and suffering from housing shortages in their 'own' areas. In many cases whites acted, for a fee, as nominees on their behalf. This led to a new generation of 'grey' areas in several major cities, especially Johannesburg. Such changes on the ground encouraged more liberal white municipalities such as Cape Town, East London, Pietermaritzburg, and Durban to seek varying degrees of change in the law. Major business firms, some of which had already bought houses for senior black executives in Johannesburg's affluent northern suburbs, also demanded changes. At the same time, the strict application of the Group Areas Act was increasingly compromised by judicial decisions.

At first the government was slow to react, given the critical role played by urban residential segregation in maintaining an 'own community life' for different population groups and the 'protection' it afforded to low-income whites in particular. Yet official hopes that the growing black middle class would form a 'bourgeois buffer' could hardly be realized if blacks were unable to move out of segregated townships. After much delay, the government's response was characteristically equivocal. In 1988 it established a mechanism for the creation of racially open Free Settlement Areas, but also sought to tighten up the operation of group areas legislation elsewhere. Exceptionally, however, it was unable to pass this legislation, a clear sign that the process of ethnic residential integration had gone too far to reverse. Prosecutions under the Group Areas Act had virtually stopped, and permits for exemption were freely and rapidly granted. President F. W. de Klerk announced his intention of repealing the Group Areas Act in 1990, and did so the following year.

With the all major enactments of urban apartheid removed, the planning process assumed new importance in determining the future shape of post-apartheid cities.

South African cities today and tomorrow

The most immediate change in post-apartheid cities has been one of governance. A complex process of transition from the racially segmented local government of apartheid resulted in local elections for new urban authorities in most urban areas in late 1995, and in Cape Town and KwaZulu-Natal in 1996. All cities now have either Transitional Local Councils (TLCs) or Transitional Metropolitan Councils (TMCs) – the latter in the largest cities such as Johannesburg and Cape Town as a second tier of councils controlling the Metropolitan Substructures (MSS). These councils are transitional in the sense that they operate under the

Interim Constitution; the changes after 1999 will include amendments to the voting system which gave a guaranteed share of representation to former white, coloured, and Indian group areas, but will otherwise be minor. The important feature of the new system is the potential which it allows for redistribution within single authorities, or within MSS: in Johannesburg (see Figure 4.7 p. 112), for instance, Soweto, Lenasia, and Eldorado Park may now benefit from the revenues of the CBD to which they are linked in the Southern MSS (Lupton and Mather 1996; Lemon 1996a). Redistribution will be no easier at local than at provincial or national level, but the role of the local state is quite crucial, because it is responsible for the delivery of services and infrastructure which are central to the hopes and aspirations of the urban majority. Progress has so far been slow, as new authorities seek to come to terms with new boundaries and responsibilities and, in the case of the TMCs, to sort out the division of powers between themselves and the MSS.

Now that all vestiges of legally enforced segregation have disappeared, the obstacles to desegregation are demographic, social, cultural, and above all economic. The very size of the black population, nearly three-quarters of the total, clearly means that only a minority could move to former white areas. Most urban blacks living in informal housing, now the majority in most cities, cannot afford access to any formal housing, let alone that in former white areas. Even among township blacks, only a minority could contemplate such a move. Moves to some former coloured and Indian areas may be more practicable economically, but the advantages may be outweighed by perceived social and cultural objections. Similar objections are likely also to deter some better-off blacks from moving to former white areas.

Desegregation will therefore affect a relatively small percentage of the urban population. Its importance goes beyond numbers, however; just as urban residential segregation was the keystone of apartheid, a measure of desegregation is a necessary (but not sufficient) prerequisite of social integration and, ultimately, nation-building. Quite simply, it is a vital means for South Africa's peoples to come to know and accept one another in a way that was not generally possible under apartheid (Lemon 1996b). It is also true that a relatively small movement of blacks is sufficient for most former white areas to undergo significant desegregation, as many already have. The speed with which this occurs depends on many factors. Desegregation of higher-income suburbs is likely to be more gradual, and social and cultural barriers may be stronger in dominantly Afrikaans suburbs.

Hart (1989) presents a scenario in which desegregation can lead to a more spontaneous process of segregation as whites move out of areas most popular with blacks, leading to the ghettoization experienced by American blacks. Using the case of central Johannesburg, where those whites who can have largely left the inner city, Hart argues that the processes occurring since the collapse of group areas accord closely with the early stages of segregation elsewhere. In other areas, however, there is much evidence to the contrary: experience in Harare, Windhoek, and Mmabatho–Mafikeng since the late 1970s, and indeed in Johannesburg itself in suburbs such as Mayfair and Bertrams, suggests a modest, incremental pattern of residential change with few signs of conflict or racially motivated out-movement by whites. Hart's model may prove applicable to some central urban areas, although the pressures found in central Johannesburg are not widely replicated.

The apartheid heritage will inevitably make South Africa's urban planners and policy-makers acutely conscious of the racial implications of the location of new low-income housing areas, the use made of former buffer strips, and the housing mix, in income terms, of new developments. There is a delicate balance between encouraging greater integration and antagonizing elements of local opinion, which may be counter-productive. It is certainly desirable that fragmented apartheid cities are gradually replaced by more compact cities which bring the poor in from the periphery. Current policies seem unlikely to achieve this, because residents of displaced urban settlements such as those in the Winterveld and KwaNdebele, at Botshabelo (outside Bloemfontein), Kanyamazane (20 km outside Nelspruit), and elsewhere all qualify for housing subsidies which enable them to upgrade their present accommodation or purchase formal homes (Crankshaw and Parnell 1996). In the face of massive need and widespread inability to contribute significantly towards housing costs, South Africa's housing debate has so far focused on the relative merits of upgrading informal homes and buildings, bottlenecks in delivery, the roles of the state and the private sector, and questions of rental or home ownership. Wider questions of urbanization policy have not been fully addressed.

Current levels of urbanization reflect the history of economic development and apartheid policies. Some 90 per cent of whites and Indians and 80 per cent of coloureds are urbanized, but the black urbanization figure is only about half the black population. Preliminary figures for the 1996 census suggest an overall urbanization level of 55.4 per cent, with considerable provincial variation (see Table 4.2, p. 106). This may be an underestimate, omitting some black settlements which are functionally urban, but it still falls below the level expected for a country at South Africa's stage of economic development, no doubt because of past controls on black urbanization. Arguably then, the state should encourage further urbanization, but the effects of present policies may have the opposite effect (Crankshaw and Parnell 1996). Granting housing subsidies in areas of tribal tenure does not provide residents with a transferable asset, and thus works against those who subsequently wish to participate in urban housing and labour markets. Subsidizing the upgrading of migrant labour hostels is also likely to perpetuate migrant labour, which is not an outcome the government itself wants.

A future urbanization strategy should allow for the continuing growth of all the country's metropolitan areas and perhaps seek to strengthen the better-located secondary cities, especially Kimberley and Newcastle (Lemon 1999). South Africa's very diverse geography demands a flexible, regionally sensitive approach which leaves the provinces free to pursue appropriate strategies within a broadly indicative national framework. At all levels, including smaller towns, local leadership, initiative, and enterprise are vital for success in generating income and employment. While the problems are daunting, most towns outside the former homelands do at least possess better communications, infrastructure, services, and human resources than do their counterparts in most of Africa and the developing world.

9 Agriculture and rural development

RODDY FOX

The current growth of the rural population of sub-Saharan Africa in relation to the production and availability of food is one of the world's major development problems. Southern Africa is no exception to sub-Saharan African conditions, although, as we will see later, South Africa presents something of a different case.

The Food and Agriculture Organization (FAO) (1998a) estimates that the world's current population of some 5.9 billion is split roughly equally between rural and urban areas. By 2020 the urban proportion should rise to 60 per cent of the total. Figure 9.1 shows that the global distribution of the rural population is uneven. Most of the world's rural population live in Asia, with Africa a distant second; in both continents the urban population is rapidly catching up with the rural but is not projected to pass it before 2020. The populations of Europe, Latin America, and North America, on the other hand, are already predominantly urban.

The last thirty years have seen world food production increase more quickly than total population (FAO 1998b). Food production per person is currently 18 per cent higher than it was at the end of the 1960s, and food available for consumption has increased from 2 300 calories per person per day to 2 700 calories. Food availability is not equally distributed around the world, however, and there are glaring disparities between the North and the South. In the late 1990s, Western Europe and North America had 3 500–3 600 calories available per person per day, while Africa had only 2 300 calories. This figure was much lower for some countries and does not reflect the distorted patterns experienced by some sectors of the population and some geographical regions. In the early 1990s there were more than 800 million undernourished people in Southern countries, 20 per cent of the people did not get enough food to meet their daily needs, and in seventeen African countries this figure reached 40–60 per cent. In these latter countries a further 30 million tons of grains were needed to meet the shortfall, a problem that is known as the *food gap*, but only 9–12 million tons were forthcoming, resulting in many deaths.

FIGURE 9.1 Rural and urban population by world regions

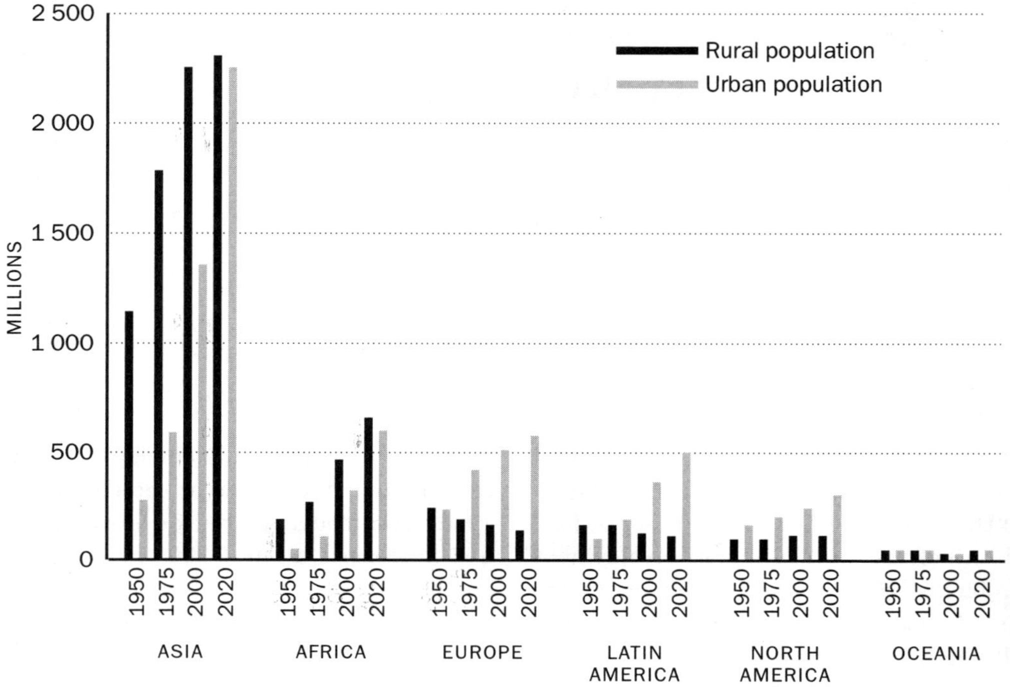

At the global scale, the required increase in food production seems relatively small: currently it is projected to grow at a rate of 1.8 per cent annually, but needs to rise to 2 per cent to bridge the food gap. The FAO estimates that the South as a whole will increase its per capita food supply from 2 520 to 2 730 calories in the period between 1990–92 and 2010. However, this will not overcome the uneven distributional patterns of food availability leaving some 700–800 million persons undernourished. The largest numbers of these people will be in South Asia and Africa; South Asia is expected to experience a relative decrease, but the numbers in Africa could increase from 100 to 300 million. The problem for African countries is where the food will be found to feed these people:

> For example, if the food were to come from the region itself, African production growth would have to be at 4.0 percent per annum for 20 years, notwithstanding increased commercial or concessionary imports, instead of 2.0 percent during 1970–1990 and 2.9 percent projected to the year 2010. Such a target may not be sustainable, economically or environmentally. (FAO 1998b: 11)

Later sections will attempt to unravel just how difficult this problem will be for southern Africa and South Africa in relation to their particular histories and environmental resources. Before then we need to turn our attention to the main environmental, political, economic, and social processes that have combined to produce the agricultural systems that characterize the world today.

Environmental factors

The principal determinants of plant growth are climatic phenomena – temperature, precipitation, water supply, and wind – in combination with soils and the topographic features of altitude, aspect, and gradient. It is possible to use combinations of these environmental factors to determine areas of optimal conditions for particular farming systems. These optimal areas will be bounded by limits determined by the diminishing returns that accrue to the farmer as the factors become less ideal. For example, as land becomes steeper, crop-growing becomes more difficult and expensive. If we combine these environmentally derived areas with historical processes such as a long tradition of agricultural specialization and political–economic features such as government subsidies, then we can begin to define areas which possess a comparative advantage over producers of the same agricultural product elsewhere (*GeographyCal* 1997). In other words, we can use the environmental features to delimit agricultural potential, although the actual agricultural system practised will depend on the interplay of environmental and human processes.

Temperature impacts on plant and crop growth because there is a minimum growing temperature and, in temperate latitudes, a minimum growing season. Growing seasons may be determined by the time of year when temperatures rise above a certain average; for example, cotton needs at least 200 frost-free days per growing season. In tropical Africa there is potentially a continuous growing season, and here altitude and rainy seasons are the important factors since they both affect moisture availability. Seasonal variations become significant as one moves into the subtropical areas of northern and southern Africa. Precipitation and water supply also have an obvious impact on farming systems. In areas with low annual totals, agriculture is likely to be based on extensive livestock grazing; high annual totals may mean that tea or sugar cane will be grown. The distribution of rainfall through the year is another important factor: coffee, for example, requires a dry period to promote the blossoming of flowers, and dry conditions are also needed during harvesting.

The FAO has produced an agro-ecological zonation for the tropics which is based on a combination of the above two environmental features (Jaetzold and Schmidt 1982). Figure 9.2 is a simplified representation of the way that agricultural potential can be identified. The rows in this diagram are temperature belts, and the columns are zones of water availability, which range from the very wet on the left-hand side to the very dry on the right-hand side. Altitude is represented in the temperature belts since temperatures fall as altitude increases (elevations in southern Africa are shown in Plate 1, colour section). The Tropical-Alpine Zone is between

FIGURE 9.2 Simplified representation of tropical agro-ecological zones

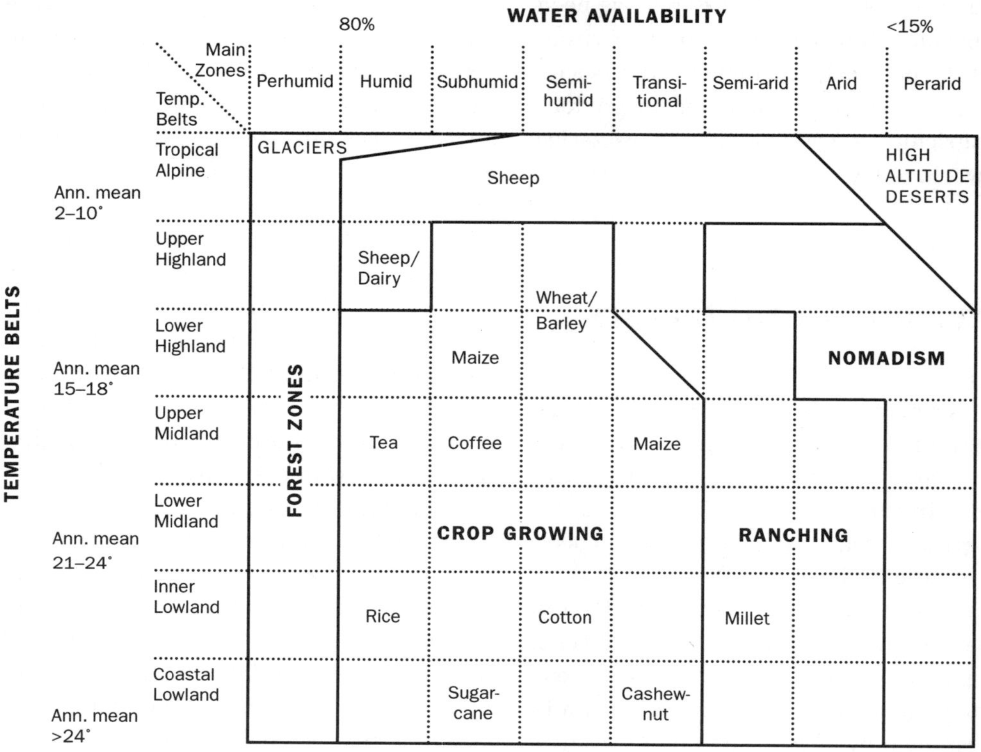

3 200 and 4 400 m, Upper Highland from 2 200 to 3 200 m, Lower Highland from 1 800 to 2 200 m, Upper Midland from 1 300 to 1 800 m, Lower Midland from 800 to 1 300 m, and the Lowland Zones are below 800 m. Water availability is calculated by dividing actual precipitation by potential evaporation; in the Upper Highland Zone, for example, conditions are cool and sub-humid, with annual average precipitation around 65 per cent of the potential evaporation (P:PET = 0.65), while in the Lower Midland Ranching Zone, where it is warm and semi-arid, annual precipitation is 25–40 per cent of potential evaporation (P:PET = 0.25–0.4). Plate 50c maps these moisture zones for the African continent.

In tropical Africa, Kilimanjaro's higher levels are an example of the Tropical-Alpine Zone, Upper and Lower Highland conditions are found in the tea- and coffee-growing regions of Malawi and Zimbabwe, and Upper and Lower Midland areas in the coffee- and maize-growing regions of Angola, Zambia, and Zimbabwe. The extensive ranching areas of Botswana correspond to the Lower Midland category, with nomadic pastoralism practised in the desert regions of Namibia, Botswana, and northern South Africa. Given good soils, the principal crop-growing zones can support rural populations of over 400 persons/km^2: Plate 10 shows areas of southern Malawi in this range. The ranching zones have very much lower densities, typically 4–40 persons/km^2, and large parts of southern Angola, Namibia, Zambia, and Botswana fall into this category. The nomadic zones have extremely low population densities, frequently supporting less than 1 person/km^2. Unfortunately, the FAO's zonation is not strictly applicable to subtropical areas such as South Africa since seasonal variations associated with higher latitudes complicate the interrelationships described for tropical areas. The Department of Environmental Affairs and Tourism's *Environmental Potential Atlas* (*Enpat* 1998) has attempted to produce a similar zonation, with biological productivity derived from measures of water availability in the soil.

Soils are the matrix upon which all farming systems depend, and, although they can be modified, their diminishing returns prevent the use of the most intractable soils. As a general rule, the heavy, wet, poorly drained nature of clay soils results in their being used for rice-growing or left as grasslands. Sandy soils are usually well drained, lighter to work, less acidic, and suitable for a variety of fruit and vegetable crops. Cereals are often grown on the dry, alkaline lime soils derived from chalk. (Soils are discussed in detail in Chapter 14.)

In addition to the impact of soil quality on agricultural productivity, the topographic features of gradient and aspect also play limiting and enabling roles. Steep slopes usually have shallower, quickly drained soils that are prone to erosion, particularly once the plant cover has been removed. Furthermore, mechanized ploughing is difficult on slopes in excess of 5–10°. Steep slopes are therefore usually not well suited to arable systems unless terracing techniques are employed. This has been done successfully in the very fertile areas with soils derived from andesite in parts of central and eastern Africa, where terracing of slopes in excess of 25° is not unusual. Aspect is another factor affecting crops, since north-facing slopes receive appreciably more radiation and sunshine in southern Africa, especially south of the tropic of Capricorn. In the northern African countries bordering the Mediterranean, the opposite is the case.

Given the role that environmental factors play in determining agricultural potential, it is not surprising that fluctuations in climate can produce devastating impacts on rural society. The 1992 El Niño–related droughts in southern Africa are just one example of this. It is particularly important to note that, as rural populations have increased in size, depending on the tenure system, landholdings have become fragmented and settlement has encroached into more marginal areas. Thus the rural population becomes more prone to disaster when rain fails to materialize.

The development of global agriculture

Le Heron (1993) has identified three stages in the development of global agriculture. The *extensive stage* lasted from 1870 to 1910, the *intensive stage* from 1940 to 1970, and the *integrated stage* from 1990 to the present. The periods between the phases were characterized by crises and changes in the organizational structure of agriculture.

The extensive stage

This period witnessed the opening up of agricultural land by European settlers in areas of 'unoccupied' territory or regions where the indigenous population could be forced off the land (*GeographyCal* 1997). Chapter 3 described how settlement and exploitation were achieved in southern Africa, where this stage continued until the Second World War. The USA, Canada, Australia, and Argentina were earlier examples of this process. European settlement and agricultural activities were promoted by governments, and, because of the comparative advantages of different regions, the temperate grassland areas of Argentina and the USA focused on grain production while the tropical areas of Africa, for example, were used for crops such as cocoa and coffee. Bulk transportation of produce was essential for this system to function and therefore dendritic networks of transport routes were set up, with inland gateway towns collecting produce before transporting it to large colonial ports and then out into the world economy (Vance 1970; Johnston 1984). Indigenous agricultural systems were destabilized by this process: the hunter-gatherers were often almost exterminated and the shifting cultivators become increasingly unable either to innovate or to produce sufficient food.

The intensive stage

Scientific research in the USA from the 1920s onwards resulted in increasing crop productivity. Hybrid crop varieties were bred, which gave higher yields when associated with inputs of fertilizers and pesticides. Furthermore, there was a far-reaching change in livestock farming when the process of meat production became dependent on inputs of animal feeds. The system changed from one in which the environment produced grass which animals ate, thus 'growing' meat, to one where the environment produced grains and soya which were crushed into compounds and fed to animals, thus *manufacturing* meat. The above developments led to the growth of industrial companies producing animal feeds, fertilizers, and pesticides, a process that has continued to this day.

Agricultural protectionism was a key policy introduced in this period to protect farm incomes and to pass on the benefits of increased agricultural production to the industrial workforce, providing plentiful, cheap food. European countries adopted this policy after World War II and it remains the cornerstone of the European Union's Common Agricultural Policy. In much of southern Africa agricultural protectionism was extended to the white, or European, agricultural sector, which was developed to the exclusion of black agriculture. The promotion of the black sector had to wait until the mid 1950s in tropical African countries such as Kenya and came later still in southern African countries such as Zimbabwe and Namibia.

The integrated stage

Five factors have been involved in the development of the current agricultural phase:

1 Market trends. Northern countries have experienced marked shifts in demand from basic foods to a wide range of fresh salad vegetables and tropical produce that are required by consumers throughout the year.
2 Technological developments. These will be discussed in more detail below, but the essential change has been the massive increase in productivity gained through the green revolution and bio-technological advances. Also, food manufacturing technology has changed, with freezing and microwave cooking becoming increasingly important.
3 Corporate finance. Funds from both private banks and organizations such as the World Bank have encouraged technological change in agriculture and necessitated new organizational structures.
4 Regulation. The internally protected markets that characterized the intensive stage have increasingly fallen into disrepute as internationally competitive agriculture has become fashionable. Signatories of the General Agreement on Tariffs and Trade (GATT) are obliged to lower their levels of protection to generate a pattern of free trade.
5 Production structure. Supply networks have become globalized as large supermarket traders have sought to ensure year-round supplies of all products. Obtaining the cheapest global sources has led to the collapse of domestic production of some commodities.

The above features can be readily understood as components of the neo-liberal growth model described in Chapter 5.

The industrialization of agriculture

There are many elements involved in the process of agricultural industrialization. The following section will investigate changes in rural social composition, increased tractor usage, agro-chemical inputs, the green revolution, agricultural intensification, and energy use. The authors of *GeographyCal* (1997) describe the industrialization process as

> [t]he application of industrial principles of management to agriculture, with increased dependence upon industrial as opposed to farm produced inputs. Industrialization is distinguished from mechanization by changes in the external relations of agriculture with the other sectors of the economy. Accompanying this is a change from farming as a 'way of life' to being a business which is part of a larger food production system.

Figure 9.3 describes this change in the farming system by comparing simplified food chains for traditional and modern farming systems. The increase in complexity in the modern food chain is the most obvious feature. Other key changes are the distancing between the farmer and the market and the falling away of the farm's role in generating its own inputs, resulting in a loss of sustainability at the local scale.

The industrialization of agriculture has altered the population composition of rural areas. A proportion of rural workers has been replaced by increasing farm

FIGURE 9.3 Food chains in traditional and modern farming systems

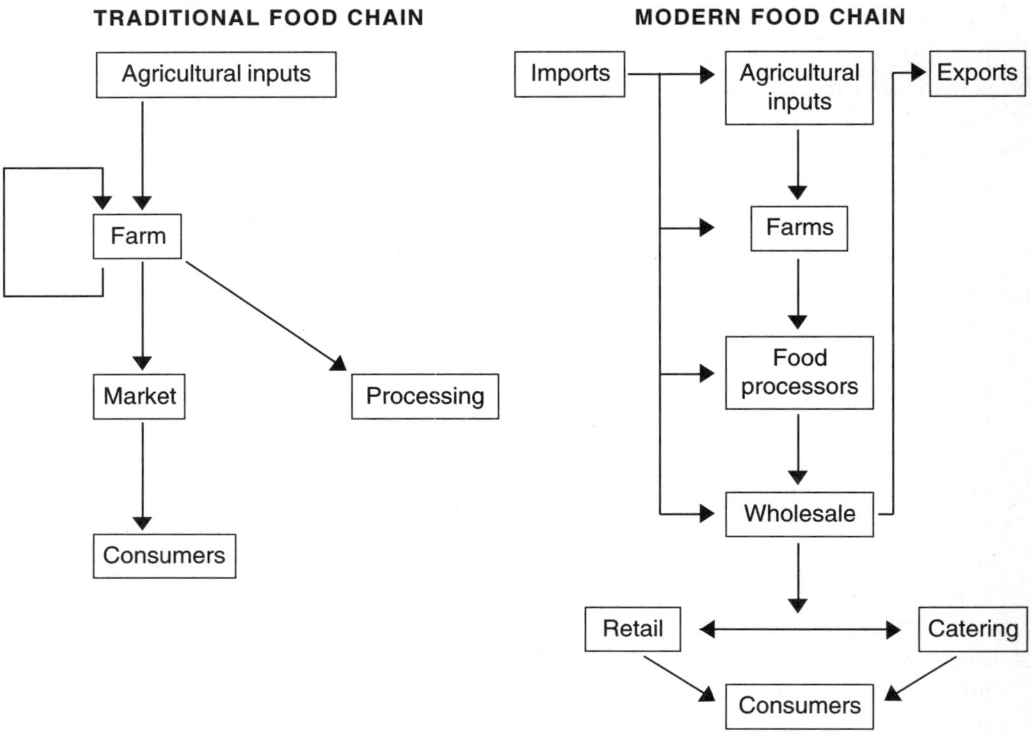

mechanization. In Great Britain the ratio of land area to rural workers was 11 hectares (ha) per worker in 1901; by the 1990s it was three times that figure, but the reduction of workers was compensated for by 0.75 tons of machinery per hectare. The declining numbers of the rural labour force are being offset by the influx of urban-employed commuters seeking a rural home environment (a trend mentioned in Chapter 8). This process has occurred in small towns in agricultural areas of countries such as England and the United States, altering their social composition.

A good index of mechanization is tractor ownership, since tractors are used in many different types of agricultural system. In the early 1990s Northern countries had 21 million tractors out of a world total of 26 million, reflecting their earlier adoption of this technology. In 1991 Europe had 42 per cent of the world's tractors, Asia had 21 per cent, the USA 19 per cent, Russia 10 per cent, South America 5 per cent, Africa 2 per cent, and Oceania 1 per cent. The South's rate of growth in tractor ownership is now much faster than the North's, reflecting its lower base and more recent adoption of mechanization. However, the high cost of tractors may well lead to debt, dependence on industrial outputs, and the use of fossil fuels, and therefore the dominant economic growth model, with its links to increasing globalization, is not necessarily going to succeed. The problems involved in the switch from ox-based ploughing to a tractor-based system is tellingly unravelled in the modern classic by Van Onselen, *The Seed is Mine* (1996). In this book we are shown that one of the problems faced by the black South African sharecropper Kas Maine was that the tractors he bought only decreased in value and needed constant fossil-fuel inputs and repairing. His oxen, however, were renewable and sustainable, and could be successfully traded because they increased their value with training.

The use of herbicides to kill weeds, fungicides to eradicate diseases, and pesticides to eliminate insects has successfully increased agricultural productivity. Some pesticides such as DDT have been banned in Northern countries as they are highly toxic to humans, but pesticide use has nevertheless increased seventeenfold in the last twenty-five years. Fungicides control plant diseases such as rust and mildew and their use has also grown spectacularly, with a twentyfold increase since the early 1970s. Herbicides kill the crop's competitors for light, water, and nutrients, and also reduce the need for labour and machinery, since weeding is less important. Their usage has grown fifteenfold over the same period. The other principal modern input is fertilizer, and in Northern countries quantities have doubled from just under 50 kg/ha in 1965 to over 100 by 1991. The *GeographyCal* (1997) team of authors comment that presently 'the application of fertilizers is usually the single most important action a farmer can take to improve yields'.

The green revolution was a spectacular improvement in agricultural productivity in parts of the South. It was the result of research and development to produce high-yielding varieties (HYV) of crops by seed companies, governments, and international agencies such as the International Rice Research Institute based in the Philippines. Wheat was experimented on in Mexico as early as the 1940s, and rice and maize were other crops that were later successfully transformed (Dixon 1990). Some countries of the South rapidly adopted these varieties, which have the capacity to double yields under farm conditions. Their attractiveness to governments and development experts can be gauged from the example of IR8 (International Rice 8), which in 1966 was producing 6 400 kg/ha under ideal conditions, in contrast to the Philippines' average rice yield of only 1 300 kg/ha. The green revolution was not only a breakthrough in plant-breeding; it also involved a package of technology, including fertilizers, insecticides, implements, and water control. The characteristics of HYV wheat are its high response to fertilizer, its short growing season, its sensitivity to moisture and temperature levels but not to the amount of light, its resistance to disease, and its short, stiff stems which prevent collapse under the weight of the wheat's ears. More recently, modern varieties (MV) of rice, wheat, sorghum, and maize have been produced which withstand drought and disease and are tolerant to poor soils. In India 55 per cent of the rice and over 80 per cent of the wheat are now MV, compared to less than 10 per cent in the mid 1960s.

Intensification of production is associated with most of the above developments and is one of the dominant characteristics of the current integrated stage of global agriculture. Intensification involves farmers adapting to the requirements of purchasing inputs, using HYV or MV crops, mechanizing production to replace labour, automating their farms, and increasing yields. Farmers who lack the capital and/or expertise to do the above have frequently been bought out by larger farmers, with the result that whole farm sectors can be dominated by only a few producers. *GeographyCal* (1997) gives the example of poultry production in the USA, where 41 per cent of production comes from only four companies and 80 per cent from twenty companies. A development frequently associated with intensification is specialization, which represents a move away from a mix of farm activities towards a single enterprise, or even a part of it. The reasons for specialization are usually to exploit comparative advantages and to derive economies of scale. In the dairy industry, for example, there is a great deal of specialization involved in milking, refrigeration, transportation, and pasteurization. Economies of scale may reduce the unit costs of production, whether these are fixed or variable, when the production process is large enough to warrant the purchasing of new machinery or large quantities of inputs at reduced prices.

The final feature to be considered in the industrialization of agriculture is the amount of energy now used in modern farming systems. Energy is frequently measured in 'oil equivalents': that is, the amount of energy applied to the land through the use of machinery, fertilizers, irrigation, pesticides, crop drying, and so on. One kg oil equivalent per worker is the same as 12.5 kilowatt hours of electricity (essentially the same as running a 1 bar electric fire for 12.5 hours). In Africa, rural workers use around 26 kg oil equivalents/ha, whereas in the USA the figure is a staggering 25 744 kg oil equivalents/ha. Southern countries typically have low energy inputs as they make more use of labour, while Northern countries have very much higher inputs, but are also more productive. Redclift (1987) has questioned the energy efficiency of using such large amounts and shows that the most efficient systems are those which combine crop rotation and fallowing with low-energy inputs (see Table 9.1). The large-income-earning systems are the least efficient.

It is now time to turn our attention to the southern African experience, which will show us that some of these broad global patterns have impacted on the region, particularly the export sector, but that we also need to examine the plight of the subsistence agriculturalists who make up the bulk of the population.

Southern Africa in crisis

Food supply and de-agrarianization

There can be little doubt that the region is in a very serious state in terms of food supply. Plate 16 (colour section) shows that in the 1994–96 period an inadequate dietary energy supply (DES) typified the sub-Saharan region, of which southern Africa is a part (FAO 1998c). DES estimates the average daily food energy available per person and incorporates food both produced and imported for human consumption. It does not allow for inequitable

TABLE 9.1 Energy efficiency of agricultural systems (kcal/ha/annum)

Agricultural system	Total energy input (A)	Energy production (B)	Energy efficiency (B/A)
Pastoralism (Africa)	5 150	49 500	9.6
Crop rotation and fallow (Mexico)	675 700	6 843 000	10.1
Estate crop production (India)	2 837 760	2 709 300	0.9
Maize (USA)	1 173 204	3 306 744	2.8
Irrigated rice (USA)	14 586 315	21 039 480	1.4
Spinach (USA)	12 800 000	2 900 000	0.23

distribution of food supplies or show how much people actually eat. A person's average energy requirement is 2 200 kilocalories per day, and it is clear from the map that many southern African countries fell well below this level: Mozambique had 1 720, Angola 1 930, Zambia 1 940, Zimbabwe 2 040, Malawi 2 050, Namibia 2 160, and Lesotho 2 170. Only Botswana, Swaziland, and South Africa had sufficient food supplies. In fact the picture is worse than this, since the FAO maintains that DES has to be considerably above the required average figure to allow for wastage and inequitable access at the national, regional, and household levels. A further problem is the dependence on cereals and tubers in the region, for these foods give inadequate quantities of nutrients, vitamins, minerals, protein, and fat. Fully 66 per cent of the diet in southern African countries is based on cereals and tubers that are deficient in these micronutrients, which are to be found in meat, fish, eggs, milk, and cheese.

Sub-Saharan Africa certainly presents itself as a region where population growth and agricultural production are not in equilibrium. The population of the region has increased by 53 per cent since 1980, while food production has risen by some 45 per cent (FAO 1998b). As a consequence, net cereal imports rose from 17 million tons in 1979–81 to 25 million tons by 1988–90, and the food aid component of these imports ranged between 10 and 40 per cent from country to country. Food supplies per capita are about the same, or slightly lower, than they were fifteen years ago. Unfortunately, national Gross Domestic Products have averaged only a 1.9 per cent growth over the last fifteen years, which translates into a 1 per cent decline per annum in real terms, given the higher rates of population growth, which have averaged 2.9 per cent. Furthermore, total merchandise imports have risen by 2.8 per cent, while exports have grown by only 1.8 per cent. Thus the overall trade balance (which might have been used to pay for food imports) has dropped from a US$7.7 billion surplus in the early 1980s to a US$6 billion deficit in 1993. Finally, although the FAO projected a growth in food production to 2.8 per cent by 2010, this is merely enough to stop the per capita food production from falling.

There have been several key changes in population and settlement accompanying the trends delimited above (Bryceson and Howe 1996). The proportion of people engaged in agriculture has actually shrunk since 1965 (see Table 9.2), with spectacular drops in Botswana, Lesotho, and Zambia, and much smaller ones elsewhere. Not surprisingly, there has also been a fall in the proportion of people resident in rural areas, an average regional decline in the order of 17 per cent, and there have been correspondingly high rates of urban population growth that can reach above 10 per cent per annum.

The trends shown in the table must be seen against the background of the zoning of the economy into rural and urban sectors, with a further division into black rural (subsistence) and white rural (commercial) areas. National development policies acted for a long time to maintain the differences between these sectors in both the colonial and post-colonial periods, although the distinctions have become more blurred in the 1990s. Bryceson and Howe describe the policy of rural zoning:

> During the colonial period, this zoning was constituted by the convergence of racial, locational and occupational categories in colonies of white settlement in particular: Africans were delimited as a race, occupationally restricted to farming and migrant labour and locationally circumscribed in rural areas…
> In fact, rural zoning remained a central feature of post-independence government policy. The nationalist phase, spanning the 1950s through the 1980s, was characterized by policies which continued to delineate rural areas as zones of agricultural work versus urban areas as zones of industrial and service-sector development. Thus urban areas were seen as areas of economic growth and new opportunities for the expanding population who had received formal education in contrast to rural areas where old agrarian occupations prevailed. (Bryceson and Howe 1996: 178)

The rapid growth of the urban population and the leakage of young, educated, and economically active males to the urban sector have been attributed to what has become known as 'urban bias'. Entry into the commercial farming sector, particularly large-scale farming, was no longer prescribed by race in the post-colonial period, but there were formidable barriers of patronage and income that ensured that only a chosen few, politically well connected and financially powerful, gained access to this sector.

Land division artificially truncated shifting cultivation systems in many countries of the region, particularly South Africa, Namibia, and Zimbabwe. In South Africa 87 per cent of the land was reserved for whites, who constituted only 15 per cent of the population, and in Namibia 40 per cent of the land was reserved for 10 per cent of the people. In Zimbabwe before

TABLE 9.2 Southern African de-agrarianization, 1960–92

Country	Population (millions) 1992	Real GDP per capita (US$)	Annual pop. growth rate	Urban pop. growth rate (%) 1960–92	Rural population (%) 1960	Rural population (%) 1992	Labour force in ag're (%) 1965	Labour force in ag're (%) 1990–92
Angola	9.9	1 000	2.3	5.9	90	73	79	73
Botswana	1.3	4 690	3.2	13.5	98	73	88	28
Lesotho	1.8	1 500	2.4	8.6	97	79	91	23
Malawi	10.3	800	3.4	6.5	96	88	92	87
Mozambique	15.1	921	2.2	9.5	96	70	87	85
Namibia	1.5	2 381	2.8	4.8	85	71	–	43
South Africa	39.9	3 885	2.6	3.2	53	50	32	13
Swaziland	0.8	2 506	2.8	10.5	96	72	–	74
Zambia	8.6	1 010	3.2	7.1	83	58	79	38
Zimbabwe	10.6	2 160	3.2	5.9	87	70	79	71

post-colonial resettlement, 47 per cent of the land was used by 7 per cent of the population, and by 1992 the figures were in the order of 35 per cent for 1 per cent of the population (Potts 1992). Shifting cultivation depends on a delicate balance between population numbers, technological input, and the availability of land to practise fallowing. Population increases in situations of land restrictions like those described above reduce the period that land can be left fallow to regenerate, eventually leading to serious land degradation. This is a well-documented process across eastern, central, and southern Africa (Zinyama 1995). For example, by the late 1980s 23 per cent of Zimbabwe's communal (black) areas were experiencing extensive to very extensive erosion and a further 20 per cent exhibited moderate erosion; the commercial farming (white) equivalents were 1.3 and 3.4 per cent respectively. Among the reasons for the low erosion rates in the commercial sector are the much lower population densities and the application of conservation measures appropriate to the agro-ecological zonation of the different farming areas. Tiffen et al. (1995) have produced compelling evidence that degradation can be overcome in black areas of the Machakos district in Kenya.

Structural adjustment programmes

Structural adjustment programmes (SAPs) have been implemented in most sub-Saharan African countries since the early 1980s in return for assistance from the World Bank and the International Monetary Fund to help them cope with balance-of-payments crises. The aims of SAPs have been to introduce fiscal discipline, to vastly reduce the role of the state in the economy, and to create the conditions by which producers will respond to the market (whether local or international). These form a very different approach to development than the modernization of the urban-industrial sector, which is based on government direction of the economy. One outcome of the implementation of SAPs has been the reduction of formal-sector employment, whether in the private or public sector, and hence much lower remittances to the rural sector from household members in urban employment.

Agricultural pricing reform has been a very common component of SAPs, coupled with the elimination of state involvement through parastatal organizations in agricultural marketing, supply, and transportation. The results of SAPs are not clear. Bryceson and Howe (1996) observe that liberalization has tended to benefit the farmers who already possess comparative advantages, particularly market-orientated smallholders with good transportation to major domestic markets and with assured agro-ecological conditions. The large-scale commercial sector that uses modern industrial methods such as irrigation and fertilizers also takes advantage of better prices. The problem is that these are relatively small sectors in the agricultural economy, and the majority of farmers depend on rain-fed systems; for them it is the presence of rain, rather than good prices, that determines output.

The three population groups in rural areas who are most vulnerable to food insecurity are the smallholder farmers (commonly over 70 per cent of the rural population), nomadic pastoralists (who may constitute over 10 per cent of the rural population), and, cutting across these two groups, the 30 per cent of households headed by women (FAO 1998b). It is difficult to see how structural adjustment alone will increase agricultural production among these groups of people. Indeed, farmers' responses to global price changes may compound the wide fluctuations in cereal production that characterize the smallholder sector.

Even in the sector producing crops for export, the impact of SAPs has been restricted by the same problem that has bedevilled African agriculturists for two decades – the declining world prices of their products. Between 1980 and 1990 all the average annual growth rates in real prices for southern African agricultural commodities were negative: coffee and cocoa fell by 11.5 per cent, tea by 0.9 per cent, tobacco (1970–80) by 1.0 per cent, cotton by 3.0 per cent, sugar by 8.4 per cent, groundnuts by 7.3 per cent, palm oil by 4.7 per cent, and rubber by 2.9 per cent. Overall, Africa's agricultural terms of trade fell from 100 in 1980 to 64 in 1991. In southern African countries the crops on which export earnings depended were as follows:

- coffee in Angola: over 150 000 metric tonnes per annum before the wars of the last decade;
- sugar (up to 184 000 tonnes), tobacco, and tea in Malawi; after the late 1980s this figure fell due to destabilization of neighbouring economies and hence transport problems;
- sugar (up to 95 000 tonnes) and cotton in Mozambique before the wars of the last decade;
- sugar (over 400 000 tonnes) and cotton in Swaziland;
- sugar (24 000 tonnes) and small quantities of tobacco in Zambia; and
- sugar, tobacco, and cotton (each over 100 000 tonnes) in Zimbabwe.

Destabilization and geopolitically inspired civil wars have wreaked havoc on agricultural productivity and transportation (as discussed in Chapter 4).

Rural development options

Before examining the sectors and approaches which may have the potential to bring about food security, we need to be very clear about the role of natural disasters and human-related calamities. In 1994 there were 5.6 million people in southern Africa, approximately one person in twenty, receiving relief assistance from the World Food Programme as a result of such dislocations (FAO 1998b). Drought in Botswana in 1985–87 and 1992 and the long-running war in Angola are just two examples of this. The problem is doubly serious, as the displaced people exert a great strain on their host communities to feed them, and at the same time they are not engaged in productive activities in their home areas. The resolution of civil wars is therefore a critical development priority for the region; fortunately this is more likely now that one of the chief protagonists of destabilization, South Africa, has undergone a democratic transformation.

Environmental impacts can be tackled on a number of fronts. Tse-tse fly infestation is a major barrier to livestock production and human settlement and hampers resource utilization over large areas, but it can be prevented by early and well co-ordinated warning and reaction. Although little can be done about the amount of rain which falls in the region, there is great scope for development in terms of the types of crops grown and the application of low-investment, cost-efficient technologies to control water.

There are a number of ways in which productivity can be increased, especially where research and development efforts are coupled with appropriate approaches to ensure social and environmental sustainability (Elliott 1994). Fertilizer use is as low as 11 kg/ha in tropical southern Africa, and this is focused on a very small area of commercial farming. A process known as soil mining has occurred, the lack of fertilizer input leading to the permanent loss of soil nutrients, which in turn leads to poor plant growth which does not adequately protect the soil from erosion. This problem can be countered by organic-matter replacement, which promotes the biological fixation of nutrients and water. African staple crops have been neglected in research efforts that have focused on cash crops for export. There are very few improved varieties of millet, sorghum, cassava, yams, cowpeas, bananas, and plantains, and not much work has been done on technologies appropriate to their usage. Drought-resistant strains need to be sought and related to diversified farming systems that can cope with insufficient and irregular rainfall.

Technical agricultural improvements can be fairly readily identified, and some examples are given above, but perhaps most critical is the approach to rural development that is adopted by development agencies. Elliott (1994) summarizes the conditions that analysts are increasingly recognizing as prerequisites for successful, sustainable local development initiatives:

1 Initiate a dialogue with the interested parties and stakeholders and be prepared to modify the aims and objectives of the 'experts'.
2 Identify and understand the priorities of the households and rural communities and make these the most important outcomes of the project.
3 Establish whether the communities will have long-term security and rights to the gains they are seeking.
4 Ensure self-help and the sustainability of the innovations by letting communities help themselves, rather than establishing dependence on outsiders.
5 Ensure that the 'outside' staff are competent, dedicated, motivated, and involved throughout the duration of the project.

This approach has been termed *neo-populism* and is associated with the techniques of Participatory Rural Appraisal (Chambers 1997). The neo-populist development agenda is frequently at odds with the neo-liberal growth model, for it stresses socio-political, rather than economic, solutions to rural development problems.

A final consideration is the highly contentious issue of land tenure and land rights, alluded to in the third point above. Virtually all of the land that was reserved for black occupation in colonial times was held in a form of communal tenure. Most of this land is still held in a system whereby the king, chief, or headman is custodian of the tribal area and responsible for allocating rights to individuals and households. Thus people gain access to residential, grazing, and arable land, but they do not own the land. Such a system is anathema both to the modernization school of development and the neo-liberals, because it lies beyond the reach of the market. Tenure reform is therefore frequently identified as critical for future economic development, even though it may be socially and politically well-nigh impossible to implement.

Agriculture in South Africa

Environmental potential and land division

South Africa is not well endowed with areas of high agricultural potential. Plate 17 is derived from the Department of Environmental Affairs and Tourism's *Environmental Potential Atlas* (*Enpat* 1998), which presents data on biological productivity that can be used to map agro-ecological potential. (This map is broadly comparable with Plate 50c, which shows the moisture availability for the whole African continent.) Biological productivity is measured as tons of biomass produced per hectare in a growing season and is calculated from the amount of water estimated in the soil. The map shows an obvious gradient in agricultural potential, with the eastern seaboard and the Great Escarpment well favoured, and a steady decline in potential to the north and west. Thus the Northern Province, North West, Free State, and Eastern Cape are all transitional, falling between areas of high and low potential. The Northern Cape is almost entirely of very low potential, while the Western Cape has zones of higher potential in the south and south-west, where the Cape Fold Mountains separate the coastal areas from the dry interior. Climate is described more fully in Chapter 12 and water availability in Chapter 16.

There is a broad relationship between the categories of biological productivity and the land-use zones delimited in Plate 18. Forestry and sugar cane cultivation are often found in the areas of highest productivity. This can be seen particularly in KwaZulu-Natal and Mpumalanga, but there are also pockets on the southern coast around Tsitsikamma, on the Amatola Mountains in the Eastern Cape, and to the west of Tzaneen in the Northern Province. Wheat and maize farming are found in the areas of moderate productivity encircling Gauteng (see Plate 18), and also in equivalent zones in the Western Cape. Cattle and game farming is found in a great arc stretching from the Kruger National Park across the north of the country and then south-westwards into the North West province and the Northern Cape. Sheep farming dominates the semi-arid Karoo. There are also extensive zones, usually of low to very low productivity, with no dominant classifiable rural land-use activity, which are categorized as 'natural areas'. Finally, there are large tracts classified as subsistence farming which are often associated with high-potential productivity along the eastern seaboard in the Eastern Cape and KwaZulu-Natal – these were the former homeland areas of Transkei and KwaZulu (see Plate 20). It is to the division of land, referred to as zoning in the previous section, that our attention now turns.

The grossly inequitable distribution of

land in South Africa has already been mentioned, but now we can take our assessment a little further using Geographical Information Systems to relate biological productivity to the political division of the country into homelands and white areas (see Figure 3.8, p. 80). Thus we can begin to unravel the anomalous situation in which black areas of high potential have low productivity and white areas of low potential have high productivity.

Table 9.3 summarizes the position towards the end of the apartheid period, showing just how small the homeland areas were in comparison to white land areas, and just how densely settled. The average amount of arable land in black areas was 0.25 ha per person in 1985, as compared to 2.95 ha in white areas. What is perhaps less well known, however, is the overall amount of low- to very low-potential land in white areas, fully 24 per cent of which was in the lowest category mapped in Plate 17. In comparison to this were the very small proportions of low-potential land in the Transkei and KwaZulu – only 5 and 6 per cent respectively – indicating that, although these two homelands possessed very little land, what they had was of high to very high potential in terms of water availability. Chapter 14 shows that when unfavourable soils, steep slopes, and rugged terrain are also taken into account, the total amount of high-potential land in the country is only 3 per cent of the land area. Former homelands such as the Transkei may appear to be well favoured in terms of water availability but the duplex soils found there are very erodible and poor in fertility. The following case study of the Transkei shows how the farming system and population growth were pushed into a state of disequilibrium and dependence on the urban sector.

The population of the Transkei rose slowly but steadily from approximately 800 000 in 1904 to 1 300 000 by the mid 1950s (Andrew 1991). Thereafter came a spectacular rise so that by 1981 there were slightly in excess of 2 600 000 people. Maize was the principal cereal grown and we could expect that, in a sustainable system, productivity would rise to keep pace with population growth by means of technological or social adaptations. This was not the case, however: although production rose from approximately 80 000 tons in 1910 to 180 000 tons in 1930, it then declined steadily to a figure of 120 000 tons in the late 1980s. Livestock numbers followed a similar trend, rising until the mid 1930s from 750 000 to 2.5 million sheep, from 300 000 to 1.5 million goats, and from 100 000 to 250 000 cattle, but then stabilized or declined slightly from those levels. Andrew's (1991) research has shown that the agriculturalists responded to the declining productivity of their arable lands by switching to cultivation of small kitchen gardens. The arable areas were abandoned as a result of a combination of processes – the soils were poor and easily eroded, fallowing periods became inadequate as farmers planted more and more frequently, fertilizer use was at low levels or non-existent, men were often absent because they worked in the urban sector, and cattle teams for ploughing became hard to establish (see Plate 21). It is not surprising that in the areas of poorer soils and lower agricultural potential there have been high rates of erosion and degradation is immediately visible. Kitchen gardens have been favoured because they can be managed by the

TABLE 9.3 Land division, biological productivity, and rural population during late apartheid

Land division	Per cent of total land area	Biological productivity < 5 tons/ha (%)	Rural population 1985	Arable land (ha/capita) 1985
'White' South Africa	87.4	69	5 600 713	2.95
Transkei	3.6	5	2 638 507	0.29
Bophuthatswana	3.0	76	1 463 587	0.34
KwaZulu	2.4	6	2 645 016	0.16
Lebowa	1.7	66	1 718 520	0.20
Ciskei	0.7	67	476 870	0.14
Venda	0.5	53	446 277	0.17
Gazankulu	0.4	100	476 924	0.17
Kangwane	0.2	0	338 483	0.33
Qwaqwa	0.04	0	160 106	0.04
Kwandebele	0.007	0	282 746	0.15

SOURCES: McGladdery 1992; Pickles and Weiner 1991.

female head of household or the wife of a migrant labourer, their fertility can be kept high by using kitchen refuse and organic fertilizer, and they can be easily fenced, keeping out sheep, goats, and cattle. The shortfall in agricultural production has been met, at least in part, by income derived from outside the homeland in the form of remittances by migrant labourers, pensions, and social welfare payouts. This transfer of funding from the urban sector constitutes, on average, some two-thirds of household income (Van Zyl et al. 1996).

Agricultural production and land policy

The various Acts which produced South Africa's land division and the subsequent underdevelopment of the black homelands are topics that have been examined by a number of authors (Christopher 1994; Lemon 1976, 1987; Van Zyl et al. 1996). As there is not space to go over this material in detail, and as they were introduced in Chapters 3 and 4, the following section will summarize the set of policies which favoured large-scale white commercial agriculture. Readers interested in the historical evolution of black farming are extremely fortunate in that two classic and highly influential texts are available: Colin Bundy's *Rise and Fall of the South African Peasantry* (1979) and Charles van Onselen's *The Seed is Mine* (1996).

The evolution of commercial farming can be appreciated in terms of policies enacted in three phases (Van Zyl et al. 1996). The first was the territorial and racial segregation of white and black farmers in the first three decades of the twentieth century. Second, in the period from the late 1930s to the early 1980s black agriculture at best stabilized while white agriculture achieved productivity levels that gave national self-sufficiency in almost all sectors. In the former homelands, commercial agriculture was promoted in large-scale development projects under expatriate management. In the white farming areas, modern mechanical and biological technologies were fostered as part of national import-replacement policies. White farmers were protected from foreign competition by import controls, received direct and indirect subsidies, had guaranteed producer prices, and benefited from carefully regulated marketing systems for all products. One consequence of these policies was that labour was replaced by capital.

The latest period commenced in the mid 1980s and it is characterized by the reversal of previous policies. Racially based spatial segregation has been repealed (for example the pass laws in 1986, the Land Acts and Group Areas Acts in 1990), thus removing the territorial barriers erected in the early part of the twentieth century. Subsidies, tax concessions, and price support mechanisms have also disappeared since the agricultural sector has been liberalized. Agricultural budgets have been cut since 1994 and the marketing boards set up by the Agricultural Marketing Act of 1968 have been abolished.

South African agriculture has the appearance of success, particularly in relation to the southern African trends described previously. Table 9.4 demonstrates that the country was self-sufficient in virtually all staple crops between 1985 and 1993, although there were exceptions in some meat commodities and sunflower oil. This is a particularly striking performance given such an inadequate agricultural resource base, described in more detail in Chapter 15. The problem, however, is that self-sufficiency has been achieved by means of extremely dualistic policies and at the expense of social equity and environmental sustainability. It has been shown, additionally, that, in spite of national self-sufficiency, 68 per cent of rural African households were living below the minimum subsistence level in 1990 (South Africa 1995).

A set of five policy instruments was responsible for the country's agricultural performance (Van Zyl et al. 1996):

- price support above market levels, particularly for maize and wheat, coupled with subsidized storage facilities, which has made possible crop production in very marginal areas such as the drier parts of the North West province and western Free State;
- subsidizing investments in irrigation facilities, which has led to an expansion into areas of marginal potential and depleted groundwater resources;

TABLE 9.4 Self-sufficiency in selected agricultural commodities, 1985–93

Commodity	Production ('000 tons)	Self-sufficiency index*
Wheat	2 242	100.4
Maize	8 019	114.4
Potatoes	1 161	101.7
Vegetables	1 776	101.2
Sugar	1 956	176.7
Beef	618	92.8
Mutton, goat's meat, lamb	176	91.2
Pork	117	100
Chicken	656	99.2
Eggs	199	101.5
Deciduous and tropical fruit	1 484	152.3
Fresh milk	2 435	100
Dairy products	2 344	101
Sunflower seed oil	121	69.1
Citrus fruits	802	217.3

* SSI = (Production/Consumption) x 100

SOURCE: Van Zyl et al. 1996.

- subsidizing agro-chemical inputs such that fertilizer use grew at the rate of 7.7 per cent per annum from 1947 to 1979 and dipping and spraying at the rate of 20.6 per cent per annum from 1973 to 1980;
- subsidizing irrigation water, leading to the perception that it was a plentiful resource that could be used inefficiently; and
- favourable taxation and interest rates, which encouraged the over-mechanization of the production process and extended production down the ecological gradient into the zones of lower moisture availability shown in Plate 17.

At the same time that these policies were favouring the white sector, the homelands were being systematically underdeveloped in a variety of ways. Enforced population movements both into homelands and within them through villagization programmes – known as betterment planning – created large-scale dislocation. Moreover, black smallholder agriculture suffered from inadequate market access and poor levels of farmer outreach and infrastructural services.

It can be appreciated that the white farming sector was enjoying very high levels of support during the intensive stage of world agriculture and has made use of many agro-industrial advances typical of the integrated stage. The challenges for South African agriculture today are quite complex, however, as the country is struggling to redress the inequities of the past, has a fragile and often degraded natural environment, and is experiencing the effects of much more open global competition. Furthermore, the whole agricultural sector, black and white, is characterized by very low employment levels which have prevented the development of local regional economies and small towns or service centres such as are commonly experienced throughout southern Africa.

Policy statements and new laws enacted by the post-1994 Government of National Unity have understandably focused on improving rights of access by the rural poor to land (South Africa 1995, 1997). The three cornerstones of the White Paper on Land Policy are:

- land restitution through the Land Claims Court for those found to have been dispossessed on racial grounds;
- land redistribution to overcome homelessness and prevent land invasions; and
- land tenure reform to overcome insecurities such as those experienced by people utilizing communal land and white farmland.

The rural development strategy was explicitly linked to the Reconstruction and Development Programme and encompassed goals such as:

- full employment and a more diverse agricultural system to gain dignity, security, and freedom from poverty;
- integrating towns and rural areas through marketing strategies;
- improving accessibility to local government and services such as water, sanitation, and fuel; and
- a healthy, productive environment.

At the time of writing it is too early to judge the impact of these policies, although first impressions are that the land policy has been slow in implementation. Furthermore, given environmental and social realities in the black rural areas, we can expect that reinventing a successful black smallholder farming community will be a daunting task, although a vital one on which many of the other goals depend. Also, it is important not to overlook the problems of the large-scale white commercial farmers with an often degraded resource base and who face escalating violence and a global market with little of the protection afforded them in the past. Clearly, South Africa's food security may well be under threat.

Conclusion

This chapter has attempted to introduce the key themes driving global agricultural change. Southern Africa has been described in terms of the precarious food security position it has occupied in the recent past, and it can be appreciated that the region is faced with daunting challenges to redress the situation. South Africa is anomalous within the region, for its use of agro-industrial practices coupled with apartheid policies has provided food security – although at a cost. The problem for the country will be to produce more than political change through successful elections, and to improve social provision such as access to water through the successful line ministries such as the Department of Water Affairs. The main challenge is one of reinvigorating black agriculture from its long-term decline and dependence on the urban sector while maintaining food security formerly derived from a protected, subsidized, and often non-sustainable white farming sector.

10 Geographies of mobility and accessibility

DAVID SIMON

Transport geographies: an introduction

Transport geography emerged as a specialized sub-discipline of economic geography in the late 1950s and early 1960s, and was concerned primarily with the economics of various transport modes and the geometry of infrastructural networks. Its dominant theoretical underpinning was – and largely remains – that of neoclassical economics and modernization theory. Its adherents claimed it to be value-free and 'objective', being concerned with the impacts and processes of modernization, that is, the diffusion and adoption of 'modern' or 'Western' technologies and methods, which were seen as a necessary precondition for broader economic development (these issues were introduced in Chapter 5).

Some perceptive 'radical' critiques emerged during the 1970s and early 1980s, but these more critical neo-Marxist and structuralist perspectives failed to gain widespread acceptance. Such studies highlighted the way in which transport was utilized by colonial and imperial powers primarily for their own ends rather than for the benefit of local people and territories, as well as the often complex interrelationships between changing transport technologies and 'development'. With the end of the Cold War and the eclipse of Marxism and other strands of socialism as alternative ideologies, neoclassical theory in its contemporary neoliberal form has become the dominant orthodoxy. There is now general agreement that the direct and indirect implications of transport innovations and policy changes for different groups of people and the environment are complex and need to be understood and addressed at the implementation stage.

Current theoretical debates concerning postmodernism, poststructuralism, and even postdevelopmentalism have to date not influenced transport geography. Readers interested in a more extensive discussion of these social scientific theories and paradigms should consult Simon (1996). The way in which they have been deployed in the southern and South African contexts will be summarized below.

The importance of transport stems from the need for people and goods (commodities) to move or be moved between locations. Although there is no direct or linear relationship, it is generally true that, as levels of economic development and disposable income increase at the household and societal levels, the extent and frequency of transport use also grows. This so-called demand for transport is both direct, for personal mobility, and indirect or derived, to gain access to an increasingly wide range of goods and services. At one extreme, poor rural or urban dwellers often suffer constrained mobility, as they are unable to afford to purchase private transport or pay to use much public transport. Walking is then the most important transport mode – and commonly witnessed throughout southern Africa – so that average distances travelled are small and, wherever possible, trips are multipurpose. Conversely, wealthy people make frequent trips, often for leisure, and use a high proportion of luxury private transport over short as well as long distances. When public transport is used, it is usually in luxury class, such as first- or business-class air and rail travel.

Transport is required at each stage of a production process, to move the raw materials and other inputs to the farm or factories where they are grown, refined, or transformed into intermediate goods and ultimately the final products which we purchase. Recent technological change has enabled the development of increasingly sophisticated products, such as consumer electronics, each comprising a myriad of components which are themselves often manufactured in several factories on different continents. These are then brought to another location for assembly, before being shipped to numerous countries in our global marketplace (see Figure 5.2, p. 119). Although the total transport requirements for production have increased, the growing sophistication of production technologies and of specialized, high-volume forms of transport over the last thirty to forty years has significantly lowered the proportion of transport costs in the final cost of a product which we buy.

As in any other policy arena, diverse and often conflicting demands and interests have to be taken into account by transport planners and policy-makers. However, the foregoing discussion implies that in any society, but especially in the highly unequal contexts of southern and South Africa, the objectives of socially responsible transport policies should be to ensure not only the existence of transport infrastructure and services to promote commercial efficiency but crucially also their *appropriateness* to local needs and conditions and their *affordability* to potential users. Two related concepts are those of *mobility*, namely the degree to which an individual or group is able to travel, and *accessibility*, which links geographical proximity to the ease (including affordability) with which individuals or groups can reach a particular facility or destination. As an integral element of any development process or strategy, transport systems should also be increasingly environmentally, economically, and socially *sustainable* in the long run.

Accordingly, too, the kind of transport geography offered here focuses as much on difference and conflict as on similarity and harmony. It is a more critical and engaged approach, in keeping with current development thinking. Put simply, there is no single (or singular) transport geography, but a series of particular transport geographies reflecting different lifeworlds, lived experiences, and analytical frames. This explains why the title of this chapter refers to 'geographies'.

Global issues and trends

Changing modes of transport

The respective modes of transport are:

- land, subdivided into road, rail, and sometimes pipeline;
- water, frequently divided into deep-sea, short-sea or coastal, and inland waterways; and
- air, both short haul and long haul.

Each mode is defined by the medium through or on which it operates, combined with a geographical and/or technological variable. There are particular technological, economic, service, speed, and cost characteristics which lend themselves to different roles and have different impacts. Most traditionally organized transport and transport geography textbooks discuss each mode in turn (for example Hoyle and Knowles 1992; Hilling 1996). A few recent texts, however, have opted for a more integrated approach, highlighting the interactions and differential development issues of various transport modes in predominantly urban, rural, and international contexts (for example Simon 1996). The intention in this subsection is to sketch some of the changing global transport patterns and processes within and between modes. In the following subsection, changing policy frameworks and approaches will be outlined.

Since roughly the 1930s in North America, the UK, and some other European countries, there has been a progressive shift of freight and passenger transport from rail to road and from sea to air, as a result of greater speed, flexibility, and falling real costs. These trends have been most marked where government subsidies and other regulatory protection of the railways have been reduced or never were substantial. Conversely, countries such as Switzerland and Germany have maintained higher subsidies and accordingly continue to run efficient and effective railway systems which account for higher proportions of passengers and freight than in most other Northern countries, despite some shift to road transport.

Similar trends from rail to road transport have occurred in most countries of the South, albeit more recently and, in many cases, to a more limited extent. State policy in terms of regulatory protection and the level of subsidy has again been a crucial source of international variation. Unlike in old industrialized states, however, railway systems in most poor countries were never extensive and even today commonly comprise only a handful of linear routes linking ports to mines, plantation cropping areas, and the principal cities. Therefore 'network' is not an appropriate description. Angola and Kenya exemplify this situation, reflecting the colonial or imperial imperatives in terms of which their railways were built. Although road transport has therefore long been the sole mode of land transport able to reach many parts of these countries, the decline of their railways has often been exacerbated by inefficient operation, underfunding, poor maintenance, and the impact of armed conflicts, as in Angola, for example. India, China, and South Africa are unusual in having extensive and well-integrated rail networks. State protection of rail transport has traditionally been strong in these three cases, although South Africa began extensive liberalization in the early 1980s (see below). Table 10.1 indicates international variation in the provision of road and rail networks. The percentage data are *relative* measures of infrastructure in relation to population, total network length, or recorded economic activity, respectively. Because of their relative basis, these indicators are more useful for comparative purposes than are *absolute* measures of, say, kilometres of road or railway track on their own.

The shift to air transport has been more recent and dramatic but also more selective, especially in terms of freight, since only low-bulk, high-value goods lend themselves to this mode. Innovations in both air and maritime transport have been dramatic in impact and near-universal in adoption. For example, the unitization (that is, containerization) of seafreight and the move towards the use of large bulk tankers for petroleum products, coal, metal ores, and grains have necessitated appropriate harbour deepening and the provision of the requisite loading facilities and equipment in every port not wishing to be bypassed by these services (Simon 1996). Traditional palletized or bagged general cargo is now found only on some low-volume coastal and inland waterway routes in poor countries.

Technological change and its impacts are not, however, solely recent phenomena. The advent of the railways and the switch from sailing to steam ships in the nineteenth century were no less dramatic in terms of speed, reliability, volumes carried, facilities required, and their implications for different groups of people and places. We will return to this theme below in relation to South Africa.

Water transport is particularly important for maritime trade between coastal countries and islands, within countries comprising island archipelagos (such as Indonesia and the Philippines) or looser groups (such as the Solomon Islands and the Maldives), and where navigable inland waterways (rivers and lakes) exist. The Great Lakes of North America, linked via the St Lawrence Seaway, form one of the principal freight corridors for the Midwestern USA and Canada. Similarly, Africa's Great Lakes provide vital international traffic routes as well as sources of protein and other resources. The same applies to major rivers such as the Congo, Nile, Mississippi, Amazon, Danube, Rhine, Euphrates, Ganges, and Yangtse.

Pipelines have grown in importance since World War II as a means of transporting oil and petroleum products, and occasionally also water and certain solids in suspension, on overland routes. Most of these link inland oilfields in Alaska, the Middle East, and central Asia to ports and refineries, often in neighbouring countries. In sub-Saharan Africa there are relatively few pipelines, and these serve to transport refined petroleum products from coastal refineries to landlocked countries, for example from Beira in Mozambique to Mutare in Zimbabwe, and from Mombasa to Eldoret (both in Kenya), where road

TABLE 10.1 Selected comparative international transport indicators

	PAVED ROADS 1988		RAILWAYS 1990	
	Road density (km/million persons)	Roads in good condition (% of paved roads)	Rail traffic (km/million US$ GDP)	Diesels in use (% of diesel inventory)
Low-income economies				
Malawi	278	56	43	77
India	893	20	593	90
Kenya	278	32	120	52
Zimbabwe	1 389	27	505	54
Zambia	751	40	294	44
Middle-income economies: a) Lower-middle				
Congo	584	50	170	56
b) Upper-middle				
South Africa	–	–	987	88
Brazil	704	30	60	62
Korea, Rep. of	236	70	–	89
High-income economies				
Australia	25 695	>85	62	–
United Kingdom	6 174	>85	66	–
Netherlands	6 875	>85	73	83
Japan	6 007	>85	144	87

SOURCE: Simon 1996.

tankers collect them for carriage to Uganda, Rwanda, and Burundi. Pipelines are quite efficient and avoid the need for large numbers of road or rail tankers, but are vulnerable to leakage if inadequately maintained and to sabotage in times of conflict.

Periodizing transport policy changes

Transport policy in most countries with a significant market sector appears to shift cyclically between periods of strong and weak state regulation. The precise duration of these periods varies but is commonly in the twenty- to thirty-year range, as it takes a considerable period for new legislation to be formulated and implemented, for its provisions to take effect, for the full range of direct and indirect implications to be felt, and for pressure for change to build up as a result of the perceived shortcomings of the status quo. Under prolonged periods of strict regulation the system lacks the flexibility to respond to changing economic and political circumstances or new technologies, and services become inefficient and unnecessarily expensive – a situation commonly known as profiteering. Pressure for deregulation and increased competition therefore mounts.

Conversely, a liberal, competitive regime often leads over time to instability and problems of road safety as cut-throat competition leads firms or owner-drivers to reduce vehicle maintenance expenditure in order to compensate for charging low fares in the battle for trade. The market may become volatile, too, with frequent insolvencies, mergers, and takeovers on the one hand, and entry by unqualified and underfunded opportunists (often called 'cowboys' in a 'free-for-all' situation) on the other. Competition may also be undermined by the emergence of oligopoly (a small group of firms) or monopoly (one firm) control in particular localities, routes, or sectors. Such trends in turn lead to pressure for greater control and regulation. Similarly, demands to protect the railways from road-based competition and to regulate the road haulage and passenger transport sectors commonly increase during periods of recession, when total traffic levels stagnate or decline and the profitability of the entire industry is threatened.

Although the precise situation varied, the Great Depression in the 1920s and early 1930s was characterized by great instability and insolvency in many countries, prompting a move towards greater regulation. During the immediate post–World War II period this facilitated reconstruction, but by the 1950s and 1960s, a period of unprecedented international economic boom, greater flexibility and liberalization were again sought. Such developments have generally occurred, although to differing extents and at different times, since then. Under the Conservative Party government of Margaret Thatcher, for example, Britain introduced rapid and profound deregulation of domestic road transport in the early 1980s, a process complemented internationally by the gradual harmonization of conditions among member states of the European Community (now Union), and culminating in the 'Single European Market' from 1992. The changing situation in South Africa will be discussed below.

Transport and development in southern Africa

Changing times, changing perspectives

As elsewhere in the former colonial world, the forms and modes of transport introduced into southern Africa by the colonial powers were generally developed in Europe rather than locally, and were utilized as tools of imperial and settler policy. After all, colonial possessions were intended principally to further the status and wealth of the 'motherland', not to be developed in their own right and for the benefit of their indigenous inhabitants.

From the seventeenth to the nineteenth centuries, land transport was based on draught animals, primarily horses and horse-drawn carriages for the personal transport of settlers, and ox-drawn wagons for goods. These required little formal infrastructure, except passes through mountainous areas, although tracks and routes emerged over time, along with wayside inns and the like. Animal transport was slow, subject to the health of the beasts, and limiting in terms of the mass or volume of goods which could be carried, but required relatively unspecialized skills to use or maintain. Wagon-riding and horse-rider messenger services employed substantial numbers of people.

The rapid expansion of the railways from the coast to the interior of South Africa after about 1870, when the Cape Town line reached the new diamond fields at Kimberley, transformed colonial fortunes and prospects. The speed and reliability of transport increased dramatically, and the ability to bring heavy mining equipment from Britain led rapidly to far more capital-intensive and deep-level mining by the few companies which could afford it. Most could not, and there then followed a period of frenetic acquisitions by the few large companies – a process which ultimately resulted in De Beers Consolidated Mining Company establishing a monopoly (Mabin 1986). This indicates clearly one of the development impacts of new transport technology.

Equally, the introduction of the railways rendered many thousands of transport riders, wagon-makers, and people in supporting activities unemployed, as their skills and often their place of residence were very different from those required on the new, more rigid, and fixed rail routes (Pirie 1982, 1993). New immigrants accounted for most of the skilled rail jobs, while the labour-intensive railway construction gangs were composed mainly of poor whites, black prisoners, and exploited migrants.

The railways were also crucial instruments in the expansion and establishment of imperial control and colonial settlement. The line was rapidly extended from Kimberley by Cecil Rhodes in furtherance of his ambitions north of the Limpopo. It reached Bulawayo in 1897 and Fort Salisbury (now Harare) in 1898, before heading into Northern Rhodesia (Zambia) (see Figure 10.1). Following the discovery of vast gold deposits in Johannesburg in 1886, the line was quickly extended there too. For Paul Kruger and his landlocked Boer state, the South African Republic, the construction of a railway line to Delagoa Bay (now Maputo) in Portuguese East Africa (Mozambique) was a prized objective in order to remove transit dependence on

FIGURE 10.1 Extension of the railway network in southern Africa

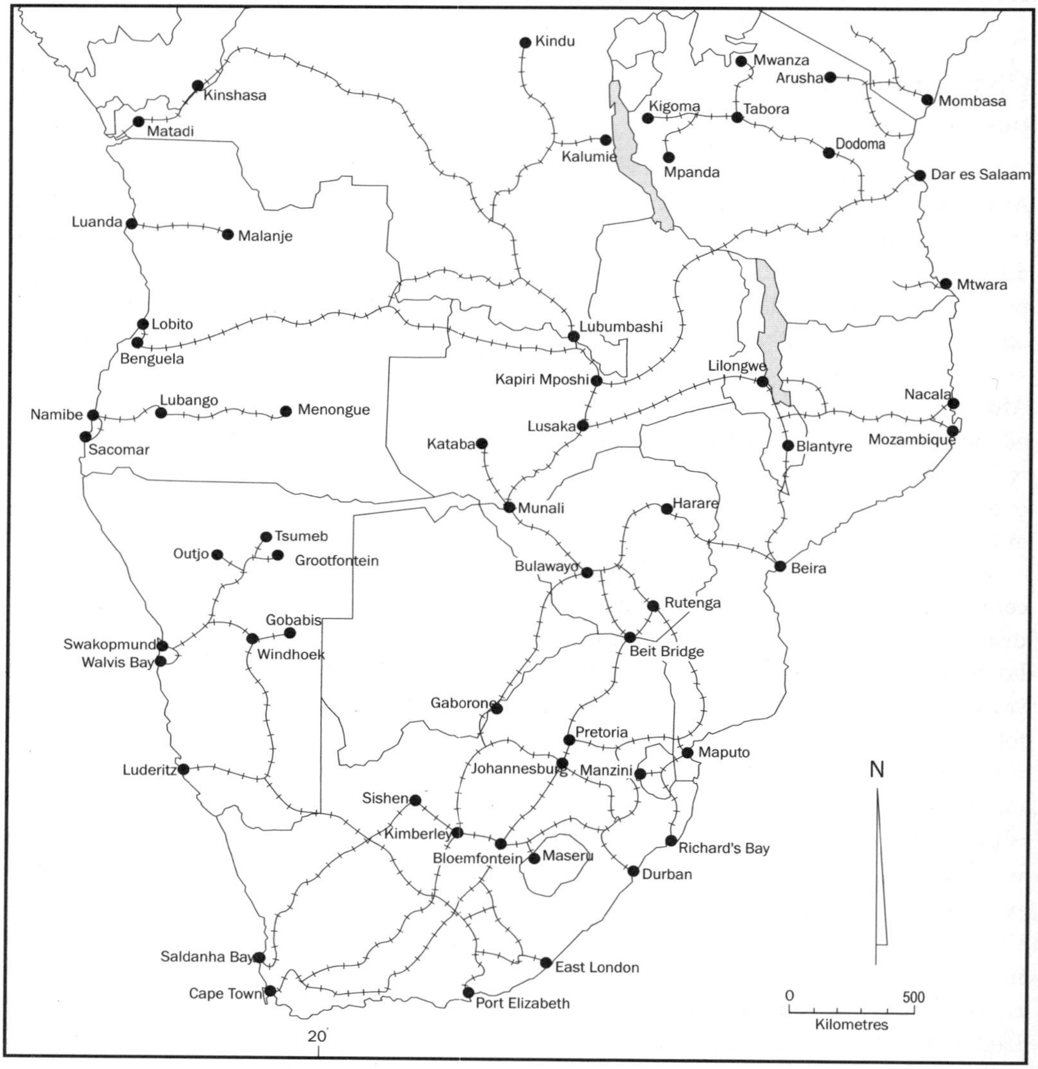

British-controlled territory. Similarly, the Benguela Railway was constructed just after the turn of the twentieth century, to enable copper exports from the Copperbelt of the then Belgian Congo, and the Trans-Zambezia Railway from Beira across Nyasaland (Malawi) served strategic British imperial interests. Britain refused to subsidize colonial development, however, requiring that all its possessions be self-financing. Accordingly, Nyasaland became severely indebted in the long term through having to bear the cost of, and subsequent debt repayments on, this line which merely traverses its territory, without gaining significant benefits (Pirie 1982).

In a more contemporary context, it was only during the 1980s that published research began to document the crucial and hitherto largely implicit role of urban and inter-regional rail and bus transport in underpinning South Africa's urban and regional planning under apartheid. Group areas, the pass laws, and forced removals all greatly exacerbated the scale, difficulties, and cost of long-distance migration and the increasingly important phenomenon of 'frontier commuting', whereby hundreds of thousands of black people were impelled to spend several hours each day travelling up to 100 kilometres each way between 'homeland' residences and urban jobs in 'white' South Africa under unpleasant conditions. Public transport operators (parastatal, local authority, and private) received major subsidies to provide these services; over time these came to represent a substantial source of profits in many cases (see Mann and Segal 1985; Pirie 1987, 1990; Khosa 1990).

Similar development implications, with some people and some areas benefiting and others losing out, can be discerned with the introduction of any new mode of transport or specific new technologies. For example, when the small, propeller-driven Dakota DC3 aircraft provided the main form of air transport between southern Africa and Europe, journeys took three days, with overnight stops en route in places such as Entebbe and Khartoum. The introduction of jetliners, initially the DC8 and Boeing 707 and from 1970 the Jumbo Jet (Boeing 747), meant that only a single refuelling stop was required. Most airlines chose Nairobi, which then developed a major industry around this transit traffic and the increasing numbers of passengers who stopped off to visit Kenya. Most of the other previous stopover points declined, as they now catered mainly to local or regional traffic. When the Boeing 747-400 entered service on the route between Europe and South Africa in 1990, non-stop flights became the norm for airlines using this aircraft, thus bypassing Nairobi and other previous stops unless flights were designed to combine passenger traffic from two or more destinations. As it turned out, this setback was not long term, since rapidly rising tourism and business travel to Kenya, and Jomo Kenyatta International Airport's hub function for East Africa, have generated steady growth in the volume of traffic into the airport. Passenger arrivals have, however, suffered sharp temporary declines as a result of political violence on several occasions during the 1990s, especially prior to the 1992 and 1997 general elections.

A truly regional-scale example of the interconnections between transport and development, and of the role of particular geopolitical and ideological strategies in relation to transport, is provided by the struggle over southern Africa's rail network during the 1980s. This case highlights efforts to avoid and enforce dependency on an historically entrenched axis of international power inequalities. For reasons explained above, the imperial system left a generally poorly integrated rail network within each country (except in South Africa), although international connections existed to link the ports with key inland centres of (mainly European) population and production. In order to broaden its options and reduce its vulnerability to line closures, landlocked Zambia joined Tanzania in the ambitious new TAZARA (Tanzania–Zambia Railway Authority) railway scheme to construct a Chinese-built railway linking the Copperbelt with Tanzania's capital and principal port, Dar es Salaam. The railway opened in 1974 but has burdened these two poor countries with considerable accumulated debt ever since (Mwase 1984, 1987; Reichardt and Duncan 1990). Justification for the foresight in building the TAZARA line came quickly when civil war closed the alternative Benguela Line from Lobito in Angola to the Copperbelt in 1975. Consequently, a substantial proportion of Zambia's foreign trade was rerouted. Within only three or four years, however, operational problems and Rhodesian sabotage rendered the line effectively inoperable, forcing Zambia to contradict its stated policy of boycotting Rhodesia under Ian Smith's illegal government by reopening the railway route through that country to South African ports.

This experience proved instructive to South Africa's neighbours, indicating their de facto dependence on the apartheid state. Equally, it showed Pretoria clearly how rail and other transport infrastructure could be manipulated to its own advantage. As the sanctions campaign and armed struggle against apartheid intensified from the early 1980s, the newly formed Southern African Development Co-ordinating Conference (SADCC) strove to reduce its members' individual and collective dependence on South Africa (these developments are discussed in Chapter 4). For the six landlocked states in particular, this meant securing trade and transport routes which avoided the need to utilize South African routes and ports. Consequently, over the twelve-year life of the organization before its reconstitution as the Southern African Development Community (SADC) in 1992, SADCC devoted roughly 60 per cent of its budget to transport and communications projects, mainly the upgrading and rehabilitation of strategic regional railway lines and harbours

TABLE 10.2 SADCC budget for the transport and communications sector, 1980–92

Sector	Number of projects	Total value (US$ m)	Foreign funding (US$ m)
Transport and communications	220	6 614	5 756
Energy	82	755	722
Fisheries, forestry, and wildlife	64	275	249
Agricultural research and training	17	272	258
Total	526	8 475	7 446

(see Table 10.2). Conversely, the apartheid government's 'total strategy' was pursued in part by the sabotage of these transport arteries, in a sometimes successful effort to enforce its neighbours' dependence upon South Africa's railways and harbours, and thereby softening their anti-apartheid stance. This formed a key element of Pretoria's regional destabilization campaign, which inflicted at least US$60 billion worth of destruction on SADCC members during the 1980s.

Attention to these complexities and differential impacts, both positive and negative, in the transport sphere came only with the political economy perspectives in the late 1970s and 1980s. In terms of modernization theory, the introduction of new transport facilities and technologies would be seen as unambiguously positive, enhancing the scope and pace of economic development. Negative implications were either dismissed as unfortunate side effects of 'progress' or even denied in extreme cases. Postmodernist perspectives are, in turn, likely to focus on alternative forms of transport, to question fundamentally the kind of hi-tech, energy- and travel-intensive solutions from new highways to airports, and to advocate more environmentally friendly, appropriate technologies, with diversity and local appropriateness as keystones.

The urban transport arena

The complexity of urban structures is reflected in their transport systems. Several modes and forms of transport coexist, sometimes by chance and sometimes by design, in order to provide an integrated service through complementarity of their respective characteristics. Examples of this would be a bus–rail–mass rapid transit (MRT) interchange and the integration of feeder bus routes through low-traffic areas using minibuses linking up with large buses on the high-volume main routes. Plate 22 shows one such main-line bus plying the dual carriageway between

SOURCE: SADCC Annual Report, 1991–2.

eign funding (%)	Local funding (US$ m)	Local funding (%)	Already secured (US$ m)	Already secured (%)
87	858	13	3 110	47
96	32	4	242	32
91	26	9	147	53
95	14	5	80	30
88	1 029	12	3 749	44

Nairobi and Thika in Kenya. Such integrated planning, however, remains the exception rather than the rule, despite several decades of formal transport and land-use planning. In cities of the North, however, it is becoming increasingly common at major airports and railway stations in an effort to promote rapid transit and reduce motor car congestion. In the South, the bus feeder services focused on mass rapid transit stations in Singapore represent a particularly good example. MRTs are expensive and disruptive to build; although now appearing in a growing number of metropolises, they are imported wholesale from the North and have been criticized as representing a form of technological imperialism. Less expensive light rail transits (LRTs), which generally run above ground, may represent a more cost-efficient alternative for cities needing a high-speed, high-volume, rail-based system.

Elsewhere, piecemeal or incremental planning is the norm, compounded by chronic shortages of expertise and funds. In rapidly growing cities of colonial origin, the challenges are particularly acute. On the one hand, the old inner core areas which predate motorized road transport are usually characterized by networks of narrow lanes and streets unsuited to bus or mass rapid transit systems despite their high population densities. On the other hand, only a small proportion of new urban developments on the urban fringe are formally planned. Most are unplanned, so-called 'spontaneous' settlements of self-help shanties. Public transport here is often non-existent, at least until some formal recognition of the settlement is obtained, unless it happens to lie on a trunk route which already has services. It is also commonly impossible for large vehicles to negotiate the winding tracks without some upgrading scheme to improve access.

There is great variety in urban transport infrastructural provision, in its quality and levels of maintenance, and in the availability and affordability of public transport services. Nevertheless, two generalizations are justified. First, urban elites and middle classes in most countries apart from some of the Asian newly industrializing countries rely heavily on private transport, especially motor cars but also motorcycles and scooters. Some Asian metropolises such as Singapore, Hong Kong, Bangkok, and Jakarta now suffer from chronic traffic congestion. The former two have implemented comprehensive traffic management schemes, including electronic road pricing, heavy entry fees, and selective bans from central areas. In the latter two cities, however, traffic jams frequently last most of the day. The situation in southern African cities has not reached such proportions, but peak-time congestion is regarded as a growing problem in metropolitan centres, especially where – as in South Africa – car ownership rates have been rising steadily beyond the traditional elite groups, albeit from a low base.

Second, public transport services are invariably inadequate in geographical coverage, frequency, reliability, and affordability. For the urban poor, walking remains the most widely used form of transport over relatively short distances, and – where cycling is culturally acceptable and the terrain relatively flat – bicycle use is very important for short to medium distances of three to four kilometres. Only beyond that distance are buses and trains the dominant modes for able-bodied people. Where these, too, are inadequate, shared taxis or other forms of intermediate transport, known as *paratransit,* commonly appear to meet demand (Mutizwa-Mangiza 1993; Simon 1996). The case of minibus taxis in South Africa is discussed below.

Apart from walking and cycling, the urban poor may utilize draught animal power to pull carts of various designs (see Plate 23). In some African contexts, donkeys and mules are relatively common, whereas in South Asia oxen or water buffalo dominate. While fairly cheap to keep, provided that access to grazing or fodder exists, and also sustainable in the sense of not burning fossil fuels or emitting polluting exhaust fumes, these animals are slow and contribute to traffic congestion. Along with rickshaws and pedal tricycles, animal-drawn carts have frequently been 'designed out' of urban areas by strongly modernist governments seeking efficiency and a modern image. Unless appropriate and affordable alternatives are found, such measures often exacerbate the plight of the urban poor.

Until quite recently, urban public transport was highly regulated almost everywhere in the developing world, with inefficient and unresponsive state, municipal, or private bus monopolies. Fares were often high, vehicle utilization surprisingly low, and maintenance levels poor – a problem often exacerbated by shortages of imported spare parts. Most of the major municipal and private bus companies in South Africa maintained their fleets well. The worldwide wave of deregulation and privatization since the early 1980s has led to improvements in many places, often through public–private joint ventures, greater attention to maintenance and financial management, and similar measures. The bus services in Harare, the Zimbabwean capital, are a case in point (Maunder and Mbara 1995).

Constrained accessibility: the rural mobility challenge

Rural dwellers face even more severe mobility and accessibility problems than poor urban dwellers. Those in wage employment tend to earn less, while un- and underemployment are generally higher. Moreover, the availability of transport infrastructure and vehicles is often more limited and geographically variable. People living close to trunk roads and railway lines are relatively fortunate, assuming that they can afford the relevant fares. However, the commercialization of public transport has often led to a reduction in off-peak frequencies and even the withdrawal of some services. For example, passenger train services have been abolished on most regional and rural lines in South Africa. Hence, even living alongside the track does not help unless one is also close to the trunk roads along which the replacement bus services operate.

Many peripheral rural regions situated away from the commercial economic cores and corridors in every southern African country lack even this access, as a glance at any map of transport infrastructure in relation to the former homelands or reserves of South Africa will reveal. The most important innovation over the last twenty-odd years, in terms of improved rural mobility for those able to afford it, is the advent of the minibus (*kombi)* taxi on inter-urban routes. Fares are generally lower than for other public transport and routes are more appropriate, but overcrowding and safety are major concerns.

Numerous villages and hamlets in all regions lie at substantial distances from the nearest access point to motorized transport. For their inhabitants, walking and the use of draught animals (perhaps with some form of cart or a more primitive sledge – see Plate 24) are the main sources of mobility and transporting goods. Walking is slow, tiring in hot weather, and limiting in terms of loads that can be carried. Walking also incurs an opportunity cost, in that other activities cannot be undertaken during this time. Accordingly, fields situated far from a homestead are generally cultivated less intensively (and can be less well protected) than those nearby, even if water is available.

By no means all rural households have access to draught animal power; indeed, the lack of such access has become a widely used proxy indicator for the prevalence of extreme poverty. Rural social inequality has commonly risen in recent decades, as a result of overcrowding and falling land productivity, differential migration, and the impact of more frequent drought and famine. Poor peasant households – with few animals to start with – are the most vulnerable to losing them all in a drought or through having to sell livestock in order to survive. Either way, this leaves them worse off and possibly destitute. Draught power for ploughing or transport then has to be rented, but this becomes less and less affordable, exacerbating the downward spiral into destitution.

In areas of flat topography, where there are not too many thorns and where cultural norms do not militate against it, bicycling may be important for short and moderate distances. Bicycles require a modest initial outlay – and hence are most often purchased by those with paid employment or those who have undertaken migrant labour – but have low maintenance and minimal running costs, and can transport remarkable loads in addition to their drivers.

The major thrust of (integrated) rural development initiatives since the 1970s in terms of improving access and mobility has been the construction of low-volume and feeder roads. The underlying principles are:

- that routes with little vehicular traffic can be built according to lower standards, thereby enabling more rapid network expansion; and
- that providing small centres and villages with at least a basic road connection will stimulate economic activity and enable the provision of other extension and development services, such as primary health clinics and schools.

The evidence is mixed. Although service access has been improved, new roads can permit the introduction of produce and commodities from outside the area just as much as facilitating the marketing of local produce elsewhere. Many previously isolated social economies have thereby been undermined, while out-migration has also become easier. One strategy pioneered by the International Labour Organization (ILO) to generate employment and wage incomes in rural areas through infrastructural provision is the construction of labour-based roads. Under the right conditions this can work well, but construction often takes longer and, by virtue of the large number of labourers involved, requires skill and patience in management. Namibia's first post-independence experiment with this approach, at Onaanda in the communal lands of the Omusati region, foundered on this account in 1992/93. Conventional mechanical construction is invariably quicker and sometimes cheaper, and various intermediate technologies such as strip roads built from different materials are also being experimented with. For further details on all these issues, see Simon (1996).

South Africa in transition: policy and practice

Transport policy in the late apartheid and early post-apartheid periods

During the decades of 'high apartheid', state policy was geared to ensuring white and especially Afrikaner control over the economy, and to the provision of a comprehensive welfare system for whites. All sectors of the economy, including transport, were subject to strict regulation and control. The parastatal South African Railways and Harbours (SAR&H) administration – under which South African Airways (SAA) also fell – held monopoly rights and engaged in rural road transport in areas lacking railway access. Strict licensing and quota systems governed the private sector's involvement in road and air transport.

By the late 1970s, technological change (such as containerized maritime cargo), the growing needs of South Africa's increasingly sophisticated economy, the onset of a severe international recession, an international trend towards deregulation, and growing pressure to isolate South Africa on account of its apartheid policies combined to force a major policy rethink. From the early 1980s, the former government commenced a process of gradual economic liberalization and deregulation, albeit one marked by frequent inadequacies and even contradictions. This period also coincided with the incremental reform of apartheid policies under President P. W. Botha. The first significant change in the transport sector was the restyling of the SAR&H as the South African Transport Services (SATS) in 1981. This was intended to create a more modern, outgoing, and commercially orientated multi-modal parastatal transport corporation. Each transport mode (rail, air, harbours, road) became the responsibility of a separate division within the SATS.

One important stimulus to debate and policy change in the early 1980s was provided by the rapid growth in unregistered or 'pirate' taxis and minibuses catering to the needs of black urban township dwellers (see below). Together with shebeens in Soweto and other townships, pirate taxis precipitated a profound official rethinking of policy towards the unregulated or informal sector. Unlike pavement hawking and other informal trades, these two activities were well organized, increasingly important, and demonstrably met popular needs unserved by – and, indeed, to a significant extent, arising directly as a result of – apartheid urban planning and administration.

By 1987, the White Paper on Privatization and Deregulation set out a clear commitment to promote a more efficient and streamlined public sector. In parallel with this, the 1987 White Paper on Transport Policy – which followed a major National Transport Policy Study – envisaged a fully deregulated transport system characterized by control over service quality instead of quantity, and the elimination of state protection for SATS commercial operations which gave them an unfair advantage over private competitors. The ultimate objective was to establish a system based on the 'user pays' principle, whereby the full infrastructural costs were to be recovered from users, cross-subsidization between different activities and modes of transport would be eliminated, and regulations would be streamlined and systematized. Existing legislation was to be remoulded by 1990 within four principal Acts: on the SATS, transport, roads, and road traffic, respectively. For various reasons, however, the schedule was not adhered to, and, although several deregulatory Acts were passed and related measures adopted, the process was then overtaken by the constitutional negotiations to end apartheid. Several key Acts were therefore delayed and then modified by the new government.

In 1990 the government acted on the 1989 legislation to replace the SATS with Transnet, with the various divisions separated into independent corporate business units operating on a strict commercial basis. Hence Portnet runs the harbours, Spoornet the railways, Autonet the road transport activities, SAA the state airline, the South African Rail Commuter Corporation the loss-making urban township rail services (with explicit state subsidy), and so forth. This restructuring followed the highly critical De Villiers Report on the SATS in 1986, which condemned heavy loss-making investments over many years, claimed that even the 'new' SATS administrative structure was anachronistic, and argued that business principles based on competitive advantage needed to be entrenched as the cornerstone of operations.

The ANC-dominated Government of National Unity elected in 1994 has retained much of the deregulatory thrust of inherited transport policies, although with one crucial change, namely an explicit commitment that transport should be developed to promote social equity. This formed part of the Reconstruction and Development Programme, the ANC's election manifesto, and reflects the overall programmatic reorientation of state policy and expenditure to redress the inequitable legacy of apartheid and other policies. The thrust of commercialization under the former government had increasingly removed 'ability to pay' and social welfare elements from policy in favour of the 'user pays' principle. The Green and subsequent White Papers on National Transport Policy published in 1996 encapsulate these new objectives, emphasizing a shift towards economically and environmentally sustainable transport policies and bottom-up planning, in contrast to the previous bureaucratic and top-down thrust (South Africa 1996a, 1996b).

Two examples of current policy direction will suffice. Deregulation of the road haulage and transport sector commenced in 1994, with the relaxation of haulage permit quotas and similar restrictions. A government bill introduced in late 1997 envisages the creation of an International Road Transport Agency to undertake the market-regulation activities (centred on cross-border permits), law enforcement, and related tasks hitherto performed by the Department of Transport. This would represent a significant privatization and streamlining of the department. Meanwhile, the National Road Traffic Act, No. 93 of 1996, which came into effect in 1997, overhauls previous legislation and standardizes provisions nationwide in the light of new provincial powers and the abolition of the former TBVC 'independent' homelands. This Act also institutes quality controls as envisaged in the late 1980s.

Many parastatal corporations are being commercialized and a programme of partial privatization is also in place. Deregulation of civil aviation in general, which has seen a rapid increase in the number of airlines and routes in recent years, has continued along the trajectory set by the previous government following a major study in 1989–90. Following the model of the BAA (formerly British Airports Authority) in the UK and similar examples elsewhere, state airports in South Africa are now run by the Airports Company, which has already made substantial improvements in upgrading and expanding the international gateways. The sale of 49 per cent of the shares in this parastatal was initiated in mid 1997. However, the Air Traffic and Navigation Services Company (ATNS), which operates air traffic control at state airports, remains wholly state-owned. In 1994, the airlines of the former so-called independent homelands were 'inherited' by the South African government. The most profitable of these, Sun Air, was prepared as the first candidate for full privatization in the transport sector, a process completed in November 1997. Black economic empowerment groups took a substantial shareholding. By mid 1998 the value of Sun Air's shares had quadrupled from the initial sale price of just over R50 million. Partial privatization of SAA – originally mooted in 1989 – was repeatedly postponed, not least because the airline is loss-making.

Changing practice: selected case studies

Minibus taxis: continuing struggles

Along with other forms of road transport, urban taxis were strictly regulated and licensed under apartheid. Restrictions were introduced whereby drivers had to operate exclusively within their 'own' group areas or between those and the central city; they could also only carry passengers of their own 'population group'. In most towns and cities, the train service (where any existed) and municipal and private bus services were inadequate for the needs of black South Africans, despite the high employer levies and state subsidies paid to underwrite commuter services from distant townships and closer settlements within the nearest homeland. Journeys were long and arduous and fares were high relative to earnings (Mann and Segal 1985). For example, bus commuters from the former KwaNdebele homeland to Pretoria endured a two-hour ride each way, leaving home at 4 or 5 a.m. and often returning only at 7 or 8 p.m. This was a classic case of the burden of apartheid borne almost exclusively by black people, a situation greatly exacerbated by forced removals under the Group Areas Act and pass laws.

Under these circumstances, illegal or 'pirate' taxis began operating in increasing numbers during the 1970s. They successfully exploited the yawning gulf in the market, providing a quicker, more frequent, and comfortable journey for smaller numbers of people at a price which, although higher than the equivalent rail or bus fares, was more than compensated for by the greater convenience. The vehicles initially used were ordinary family sedan/saloon cars (see Plate 25) such as the Chevrolet Constantia and Chrysler Valiant, which could be bought relatively cheaply second-hand and were not conspicuous. Gradually, however, minibuses became the vehicles of choice on account of their larger carrying capacity. The most common such vehicle at the time was the Volkswagen Kombi – from which *kombi-taxi*, the popular name for minibus taxis, is derived.

Initial reaction to this new black-owned industry by organized urban transport interests was predictably negative. Bus, taxi, and rail operators claimed unfair competition, saying that their own services would be undermined. Police clampdowns failed to curb the growing numbers of minibuses countrywide, however. At the same time, this informal mode of transport, together with illegal shebeens in private homes within the townships, became the subject of fierce policy debates between vested industry interests seeking their suppression and those who increasingly perceived these developments as the vanguard of a black business sector, showing precisely the initiative and innovativeness required to make inroads into poverty and unemployment (Khosa 1992). Eventually this debate crystallized into a more positive attitude and policy towards the 'informal' economic sector and black business as a whole, but not before the Welgemoed Commission in 1983 had recommended outlawing *kombi-taxis*.

By the mid 1980s, the South African Black Taxi Association (SABTA) had been embraced by the state and organized commerce, serving as the vanguard of reformist initiatives to foster a black business elite and middle class through economic empowerment. Thereafter, the industry grew dramatically in scale and complexity, stimulated by rising demand as some people's real incomes rose and also by the increasing cost and unreliability of rail and bus services – which in turn became the subject of frequent boycotts (Khosa 1995). Wealthier owner-drivers and professional people, along with some white and corporate investors, purchased fleets of minibuses (the majority of which are now Nissans and Toyotas) and hired drivers (Khosa 1991a; Noah 1995). Interurban services were introduced and fell under a separate body, the South African Long Distance Taxi Association (SALDTA). These associations negotiated bulk discounts or favourable hire-purchase terms from vehicle, tyre, and spare parts suppliers and petrol companies.

Inevitably, however, there have been immense problems. Minibus taxis have become notorious for poor safety standards, reckless driving, and disproportionately high accident rates, often with multiple fatalities. In 1994, the Government of National Unity immediately announced its determination to address these and other problems facing the industry.

By the late 1980s, rivalries and competition over routes and taxi ranks spawned a number of new associations, some defined by geographical area, and others as breakaways from or direct competitors with existing bodies. Examples are the Transkei Taxi Owners' Association (TTOA), the Border Alliance Taxi Association (BATA), and the Federated Long Distance Taxi Association (FLDTA). Increasingly violent conflicts, known as 'taxi wars', resulted (Khosa 1991b). Ambushes of minibuses in transit and gunfights at taxi ranks are the most common incidents, but planned murders and sabotage of vehicles also occur. Mediation efforts have seldom had more than a temporary impact, while new conflicts frequently arise in previously peaceful areas. During the late apartheid years, political affiliations also played a role; in some areas, especially KwaZulu-Natal and Gauteng, this sometimes continues to be the case. A decline in violence immediately after the new government took office proved short-lived and all the efforts of the Minister of Transport and provincial leaders to intervene have proved fruitless.

Table 10.3 illustrates both the magnitude of the problem countrywide and the significant regional variations, with the Eastern and Western Cape together accounting for virtually two-thirds of all incidents and over half the fatalities in 1996. Importantly, rivalry between the ANC and the Inkatha Freedom Party (IFP) is not a factor in either of these provinces, suggesting that conflicts within the taxi industry lie at the heart of the problem. Newspaper reports continued to document frequent incidents with high casualties around the country in 1997 and 1998. One of the most audacious was an ambush outside the Durban High Court in May 1998 in which ten men armed with AK-47s and handguns attacked rivals attending two taxi-linked hearings in the court. Three people were killed and nine others wounded.

An in-depth study of the Eastern Cape taxi industry (Noah 1995) has supported the earlier findings by Khosa (1994) that increased competition and declining profitability – partly because of the competition but also because of rising fuel and vehicle replacement and repair costs – have intensified rivalry between competing associations over particular routes, taxi ranks, and fare levels, and the violence commonly associated with these. In addition, the safety of vehicles has deteriorated markedly during the 1990s as a result of competitive pressures and rising costs relative to fare revenues. Maintenance is delayed or carried out inadequately, inferior spare parts and inappropriate or non-matching tyres are fitted, unskilled mechanics are utilized, untrained or unlicensed drivers are used, and so forth. These pressures, often exacerbated by exploitative practices and revenue demands on drivers by fleet owners, also lead to dangerously fast and reckless driving, the working of long shifts without adequate rest and sleep periods, rapid turnaround times for vehicles, and the delaying of maintenance. All these factors contribute to the rising toll of death and injury, compounded in the last few years by reduced enforcement of traffic regulations by the traffic police.

In an effort to address the industry's problems in a comprehensive and consultative manner among the various stakeholders, the Minister of Transport established the National Taxi Task Team

TABLE 10.3 Taxi-related violence in South Africa

Province	Persons killed	Persons injured	Incidents reported	Proportion of total incidents
Eastern Cape	104	243	353	40.0%
Western Cape	76	164	225	25.5%
Gauteng	74	104	120	13.6%
KwaZulu-Natal	43	60	61	6.9%
North West	22	63	38	4.3%
Mpumalanga	8	16	37	4.2%
Northern Province	6	56	26	2.9%
Free State	5	2	20	2.3%
Northern Cape	0	4	3	0.3%
National Total	338	712	883	100%

SOURCE: South African Institute of Race Relations, *Fast Facts*, June 1997, Johannesburg.

(NTTT) in early 1995. Its interim report was tabled late that year and the final version in 1996. The report highlighted the extent to which minibus taxis have come to dominate the public transport market, carrying an estimated 2.2 million passengers per day, far more than any other mode and approximately half the total of public transport users. The pleas encountered in the NTTT's hearings and reflected in its recommendations focused on enhancing safety, quality, professionalism, and greater profitability in a situation of greater peace and order. Government recognition of the industry and appropriate training, regulation, and enforcement of enhanced safety standards and procedures were therefore central issues. Many of these were compatible with the overall objectives of the Department of Transport, and were embraced by the Green and White Papers on National Transport Policy in 1996 (South Africa 1996a, 1996b). A system of regulated competition is envisaged; this will require further formalization of the industry. Measures to enhance economic viability will include financial and technical assistance and the issuing of permissions to operate on particular routes/networks on the basis of demand as estimated by the local passenger transport plan. As mentioned above, the White Paper provides the new government's strategic forward view; the minibus taxi industry in particular and black economic empowerment in the sector in general now form key elements in the plans for a more equitable and sustainable transport system.

South Africa in the world: seafreight

South Africa's return to the international fold, a process completed in mid 1994 but begun a few years earlier, has heralded many new challenges for the country's international transport industries, both air and sea. Passenger and freight volumes have increased markedly, as have the numbers of flights and sailings, but this, together with progressive deregulation and partial privatization in the sector, has greatly increased the levels of international competition facing South African airlines and shipping lines previously insulated to some extent by the country's isolation during the apartheid years. In addition, technological change and restructuring of the maritime sector worldwide are impacting simultaneously. These new challenges are being met in a number of ways; here the focus is on seafreight, and especially the growth of unitized (containerized) freight.

Portnet, the harbour arm of the parastatal transport utility Transnet, which was commercialized in the early 1990s, has been forced to undertake rapid expansion of capacity at the container terminals in the main ports of Durban and Cape Town as well as Richard's Bay (see Plate 26). The last-mentioned, situated to the north of Durban in KwaZulu-Natal, is somewhat smaller but has diversified away from its original sole function as a bulk coal export terminal. By 1995, the growth in container traffic (measured in terms of twenty-foot equivalent units (TEUs)) had greatly exceeded Durban's capacity, leading to significant delays which were exacerbated by labour disputes. The situation in mid 1997 was considerably better, following the investment of R654 million in 1996/97 and R1.4 billion in 1997/98 in harbour development countrywide. Nevertheless, if the current 6–8 per cent annual growth rate of TEUs continues, Durban's maximum capacity of 1.1 million TEUs per annum will be reached some time in the period 2000–2005. Cape Town's current capacity of 450 000 TEUs is likely to be reached in the year 2003, although the potential exists to expand this to 850 000 TEUs (*Lloyd's List* 1997).

Uncertainty over a longer-term national capacity expansion strategy remained in mid 1997. Certain interests favour building a new container port in Durban, to capitalize on existing infrastructure and demand patterns. However, Richard's Bay is seeking to attract that dedicated facility on account of its room for expansion and its location within a poor region. The recent announcement of a major transport and industrial corridor linking Maputo in Mozambique to Gauteng at a cost of over R400 million, however, suggests that a growing proportion of Gauteng traffic might be routed through Mozambique's main port, a move which could jeopardize Richard's Bay's prospects. Additionally, Port Elizabeth in the impoverished Eastern Cape has been used as a back-up landing point for Durban during the latter's recent overcrowding, and long-term expansion in traffic there would make a substantial contribution to the local economy. Moreover, environmentally controversial plans to construct a new bulk port from scratch at Coega, 20 km along the coast of Algoa Bay from Port Elizabeth, to refine and export zinc and other industrial products, have recently been approved without a full environmental impact assessment. This project will form the core of the Fish River Spatial Development Initiative (see Figure 5.5, p. 136), one of a series of new geographically defined public–private partnerships designed to target employment and development in depressed and peripheral regions, especially those, like the Eastern Cape, suffering from a particularly severe apartheid legacy.

Meanwhile, rates and commissions on most categories of unitized and bulk cargo are falling, putting pressure on agents, freight forwarders, shipping lines, specialist repair, salvage, and pollution control operators alike. In addition, the trend towards the use of larger container vessels and the development of hub-and-spoke distribution and collection systems in each continental region are leading to selective growth and failure of firms. Consequently, South African firms, many of which are specialized but small by global standards, are forging new strategic alliances with foreign firms in order to strengthen their effective leverage, to position themselves better in relation to particular geographical or specialist markets, and to exploit new opportunities (*Lloyd's List* 1997). The future is almost sure to be governed by growing internationalization, increasing competition for routes and loads, rationalization and increasing specialization of both firms and ports, and the need for flexibility in times of rapid change. This is true within South Africa, in the southern and central African region, and globally (Hoyle and Charlier 1995; Iheduru 1996; Simon 1998).

Conclusions

This chapter has outlined some of the principal dimensions and policy debates surrounding the provision of transport infrastructure and services. Technological change and processes of globalization may be 'shrinking the world', but this development is intensifying the reliance on new forms of transport and telecommunications. Internationally, policies have tended to undergo cycles of greater and lesser regulation and state control. South Africa has not been immune to such shifts, even though it was relatively isolated in the apartheid era.

Salient contrasts between prevailing conditions and policy issues in urban and rural areas were summarized as a prelude to more detailed consideration of conditions in southern and South Africa. The historical importance of technological innovation, such as the introduction of the railways, against the backdrop of colonial conquest and policy in the region, was emphasized to facilitate an appreciation of contemporary issues. Equally, current policies in post-apartheid South Africa are designed explicitly to promote a more efficient as well as equitable transport system which relieves the burdens of the immediate past. Finally, the South African minibus taxi and seafreight industries were examined as case studies of domestic road and international maritime transport trends and debates. The interrelationships between transport and development are complex. While the former may well be a prerequisite for promoting the latter, the presence of adequate transport is certainly no guarantee that development will necessarily follow.

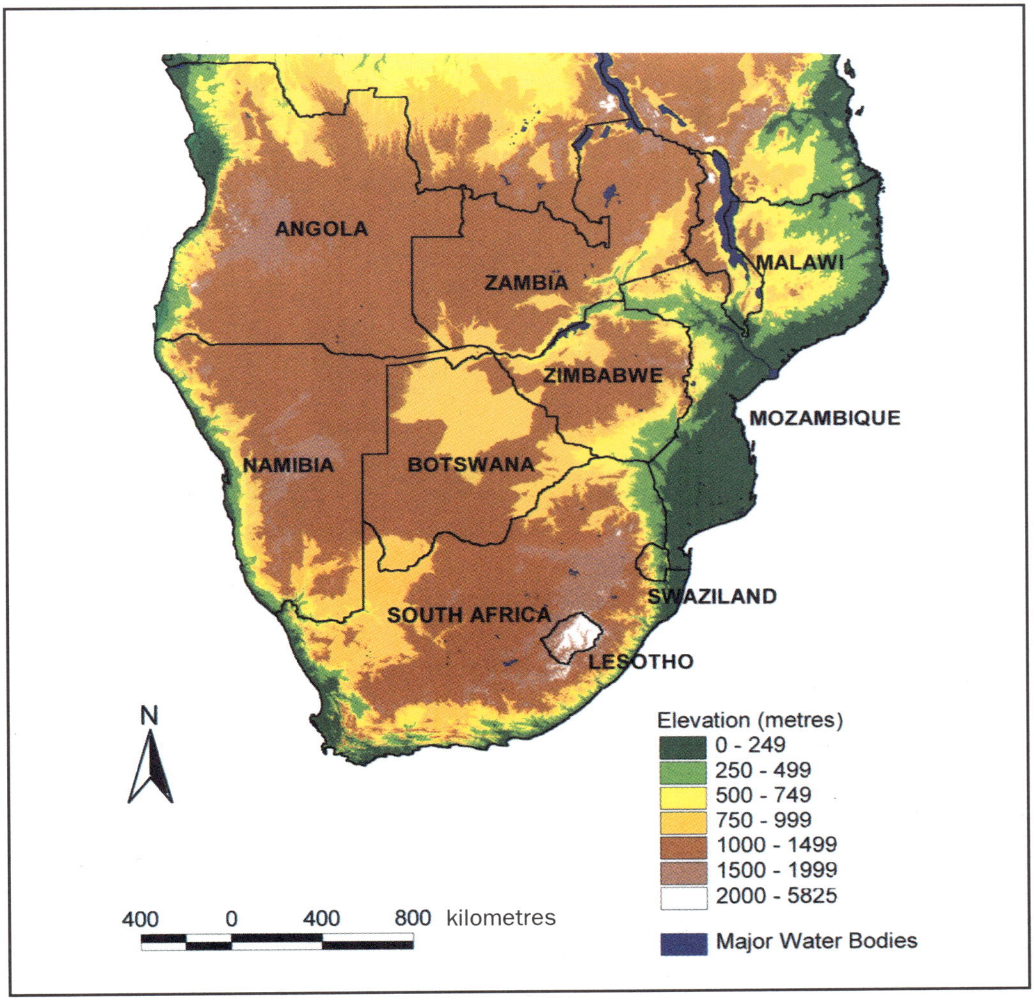

Plate 1 Southern Africa

Plate 2 Donga erosion in the Transkei. Professor Tim Partridge is pointing at Stone Age tools exposed in the gully wall

PHOTO: K. M. Rowntree

Plate 3
Cruikshank's cartoon

Plate 4 Possibly the easternmost example of the Cape house form, Smithfield in the Free State

PHOTO: B. Dodson

Plate 5 Typical wine estate homestead in 'Cape Dutch' style, Overgaauw near Stellenbosch

PHOTO: B. Dodson

Plate 6 Union Buildings, Pretoria, with the 'new' South African flag, April 1994

PHOTO: B. Dodson

Plate 7 St George's Cathedral, Grahamstown, during the National Arts Festival

PHOTO: R. C. Fox

Plate 8 Queen Victoria outside the Houses of Parliament, Cape Town

PHOTO: B. Dodson

Plate 9 Muslim shrine, Robben Island

PHOTO: B. Dodson

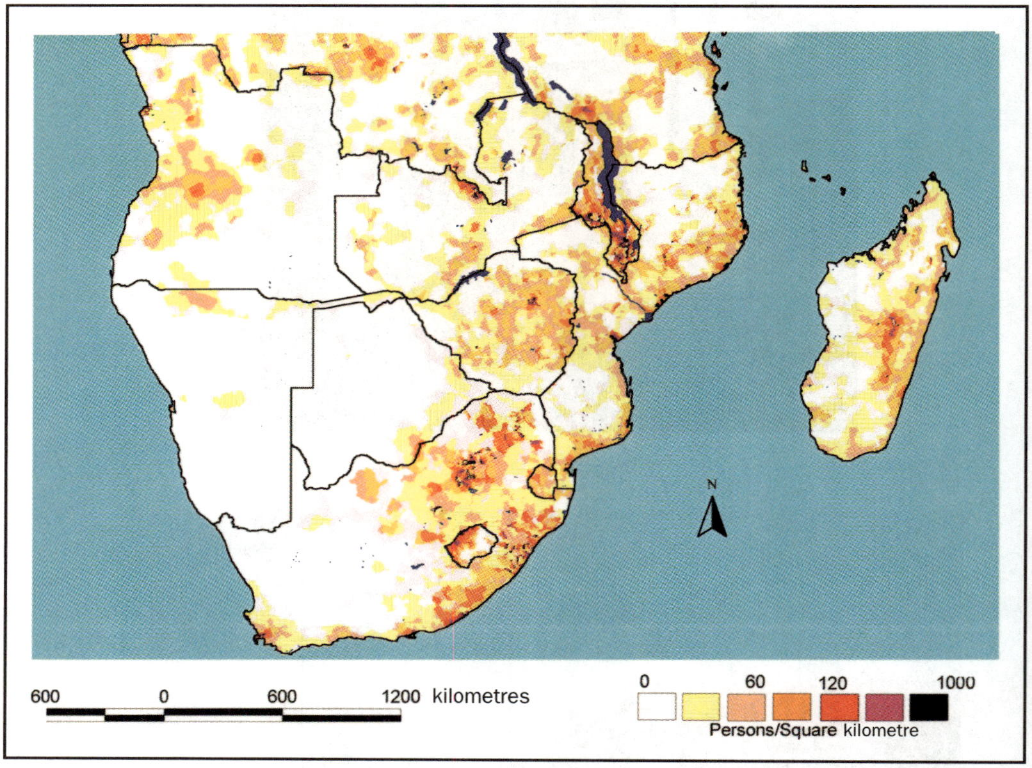

Plate 10 Southern Africa: population distribution, 1990

Plate 11 Dutch Reformed Church, Graaff-Reinet

PHOTO: R. C. Fox

Plate 12 Graaff-Reinet from Spandau Kop PHOTO: R. C. Fox

Plate 13 Indian group area, Durban PHOTO: R. C. Fox

Plate 14 Black township, informal and formal housing, Mdantsane.

PHOTO: R. C. Fox

Plate 15 White group area, Durban

PHOTO: R. C. Fox

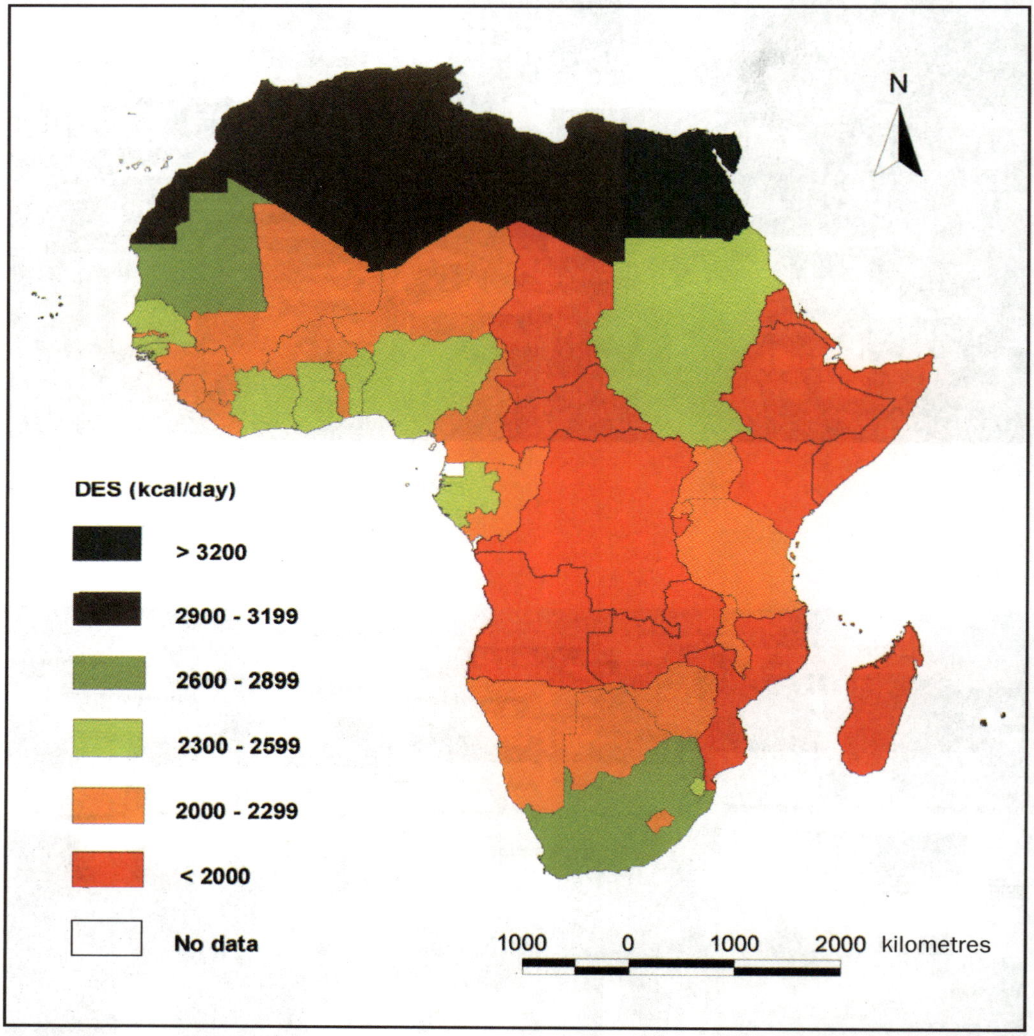

Plate 16 Africa: dietary energy supplies

Plate 17 South Africa: biological productivity

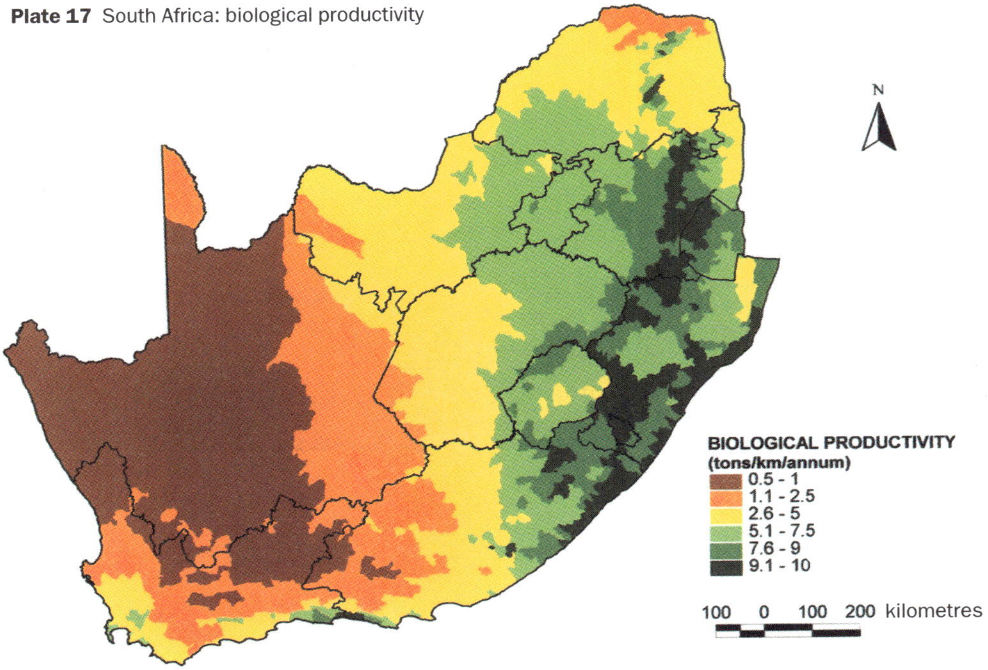

Plate 18 South Africa: land-use zones

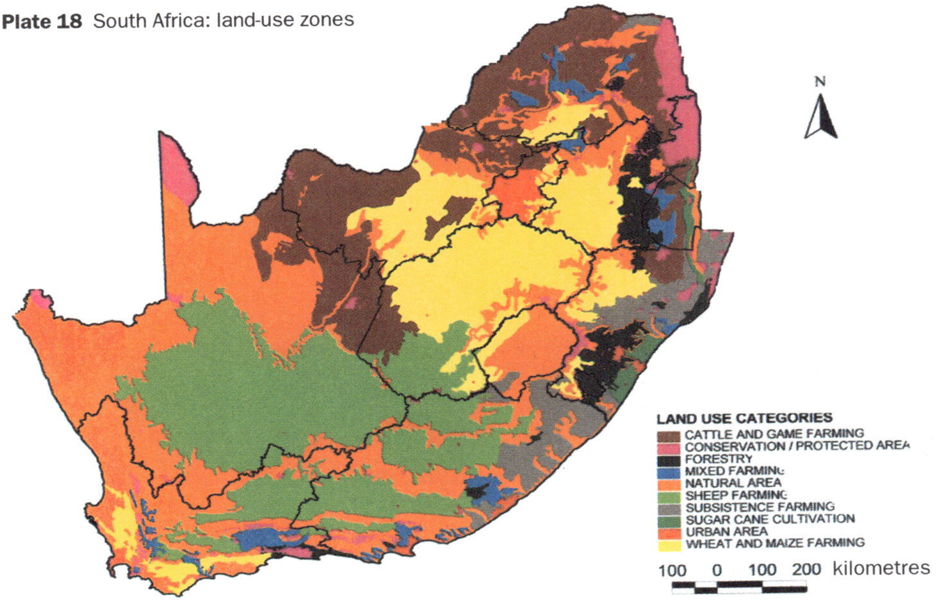

Plate 19 Centre-pivot irrigation, crop-growing region around Estcourt, KwaZulu-Natal PHOTO: R. C. Fox

Plate 20 Rural KwaZulu-Natal below the Drakensberg PHOTO: R. C. Fox

Plate 21 The neglected plough, Transkei Wild Coast

PHOTO: R. C. Fox

Plate 22 Express bus service, Nairobi–Thika dual carriageway, Kenya

PHOTO: R. C. Fox

Plate 23 Ox cart at Wartburg market, Eastern Cape

PHOTO: R. C. Fox

Plate 24 Rural sledge in the former Transkei homeland, Eastern Cape

PHOTO: R. C. Fox

Plate 25
Empty taxi at a deserted long-distance taxi rank, Grahamstown, election day, April 1994

PHOTO: R. C. Fox

Plate 26
Richard's Bay, bulk minerals carriers

PHOTO: E. L. Nel

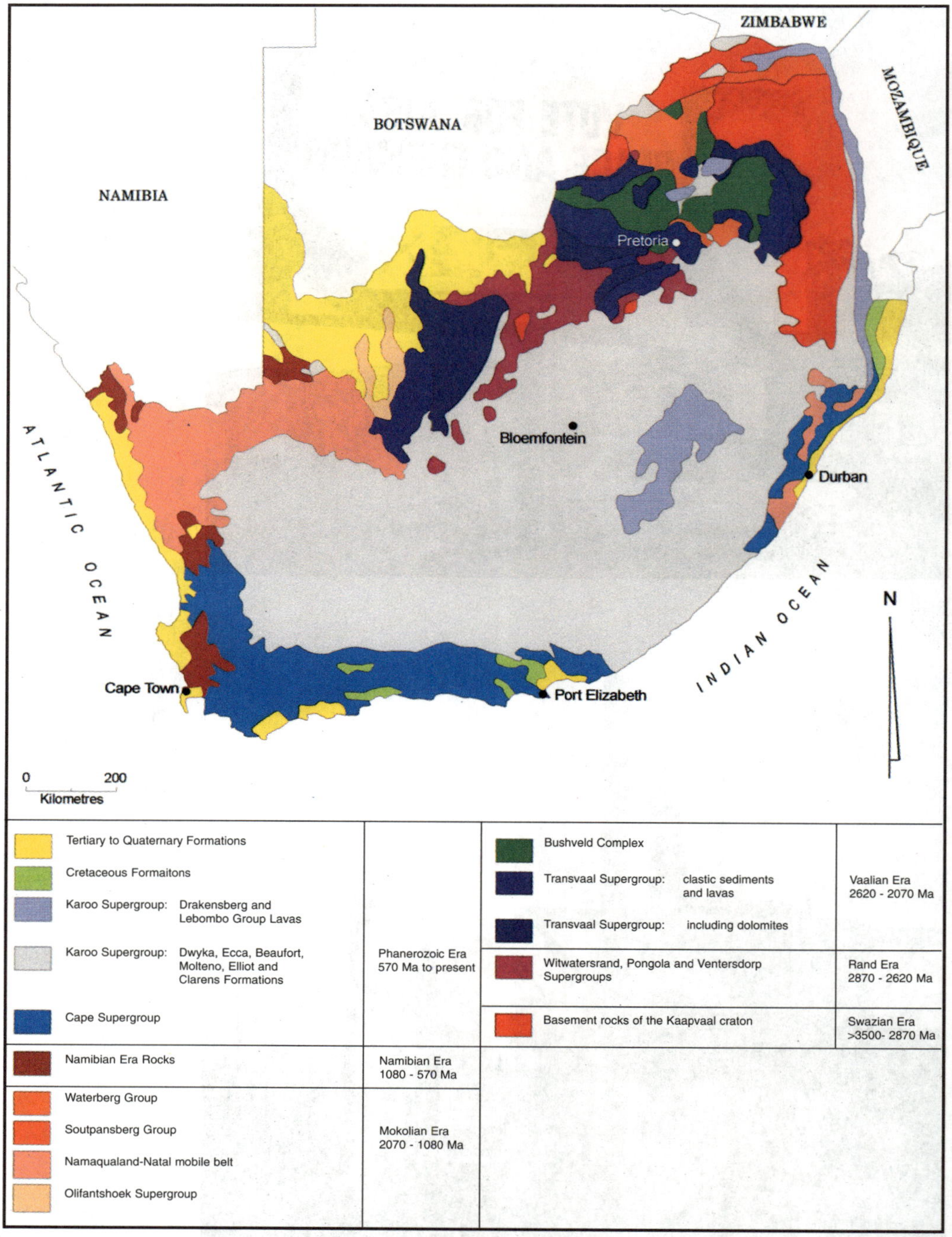

Plate 27 Geological map of South Africa

This map shows major units in geological succession, including those mentioned in the text. Note also the South African stratigraphic time scale, encompassing the Swazian era (> 3 500–2 870 Ma), the Randian era (2 870–2 620 Ma), the Vaalian era (2 620–2 070 Ma), the Mokolian era (2 070–1 080 Ma), the Namibian era (1 080–570 Ma), and the Phanerozoic era (570 Ma to present)

Plate 28 The Augrabies Falls on the Orange River formed in the basement rocks of the Namaqualand mobile belt

PHOTO: K. M. Rowntree

Plate 29 Folding in the Swartberg Pass, Cape Fold Mountains

PHOTO: K. M. Rowntree

Plate 30 Kaolinite formed by deep weathering of underlying silcrete, Grahamstown, Eastern Cape

PHOTO: K. M. Rowntree

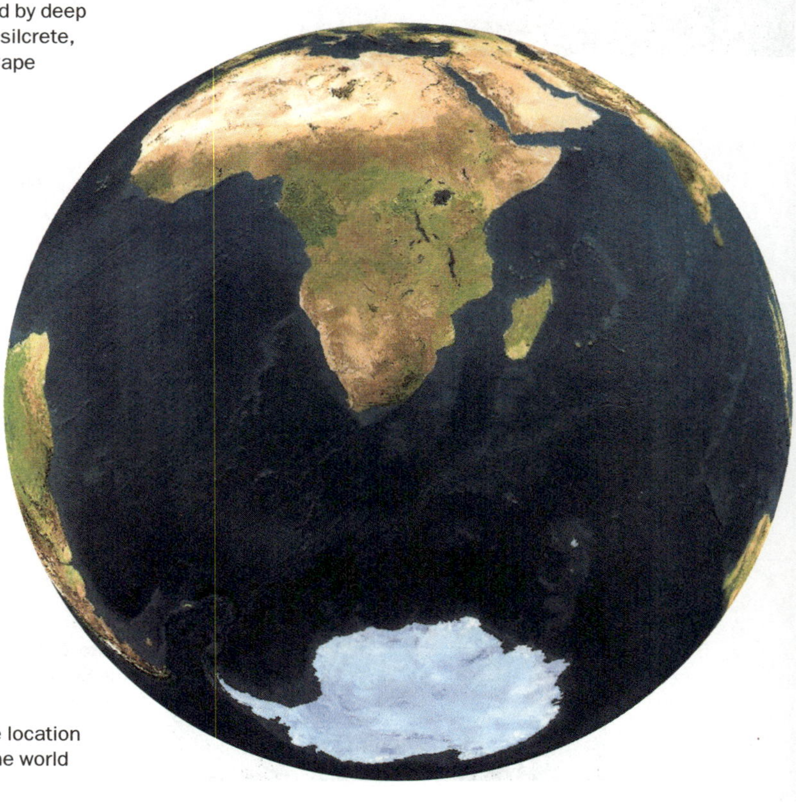

Plate 31 The location of Africa in the world

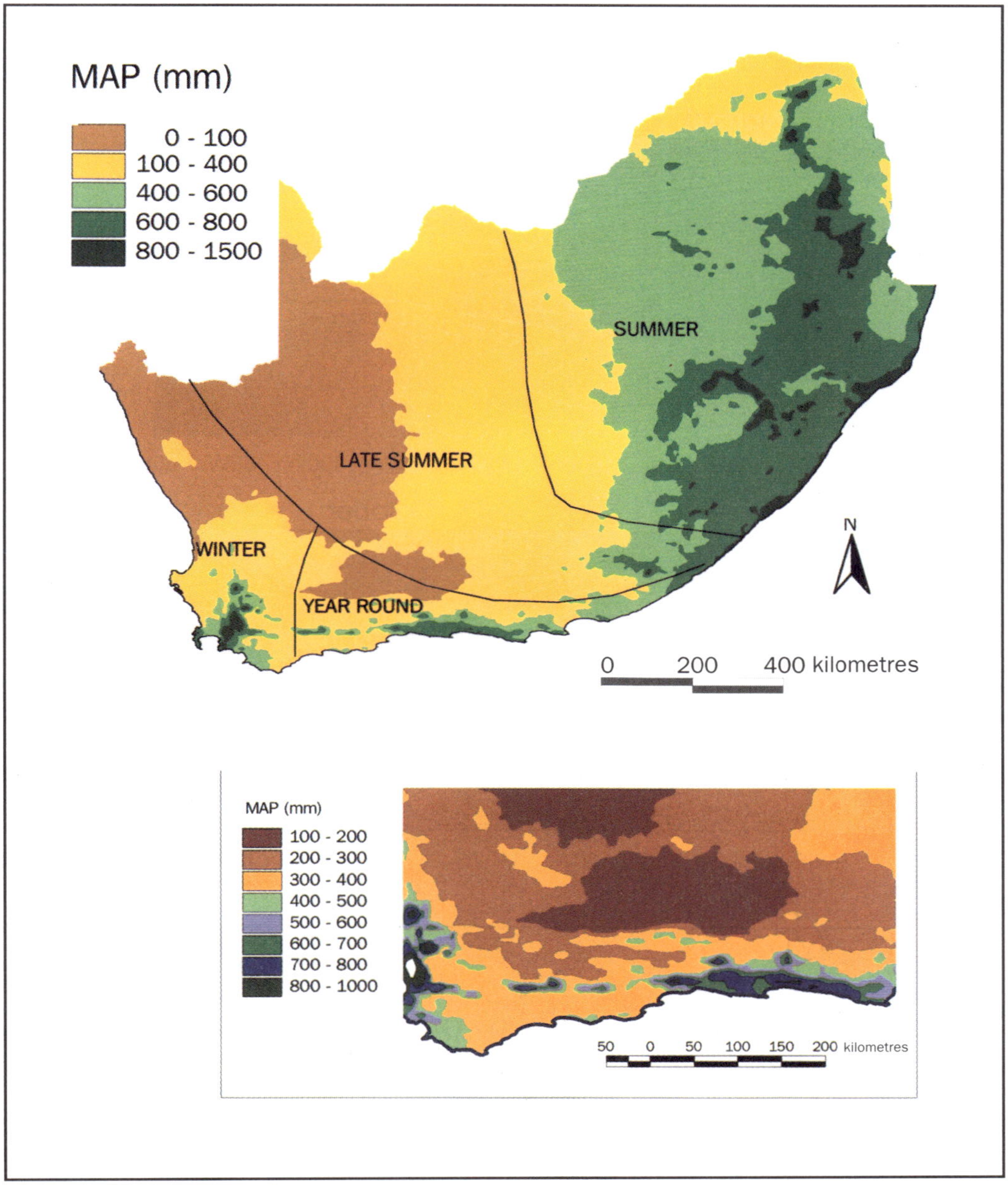

Plate 32 Spatial distribution of mean annual rainfall in South Africa

Inset: Orographic rainfall over the Cape Fold Mountains

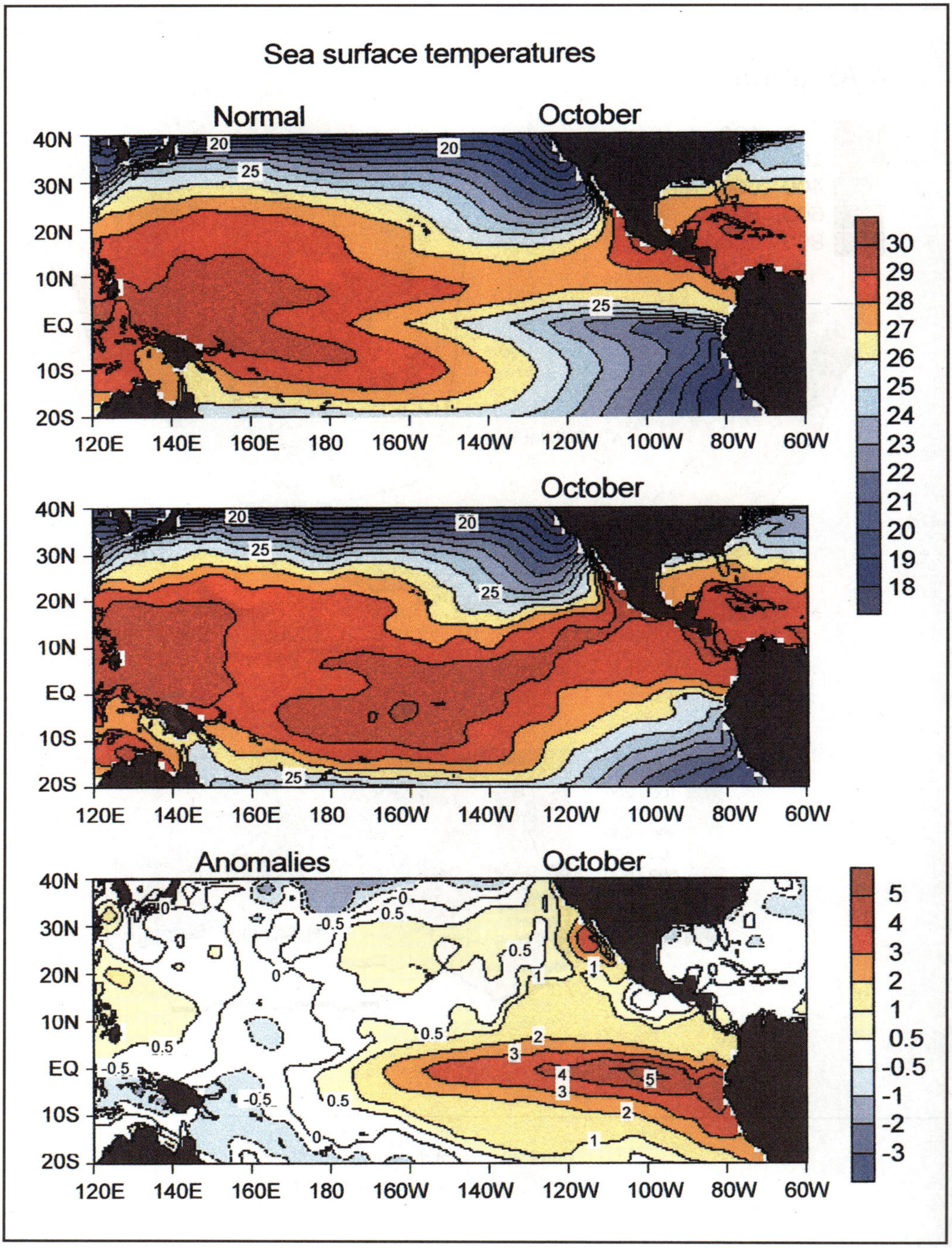

Plate 33 Sea-surface temperatures during the El Niño event of 1998

Plate 34 Drought in the Sneeuberg, Eastern Cape, causes the drying out of a farm dam

PHOTO: K. M. Rowntree

Plate 35 Relic boulderstream in the Lesotho highlands formed under previous colder conditions

Plate 36 Relic screes in the Amatola Mountains

Plate 37 Surficial landsliding in the Injasuti Valley, Drakensberg

(telephone pole in foreground for scale)

Plate 38 Soil erosion caused by multiple footpath development in the Drakensberg

Plate 39 Weathering in the Clarens Formation sandstone causing deterioration of rock art

Plate 40 Angular basalt clasts: structural control or cryogenic weathering?

(camera case in foreground for scale)

Plate 41 Typical open savanna woodland, bushveld in South Africa

Note the continuous cover of grasses. The trees are deciduous during the winter dry season and are also fire-resistant. The herbivores are red hartebeest

PHOTO: M. E. Meadows

Plate 42 Karoo shrubland near Aberdeen

Many dwarf Karoo bossies are deciduous as a drought adaptation. Their spacing suggests competition for a scarce resource – water

PHOTO: M. E. Meadows

Plate 43 Dunes in the Namib Desert

Both the intense aridity and the unstable nature of the substrate make this a difficult place for plants and animals to survive PHOTO: M. E. Meadows

Plate 44 *Welwitschia mirabilis* in Namibia

This remarkable plant has specialized adaptations, which include a tap root and hard leaves with sunken stomata, to enable it to grow in the stony desert PHOTO: M. A. N. Müller

Plate 45 Afromontane forest

A patch of forest with *Podocarpus* (yellowwood) and *Olea* (ironwood) trees surrounded by fire-prone grasslands

PHOTO: M. E. Meadows

Plate 46 Fynbos

This mosaic of many species is unique to the south-western Cape of South Africa and is one of the biodiversity jewels of the continent

PHOTO: M. E. Meadows

Plate 47
Human impact on the landscape

The fynbos has been fragmented by urbanization, agriculture, and invasion of alien trees. Here in the Franschhoek valley, all three elements of disturbance are evident

PHOTOGRAPH: M. E. Meadows

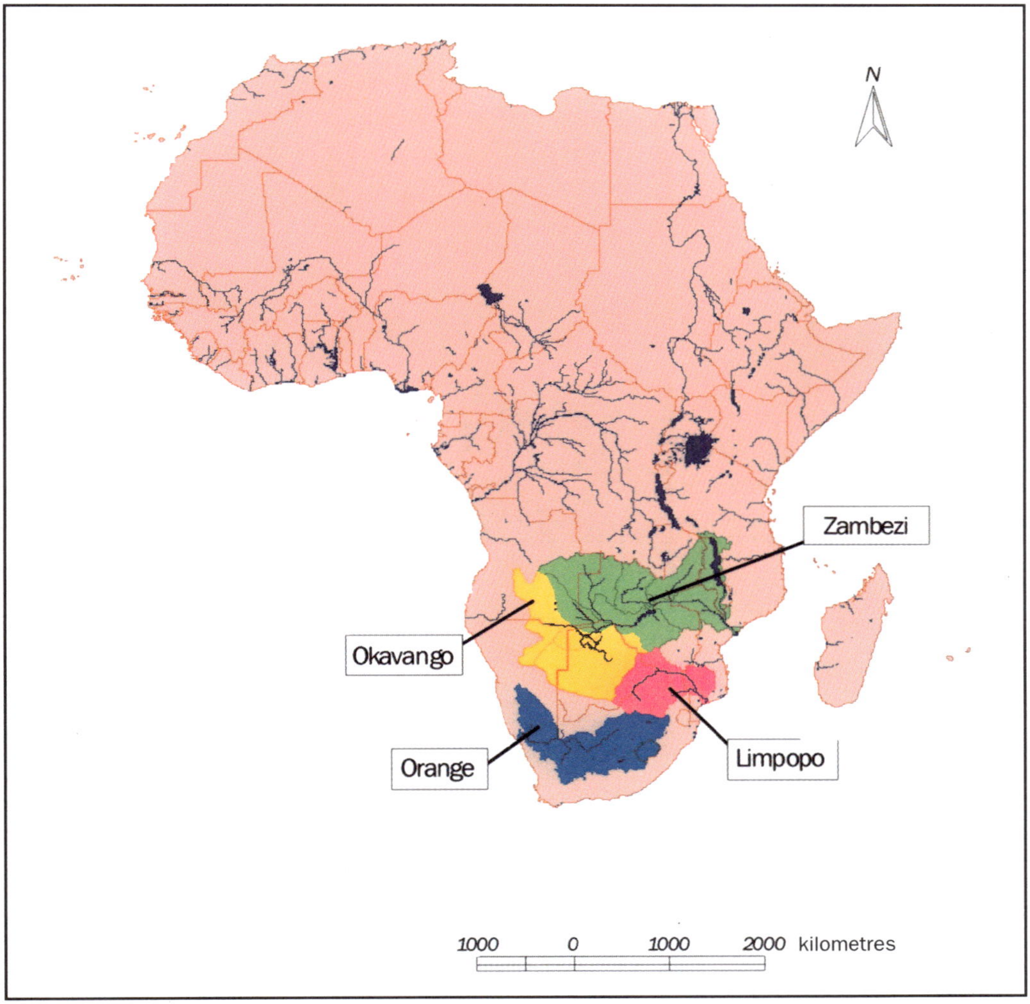

Plate 48 The major drainage basins of southern Africa and rivers of Africa

Plate 49 Contrasting land use on either side of a drainage divide in KwaZulu-Natal PHOTOS: K. M. Rowntree

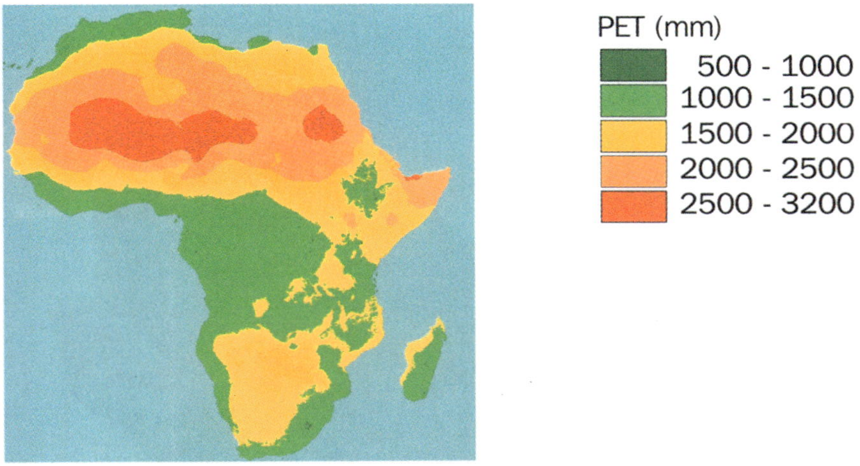

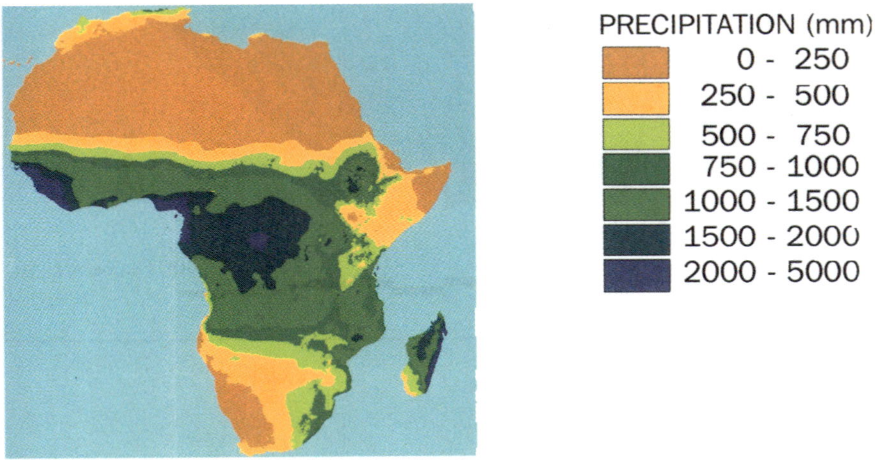

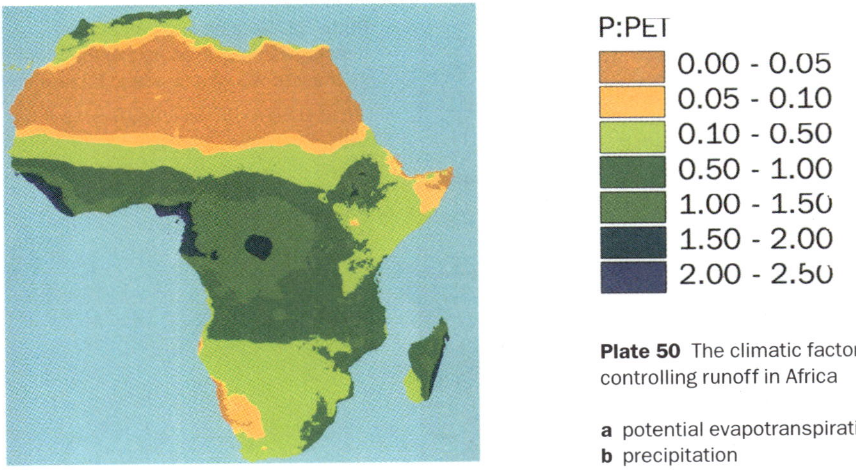

Plate 50 The climatic factors controlling runoff in Africa

a potential evapotranspiration (PET)
b precipitation
c ratio of precipitation to evapotranspiration (P:PET)

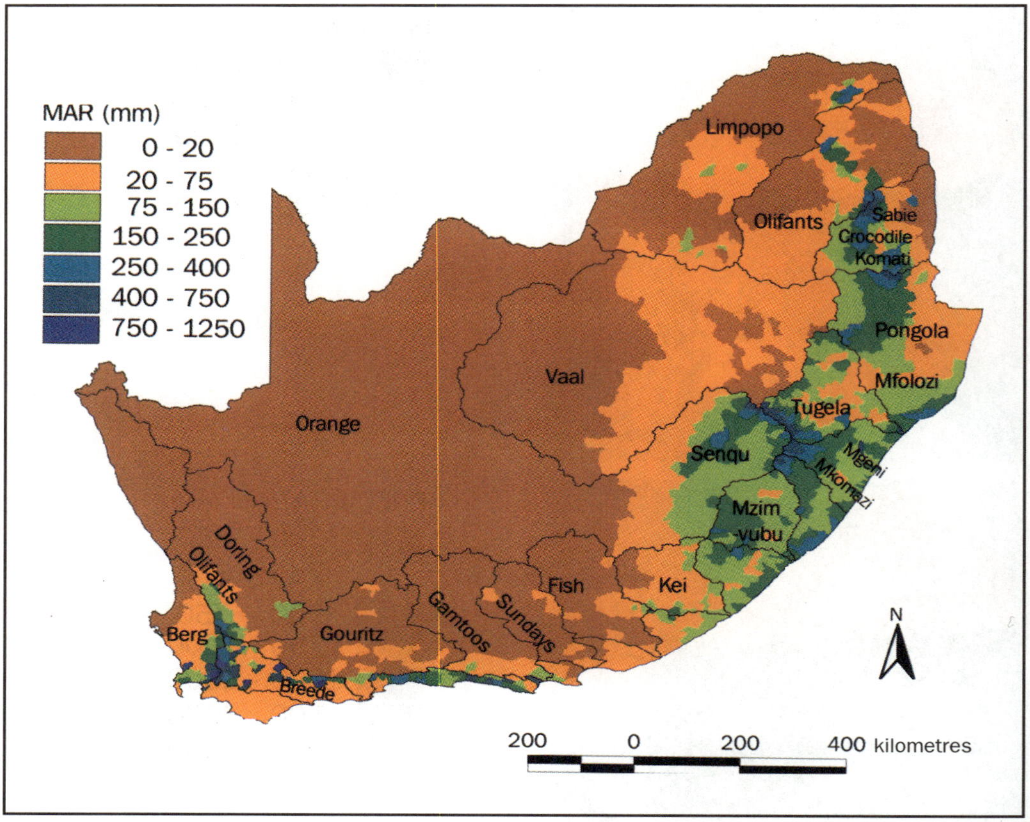

Plate 51 Distribution of mean annual runoff for South Africa

Plate 52 Clearing alien vegetation near Grahamstown as part of the DWAF-sponsored Working for Water Programme

Left-hand inset: sign board; right-hand inset: water seeping off a cleared slope

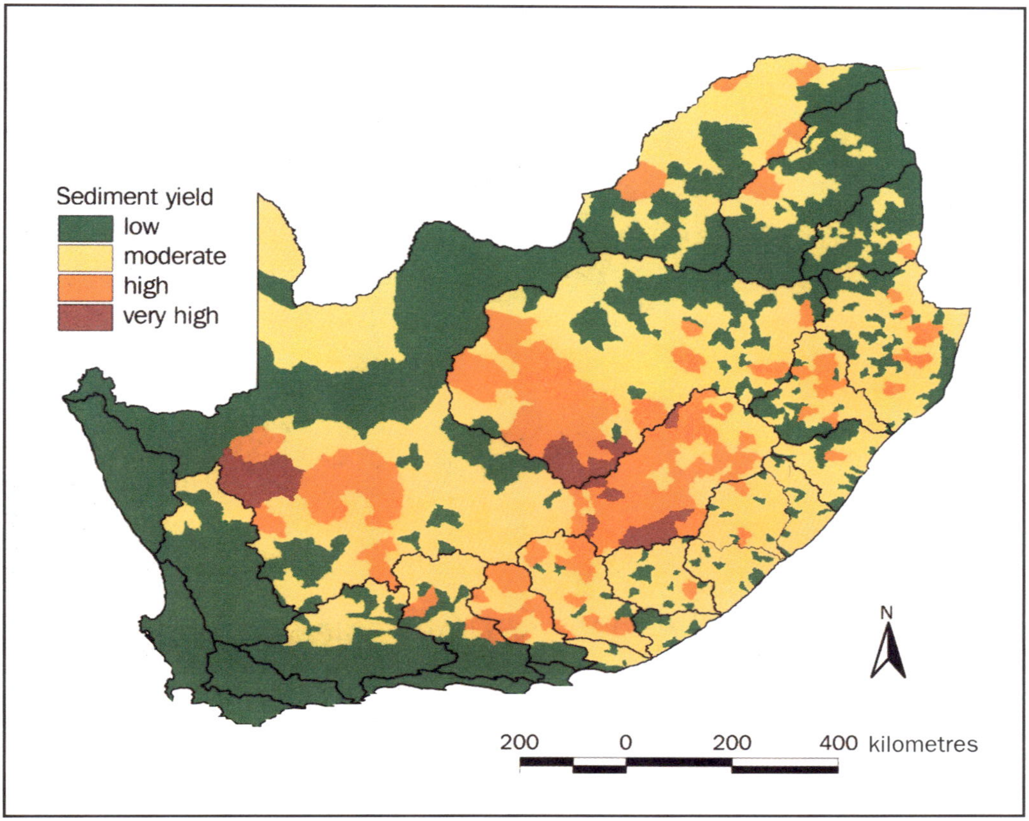

Plate 53 Distribution of sediment yield in South Africa by quaternary catchment

Plate 54 Riverside recreation in the United States of America

PHOTO: L. M. Magi

Plate 55 Riverside recreation in South Africa

PHOTO: L. M. Magi

SECTION III

Physical geography

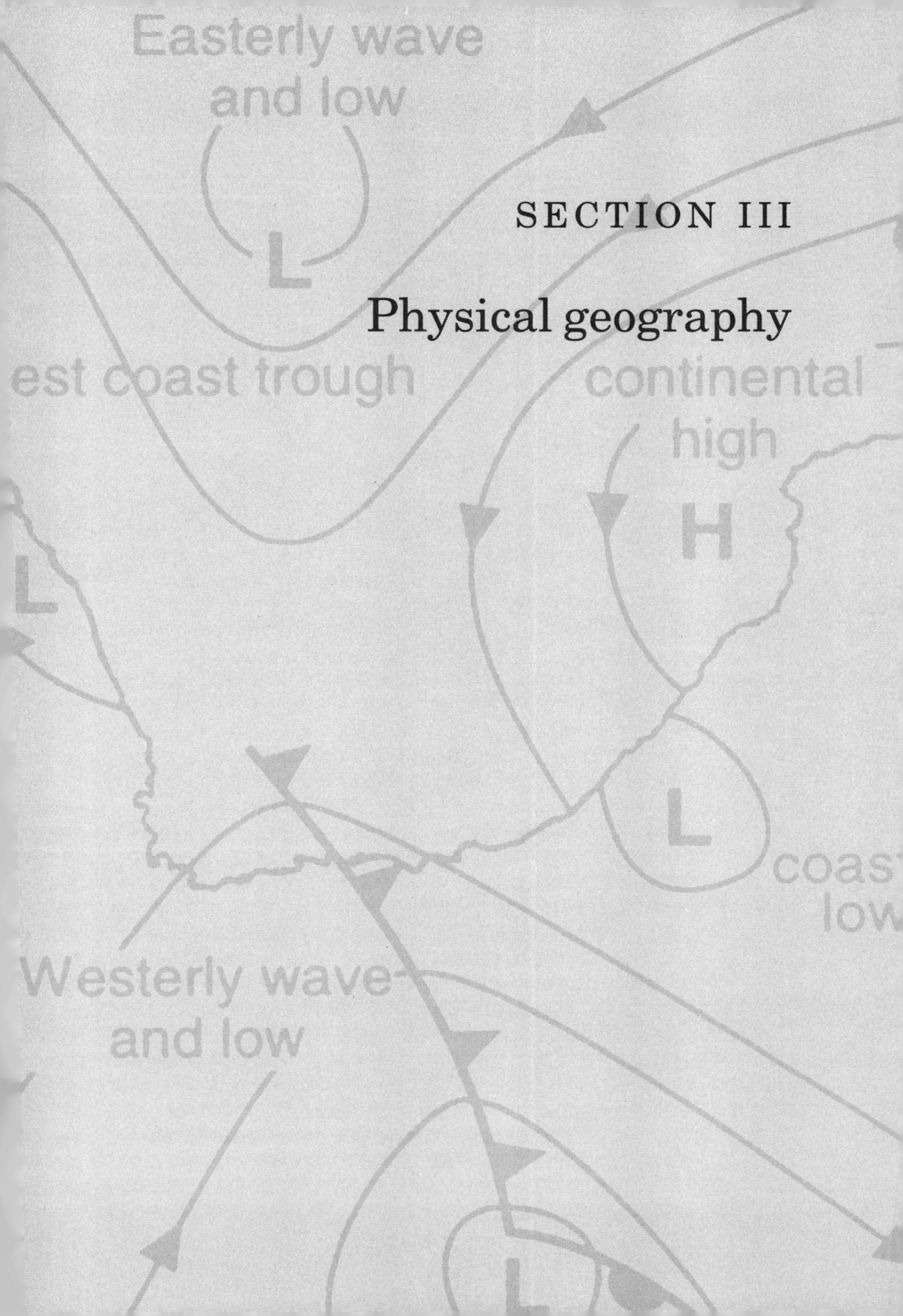

11 The geological template

PATRICK ERIKSSON

A knowledge of geology is important for an understanding of the geography of any country or region. The macro-scale landscapes are a direct result of geological processes, while at a local or meso scale landforms reflect the varying weathering and erosion rates of different rock types. Through geological time, macro-scale geological processes, the configuration of the continents and oceans, the composition of the atmosphere, and the world's climate have all been closely interlinked. Geological processes also determine the distribution of mineral resources and groundwater, as well as the incidence of natural hazards such as earthquakes, volcanic eruptions, and landslides. This chapter presents an outline of the most important geological events and processes on both the global and southern/South African scales. Perhaps because of the economic importance of its mineral wealth, South Africa has produced some world leaders in geology. The contributions of some of these scientists are noted where appropriate.

In a chapter such as this, the vast subject of geology and its relationship to landscape development in South Africa can only be touched upon. The reader is referred to a number of basic textbooks on the material covered here: Truswell (1977), South African Committee for Stratigraphy (1980), Strahler (1981), Tankard et al. (1982), Tarbuck and Lutgens (1987), Hamblin (1992), and Snyman (1996). In addition, many issues of the *National Geographic* magazine, published by the National Geographic Society in the USA, contain extremely well-illustrated articles on relevant aspects of global geology.

Table 11.1 provides a very simple rock classification scheme, outlining terms which are used in this chapter. *Igneous* rocks are products of molten magma derived from deep within the earth, while *sedimentary* rocks comprise debris weathered from existing rocks and transported by erosive surficial processes such as running water (river, waves, and tides), moving ice, wind, and gravity. *Metamorphic* rocks, the third major grouping, reflect the results of temperature and pressure on pre-existing lithologies (rock types); as a result of such processes,

TABLE 11.1 Simple rock classification scheme

Igneous (magmatic) rocks – crystallized directly from molten magma.

Can be subdivided according to two criteria: composition and texture (i.e. crystal size).

Texture:	
plutonic rocks	crystallize slowly deep in the crust and are coarse-grained
hypabyssal rocks	form at intermediate depths and are medium-grained
volcanic rocks (lavas)	solidify rapidly at surface and are fine-grained
Composition:	
acid rocks	typically are light-coloured, have significant amounts of quartz; feldspar minerals are predominantly of the potassic variety
intermediate rocks	have less quartz; potassic feldspar and plagioclase feldspar are about equal
mafic rocks	are darker and denser, have less quartz; plagioclase feldspar > potassic feldspar; in addition, dark pyroxene group minerals become important
ultramafic rocks	have the least quartz, minor feldspar and predominant pyroxenes, and, commonly, the mineral olivine

By combining these two criteria, rock classification and naming are achieved:

basalt	volcanic, mafic rock
andesite	intermediate, volcanic rock
tonalite and *trondhjemite*	intermediate, plutonic rocks
granite	plutonic, acid rock

Sedimentary rocks – transported weathering detritus which becomes lithified (clastic sediments), or chemically–biochemically precipitated material (chemical–biochemical sediments)

Clastic sediments:	
conglomerate	course-grained
sandstones	medium-grained
shales, mudstones	fine-grained
diamictites	composed of a predominantly muddy matrix carrying angular large clasts therein. A *tillite* is a diamictite with evidence of glacial transport of the large clasts (e.g. striations and facets thereon)
immature sediments	composed of easily weathered particles and course and poorly sorted grain sizes, implying short transport distances.

Chemical–biochemical sediments:

banded iron formation (BIF)	alternating cherty (silica) and iron mineral laminae/bands, formed most probably by a combination of subaqueous volcanism, chemical or biochemical precipitation, and physical sedimentation processes within early-Precambrian oceans as the earth's atmosphere became more aerobic.
dolomites	limestones (calcium carbonate), often precipitated by blue-green algae or cyanobacteria in the Precambrian era which, after burial, are dolomitized by the partial replacement of $CaCO_3$ by $MgCa(CO_3)_2$

Metamorphic rocks – result from pressure and/or temperature applied to existing rocks. Divided into the products of three types of metamorphism:

cataclastic	predominantly pressure; e.g. a breccia
thermal or *contact*	where heat from an igneous intrusion affects wall rocks
regional	combination of pressure and temperature

In addition, low to high grades are recognized in each of these categories:

gneiss	high-grade regional metamorphism (banded rock with dark and light bands)
amphibolite	medium- to high-grade regional metamorphism of a mafic rock; dark-coloured
quartzite	low-grade thermal or regional metamorphism of a quartz-rich sandstone

primary minerals may be recrystallized or new minerals may form, but the rock maintains most of its chemical character. Metamorphic rocks typically form when plate tectonic collisions (discussed below) occur.

Standard international stratigraphic terminology is used when referring to specific portions of the South African geological record: *supergroup*, *group*, and *formation*. These terms, ranked in descending order of magnitude, reflect stacked successions of layered or stratified rocks of sedimentary and/or volcanic composition. Igneous intrusions are generally termed *complexes*, while assemblages of metamorphic rocks are known as *suites*, *complexes*, or, if very large, *provinces*.

The global perspective

Beginnings

The present-day landscapes which we may observe anywhere on the earth are impermanent features (Strahler 1981).

Within the scale of geological time, the human chronological perspective is infinitesimally small. The geological record preserved on the earth goes back to about four billion years before the present (expressed normally as 4.0 Ga; Ga = giga-annum). The earth, the sun, and our solar system probably formed about 5.0 Ga ago, when cosmic debris condensed under the influence of gravity (Tarbuck and Lutgens 1987). Radioactive heat trapped within the early earth, combined with massive meteorite bombardments until about 4.0 Ga ago or even more recently, melted this material, resulting in an internal structure comprising, from the centre outwards, a solid nickel–iron core, a liquid nickel–iron outer core, the hot and convecting mantle, and, outermost, a relatively thin, essentially rigid crust, about 3–70 km thick (see Figure 11.1) (Hamblin 1992). It is on this outer crust that earthscapes have evolved, as will be outlined below.

Plate tectonics, mantle plumes, and volcanism

The theory of *plate tectonics* dates back to at least 1960, when Harry H. Hess of Princeton University first formally sug-

FIGURE 11.1 Internal structure of the earth

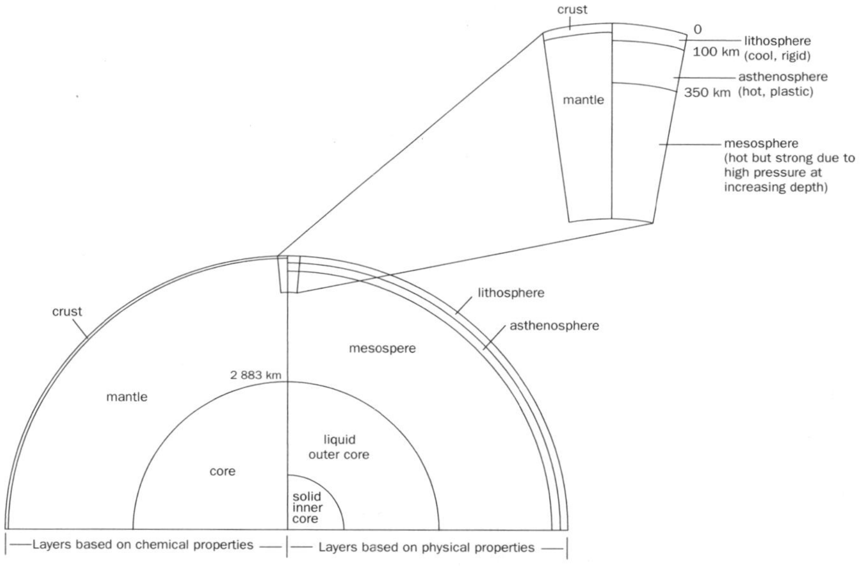

Note that the uppermost part of the generally viscous mantle (known as the asthenosphere) is more plastic and thus more mobile during mantle convection and that the lithosphere comprises both the rigid crust and the immediately underlying more solid portion of the asthenosphere. Mantle plumes ascend from the core–mantle boundary to lithospheric levels

SOURCE: adapted from Hamblin 1992.

gested the now widely accepted mechanism of sea-floor spreading. Among the very few earlier scientists who had supported the then-ridiculed idea of continental drift (a theory similar to plate tectonics) was Alexander du Toit, one of South Africa's most renowned geologists, who published a book on the subject in 1937. Modern research has elucidated how, through the process of plate tectonics, reasonably rigid crustal plates, which may be either continental or oceanic plates, move and collide with each other on the surface of the earth (see Figure 11.2). This process also extends deeper into the earth, as subducting oceanic crustal plates, which dive beneath other tectonic plates during collision, appear to sink slowly to the core–mantle boundary. Convective heat, returning from this same zone, brings hot mantle material to crustal layers in the form of large mantle plumes or smaller, more linear systems of convective upwelling. Plumes may impinge on ocean plates, continental plates, or their contacts. The interaction of the magmatic mantle processes and the crustal plate tectonic processes are the main large-scale forces shaping the landscapes on the surface of the earth. The importance of plumes has only recently become apparent. In 1997 a large international conference examining their nature and interaction with plate tectonics was held in South Africa ('Plumes, Plates and Mineralisation '97', University of Pretoria). Most of the leading scientists involved in this research were present, and the proceedings of the

FIGURE 11.2 The process of plate tectonics

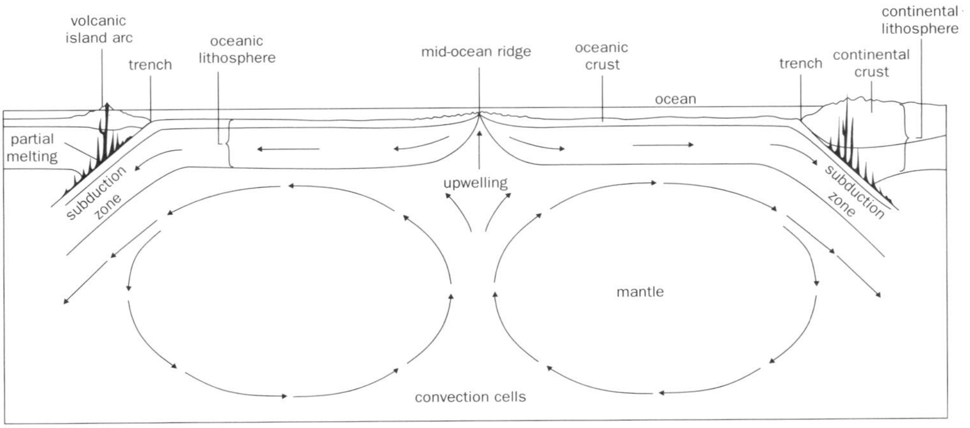

Note the upwelling of hot mantle, forming a mid-ocean ridge, and the descent of cooling oceanic crust beneath a continental plate (right-hand side) and an oceanic plate (left-hand side), with development of a subduction zone in both cases. Partial melting of the descending subducted plates produces magmas, which intrude into the continental crust and may also form extrusive lavas at the surface. Lavas above the oceanic plate subduction (left-hand side) extrude to build emergent island-arc chains

SOURCE: adapted from Tarbuck and Lutgens 1987.

meeting are published in the centennial volume (100/4) of the *South African Journal of Geology*.

Ultimately, the rocks making up the surface of the earth, the atmosphere that enables us to breath, and the waters covering about two-thirds of the earth were all derived from the cooling mantle during early geological time. An original helium–hydrogen atmosphere was lost to a 'sun-wind' (radiation). Due to the absence of a dense atmosphere, bombardment of the earth by comets, asteroids, and meteorites was common until about 4.5–4.0 Ga ago. Complex organic molecules arriving on the earth from these collisions most likely contributed to the origin of life. As the early earth cooled, a crust formed in the uppermost part of the mantle; this was most likely of basaltic composition (a fine-grained mafic or dark-coloured lava, derived directly from the mantle, which is predominantly ultramafic in its composition). The formation of this early basaltic crust was associated with large-scale volcanism and concomitant degassing of the mantle. The composition of the crust was equivalent to that of oceanic plates or ocean-floor crust in the younger plate tectonic systems which had yet to develop.

The main gas emitted by the volcanoes was water vapour (approximately 70 per cent), with lesser amounts of carbon dioxide, sulphur dioxide, and a host of other gases in small quantities. The water vapour collected and condensed from volcanic steam over many millions of years, forming the world's oceans; about 60 per cent of present ocean volume had accumulated by about 4.2 Ga ago. A second atmosphere had developed by about 3.8 Ga ago from the degassing, comprising mostly carbon dioxide, nitrogen, perhaps methane, and small amounts of other gases. At this stage, about 3.8 Ga ago, the basaltic crust was covered by the emerging oceans, above which the CO_2-rich atmosphere existed (Eriksson 1995). The latter was a 'greenhouse' atmosphere, trapping incoming solar radiation much as a modern greenhouse does for plants. This, combined with the large oceans, produced a climatic milieu characterized by warm, humid conditions with the potential to attack and weather any emergent rocks at a very high rate.

It is uncertain exactly when the process of plate tectonics began, nor how the early and thin basaltic crust reacted to the convective and plume-related magmatic processes in the underlying hot mantle. It is thought that by about 4.0 Ga ago a primitive form of plate tectonics had emerged, whereby convective upwelling from the mantle formed mid-ocean ridges within the crust, along which hot mantle material was erupted and long and high-standing mountain chains formed on the ocean floor. Due to a lower ocean volume than at present (about 60 per cent, as already noted), interaction between these high-standing ridges and water was unlikely to be common. As plate tectonics requires hydrated ocean crust to operate, further time was needed for more water to be released from the mantle and raise ocean levels, thus covering the mid-ocean ridges. Once this had occurred, by about 3.8 Ga ago, plate tectonics became fully operative, as shown in Figure 11.2. This model of the onset of plate tectonics is described by De Wit and Hynes (1995). Maarten de Wit, currently Professor of Geology at the University of Cape Town, is an international leader in modern research in this field.

Initially, the process involved the formation of discrete ocean plates as a result of mantle upwelling along the mid-ocean ridges. As upwelling continued, new hot oceanic crust thrust up at the ridges forced older and cooler basaltic crust away from these ridges. As this process was spread across the globe, and as ocean crust thus became subject to movement stresses, collisions between discrete moving ocean plates began. Amphibolitic rock types were formed through the process of collision of hydrated basaltic plates and their subjection to pressure, overriding, and burial. As the collisions continued, ocean plates became forced beneath other ocean plates, a process called *subduction*. The descending basaltic–amphibolitic plates were partially melted as they descended under pressure into the hot mantle. Magmas formed from this melting were extruded as lavas in volcanic island chains (island arcs) along the subduction zone contacts between colliding ocean plates, as shown in Figure 11.2. With time and continuation of these combined processes, tonalitic rock types formed at depth in the mantle and intruded into the island arcs; these more acidic rocks represent the beginning of the formation of continental crust. Subsidence and melting of tonalitic–amphibolitic crust formed granitic rocks, which intruded into the small continental plates, strengthening and enlarging them and making them more rigid. Small continental plates, known as *cratons*, were thus formed. The remains of the original island-arc terranes, preserved as deformed and metamorphosed remnants within the growing cratons, are called *greenstone belts* (De Wit and Ashwall 1996), due to the characteristic colour of the rock types preserved therein.

Cratons and continents

These emerging cratons, built up predominantly of granites, greenstones, and tonalitic gneisses, were less dense than the oceanic crust (plates). Oceanic crust, in turn, is lighter than the underlying upper mantle. Both continental and oceanic crustal plates thus effectively float on denser mantle rocks. The underlying mantle operates as a viscous fluid, with movement and displacement of mantle occurring over long geological time periods, rather than rapidly, as would be the case with a fluid such as water. The buoyancy of the crustal plates is affected by processes causing loading (addition of weight) and unloading (loss of weight). If a large and dense mafic intrusion, for example, is emplaced within a craton, the continental plate will subside due to the additional weight. A similar effect results from glaciation of a continent such as Antarctica due to the weight of the ice burden. Unloading resulting from long-term erosion of mountain masses leads to uplift of the continental plate. This buoyant adjustment is known as *isostacy*.

After 3.8 Ga and continuing until about 2.5 Ga ago, small initial cratonic nuclei gradually grew into larger continental plates, by accreting newly formed island arcs and by colliding with each other. The formation of continental crust peaked around 2.5 Ga ago, at the boundary between the preceding Archaean geological period and the following Proterozoic period (2 500 to 570 million years, or Ma, ago). This rapid crustal growth, together with an increased ocean volume (about 90 per cent of present ocean volume was achieved by about 3.8 Ga ago), caused ocean levels to rise. At the same time, aggressive weathering occurred, as a result of the greenhouse atmosphere. The late-Archaean–early-Proterozoic continents, emergent as a result of isostacy, were thus subject both to eustatic flooding and to down-wearing weathering, quite apart from the normal erosion processes caused by the action of running water, gravity, wind, and ice (see Figure 11.3). Under greenhouse conditions and a humid climate, it was probably too wet for the formation of true arid deserts, as found today, and, apart from a single global episode around 2.4–2.3 Ga ago, large-scale glaciation was unknown until near the end of the Proterozoic period. The first known desert sand seas (ergs) developed only about 1.8 Ga ago, once the greenhouse atmosphere had begun to change. These ideas are well summarized in a publication by K. A. Eriksson (1995), who, after studying at the University of the Witwatersrand, is now a professor at Virginia Polytechnic in the USA, and who is one of the editors of the journal *Precambrian Research*, the leading international publication on Precambrian geology.

Eustatic change is a worldwide change in sea level due to a rise or fall in the ocean, as distinct from uplift or subsidence of the land due to isostacy.

Early climatic change and supercontinents

Atmospheric change was largely the result of two factors, both related to the growth of continents. Marine flooding of

FIGURE 11.3 Processes acting on the early earth

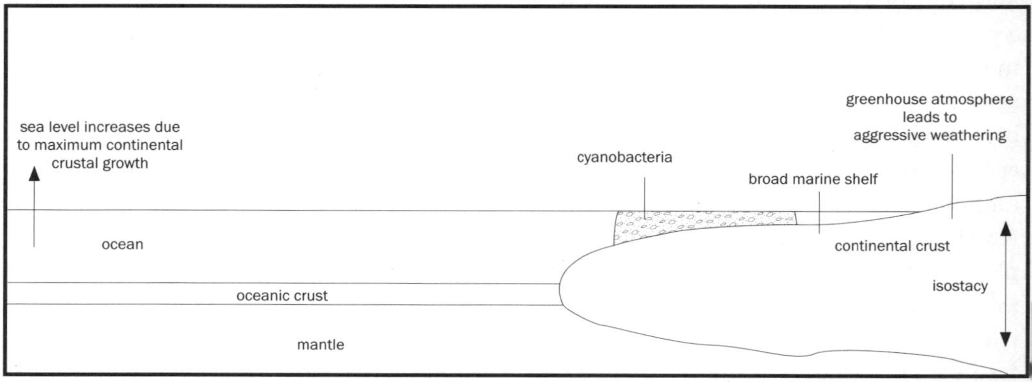

Approximately 2.5 Ga ago, with continental crustal growth being at a maximum, sea level rose and flooded large parts of isostatically emergent continents, forming broad continental-shelf environments. On the latter, large colonies of cyanobacteria became established and their photosynthetic consumption of carbon dioxide, allied to the production of oxygen, began the process of change from a greenhouse atmosphere to a more aerobic one. The greenhouse atmosphere aggressively attacked emergent rocks of the continents the early continents due to enhanced sea levels around the Archaean–Proterozoic time boundary promoted the growth of enormous colonies of cyanobacteria (blue-green algae), primitive early life forms which formed organoclastic mounds, consisting of clastic sediment grains trapped by these fibrous organisms (see Figure 11.3). These mounds, known as stromatolites, consisted predominantly of calcium carbonate. Their formation removed large quantities of CO_2 from the greenhouse atmosphere. At the same time, the cyanobacteria were photosynthetic, producing oxygen as a by-product of this natural nutritional process (Schopf 1983, 1992). (Bill Schopf, from the University of California, Los Angeles, is an international leader in research into the evolution of life; he follows in the footsteps of his equally famous father, who pioneered much of this early work.) Further drawdown of CO_2 from the atmosphere occurred on a large scale due to the aggressive weathering regime under the greenhouse climatic–atmospheric system then operative (see Figure 11.3). By about 2.0 Ga ago the atmosphere had begun to change significantly, gradually becoming more aerobic (carrying free oxygen). Climates during much of the subsequent Proterozoic period, until about 1 100–1 000 Ma ago, were dry and even partly arid. The first global glacigenic (glacier-forming) event, about 2.4–2.3 Ga ago, is poorly understood, but probably relates to a combination of loss of greenhouse CO_2, as described above, and upraised large continental land masses, or even the first supercontinents, which became elevated due to mantle plume activity. A recent review publication on early-Precambrian sedimentology provides an overview of the interacting variables affecting the early evolution of the earth (Eriksson et al. 1998).

Throughout geological time, world climates have largely been controlled by the proportion of land and sea (about 1:2) and by the arrangement of land masses in discrete continents surrounded by the various oceans. Wind and oceanic circulation patterns, critical for the development of weather patterns and thus also for landscape development, are subject to radical change by any alteration in the arrangement of land and ocean masses (Strahler 1981). Due to the ongoing process of plate tectonics, modified by the operation of mantle plumes, the continental plates are constantly subject to change over geological time periods. As smaller continental plates collide, supercontinents are formed, driven by processes along the mid-ocean ridges. Once a large supercontinent has formed, mantle heat becomes trapped beneath the large land mass, and new spreading centres develop due to mantle upwelling or plumes. The supercontinent thus breaks up again into discrete smaller continents, only to start the reassembly process once more (Hamblin 1992). The volume of continental crust has been approximately stable since about 2.5 Ga ago (Eriksson 1995) and, because of its low density, this material cannot be subducted back into the mantle. Existing continental crust is continually reworked, by tectonic collisions forming mountain belts or by mass-wasting processes leading to sedimentation within basins on land and in the sea. Unlike the oceanic crust, however, it is not lost to recycling in the mantle.

Although there is evidence of possible supercontinents during the late Archaean and earlier parts of the Proterozoic period, the oldest truly global-scale supercontinent was Rodinia, formed in the late Proterozoic. Rodinia began fragmenting about 750 Ma ago. The cycle was soon complete again, with the formation through plate tectonic collisions of the next supercontinent, Pannotia, about 600 Ma ago. With the break-up of Pannotia, coupled with increasing mid-ocean ridge development and enhanced spreading rates affecting ocean volumes, the sea level rose. The resultant shallow, warm seas which developed in many areas provided ideal conditions for an explosion in life forms about 600–500 Ma ago. During this time most of the predecessors of modern marine invertebrates emerged.

The fragmentation of Pannotia resulted in a large southern land mass, Gondwana, and scattered northern fragments. Africa formed part of Gondwana, which initially lay over the South Pole and which later drifted, through plate tectonic movement, into more temperate latitudes. Conglomeration of the northern fragments began, while Gondwana enjoyed a separate history. Around 300 Ma ago the next global supercontinent assembled, known as Pangaea. This stretched from pole to pole, with an internal ocean, the Tethys (later the Mediterranean), and was surrounded by the forerunner of the Pacific Ocean. The formation of the supercontinent increased ocean volume, causing sea levels to fall. Plants began to colonize the land. Falling sea levels, combined with changed ocean circulation patterns and most likely also with large-scale volcanism, caused an enormous-scale extinction event about 250 Ma ago, largely affecting marine life forms. Surviving reptiles on land adapted and flourished as the subsequent dinosaurs. Between about 200 and 140 Ma ago, Pangaea and its smaller Gondwana portion broke up, forming the discrete continents known today (this process is illustrated in Figure 2.2, p. 43).

Environmental implications

On a global scale, the development of macro-landscapes at any particular time is a result of plate tectonics and the concomitant distribution of land and sea masses. Uplift of the land is largely caused by tectonic collisions, either between two continental plates or between island-arc terranes and continents. The former process results in massive mountain chains such as the Alps and the Himalayas. Collision between island-arc terranes and continents usually forms more linear mountain belts and many active volcanoes, such as the circum-Pacific belt, which is characterized by seismically unstable islands (such as Japan and the East Indies). Volcanism and associated earthquakes also characterize most subduction zones, oceanic crust being thrust beneath continental plates, as illustrated in Figure 11.2. An example of the latter situation is found on the west coast of North America, where several ocean plates either are being subducted beneath the continent or are moving past each other with a strike-slip type of motion, as exemplified by the well-known San Andreas fault system passing through California. Earthquakes and their cataclysmic effects on human habitation are well known from San Francisco early in the twentieth century (1906), and in Los Angeles in recent times.

Further north, in the states of Oregon

and Washington and in Canada, similar potential for major geologically related disasters looms. In this region a complex system of faults and associated volcanoes exists, related to the descent of ocean crust beneath the continent. Movement rates of several centimetres per year mean that, on a geological time scale, earthquakes, volcanic eruptions and explosions, and tsunamis (tidal waves) will be common events in the region. From a human time perspective, such events are likely every several hundred to several thousand years. Apart from the obvious economic impact of such disasters, there are potentially larger effects for world climates. Enormous quantities of dust enter the atmosphere during volcanic eruptions, often blocking out sunlight and thus cooling climates. In addition, vast amounts of toxic gases, mostly sulphurous, enter the atmosphere. The 1991 eruption of Pinatubo in the Philippines, for example, added some 20 million tons of sulphur dioxide to the atmosphere. The circulation of this material in the stratosphere cooled the earth for two years.

The real impact of human-generated pollution on atmospheric processes and the world's climate is an as yet uncertain variable in natural systems. It would appear to have been significant since the Industrial Revolution began, but data are not yet adequate to provide a reliable quantification of our contribution to an enhanced greenhouse effect (see *National Geographic*, May 1998). The scale of natural disasters such as volcanic eruptions and the much longer-term natural carbon cycle on the earth may, on geological time scales, easily outweigh the effects of human activities. Nevertheless, the impact on life on earth could well be catastrophic within a short time interval.

Another factor affecting global climatic change is the variability in the earth's orbit known as the Milankovitch cycles (see Figure 11.4). These take three forms: a 100 000-year cycle of variation in the earth's orbit around the sun, a 41 000-year cycle of variation in the tilt of the earth's axis, and a 23 000-year cycle of variation around the rotational axis of the earth (a wobble effect, much like that of a spinning top). The mutual interference of these cycles affects their combined influences, which are further superimposed on the much larger-scale influences of plate tectonics and supercontinent amalgamation and separation. The unravelling of the earth's climates is thus a highly complex process.

The operation of climate on natural mass-wasting processes such as weathering and erosion is subject to much change over geological time. Landscape-forming processes such as desertification or glaciation, too, have not been constant variables. Prior to 2.4 Ga ago, no large-scale glaciation had yet occurred, and large deserts were apparently impossible before 1.8 Ga ago. Our present landscape and its potential for change on the human time scale are thus of limited significance when seen from a geological historical perspective.

The southern African perspective

The Kalahari craton

The African continent, the second largest on earth, is the most tropical of the land masses, and much of the geological record

FIGURE 11.4 The variability of the earth's orbit around the sun, referred to as Milankovitch cyclicity

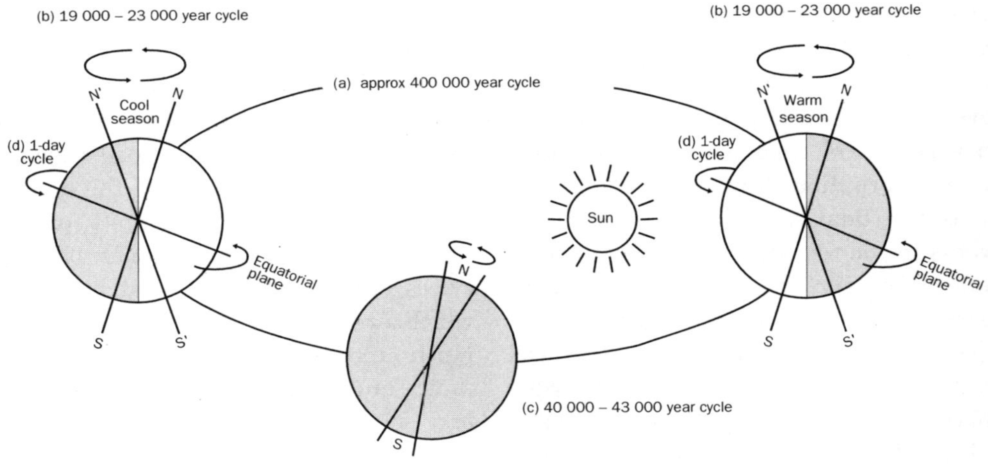

SOURCE: Pickering and Owen 1994.

is consequently buried beneath a covering of deeply weathered material and/or a verdant vegetation cover; in drier areas, planation by erosion has been widespread (Petters 1991). The Precambrian geology of Africa was characterized by well-defined cratonic terranes – the Kalahari, Tanzania, Congo, and West African cratons – as well as the poorly exposed East Saharan or Nile craton (see Figure 11.5). These cratonic terranes, comprising the so-called granite–gneiss–greenstone rock assemblage discussed above, are separated by what are known as *mobile belts*. These are younger terranes caused by plate tectonic collisions, characterized by highly deformed (folded, faulted, and sheared) and highly metamorphosed lithologies, and forming mountain belts prior to erosive downcutting (Hamblin 1992). Where continents are well exposed, the granite–gneiss (light-coloured rocks) and deformed greenstone remnants (dark-coloured rocks) in between are visible from satellite images, as are the mobile belts bounding and cutting across cratons.

Within southern Africa, the earliest cratonic plates were the Archaean Kaapvaal and Zimbabwe cratons, both of which began forming about 3.5 Ga ago and which were built up by the accretion of island-arc terranes with originally small cratonic nuclei (McCourt 1995). The Kaapvaal and Zimbabwe plates collided along the Limpopo mobile belt about 2.7–2.6 Ga ago, forming the larger Kalahari craton (see Figure 11.5). The Limpopo belt thus cuts across the middle of this early stable continental land mass, which is bound on all sides by younger mobile belts (see Figure 11.6). The Archaean plates that make up the Kaapvaal and Zimbabwe cratons are usually very rich sources of minerals of economic value. A number of late-Archaean–early-Proterozoic sedimentary basins were deposited on the Kaapvaal

FIGURE 11.5 Cratons and mobile belts of Africa

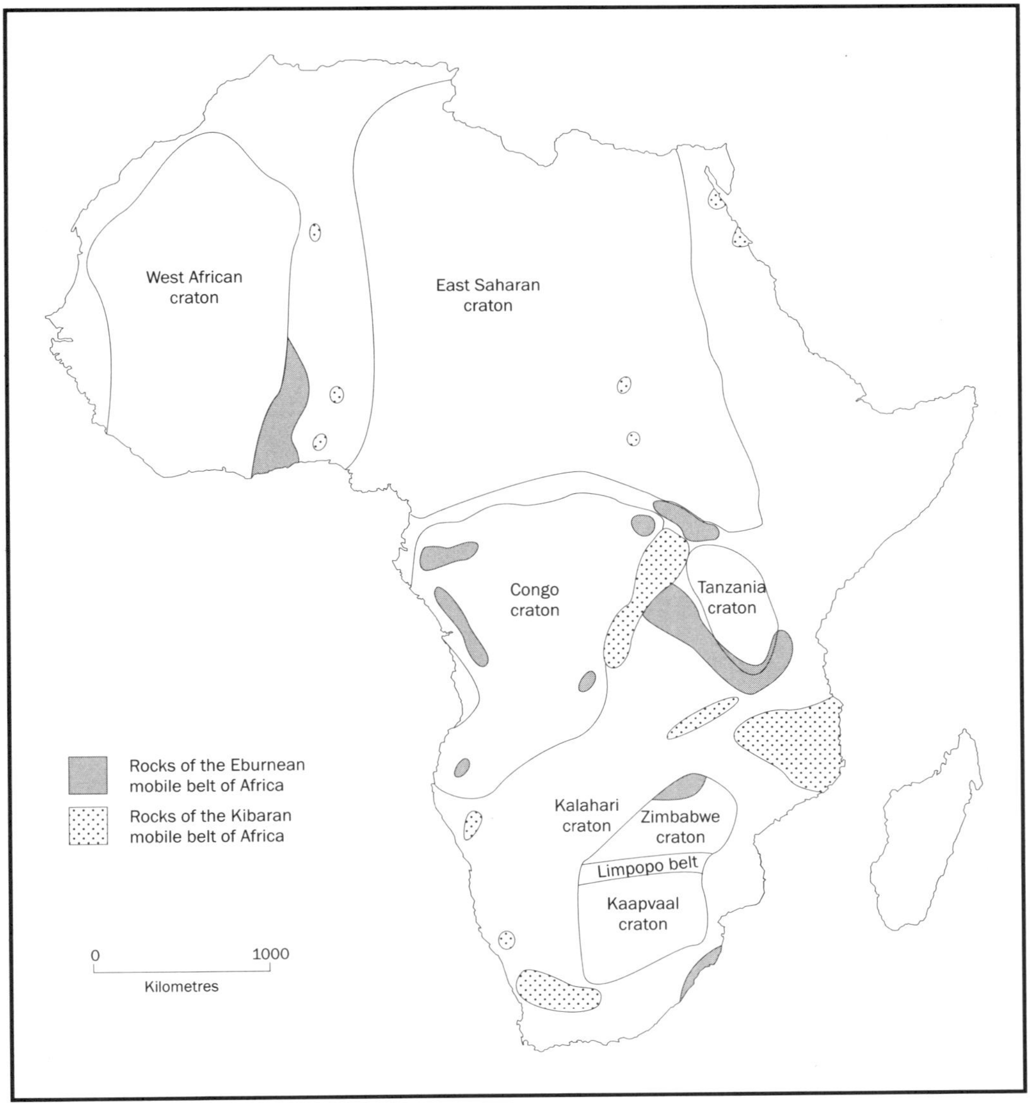

Note also the Eburnean and Kibaran rocks, reflecting mobile belt (tectonic collisional) events that affected large parts of the African continent. These cratons are shown in their present positions relative to modern Africa, but due to continental drift were not always in the same relative and absolute positions over preceding geological time

SOURCE: adapted from Petters 1991.

FIGURE 11.6 Crustal architecture of southern Africa

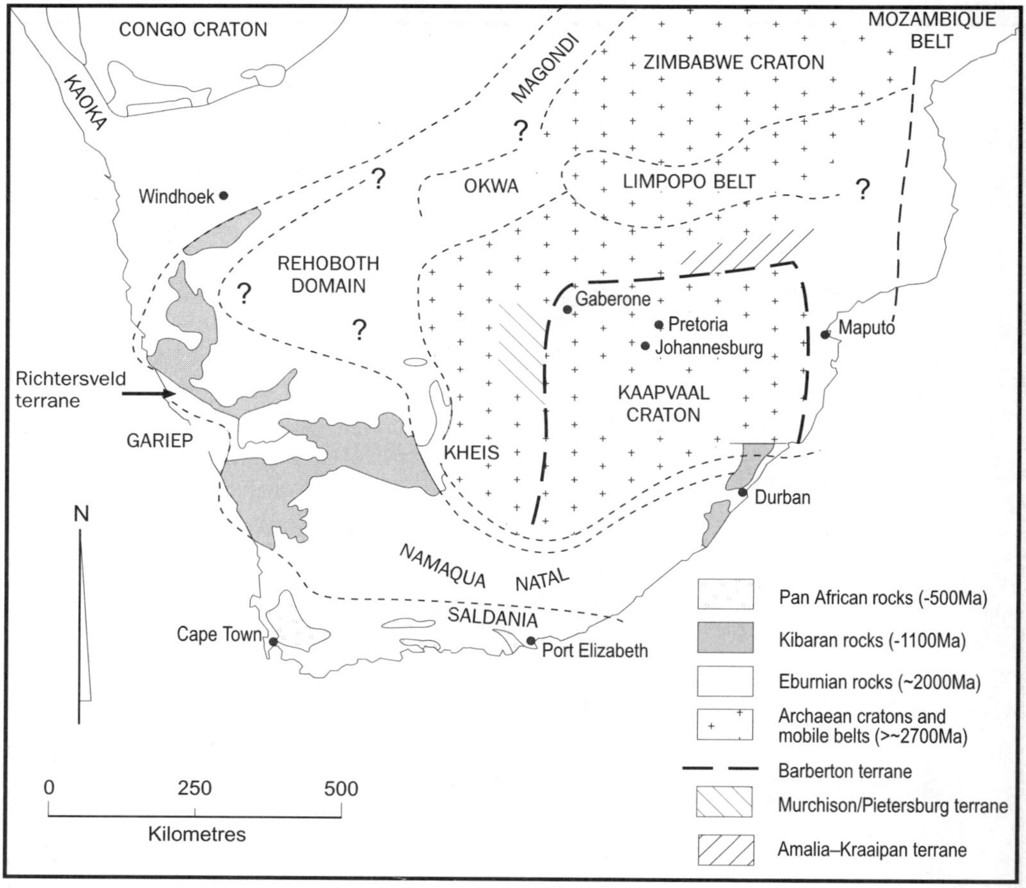

Note the Zimbabwe and Kaapvaal cratons which, together with the Limpopo mobile belt, comprise the Kalahari craton. Note also the ancient core of the Kaapvaal craton (the Barberton terrane) onto which the Murchison, Pietersburg, and Amalia–Kraaipan greenstone terranes were accreted about 2.9–2.8 Ga ago. Younger terranes were added later in geological time, including the Eburnean-aged (2.0 Ga) Richtersveld terrane, the Kheis mobile belt, and the Magondi mobile belt, as well as the Kibaran (about 1.1 Ga) Namaqua–Natal mobile belt

SOURCE: adapted from McCourt 1995.

cratonic terrane from about 3.0 to 1.7 Ga ago. These include the world's richest gold deposit, the late-Archaean Witwatersrand basin. In addition, enormous layered intrusive bodies of igneous rocks were emplaced into both the Kaapvaal plate (the Bushveld Complex, around 2.05 Ga ago) and the Zimbabwe plate (the Great

Dyke, around 2.85 Ga ago). These two intrusive bodies contain the world's greatest reserve of platinum and chrome, as well as significant quantities of vanadium and many other minerals.

Growth of the Kalahari craton

Figure 11.6 shows the geological architecture of South Africa's major geological elements which arose from the growth of the Kalahari craton. The Murchison, Pietersburg, and Amalia–Kraaipan greenstone terranes were accreted onto the ancient core of the Kaapvaal craton (the Barberton terrane) about 2.9–2.8 Ga ago. Younger terranes were added later in geological time, including the Eburnean-aged (2.0 Ga) Richtersveld terrane, the Kheis mobile belt, and the Magondi mobile belt, as well as the Kibaran (about 1.1 Ga) Namaqua–Natal mobile belt. The Pan-African orogenic (mountain-building) event (about 0.9–0.6 Ga ago) resulted in the addition of the Saldanian belt, south of the growing Kalahari craton, the small Gariep belt along its western edge, and the Mozambique belt to its east, and also reactivated many earlier mobile belts.

The Gondwana supercontinent

By about 600 Ma ago, as the supercontinent Pannotia broke up, the southern large land mass of Gondwana remained. This comprised Africa, South America (to its west), India and Antarctica (to its east-south-east), and the Falkland (Islands) plate (to its south) (see Figure 11.7). Following the Pan-African orogenies, widespread tectonic stability was attained over much of Africa. Subsequently, the Cape Supergroup and younger Karoo Supergroup basins formed in South Africa as a result of renewed tectonic activity in southern Gondwana. These rocks are preserved in many parts of the Gondwana supercontinent. The break-up of Gondwana between about 180 and 135 Ma ago (Thomas et al. 1993) led to the formation of the discrete continents known today, separated by vast oceans. Southern African landscapes owe a great deal of their development to this event, and modern climates are strongly related to the present distribution of the land masses and oceanic areas in between.

Seismic activity

The break-up of Gondwana initiated many processes in southern Africa, including seismicity, faulting, hot springs, uplift of the margins of the subcontinent (due to isostatic adjustment after break-up), and slumping of sediments offshore of the newly formed continent (Thomas et al. 1993). The East African Rift, a feature as ancient as the Precambrian era but more active in post-Gondwana times, extends southwards into a zone of active seismicity, running along the eastern margins of the Zimbabwe and Kaapvaal cratons to the escarpment area around Swaziland; a southern seismically active zone coincides with the southern margin of the Kaapvaal plate (Thomas et al. 1993). These seismically active zones pose no significant known threat to modern society, and non-mining-related seismicity is generally low in southern Africa. Isostatic adjustment and denudational unloading of the margins of the subcontinent, however, have led to localized seismic activity and even earthquakes, as for example at Ceres on 29 September 1969 (Partridge and Maud 1987; Snyman 1996).

FIGURE 11.7 An approximate reconstruction of the Gondwana supercontinent, showing only major component terranes

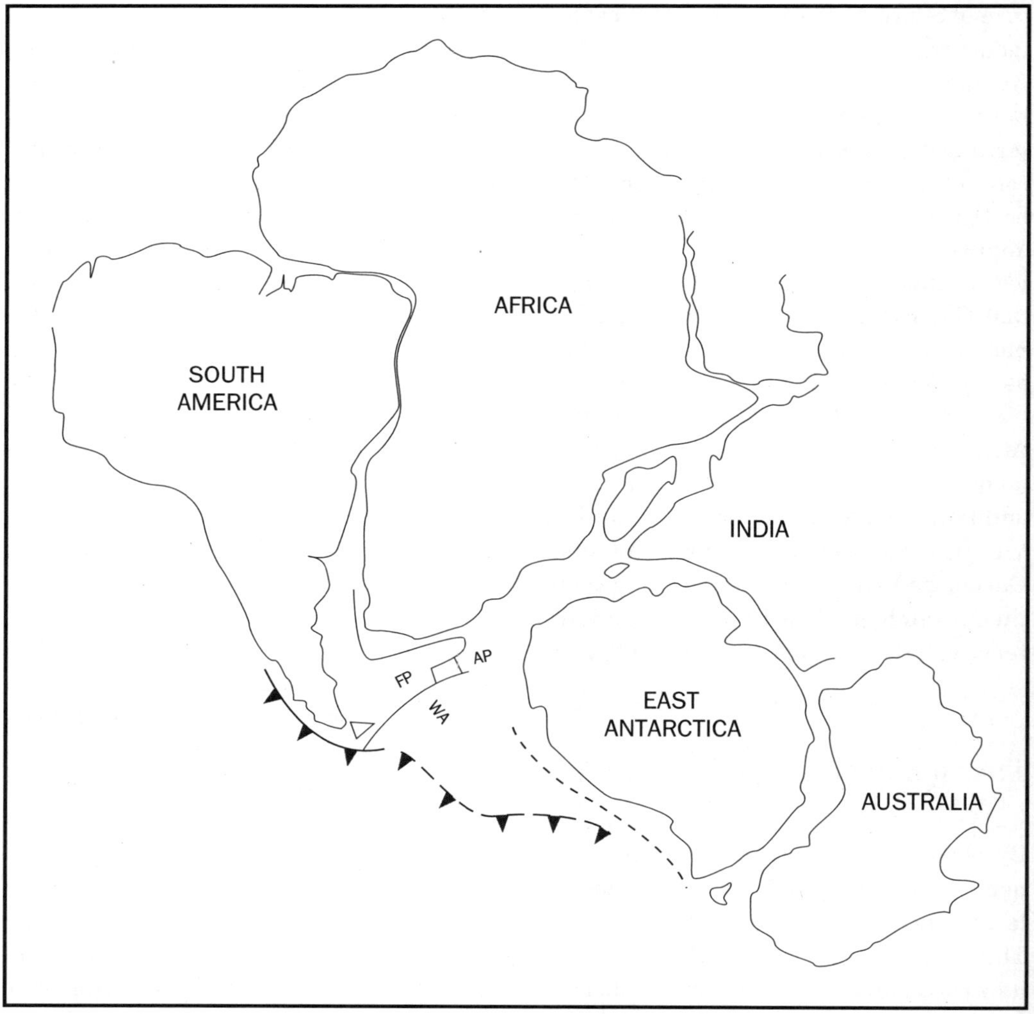

SOURCE: adapted from Partridge and Maud 1987, Figure 15.

The South African perspective

A brief review of the geological history of South Africa is offered below, noting predominant rock types, their genesis, and their relation to landscape formation. Associated economic aspects and environmental problems are discussed after this. The reader is referred to Plate 27 (colour section), which shows the geology of South Africa.

Formation of the continental crust

As noted previously, the geology of South Africa began with the formation of the Kaapvaal craton, composed predominantly of gneissic rocks (trondhjemitic–tonalitic, light-coloured, frequently banded crystalline rocks), with highly deformed greenstones (metamorphosed lavas and sediments representing remnants of island-arc systems) and intrusive granite plutons (roughly mushroom-shaped masses of granitic rock intruded at a relatively high crustal level). Together, these lithologies comprise a continental crustal basement. The core of this craton, shown as the Barberton terrane in Figure 11.6, formed as a result of plate tectonics between about 3.7 and 3.2 Ga ago. Tectonic stability of this terrane was established by about 3.1 Ga ago. Between 3.1 and 2.65 Ga ago, collisional additions increased the size of the core terrane, forming the larger Kaapvaal craton. Around 3.0–2.9 Ga ago the Murchison and Pietersburg greenstone terranes accreted onto the northern part of the core shield, while the western Amalia–Kraaipan greenstone terrane collided with it about 2.8 Ga ago. These accretionary events were followed by extensive granitic intrusions which thickened and strengthened the Kaapvaal craton (McCourt 1995). Between about 2.7 and 2.6 Ga ago, during the Limpopo orogeny, the Kaapvaal and Zimbabwe cratons collided and became sutured, resulting in the Limpopo mobile belt. This belt comprises highly deformed and very high-grade metamorphosed basement rocks, overlying shallow marine sediments and intrusive complexes of diverse composition. The basement rocks of the Kaapvaal craton (see Plate 27) and Limpopo mobile belt tend to be deeply weathered, despite being composed predominantly of hard crystalline lithologies which are resistant to weathering. Their landscapes are thus mostly low-lying, with resistant greenstone-belt rocks commonly forming either low ridges or, as at Barberton, a mountainous landscape. The granites commonly form rounded hills covered by large boulders.

Earliest basins – Witwatersrand, Pongola, and Ventersdorp

During the final accretion and stabilization of the Kaapvaal craton, the oldest supracrustal rocks belonging to the Witwatersrand and Pongola Supergroups were formed, around 3.1–2.8 Ga ago. Both were formed by early thick lavas, followed by shallow-water marine sediments, with evidence of fluvial and deltaic coastline sedimentation. The two supergroups likely represent the remnants of a single large epeiric basin, formed as the sea transgressed onto the still-evolving craton (Stanistreet and McCarthy 1991). The collisions of greenstone terranes noted above (between 3.0 and 2.8 Ga ago) affected this large basin, leading to an uppermost regressive fluvial braidplain environment, during which the famous auriferous (gold-bearing) quartzites and conglomerates of the upper Witwatersrand Supergroup were laid down. Numerous South African geologists have attained international recognition for their research on the Witwatersrand geology.

An **epeiric basin** is a shallow sea located on a continent and linked to a large ocean.

A **fluvial braidplain** is a sandy river plain characterized by a braided channel pattern.

Thermal subsidence of the epeiric basin, coupled most likely with the impingement of a mantle plume on the base of the central Kaapvaal craton, led to rifting and the vast outpouring of the Ventersdorp andesitic lavas approximately 2.7 Ga ago. Graben sediments (sediments deposited within a graben or rift valley) also formed within the Ventersdorp rift systems, with semi-arid deposition on alluvial fans and fluvial braidplains and in small ephemeral lakes (Van der Westhuizen et al. 1991). The Ventersdorp lavas erupted mainly to the west-south-west of the former Witwatersrand basin, as shown in Plate 27. The ancient rocks of the Witwatersrand, Pongola, and Ventersdorp successions generally outcrop in limited areas and tend to form subdued ranges of hills. The Witwatersrand quartzite ridges dominate the local topography within the Greater Johannesburg region, and the Pongola rocks form broken country in Swaziland and northern Kwazulu-Natal.

Transvaal sedimentation and Bushveld intrusions

The sedimentation of the Transvaal Supergroup began sometime between 2.7 and 2.6 Ga ago, contemporary with the lava eruptions of the late-Ventersdorp and Limpopo mobile belt collision. A large proportion of the sediments of the Transvaal Supergroup were laid down in a single sedimentary basin, but two million years of weathering and erosion have reduced this to three remnants, namely the Transvaal and Griqualand West basins in South Africa and Kanye in Botswana. Today the Transvaal Supergroup attains the maximum preserved thickness of about 15 km.

The sediments of the Transvaal Supergroup represent a long and varied sequence of events. Initial rift basins related to the Limpopo event formed at the base of the Transvaal depository (Eriksson and Reczko 1995). These were filled with lavas and immature fault-related sediments. Planation and concomitant thin fluvial sheet sandstones in the Transvaal basin were approximately coeval (contemporary) with shallow marine coastal deposits and minor lavas in Griqualand West. Following this, about 2.5 Ga ago, elevated sea levels due to continental crustal growth and thermal subsidence, in part related to Ventersdorp fault systems and the Limpopo collision, together enabled transgression of the Transvaal epeiric sea over much of the Kaapvaal craton. Warm and shallow marine conditions led to vast colonies of cyanobacteria, forming stromatolitic (algal-laminated) dolomites, succeeded by banded iron formations (BIF) about 2.43 Ga ago. BIF are iron-rich sedimentary rocks formed by the unique atmospheric conditions then pertaining (Eriksson and Reczko 1995; Eriksson et al. 1998). These rocks provide vast reserves of iron ore in South Africa. The Transvaal dolomites have been studied by many renowned international scientists such as W. J. Schopf and W. Altermann, who carried out post-doctoral studies at Stellenbosch University, as well as well-known South African geologists such as K. A. Eriksson and N. J. Beukes. South African research on the Transvaal carbonate rocks and BIF has contributed significantly to internationally accepted models of early-Precambrian chemical sedimentation, atmospheric change, and the evolution of early life.

Sandstones, shales, and andesitic lavas succeeded the carbonate–BIF deposits in both basins after an erosional hiatus. Localized glaciation, associated with the first such global event between 2.4 and 2.3 Ga ago, was concentrated in the Griqualand West basin (Eriksson and Reczko 1995). In the Transvaal basin, an alternation of transgressive marine (epeiric) environments and continental fluvial braidplains persisted until about 2.1 Ga ago (Eriksson and Reczko 1995).

The sedimentation of the Transvaal Supergroup was terminated when a second major mantle plume beneath the Kaapvaal craton resulted in the emplacement of the world's largest mafic layered intrusion, the Bushveld Complex, into the Transvaal basin about 2.05 Ga ago (Von Gruenewaldt and Harmer 1993). Acid lavas and vast granitic intrusions were associated with this event, as well as an analogous intrusive complex in Botswana: the Molopo Farms Complex. The layered rocks of the Bushveld Complex contain the world's single richest source of platinum-group minerals, chromite, and vanadium. The Bushveld Complex is thus unique in a global context. As a result of research into this complex by such South African luminaries as J. Willemse and G. von Gruenewaldt, the Department of Geology at the University of Pretoria has long been acknowledged as a centre of world-class igneous petrological research.

The rocks of the Transvaal Supergroup dip inwards beneath the saucer-shaped Bushveld Complex, and resistant Transvaal quartzites and mafic lithologies of the Bushveld form uplifted ridges around a central basinal landscape, stretching for about 350 km east–west and 150 km north–south. The central granites are resistant to weathering and have formed the Nebo Plateau in the east of this basin. Further east in Mpumalanga, westerly dipping Transvaal and Bushveld lithologies have been deeply eroded, forming an impressive escarpment. In Griqualand West, the topography is generally much more subdued, forming relatively low-lying ridges of Transvaal rocks. Folded ranges of hills occur in the south-west, in the Prieska region.

Waterberg Group and Namaqua–Natal mobile belt

Approximately 2.0 Ga ago, the Richtersveld cratonic terrane and the Kheis mobile belt were accreted onto the western Kaapvaal craton during what is known as the Eburnean event. Plate tectonic collisions associated with this event also had effects within the Kaapvaal craton. Transmitted collisional stresses formed two rifted basins, one north-west of the Bushveld Complex (the Main Waterberg basin) and one to the east of Pretoria (the Middelburg Waterberg basin). The approximately coeval Soutpansberg rifted basin formed to the north-east of the Main Waterberg basin. To the west, the Olifantshoek succession of shallow marine sediments and volcanics, which were subsequently metamorphosed, formed over the previous Griqualand West basin. The Olifantshoek rocks were locally tectonically thrust over earlier andesitic lavas of the Griqualand West basin, and contain enormous manganese deposits.

Waterberg rocks comprise predominantly red-coloured sandstones and conglomerates, associated with a range of sedimentary environments including rivers, alluvial fans, lakes, and the first large deserts in South Africa (Callaghan et al. 1991). The red colouration of these rocks reflects the change to an oxidizing atmosphere by about 1.8 Ga ago, and the changing global climate enabled the formation of the first ergs. The Soutpansberg succession is analogous, but also includes large volumes of lava. The Main Waterberg basin forms a resistant emergent plateau landscape in the north of the country near Elisras, north of Pretoria. The Soutspanberg rifted basin forms a prominent mountain range to the north-east of this area.

Around 1.2 Ga ago, major accretion along the southern margin of the Kaapvaal craton and Richtersveld terrane occurred, resulting in the formation of the Namaqua–Natal mobile belt (Thomas et al. 1993). This assemblage of highly deformed and metamorphosed rocks (altered sediments, lavas, gneissic basement, and intrusive granites) builds the basement to the Namaqualand and Kwazulu-Natal regions (see Plate 28). Deeply eroded valleys have been cut into the latter, while the former comprises rolling hills in the south and a more broken and rugged terrain in the north. During the Pan-African orogenic event of 900–600 Ma ago, the accretionary Saldanian belt in the south, the Mozambique belt in the east, and the Gariep belt in the west completed the assembly of the South African basement rocks.

Gondwana supercontinent: Cape and Karoo Supergroups

As previously discussed, South Africa formed part of the Gondwana land mass by about 600 Ma ago. From about 450 to 370 Ma ago, the Cape Supergroup was laid down in the Eastern and Western Cape provinces, extending eastwards along the coast into Kwazulu-Natal. This supergroup extended also into adjacent parts of the Gondwana supercontinent and comprises predominantly quartzitic sandstones and shales, laid down in littoral and continental environments. During subsequent Karoo times, and due to the development of a low-angle subduction

> A **Littoral zone** is the area of a shoreline between the high- and low-water marks; deposits within the littoral zone generally consist of sand, shells, and shingle.

zone south of Gondwana, compressional stresses were transmitted inland into the south of South Africa (Thomas et al. 1993), forming the Cape Fold Belt and its mountains (see Plate 29).

From about 300 Ma to approximately 190–180 Ma ago, the Karoo Supergroup was laid down over vast parts of South Africa and neighbouring southern Africa (Smith et al. 1993), as well as over much of the rest of Gondwana. A number of separate basins are preserved in South Africa. The Karoo Supergroup dominates the geological map of South Africa, covering a very large proportion of the country. This reflects its relatively young age and the as yet limited time there has been for its reduction by weathering and erosion. Earlier basins of sediments, such as the Transvaal Supergroup, probably attained much the same size as the Karoo depository, but have not survived the passage of time at their original scale of formation.

The Karoo Supergroup, a 9 km-thick succession of alternating sandstones and shales, began with glacial diamictites, formed when this part of Gondwana lay near the South Pole. The so-called Dwyka Glaciation was related to ice sheets much larger than the Karoo basin itself, surrounding it to the north, west, and east. The glaciation was essentially glaciomarine in the south of the Karoo basin, glacifluvial in the north, with tillites in between reflecting mainly sub-glacial deposition. The sedimentary rocks associated with this glaciation are known as the Dwyka Group. The glaciated landscape left behind by the retreating ice was characterized by a shallow basin in the south and incised glacial valleys in the north of the Karoo depository.

With tectonic plate movement (continental drift) away from the polar location and towards increasingly temperate latitudes, Karoo environments changed gradually from locally shallow marine and more common fluviodeltaic settings to semi-arid floodplains, uppermost braided river systems, and arid erg deposits (Smith et al. 1993). The shales and sandstones of the Ecca Group were deposited initially under shallow marine conditions in the south, in the remnant Dwyka basin, with subsequent deeper-water suspension deposits and even turbidites following. Towards the north and north-east, advancing river and deltaic complexes formed large coal swamps, partly controlled by pre-existing glacial valley palaeotopography. This Ecca sedimentation graded upwards into the Beaufort Group, whose shales and sandstones were deposited on enormous semi-arid river

Glaciomarine processes and landforms are related to glaciers melting in a marine environment.

Glacifluvial processes and landforms are related to the action of channelized glacial meltwater.

Fluviodeltaic sediments are laid down in associated river and delta depositional environments.

Turbidites are sediments deposited in water by a turbidity current – a submarine flow of a dense slurry of water and sediments which can transfer large volumes of sediments into the deeper zones of the oceans. They may also occur in deep lakes. A greywacke is an example of a turbidite.

plains, subject to strong seasonal variations and populated by large numbers of reptiles; uplifted fluvial source areas lay to the west, south-east, and north-east of the basin (Smith et al. 1993). Subsequently, the Karoo basin shrank, with the depositional centre migrating north-eastwards. Molteno Formation sandstones were laid down, distal to fan systems developing along a southern uplifted mountain range (related to the Cape Fold Belt). As aridity increased, this environment changed to one characterized by river systems and dry floodplains subjected to wind deflation (the Elliot Formation), followed by proper desert conditions for the overlying Clarens Formation (Smith et al. 1993). David K. Hobday, who obtained his early training at the University of Natal, Durban, and who did much pioneering research on these Karoo Supergroup sediments, was largely responsible for the introduction of modern sedimentology to South Africa.

The thick Drakensberg basalts (mafic lavas in the main Karoo basin, largely acidic lavas in the correlated Lebombo basin – see Plate 27) terminated Karoo sedimentation and were associated with the break-up of the Gondwana supercontinent. Kimberlite pipes, some of them diamondiferous, were a related deep-seated magmatic effect of the fragmentation of Gondwana in the subsequent Cretaceous period (144 to 66 Ma ago).

Post-Gondwana geology

South Africa existed as part of a discrete African continent after the fragmentation of Gondwana. Break-up along the east coast of the country occurred about 180–160 Ma ago. South America on the west coast separated approximately 135 Ma ago, and strike-slip rifting occurred along the southern coast of South Africa at this time (Thomas et al. 1993), forming a series of offshore fault-bounded basins on what is today the Agulhas Bank (continental shelf). Isostatic uplift of the country resulted in a high-lying continental interior, and rifting and readjustment along the emergent coastlines created faulted basins during the Cretaceous period. These became filled with sediments of continental and shallow marine origin and minor lavas. The post-Gondwana uplift was a major geomorphological event, leading to, among other things, the formation of the Great Escarpment of southern Africa (Partridge and Maud 1987). The latter, largely comprising deeply eroded Karoo rocks along the Drakensberg ranges and Lesotho highlands, extends further north to encompass the escarpment along the eastern margin of the Transvaal–Bushveld basin, and even further north into eastern Zimbabwe. The development of the large interior Kalahari basin and the major river drainage systems of South Africa took place in the Tertiary and Quaternary periods, which followed after the Cretaceous period. (For an outline of the geological time scale and associated environmental changes in southern Africa, see Table 2.4, p. 42.)

Much of the present-day landscape of South Africa thus owes its origin to geological and geomorphological events in the post-Gondwana period. Older landscapes, however, remain partially preserved within the rock record of the country, and, where fortuitously exposed by denudation, they will become visible to the trained eye.

Examples of pre-Karoo landscapes include the Kuruman hills and Langeberg mountains in the Northern Cape province, as well as the Witwatersrand (Supergroup) quartzite ridges around Johannesburg and the ridges of Magaliesberg quartzite (Transvaal Supergroup) in Pretoria (Snyman 1996).

A number of geomorphological cycles are recognized in South Africa from post-Gondwana or post-Karoo times, the first such landscape being about 190 Ma old. Each cycle began with a tectonic–isostatic upliftment (and/or deformation) event, leading to renewed denudation and the formation of planated internal landscape surfaces, with enhanced erosion and valley formation along the margins of the internal plateau and concomitantly thicker deposits of the resultant detritus along the low-lying coastal margins (Partridge and Maud 1987).

Although there has been, and indeed still is, considerable controversy about these cycles, a widely accepted summary of post-Gondwana geological events and their geomorphological results is presented by Partridge and Maud (1987) in a seminal paper in the *South African Journal of Geology*. The Great Escarpment, formed after the rift-related break-up of the Gondwana supercontinent, survived to the present due to the high-standing nature of the African plate and, particularly, of southern Africa. This most likely reflects not only a thick continental crust, but possibly also enhanced heatflow in the underlying mantle. Within the general geomorphology of an uplifted interior, passing through the escarpment into lower-lying coastal plains, landscapes at different levels yet approximately coeval came into being. Partridge and Maud (1987) interpret a single erosional cycle from the break-up of Gondwana until the early Miocene period (24 to 5 Ma ago), resulting in the gentle African surface pediplain which extended across much of southern Africa at elevations of about 500–600 m.

Most of the erosion and scarp retreat occurred soon after the fragmentation of Gondwana, as isostatic adjustment fractured the coastal margins of the newly emerging African continent. The products of this erosion produced the thick Cretaceous sedimentary fills of fault-bounded coastal basins such as the Algoa basin in the Port Elizabeth area. Partridge and Maud (1987) infer that by about the Cretaceous–Tertiary boundary (66 Ma ago) South Africa comprised a vast undulating plain, mostly 500–700 m above sea level, with infrequent erosional residuals and elevated massifs. A second cycle of erosion, known as the Post-African I, followed uplift of approximately 150–300 m during the Miocene. This continued into the late Pliocene period. A new planation surface formed at no more than 100–300m below the African surface, but was not fully developed due to the short duration of this second cycle. Offshore sedimentation rates were significantly below those attained during the Cretacious period. Major late-Pliocene uplift initiated the Post-African II cycle, with up to 900 m of elevation of the eastern interior of southern Africa contrasting with subdued uplift along the western and central interior regions. Large-scale monoclinal warping affected

A **massif** is a clearly defined mountain mass with fairly uniform geology and structure.

Kwazulu-Natal, and deep incision of interior rivers and coastal hinterlands occurred; pre-existing surfaces became highly dissected, especially in the south-east of South Africa, but planation was limited and occurred mainly near the coast.

African landscape surfaces are commonly marked by deep weathering associated with kaolinization of underlying bedrock where the latter contains feldspar minerals (see Plate 30). The upper surfaces of these deep weathering/soil profiles are often marked by hard crusts known as duricrusts, normally either calcrete or silcrete, and less often ferricrete (Partridge and Maud 1987). These duricrusts reflect seasonal variation and dry conditions: groundwater dissolves rock components and deposits them at the top of soil profiles when the water-table is high in the wet season. In the dry season, with a lowering of the water-table, these minerals are left behind, forming a crust on the ground surface, generally highly resistant to erosion. Calcrete, silcrete, and ferricrete reflect solution and precipitation from the groundwater of calcite, silica, and iron minerals respectively.

South Africa's major river system, the Orange, predates the onset of the uplift and rifting related to the break-up of Gondwana, and maintained its general westward course during this upheaval (Partridge and Maud 1987). Previous work by Partridge suggests that the Cretaceous Orange River comprised two separate systems: an upper Orange entering the Atlantic via the lower course of the Olifants River, and a discrete lower Orange system. Miocene uplift, already described above, led to capture of the upper Orange–Olifants system by the lower Orange, via the Koa valley; the modern Orange River course dates only from about 2 Ma ago or less, reflecting further uplift and capture near Prieska.

A **monoclinal fold** is a gentle bend or flexure with a step-like form in horizontally bedded or gently dipping rock strata.

Kaolinization is a chemical weathering process occurring under acid leaching conditions which produces the white clay known as kaolin.

Economic aspects of South Africa's geology

South Africa is the dominant source of many of the world's most important mineral resources, on the basis of proven reserves and of current production. An up-to-date review of the details and importance of the country's mineral resources is given by Vermaak (1997), in a special issue of the journal *Mineralium Deposita*. The papers in this issue detail many other specific mineral deposits. As is well known, South Africa is the world's leading gold producer, although the importance of this metal both in a world context and within the country's economy is decreasing and will most likely continue to do so. South Africa also holds most of the globe's known reserves of platinum-group metals (55.7 per cent), chromium (68 per cent), vanadium (44.8 per cent), and manganese (81.4 per cent), as well as having significant quantities of diamonds, coal, iron ore, titanium, fluorspar, and many other minerals. The only major mineral commodity for which significant known reserves are as yet lacking is oil, although large gas fields have been located offshore, as well as small finds of light crude oil; exploration continues.

In terms of statistics since 1980, gold mining has accounted for an average of 43.3 per cent and other mining for 33.9 per cent of South Africa's trade earnings (Vermaak 1997). In 1994, minerals contributed 8.5 per cent of the country's GDP. Mineral exports in the same year accounted for 48 per cent of foreign earnings, with the relative proportion contributed by gold having declined regularly since about 1980 (Vermaak 1997). It should be emphasized that a large number of secondary and tertiary industries also owe their existence to the primary mining-related products.

In a generally dry country such as South Africa, which also suffers from common droughts like the rest of the continent, groundwater resources are another important geological resource. Presently, although it is estimated that groundwater volumes are about thirty-three times larger than surface waters on land, only about 10–15 per cent of South Africa's water needs are met from groundwater (Snyman 1996). For about 65 per cent of the country's surface area, groundwater is the main or only source of water for humans and animals (Snyman 1996). Average annual rainfall contributions to the groundwater reserves are calculated at about 13 100 m^3/km^2, which is a low figure (Snyman 1996), and a significant portion is lost to evaporation, seepage, and evapotranspiration. In addition, the permeability of many South African rocks is very low, making borehole yields small. In many areas, due to poor rains and overutilization, there is a real danger of lowering the water-table and ruining this scarce resource for long periods to come. Generally, the more unconsolidated sedimentary (surficial) deposits such as river and aeolian (wind-borne) sands provide good potential water sources where there has been adequate infiltration from precipitation. Poorly consolidated and incompletely cemented sediments, particularly from younger Tertiary, Cretaceous, and even Karoo rocks, also often provide good water potential. In older sediments and in hard crystalline igneous and metamorphic rocks (see Table 11.1), factors such as weathering, faulting, jointing (regular crack systems in rocks), shearing, metamorphism, and any fragmentation which may result from such processes improve water-holding capacity and its utilization. Where igneous intrusions are emplaced, there is often an improvement in both porosity and permeability for groundwater development along the contact zones. For more details the reader is referred to the relevant chapter in Snyman (1996: Chapter 28).

Environmental aspects of South Africa's geology

A special issue of the *Journal of African Earth Sciences* (vol. 16/1–2, 1993) examines many of the relationships between geology and development in southern Africa. These include not only economic aspects, as discussed in the preceding section, but also some of the environmental issues arising from the geology of South Africa.

In general, the country is tectonically stable and subject only to minor seismicity. Such seismicity is mostly mining-related, specifically to the deep Witwatersrand gold mines, already active down to depths of 4 km, with planning for 5 km being actively pursued at present. Relatively subdued seismic activity associated with the filling of large dams the world over is a well-known phenomenon, and once stability is achieved no danger is posed by these events. Initially, some property damage may ensue, for which compensation is the norm; recent press reports about grave seismic dangers associated with the Katse Dam in Lesotho are overstated.

With the Great Escarpment in the east, thin soil and vegetation cover, common overgrazing, and infrequent torrential rains in South Africa, flooding remains an ever-present hazard, particularly in Kwazulu-Natal (see *Journal of African Earth Sciences* 16/1–2, 1993). Floods are dealt with in more detail in Chapter 16.

Unstable foundation conditions for large and small human structures are common where there occur mafic igneous rocks which weather to residual material rich in smectite clays; this problem is particularly pertinent in urban areas underlain by Bushveld rocks, such as northern Pretoria. Within the Greater Witwatersrand area and the Far West Rand, where large developed regions are underlain by Transvaal Supergroup dolomites, there are serious geological hazards. Being easily dissolved by acidic surface water and groundwater, the dolomites tend to have a complex and well-developed system of underground caves and pathways. Where weathering breaks through to the surface, narrow sinkholes are commonly filled with collapsible soils and rock debris, below which are large caverns filled by groundwater. The potential for collapse of the sinkholes becomes greatly enhanced when groundwater is withdrawn from such systems, particularly where large-scale dewatering related to deep gold mining is undertaken. Dolines, much larger and shallower subsidences in dolomitic lands which are generally related to compaction phenomena, also pose serious environmental threats to human life and property. In addition, in industrialized regions, there is a high potential for pollution of dolomitic (and other) groundwater from effluents. Many of these issues are discussed by Buttrick et al. (1993).

Conclusion

The geological history of South Africa forms part of the broader global processes which formed the earth and its surficial continents and oceans between about 5 and 2.5 billion years ago. South Africa's geology can be understood only in the context of plate tectonics and mantle plumes. Geological processes responsible for landscapes in the modern world were also directly involved in forming and altering the earth's atmosphere and hydrosphere, and, through repeated supercontinent formation and dispersion, exercise a great degree of control on the world's climatic systems and their changes over long time periods.

This chapter has outlined the main geological events which have brought about the present-day landscape of South Africa. The country is underlain by one of the world's oldest continents, and extending deep below the surface are some of the world's oldest rocks. These are now the source of many important minerals which are such an important part of South Africa's economy. More recently, events subsequent to the break-up of Gondwana have led to the present-day macro-scale geomorphology and associated climate.

The physical geography of South Africa, as well as its economic base, depends to a large extent on the regional geology. Weathering, soil formation, and surface processes are all to a large extent controlled by the underlying rock types; mining, as we have seen, depends on the whole spectrum of rocks found at depth. It is hoped that this chapter has provided a better appreciation of the importance of geology.

12 Climate and climatic change: causes and consequences

COLEEN VOGEL

The global picture

The climate of a country has a major impact on its landscapes and its people. The geomorphology, biogeography, hydrology, and soils of a region are all an expression of the present and past climates. Climate also plays a major part in determining the resource base for economic development, whether through agriculture, tourism, or water-dependent industry. Extreme events such as floods and droughts result in widespread economic damage, possible loss of life, and poor living conditions.

Weather and climate take place within the atmosphere and are the result of the atmosphere's interaction with solar energy and the surface of the earth, both land and ocean. *Climate* is the term given to the long-term weather conditions experienced at any one place. Climatologists study both long-term averages and the degree of variability expected, which affect the long-term planning of human activities. *Weather* refers to the atmospheric or meteorological conditions experienced at a particular time and place. Weather changes from day to day (and even from hour to hour) and affects day-to-day activities. Weather is what climatologists measure to build up a long-term picture of the climate.

Of all the branches of physical geography, climatology (the scientific study of world climates) has probably made the biggest scientific breakthroughs over the last fifty years. Climatologists' main concern is to try to understand and explain the physical processes underpinning the patterns of climate both in time and space. Until the 1930s, climatologists depended on data collected primarily on the land. The standard distance above the ground for the measurement of air temperature and humidity was 1.016 m (4 feet), wind speed was measured somewhat higher, but upper-atmospheric observations could be made only by looking at the clouds from below. Weather phenomena, however, usually take place within the lower 10 000 m of the atmosphere, the troposphere, so a true understanding of the

> The **troposphere** is the lowest layer of the atmosphere. It contains almost all the water vapour and aerosols and three-quarters of the total gaseous mass of the atmosphere.

world's climatic system could come about only once scientists and their instruments could go up higher into the atmosphere.

The advent of aeroplanes in the 1930s gave climatologists their first three-dimensional view of the atmosphere. As aeroplanes flew through and above the clouds, the phenomenon of jet streams and other upper-atmospheric airflows were experienced. In the 1960s the stage was set for a second revolution in climatological thinking when satellites began to circle the earth on a regular basis. For the first time it was possible to see weather systems at the scale of global circulation patterns. Areas for which ground data were previously lacking, such as many Third World countries, areas of low population density, and the oceans, could now be monitored as regularly as the data-rich parts of the world. This has important implications for southern Africa, where vast areas are sparsely populated and where the financial and other resources necessary for ground monitoring are not always available. Furthermore, with the ability of satellites to measure ground-level temperatures, the great importance of the oceans as a major storage and conveyor of energy has become apparent. Ocean–atmosphere linkages have therefore become an increasing focus of climatological research. For southern Africa, situated as it is between three oceans (see Plate 31, colour section), this presents an important breakthrough in understanding local atmospheric dynamics.

The new three-dimensional global picture brought about a realization that weather events throughout the world are connected through teleconnections. *Teleconnections* are the processes by which a meteorological event originating in one part of the world impacts on weather systems over a much wider area. One such phenomenon which has been researched increasingly over the last fifteen years is the El Niño Southern Oscillation (ENSO), which has its origins in the eastern Pacific, but whose effects are felt throughout the world in the form of drought or floods. A particularly severe ENSO event in 1982/83, for example, triggered droughts in many parts of the world, such as the eastern United States, southern Europe, Indonesia, and much of Africa, including southern Africa. In 1997/98 an event of even greater severity brought drought to Indonesia and floods to California and Kenya. ENSO and its impact on the climate of southern Africa will be discussed in more detail below.

As climatologists became more aware of the far-reaching effects of climatic teleconnections, they also became concerned that local-scale impacts of human activities could have global expression through climatic disturbance. The enhanced greenhouse effect and global warming became important global issues and resulted in a number of international conventions and resolutions. Ozone depletion is closely related to the dynamics of the recently detected hole in the ozone layer over the Antarctic. South Africans are especially vulnerable to the harmful effects of ultraviolet radiation, increases of which are related to variations in ozone concentrations.

With our increased understanding of

global dynamics it is now becoming possible to simulate future climatic scenarios by means of global climatic models. South African scientists are particularly active in this field. The possibility of future human-induced change has also redirected researchers to unravelling natural climatic changes in the past. Of particular importance have been the research stations established on the Antarctic continent, where new technological developments have helped scientists to unlock the secrets of past temperatures and atmospheric composition bound up in the huge depths of ice accumulated over thousands of years. South African scientists continue to play an important part in this research. A discussion of past climatic changes is not included in this chapter, but can be found in the relevant sections of Chapter 2 (on environmental change) and Chapter 11 (on geology).

World climates are now seen as dynamic, interconnected systems which are characterized as much by their variability and propensity for change as they are by their 'normal' conditions. The geography of a country may be determined more by the extremes of climate than by long-term averages. This is certainly true of southern Africa, where oscillation between drought and flood is characteristic. Climatic variability and climatic change in the subcontinent therefore provide the focus of this chapter.

The climate of southern Africa

The atmosphere is a fluid and has no clear horizontal boundaries. It is therefore difficult to define an area that 'climatologically' is southern Africa. A convenient northern boundary is the tropic of Capricorn at 23.5° S, which cuts through the north of South Africa. North of this line the climate becomes tropical, grading into equatorial. South of the tropic the climate is subtropical, trending towards temperate along the southern coast. No distinction is made in this chapter between southern Africa and South Africa; the two areas will be considered together. Within the region, two distinct landscape areas with distinct climates can be identified: the continental interior or high plateau, and a narrow coastal plain (see Plate 1). The altitude of the interior plateau is relatively high, ranging from around 500 m in the west to over 1 000 m in the east. It is separated from the coastal plain by a distinct escarpment. The main weather systems and climatic characteristics of each area are outlined below. A comprehensive treatise on the atmospheric systems and weather of southern Africa is given by Preston-Whyte and Tyson (1988) and Lindesay (1998).

Weather systems affecting the climate of South Africa

The weather and climate of South Africa are strongly influenced by the position of the country relative to the major global circulation patterns (see Figure 12.1). A belt of subsiding air from the Hadley Cell dominates much of our weather and climate all year round. High pressure and north-easterly inflow of tropical air characterize the climate of the interior. Systems approaching from the west (cold fronts) and those from the east (tropical systems)

FIGURE 12.1 Some important surface atmospheric features over southern Africa

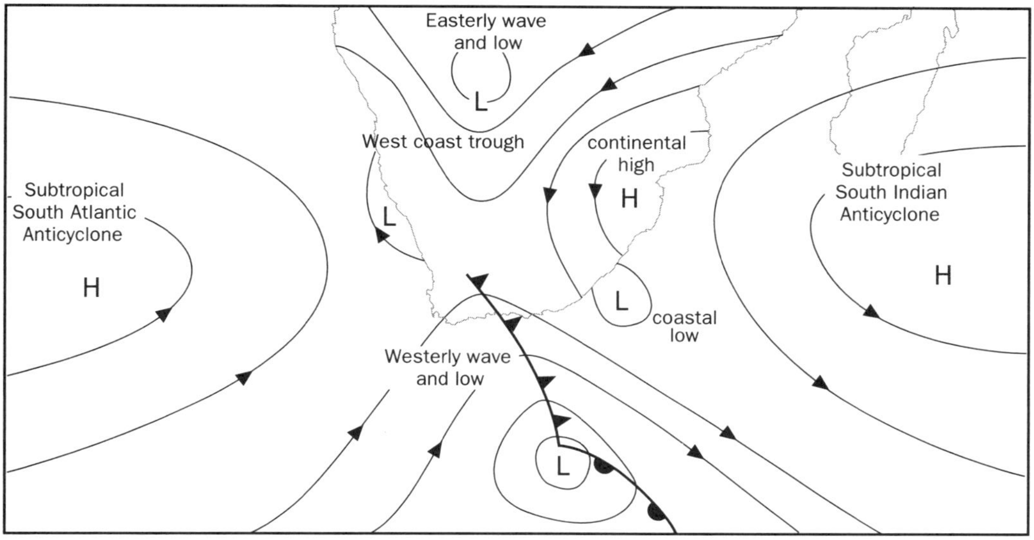

SOURCE: Preston-Whyte and Tyson 1988.

> The **Hadley Cell** is a circulating cell of air which rises above the equator and subsides in the subtropics.

also influence the day-to-day weather. These oceanic influences play an important role in influencing the progressions and strength of weather systems across the country. The distribution of rainfall over South Africa, shown in Plates 32 and 50b, can be related to these weather systems.

Over the interior there is a distinct east–west trend in rainfall. North-easterly airstreams affecting the eastern highveld bring annual rainfall totals of around 800 mm, concentrated in the summer months. Rainfall totals decrease to below 125 mm in the arid west, bringing desert conditions to the Kalahari and southern Namibia. The warm summer months in the highveld are punctuated by intense thunderstorms. These storms may be accompanied by hail, which can cause extensive damage to buildings, vehicles, and crops, and injury to people and animals.

The temperatures over the interior of South Africa are strongly linked to the high-pressure field over the region. Subsiding air from the large semi-permanent cells of the subtropics, associated with the descending limb of the Hadley Cell of the general circulation, usually brings clear conditions for much of the year. Winter weather in the interior is normally sunny and cloudless, with frost occurring in the early mornings if temperatures drop sufficiently during cloudless, calm nights. The anticyclonic systems can, however, vary in intensity and position with the seasons. Interruptions of the usual calm, clear conditions are usually the result of a passing cold front or the influx of warm, moist tropical air that may result in thunderstorms. During the months from December to February the highest near-surface temperatures occur in the tropics, when the thermal equator is in the southern hemisphere. Above-average summer temperatures occur during anomalously dry years over the region (Hulme 1996), mainly due to a predominance of anticyclonic conditions.

The **thermal equator** marks the line encircling the low latitudes of the globe which exhibits the highest mean surface air temperature. The thermal equator moves north and south following the apparent movement of the sun between the tropics. In December it lies approximately over or just to the south of the equator.

Temperatures over southern Africa have not been constant over the recent past, and there may be some evidence to support the theory of global warming. Researchers such as Hughes and Balling (1997) note evidence of a warming trend over the last century of more than 1°C over South Africa and of 0.5°C over southern Africa as a whole. Most of this warming has occurred during the last three decades, but the trend is not ubiquitous and may be a local urban effect.

The coastal climate comes under a strong oceanic influence, depending on the ocean and the predominant winds. The east coast is dominated by the warm Agulhas Current, which brings equatorial water as far south as the Cape. Conversely, the cold Benguela Current brings cold, Antarctic water up the west coast. The south coast becomes a mixing ground for the two currents, and its weather can be significantly affected by shifts in ocean temperature. The coastal areas can be strongly affected by continental high-pressure systems, with hot föhn-type berg winds blowing off the escarpment. They are also frequently affected by the frontal systems of temperate westerly depressions, which move eastwards over the country bringing with them cooler weather and the potential for rain. The coastal areas are thus characterized by strong swings in weather, alternating from hot, dry, berg-wind conditions to cold, wet, frontal conditions.

Rainfall along the coastal margins is the result of a number of mechanisms, including frontal systems and ridging anticyclones which enhance onshore airflow. Orographic effects of the Cape Fold Mountains and the Drakensberg escarpment result in high rainfall in these areas. The highest annual rainfall totals have been recorded in the Hex River Mountains of the Western Cape (locally over 3 000 mm) and the Drakensberg escarpment of Natal (over 1 500 mm). Distinct rain-shadow effects also occur in the Cape Fold Mountains, causing locally dry conditions in the lee of mountain ranges. The Little Karoo in the vicinity of Oudtshoorn is a particularly good example (see Plate 32, inset). The seasonality of rainfall along the coast varies from west to east. The Cape Peninsula has a strong winter rainfall maximum, the south coast an all-year rainfall pattern, and from the Eastern Cape to KwaZulu-Natal the rainfall trends towards a summer rainfall maximum. Towards the east, June and July tend to be consistently dry, but heavy precipitation in early spring and late autumn is not

A **ridging anticyclone** occurs when a ridge of near-surface high pressure lies along the south coast of South Africa, underlying an upper-air trough of low pressure. Surface-level anticyclonic circulation results in onshore flow. Upper-level divergence in the westerly wave simultaneously produces uplift. Together they produce widespread rainfall.

Orographic precipitation occurs as a result of uplift induced by a topographic barrier such as a mountain chain.

A **rain-shadow** is the area on the leeward (sheltered) side of a topographic barrier, where precipitation amounts are less than those on the windward side.

uncommon, even in the more easterly regions, and significant snowfall may occur over the Drakensberg mountains.

Cold frontal systems are an important source of rainfall to the coastal areas. Cold fronts also on occasion penetrate into the interior, bringing unseasonally heavy rainfall. They are the product of different air masses: one containing warm, moist air (usually from the north-east of the country), the other containing colder, drier air (associated with cold sub-polar air from the south-west of the country). When these two air masses meet, a zone or 'barrier' is observed, characterized by changing pressures, wind directions, and temperatures. This zone is referred to as a 'front'. The leading surface of warm air is known as a warm front, and the leading surface of cold air a cold front. As the front becomes enhanced in its development, a depression or low-pressure cell is formed. Unlike in northern hemisphere countries, warm fronts are seldom experienced over southern Africa, as they tend to form to the south of the country where they characteristically take up a west–east alignment. Cold fronts, in contrast, have a north–south orientation and frequently cross the coastal areas, occasionally penetrating inland (see Figure 12.1).

Not only are fronts caused by the meeting of two different air masses at ground level, but they are also influenced by the upper-air circulation, as shown in Figure 12.2. A circumpolar vortex of fast-flowing westerly air dominates the circulation of the atmosphere in the mid-latitudes. This vortex of fast-moving air usually comprises a wave-like pattern of airflow. The position and shape of these waves in the upper air play an important role in influencing the flow of air either upwards into the atmosphere or downwards towards the surface. Upward airflow usually enhances the development of low pressure at the ground surface; downward flow usually enhances ground-level high pressure. Conditions suitable for instability (usually uplift) or stability (descending air) are thus influenced by these wave patterns.

A **circumpolar vortex** is a vortex of spiralling winds that are concentrated around polar and sub-polar regions, carrying with them weather systems such as low pressures.

The origin of the waves in the westerly airflow results in the low-pressure systems or *cyclones* and frontal features usually traversing the country in an easterly direction, bringing rain to the south-western and southern coasts. Often the

FIGURE 12.2 The development of depression and frontal systems

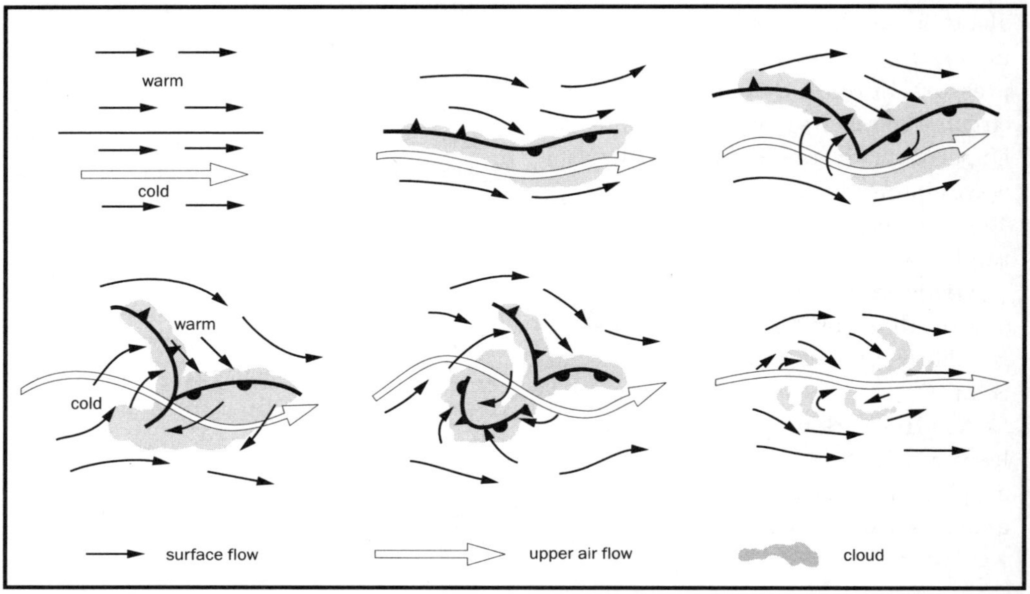

Note the interaction between surface and upper-air controls

SOURCE: Preston-Whyte and Tyson 1988.

presence of a strong, blocking high-pressure system over the Indian Ocean may preclude the northward extension of the system, which is then deflected to the south-east.

Another important weather system affecting the east coast is that of *tropical cyclones*. These originate in the Indian Ocean to the west of Madagascar and move in south-westerly tracks onto the mainland, where they bring heavy rain, flooding, and destructive winds. Tropical cyclones seldom penetrate south of 30° S, but on occasion affect the north Natal coast. Cyclone Demoina, for example, caused extensive flooding in northern KwaZulu-Natal in 1984.

Rainfall variability in South Africa

Southern Africa is a predominantly dry region, with rainfall highly erratic both in time and space. As noted above, this variability has important consequences in terms of extreme events such as floods and droughts, so that the geography of the subcontinent is determined as much by climatic extremes as it is by climatic 'norms'. Much climatological research has focused on the short- and long-term variability of rainfall, recently extending its scope to include the social and economic consequences of extreme events such as droughts. This research is reviewed below.

Active research on rainfall variability is ongoing in the country and the region. Debates as to whether South Africa has undergone desiccation, for example, date back at least 100 years (Wilson 1865) and continue today (Tyson 1996). A number of researchers have identified several notable patterns of rainfall, in particular the distinctive annual oscillations over southern and South Africa both temporally and spatially (Tyson et al. 1975; Ngara et al. 1983; Tyson 1986; Mason and Jury 1997). A ten- to twelve-year oscillation accounts for over 30 per cent of the inter-annual rainfall variance along the south coast of the country. Of greater importance is an approximately eighteen-year cycle that is evident over much of the summer rainfall area (see Figure 12.3) and which extends into Zimbabwe and Botswana. In addition to these oscillations, a broad band of variability in the range of three to six years occurs throughout southern Africa and appears to be associated with the El Niño Southern Oscillation phenomenon discussed in more detail below.

Several wet and dry years have been identified from meteorological records. As is evident from Figure 12.3, of the following dry spells – 1906/07–1915/16, 1925/26–

FIGURE 12.3 Spatially averaged rainfall series for October–September rainfall in the summer rainfall region

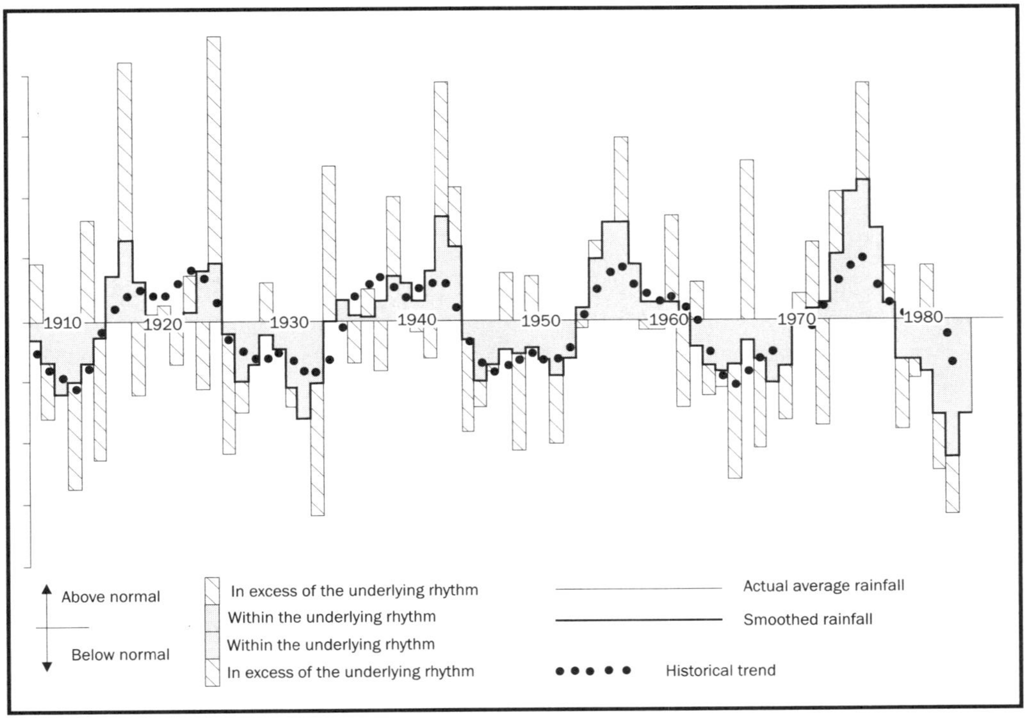

SOURCE: Tyson 1986.

1932/33, 1944/45–1952/53, 1962/63–1970/71, 1980/81–1990 – the most consistently dry period was 1925/26–1932/33. Available evidence suggests that this pattern of wet and dry years has generally persisted since at least 1840. There is marked variability in the spatial distribution of wetter and drier conditions which changes year by year; drier spells, however, tend to exhibit a greater areal extent and spatial homogeneity than do wet spells.

Causes of rainfall variability

Much of the research on rainfall variability has focused on the relationship between rainfall and circulation parameters such as changes in the intensity and positioning of pressure systems. This work is summarized by Tyson (1986), Mason and Jury (1997), and Lindesay (1998). Rainfall variability for the country has been associated with a number of local circulation patterns, including changes in the strengths of the tropical easterlies and westerlies, the interactions between tropical and temperate systems, and the variation in pressure patterns and atmospheric long waves over Marion and Gough Islands. During warmer summer months, for example, the pressures over the continent are usually much reduced in the lower layers of the troposphere, partly as a result of surface heating and of shifting pressure belts associated with the seasonal movement of the sun. These lower pressures, together with the shifting position of the Inter-Tropical Convergence Zone (ITCZ) and the movement of moist air onto the continent from the east, allow for the development of thunderstorms, which are a characteristic feature of the summer months in the northern and central parts of the country.

Changes in rainfall patterns over southern Africa are also related to adjustments in the location of both cloud bands and ocean temperature. The oceans form significant stores of energy, while cloud bands, usually aligned from north-west to south-east over the country, represent important conduits of energy and momentum into the subcontinent. Important releases of latent energy over tropical Africa, for example, influence the Hadley Circulation of southern Africa, which ultimately determines the rainfall receipts for the interior of the region. Troughs of low pressure, linking tropical and temperate circulations and their associated cloud bands, are significant contributors to South Africa's summer rainfall. They form when a tropical low is coupled to a temperate westerly wave. Harrison (1984, 1988) has shown that over the plateau these tropical–temperate troughs and cloud bands contribute 39 per cent of annual eastern Free State rainfall. Over the period 1967–77 they contributed between 50 and 90 per cent of observed January rainfall.

Shorter periods of persistent rainfall are produced by *cut-off lows*. Cut-off lows develop in conjunction with upper airflow in the westerlies and an anticyclonic ridge to the south-east of the subcontinent which provides onshore flow from the warm Agulhas Current. Centres of low pressure hence become 'cut off' from their upper-air 'parent circulation'. Cut-off lows are usually most frequent during the transitional seasons and are important contributors of early- and late-season rainfall (Taljaard 1985, 1986). Cut-offs are occasionally associated with major floods, as happened in January 1981, causing the Laingsburg floods.

Teleconnections and South African rainfall

Recent research has focused on the global-scale forcing mechanisms of rainfall and circulation changes across the southern hemisphere. Two major factors accounting for the variations in atmospheric circulation and rainfall over South Africa are, firstly, the Southern Oscillation and, secondly, the role of sea-surface temperatures, particularly those of the South Atlantic and South Indian Oceans. The Southern Oscillation is of particular interest because it is related to the more familiar El Niño phenomenon, which has been evoked to explain short-term changes to the world's weather patterns in the form of regional drought and flooding. The dynamics of the Southern Oscillation are described in more detail by Preston-Whyte and Tyson (1988).[1] A brief introduction is provided below.

The *Southern Oscillation* refers to the fluctuations in inter-tropical pressure, wind, sea-surface temperature, and rainfall that are related to an exchange of air between the south-east Pacific subtropical high and the Indonesian equatorial low. Associated with the Southern Oscillation is the *El Niño* phenomenon or ENSO (El Niño Southern Oscillation). El Niño is associated with changing weather patterns that are a result of unusually warm waters off the coast of Peru and Chile in South America. As a result of the warmer waters and a consequent lowering of pressure, large amounts of moisture are pumped into the atmosphere. This condition usually occurs around Christmas time, hence the Spanish name El Niño (Boy Child). La Niña is the term used for the opposite of El Niño conditions and is related to abnormally cooler waters off the eastern Pacific coast. Sea-surface temperatures and temperature anomalies in the Pacific during the 1997/98 El Niño are shown in Plate 33.

One mechanism by which the Southern Oscillation signal is transmitted to southern Africa is via the *Walker Circulation*. The Walker Circulation, described by Preston-Whyte and Tyson (1988), is a thermally driven longitudinal or zonal cellular circulation extending across the Pacific Ocean from Indonesia to the Peruvian coast (see Figure 12.4). It forms an important component of the Southern Oscillation. Low-phase years (El Niño) are accompanied by reductions in heat release and convection over tropical southern Africa which in turn are caused by zonal adjustments in the Walker Circulation and a diminished number of tropical–temperate troughs across South Africa (see Figure 12.5). Fewer cloud bands occur and rainfall decreases. The reverse occurs for the high phase of the oscillation, or wetter years (La Niña). Thus high- and low-phase years of ENSO influence rainfall over southern Africa, with El Niño events usually associated with dry conditions. This general association between the Southern Oscillation phase changes and South African rainfall has been shown to exist in present and pre-instrumental rainfall records from at least 1820 (Lindesay and Vogel 1990).

The Southern Oscillation is thus a major forcing mode (though not the only one) of the inter-annual variations in atmospheric circulation over southern Africa, often enhancing local weather and climate conditions. The 1982/83 ENSO event, for example, served to exacerbate the prevailing dry conditions in much of

FIGURE 12.4 Schematic representation of the anomalous Walker Circulation

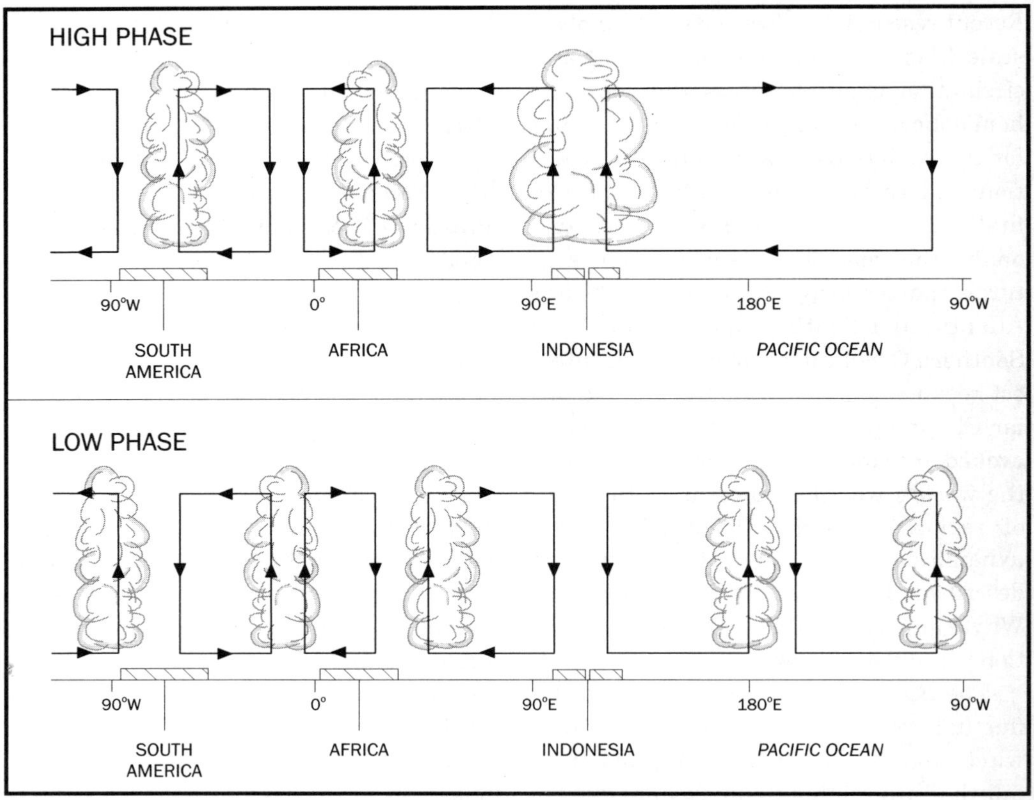

SOURCE: Preston-Whyte and Tyson 1988.

the subcontinent (Dent et al. 1987). The rainfall-producing systems were displaced eastwards during this time, with anomalous equatorward rather than poleward airflow and momentum fluxes over the region combining with a northward shift in the mid-latitude westerly circulation (Lindesay 1988; Muller and Tyson 1988).

An integral part of the Southern Oscillation is the role that sea-surface temperatures play in modulating the occurrence of low- and high-phase changes and associated atmospheric circulation interactions. Correlations between local sea-surface temperature anomalies, atmospheric circulation, and rainfall have been shown to exist in several instances (Mason 1995; Walker et al. 1984; Tyson 1996). Walker (1990) and Mason (1995) have shown that the temperatures of the oceans around South Africa exert an influence on rainfall over the subcontinent. Rainfall over southern Africa has been shown, for example, to be anomalously high when

FIGURE 12.5 The Walker Circulation during high and low phases of the Southern Oscillation

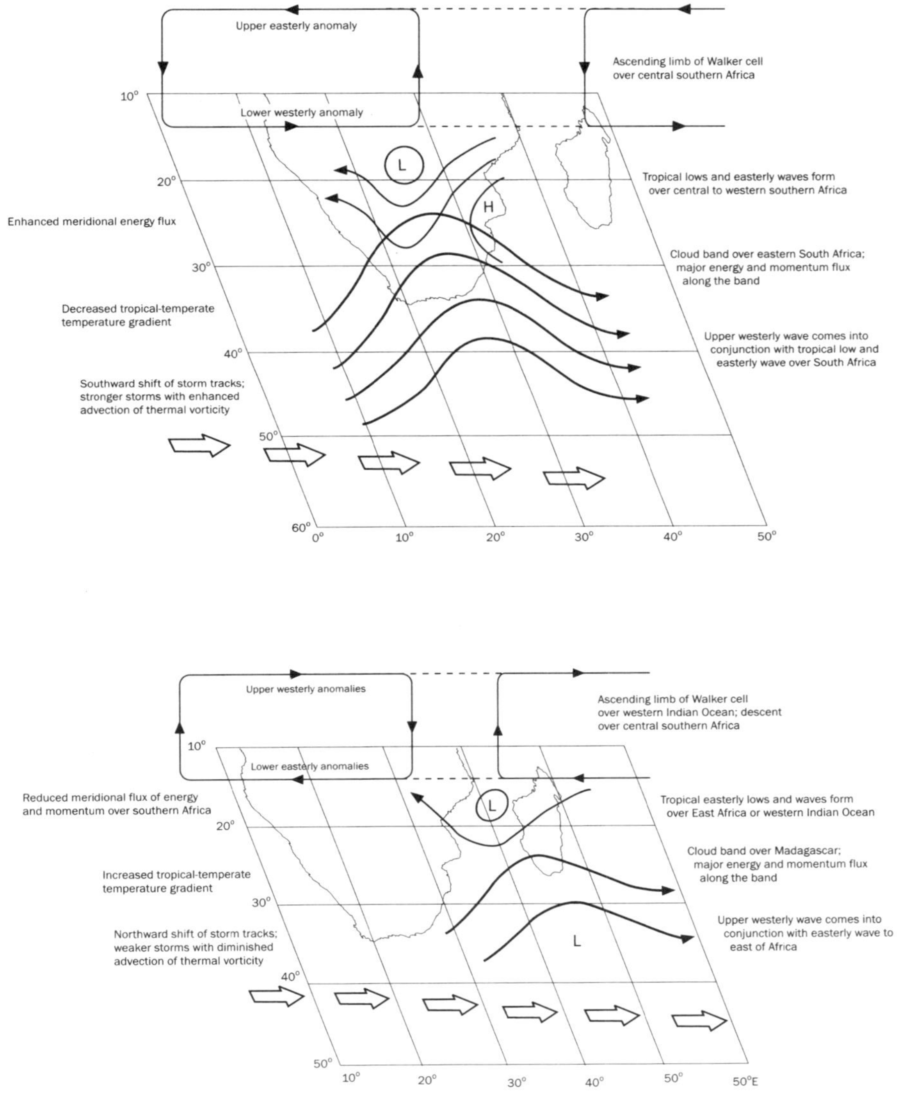

The Walker Circulation and derived flow produce above-normal rainfall over southern Africa during the high phase of the Southern Oscillation (top), and the circulations during the low phase (bottom) result in below-normal rainfall. Light and dark lines denote surface and upper-tropospheric conditions

SOURCE: Preston-Whyte and Tyson 1988.

sea-surface temperatures are above normal. More recent detailed examination of sea-surface temperatures has shown that the situation is indeed complex, with temperatures in the northern Indian Ocean associated with decreased rainfall over southern Africa in the summer (see Walker 1990, Mason 1995, Mason and Jury 1997, and Lindesay 1998 for further discussion).

In summary, rainfall variations over the country are influenced by a number of interactive mechanisms which are the subject of continued research. The influence of teleconnections and global-scale circulation forcing mechanisms is emerging as one of the key determining factors of local climates. Much of the recent work also points to an increase in inter-annual rainfall variability, particularly of extreme events. These findings are in keeping with international research which suggests that tests for climate change should focus on changes in the frequency of extremes rather than on the changes in climatic means (Von Storch and Zwiers 1988). The study of extreme events has therefore become an area of increased research interest, examining both the causes of such events and their consequences.

Extreme climatic events: droughts and floods

Droughts and floods are recurrent phenomena in Africa due to the variability of rainfall and the large areas of marginal land often settled by people lacking the political or economic empowerment to control their own lives, thus making them more vulnerable to such hazards. The prevalent atmospheric pressure system over much of the southern African region is anticyclonic, with persistent sinking air (subsidence) associated with dry conditions. Anything which enhances this anticyclonic activity and extends its influence in both time and space will decrease rainfall totals below the average and create drought conditions. The forcing mechanisms for droughts are well documented for the region and include the El Niño Southern Oscillation referred to above. Heavy, flood-producing rainfall is another feature of the South African climate. Some of the main causes of flood-producing rain are severe local thunderstorms, tropical cyclones, and cut-off lows (Preston-Whyte and Tyson 1988). Periods of extended heavy rainfall are also associated with tropical–temperate troughs, which may give rise to flooding (Lindesay and Jury 1991).

Although the primary cause of a flood is climatic, floods are hydrological phenomena controlled by the interaction of climate and drainage basin processes. A full treatment of floods is therefore given in Chapter 16 (on river basin systems). The present chapter will focus on droughts: their nature, causes, and consequences.

What is drought?

It has proved difficult to establish a precise, yet widely applicable, definition of drought. One reason is that drought can occur within a number of different contexts; one can distinguish between meteorological drought (rainfall totals well below average), agricultural drought (rainfall insufficient to support the established agricultural practices), and hydrological drought (rainfall insufficient to replenish normal streamflow or groundwater storage) (Dent et al. 1987; Rowntree 1988). Droughts are also highly variable in time and space, and severe drought may often characterize some areas of the country during a specific period while other areas experience normal or even above-normal rainfall (for example 1982/83 and 1991/92). It should therefore be noted that drought is a relative phenomenon, and definitions must relate it to deviations from the long-term mean or accepted conditions.

Many drought definitions are based on cumulative percentage curves of rainfall or some other drought measure. To derive a cumulative percentage, the values are ranked so that the percentage of time that a given value is exceeded can be calculated. These percentages can be divided in to deciles (10 per cent classes), with drought then defined where values fall into the lowest deciles. Drought is thus given a statistical definition based on the long-term record.

The intensity of short-term drought, over a period of a few weeks, has been described by the Palmer Drought Severity Index (PDSI), a single numerical value that includes inputs of soil moisture, runoff, evaporation, and temperature. For longer periods, Bruwer (1989) has defined meteorological drought as those periods when less than 70 per cent of normal precipitation is received; a severe drought occurs if such a situation extends over two consecutive seasons.

A recently developed index of agricultural or ecological drought is the Normalized Difference Vegetation Index (NDVI). This is derived from satellite data and is a measure of the photosynthetic productivity of the land surface and therefore of active vegetal cover. The Institute of Soil, Climate, and Water in Pretoria has developed a twelve-year databank of these NDVI values which has been used in regional-scale drought research. NDVI data provide a useful means to assess the impact of rainfall variability on vegetation. Jury (1993), for example, has assessed the relationship between vegetation change and rainfall over the south-western Cape from 1985 to 1988. His study shows that during 1987 a weak El Niño event reduced rainfall to below normal, resulting in drought-stressed vegetation, poor crop yields, and inadequate grazing across the Karoo as far as Kimberley and Upington.

Single drought indices clearly have their limitations, because they cannot encapsulate the complex nature of drought. The future for drought identification appears to lie in devising integrative systems based on a collection of indices such as rainfall, soil moisture, crop yields, and vegetation cover which can be used to assess the extent and severity of drought.

Consequences of extreme events

Droughts and floods are natural events which result from the inherent variability of the climatic system. They become hazards and disasters when given a human dimension – that is, when they impact on normal human activity (Vogel 1998). The consequence or outcome of the hazard also depends on the coping mechanisms and resilience that an affected community possesses. In this sense drought or flood hazards become as much a human-induced phenomenon as a natural one. Moreover, humans can exacerbate the natural variability of the climatic system by their effect on atmospheric dynamics, for example through the enhanced greenhouse effect. These various issues will be explored in the following sections.

Drought impacts in southern Africa

Drought usually impacts negatively at both the national level and at the local or household level. At the national level it can affect several sectors of the economy, particularly those relating to agriculture or those making heavy use of water, as well as on national food security. At the local level it effects household food security, particularly for the rural poor, who commonly have few reserves to cope with food shortages.

Reduced agricultural production and increased malnutrition, for example, were the most important consequences of the 1980s' drought in southern Africa. The impact on agriculture and food production in the SADC countries during this drought has been studied by Mapuwose (1985). Aggregate cereal production in 1984 was in the order of 3.4 million tonnes, compared with 4.8 million tonnes in a normal year. Economic and social conditions among the SADC countries deteriorated, and real regional output was estimated to have fallen by 1.3 per cent in 1981, stagnating further in 1983 and 1984. National cattle herds, a source of wealth and draught power in Zimbabwe, Zambia, Botswana, Lesotho, and Mozambique, declined to 30 per cent of the norm, with estimates of one million cattle dying in the region. Other consequences of the 1980s' drought included increased crime, famine, and social dislocation (Ballard 1986; Vogel 1994).

In South Africa, too, drought has brought about serious national and regional problems. Agriculture's share of the Gross Domestic Product (GDP) has declined from over 17 per cent in the 1950s to around 4 per cent in the 1990s. Frequent droughts have accelerated this trend and have had negative consequences for this and associated sectors, as well as for the broader economy (Van Zyl 1993). Drought shocks to the country include agricultural-related job losses, farm labour retrenchments, farm owners mortgaging their farms, and possible migration. During the severe drought of the early 1990s it was estimated that 50 000 jobs would be lost in the agricultural sector and a further 20 000 in related sectors, and that about 250 000 jobs in total would be affected (Van Zyl 1993). The declining agricultural yields associated with the droughts of the 1990s negatively affected GDP growth by between 0.5 and 2 per cent. Other impacts such as migration are difficult to quantify, but indications are that permanent urban migration during droughts is an option that is considered only once a range of other coping strategies has been employed (Vogel 1994; Jallow 1995).

Drought is not always the underlying cause of declining agricultural productivity, but drought periods may enhance other forms of degradation of the landscape. Overstocking of farmland, for example, is regarded as the major problem hindering farming, with drought impacts only secondary. Local farmers are well aware of these issues, but often fail to manage their farms correctly. Estimates of overstocking indicate that 50–60 per cent of the country's veld may be overgrazed (Bruwer 1989). The adverse impacts of drought on natural vegetation can, however, be reduced by judicious management, by periodically withdrawing the veld from grazing so that the species composition can recover.

One of the major impacts of drought in southern Africa is that relating to surface and underground water. There have been a number of hydrological droughts in South Africa since 1920, including 1930–32, 1950–53, 1967–72, 1978–86, and 1989–91. Reservoir water levels in several areas of the country were reduced during the 1980s and 1990s. In certain cases, levels dropped so notably during the 1990s' drought that it became necessary to augment water in some dams such as the Vaal Dam. These dams are essential for supplying water to the industrial areas of Gauteng, and water shortages can therefore have a serious effect on industrial productivity and national power generation.

While the levels of dams such as the Vaal evoke national concern, the levels of other smaller dams, often located in remote rural areas, are also drastically reduced during drought periods, causing grave hardships (see Plate 34). Other sources of water, such as underground reserves (groundwater) that serve boreholes and pumps, are also placed under strain. Several dams in the former homelands were particularly low during the 1990s and numerous cases of acute water shortages were noted (National Consultative Forum on Drought, 1992). Recognizing this important problem, an initiative known as the Water Supply Task Force of the National Consultative Forum on Drought actively monitored the water situation in rural areas and provided assistance where possible. It soon became evident that the severity of the drought impact in these areas was a result not only of reduced rainfall, but also of a lack of effective water management, a lack of reliable data on water resources, poor maintenance of water infrastructure, and inadequate supply. Drought must therefore be viewed as an integrated phenomenon that is the product of a lack of rainfall and of proactive, mitigatory measures, as well as poor or inadequate infrastructure development. This is particularly true in the African context.

Drought is thus not the fundamental problem in sub-Saharan Africa. Rather, it often lays bare the African development crisis. In the South African case, droughts have exacerbated poverty and have exposed the legacies of previous political policies. Such polices led to the lack of basic infrastructure and services in many poor, black, rural areas of the country.

Few detailed studies of disaster impacts at the rural-poor household level have been undertaken in South Africa (Adams 1993; Freeman 1984; Vogel 1994). More attention is being given, however, to enabling those who normally experience the most severe impacts of drought to voice their concerns and incorporate their experience into planning and mitigation strategies. Through such processes it has become evident that drought-coping responses include a broad array of activities (see Table 12.1). Much more research and understanding of these complex coping strategies is needed if effective mitigation and management of droughts is to occur. Technical fixes cannot be the only measures used to address droughts, and the understanding of household and community coping is essential to assist in 'drought-proofing' at the local level. This can then be fed into and complement drought management at the national level.

TABLE 12.1 Some community coping strategies in the North West and Northern Province

TYPE OF COPING STRATEGY	DESCRIPTION OF STRATEGY
Drinking of traditional beer	Believed to reduce daily food consumption
Saving water	Reduction of water used during a drought; watering of gardens is curtailed
Social networks	Taking children to relatives; taking credit from local shop owners; receiving remittances
Food storage Buy water	Store foods as reserves Households buy additional water for household chores
Sell livestock	Sales of livestock to avoid losing stock due to poor water and grazing conditions. Farmers often, however, reluctant to do so for livestock is a form of wealth and has other values
Micro-enterprises	Households embark on small business enterprises, e.g. weaving, selling products, sewing, etc.

SOURCES: Vogel 1994; Spio and Nchabeleng 1997.

Future climates and the role of climate forecasting

Having outlined some of the causes and consequences of climate variability and change, we shall now examine ways that society can prepare to mitigate and cope with possible climatic changes. One important development has been the advent of climatic modelling, enabling climatologists to make predictions about future changes. Possible changes in weather and climate are much-debated areas of science and policy, and are the focus of several international research teams and initiatives, such as the International Geosphere and Biosphere Programme (IGBP) and the International Panel on Climate Change (IPCC). The science required to respond to future changes must be interactive and interdisciplinary. Thus, while the need to understand climate change is important, so too is the need to understand the associated environmental and human dimensions. Global climate modelling and forecasting, together with an improved understanding of the vulnerability of landscapes and people, are becoming important areas of research both locally and internationally. Examples of the concerted, combined international activity include the International Human Dimensions Programme on Global Environmental Change (IHDP), the project on Land Use and Cover Change

(LUCC), and the International Decade for Natural Disaster Reduction (IDNDR). Examples of local initiatives that link into the wider research foci are described below.

Changes in atmospheric circulation patterns have been presented in conceptual models of the circulation features exerting a control on the weather of South Africa (Tyson 1986). It is now generally accepted that extended wet spells in the summer rainfall region are caused largely by an invigorated tropical-induced circulation forced by the tropical easterlies. Conversely, an expansion of the circumpolar vortex and an increased occurrence of westerly disturbances bring extended summer dry spells to the interior, but wetter winters. As a result, drought in the interior often coincides with increased wetness in the winter rainfall region of the Cape, and vice versa. This model has been used to extend our knowledge of past climate change in the country (Muller and Tyson 1988).

The climatic models for the southern African region are always under review, and several are currently being examined as a basis for numerical weather prediction (Mason and Jury 1997). In South Africa, seasonal forecasting is a growing scientific activity exhibiting varied degrees of success. Most forecasting skills relate to predicting rainfall variability attributable to the tropical circulation, including ENSO-related anomalies. Reasonably successful hindcasts ('predictions' of past events) of rainfall for extreme drought years over southern Africa have been produced using general circulation models (GCMs) based on forecasted and observed sea-surface temperatures. Such rainfall simulations generally reproduce the broad-scale features of spatial and seasonal variability, but have significant weaknesses. These lie in the inability of the general circulation models to effectively capture important local climate influences such as topography. A possible solution is to 'downscale' the models or to nest the regional models within the larger global circulation model environment (Joubert and Hewitson 1997).

Probable changes in the occurrence of dry years under doubled CO_2 conditions have also been examined, for example by Joubert and Mason (1996). Their results indicate that, while changes in the annual mean rainfall are not expected to be significant, an increase in the probability of dry years in the tropics as well over the western and eastern parts of South Africa is likely. Increases in the frequency of midsummer droughts are also predicted over the eastern part of the subcontinent south of 20° S. The possibility of increased magnitudes and frequencies of flood events has also been investigated by Mason and Joubert (1997). Their simulations suggest that the result of increased CO_2 will be to increase the severity of flood events rather than to change the long-term means. The largest flood increases are shown to be most likely over the winter rainfall area. The intensity of other flood-producing systems such as tropical cyclones may also be expected to increase under enhanced greenhouse conditions as a result of warmer sea-surface temperatures. The largest changes associated with enhanced CO_2 will thus probably find their expression in the most extreme events: droughts and floods.

Future changes in climate must not only be viewed within the context of the physical dimensions of these systems, however. The influence of people on climate has increasingly been explored, and it must be remembered that human activity will continue to have an influence on the climates of the future. Regional and local changes therefore may well differ in magnitude from the broader global changes observed. For these reasons, greater understanding of the complex interactions and feedbacks of the climate–human system is constantly being sought.

Changes in climate and associated impacts are being extensively monitored in the region. Recent integrated research submissions, several meetings on drought such as the High Level Policy Seminar on Drought in Southern Africa (Botswana, 1997), and the presence of South Africa at the important meeting on Global Change in Kyoto, Japan (December 1997), point to the concern with which local scientists and policy-makers view current and future climate change. While some ideas of possible future changes and variability within the climate system are emerging, scientific results are not yet conclusive and much still needs to be done to improve our understanding of how the present climate system operates, as well as how it could change in the future. The results of simulations must always be viewed with caution, being based on models and imperfect knowledge of climatic systems. Nevertheless, they present some interesting scenarios and enable climatologists and meteorologists to enhance their understanding of the complexities of climate.

Conclusions

Aspects of the changing weather and climate of South Africa over varying spatial and temporal scales have been presented in this chapter. While much is known about the local conditions, a great deal is still to be discovered and understood, particularly the interactions between the local, regional, and global atmospheric forcing mechanisms that ultimately influence climate and weather.

Climate is a complex system, changing on several spatial and temporal scales. The position of South Africa in the subtropics means that, while the country comes under the predominant subtropical high-pressure systems, it is also influenced by warmer, moister tropical atmospheric systems from the north-east as well as cooler temperate systems from the south-west. Changes in features that are global in scale, associated with temperature and pressure oscillations such as ENSO, also have an effect on local climate. The country is prone to periods of drought and other climate extremes. Future climate change is an area of much research and here the use of global circulation models has begun to produce results.

The related aspects of such research, including the broader biophysical and social interactions of the larger system, must also be borne in mind. Effectively coping with and responding to extreme events requires a concerted, combined response that includes effective mitigation during 'normal periods' through appropriate development strategies, forecasting by means of meteorological and other early-warning systems, an understanding of who the vulnerable are (for example through nutritional surveillance and delimiting vulnerable landscapes), as well as a well-planned strategy that can effectively manage the outcomes of events when they heighten into 'disasters'.

Geographers bring a wealth of skills and expertise to the study of climate and associated changes. Improved understanding of the mechanics of the climate system, together with a better appreciation of our influence on the atmosphere–biosphere system, should enable us to 'manage' the environment better. Hopefully our imprint on these systems will be made sensitively, thereby ensuring their sustainability for future generations.

Notes

1 An up-to-date explanation of ENSO and details of current ENSO events can be found on the Internet at sites such as http://www.pmel.noaa.gov/toga-tao/el-nino-story.html or http://ww2010.atmos.uiuc.edu/(Gh)/guides/mtr/eln/home.rxml.

13 Landscape evolution in a changing environment

PAUL SUMNER AND IAN MEIKLEJOHN

Geomorphological processes are directly influenced by the environment within which they operate. Landforms and landscapes, in turn, reflect the environmental conditions that existed during the period of their formation. The concept of distinctive landscapes or landscape assemblages associated with specific climates has been a dominant philosophy within international geomorphology and has given rise to the field of 'climatic geomorphology' (Thorn 1988). Central European geomorphologists have long been proponents of the idea that regional geomorphology is a response to climate.

In a time when predicting environmental changes is recognized as important for the well-being of society (see Chapter 2), understanding the geomorphological response to environmental change is becoming increasingly relevant. Geomorphological features can be useful indicators of past environments. In order to predict possible future landscape changes, it is first necessary to understand the geomorphological response to previous environmental change.

Understanding the past and present to predict the future is a difficult task. One of the greatest challenges facing geomorphologists in attempting to interpret the physical landscape is the development of an understanding of processes, particularly the identification of the type and rate of processes both within present-day conditions and those occurring during previous climatic fluctuations. One of the most important fields of international geomorphological research involves the use of relic landscape features to determine certain palaeoenvironmental variables such as temperature and moisture conditions. A serious problem hindering the interpretation of landscape features is the fact that relic landforms may have been obliterated or their features reworked by other processes. Evidence for marginal glaciation, for example, may have been reworked by modern fluvial processes. In other situations the climatic influence appears to be less important. For example, on rock-dominated slopes in southern Africa, such as the Drakensberg Main Escarpment, slope gradients are maintained in equilibrium with their underlying rock mass strength, which is determined by the rock type and structure (see Moon 1990).

Notwithstanding this, certain landforms can be attributed to specific climatic conditions and can be used as environmental indicators. A number of these are described in Table 13.1. For example, glacial environments require specific temperature and precipitation conditions conducive to substantial ice-mass build-up. Protalus ramparts require lower snowfall conditions and the existence of snowbeds adjacent to cliff lines where weathered material from the cliff moves across the snow surface and is deposited downslope (Shakesby 1997). Ice wedges, deep vertical cracks in the soil caused by thermal contraction and filled with ice, form in moist periglacial conditions with annual air temperatures below –6°C. Investigations of landforms have the potential to provide an indication of general climatic conditions. Some of these features found in southern Africa and used as palaeoenvironmental indicators are discussed below. Together with additional evidence from other disciplines, such as the palynological evidence derived from the pollen record preserved in certain sediments, geomorphological investigations can make a meaningful contribution to palaeoenvironmental reconstruction.

Meadows (1988: 307) summarizes the array of geomorphological features that have been used as evidence for various environmental conditions in the Quaternary period. These include alluvial and colluvial sediments (river and hillslope deposits respectively), lake levels and lake sediments, pans, dunes, marine and estuarine sediments, archaeological evidence, palaeosols, duricrusts, and periglacial and glacial features.

Due to the relative scarcity of continental land masses in the southern hemisphere, southern Africa is well located to assist with the worldwide interpretation of landscape evolution under different palaeoenvironmental conditions. While climatic change and landscape response are reasonably well documented for the northern hemisphere (for example the extent of the Pleistocene ice sheets), few definitive geomorphological palaeoenvironmental data exist for southern Africa. Understanding the geomorphic response in the region, both past and present, is thus of international interest.

Geomorphic processes are generally more active in high-altitude environments. This is due to the high slope gradients, generally higher precipitation, rapid changes in temperature, poorly developed soils, and sensitive vegetation found in mountainous regions. As a consequence, any alteration of the climate or the environment will rapidly manifest itself in changing geomorphological processes. For example, a decrease in global temperatures and increased precipitation could result in a lowering of the snowline and associated nival (snow-related) processes in mountainous regions, as well as alpine glacial advance. Mountainous areas thus offer ideal locations for interpreting geomorphic response to climatic change. This chapter will focus on both the current geomorphic landscaping processes and long-term landscape evolution in the mountainous environments of southern Africa (see Figure 13.1). These include the escarpment of the Drakensberg mountains, the Lesotho highlands, and the Eastern and Western Cape mountain regions.

TABLE 13.1 Brief definitions, general location, and main references for relic Quaternary periglacial features found in southern Africa

RELIC FEATURE	DEFINITION	LOCATION
Blockfield	Open-work (no interstitial fines) block accumulations generally on low-gradient slopes formed by *in situ* weathering (often suggested as freeze–thaw)	Lesotho highlands
Blockstream	Open-work block accumulations that have a downslope stream-like appearance.	Drakensberg/ Lesotho highlands
Glacial striae	Scratches on a rock surface characteristic of glacial movement	Eastern Cape Drakensberg
Glacial moraine	A distinct landform caused by deposition of glacially transported material (*moraine* refers to the landform and *till* to the sediment)	Lesotho highlands, Eastern Cape Drakensberg
Nirvation cirques	A hollow, open downstream but bounded upstream by a cliff or slope and formed by nival (snow-related) processes	Lesotho highlands, eastern Free State
Niche glacial deposits	Deposits from small glaciers found in protected sites	Drakensberg Main Escarpment
Rock glacier	Thick lobate mass of coarse material moving downslope through deformation of internal ice or frozen sediments	Eastern Cape Drakensberg
Protalus ramparts	Ridge of predominantly coarse deposit formed parallel to the slope at the base of a snow patch	Eastern Cape Drakensberg
Stone-banked lobes (terraces)	Stone-fronted lobes (terraces) formed though solifluction and/or sorting by freezing and thawing	Drakensberg
Screes	Open-work block deposits (of cryogenic origin) on slopes below a cliff line	Amatola Mountains, Eastern Cape
Valley asymmetry	Valleys with different valley-side gradients caused by different (aspect-related) slope processes	Drakensberg, Lesotho highlands

See Boelhouwers 1995, 1998b.

FIGURE 13.1 Locations of recorded relic cryogenic landforms in southern Africa

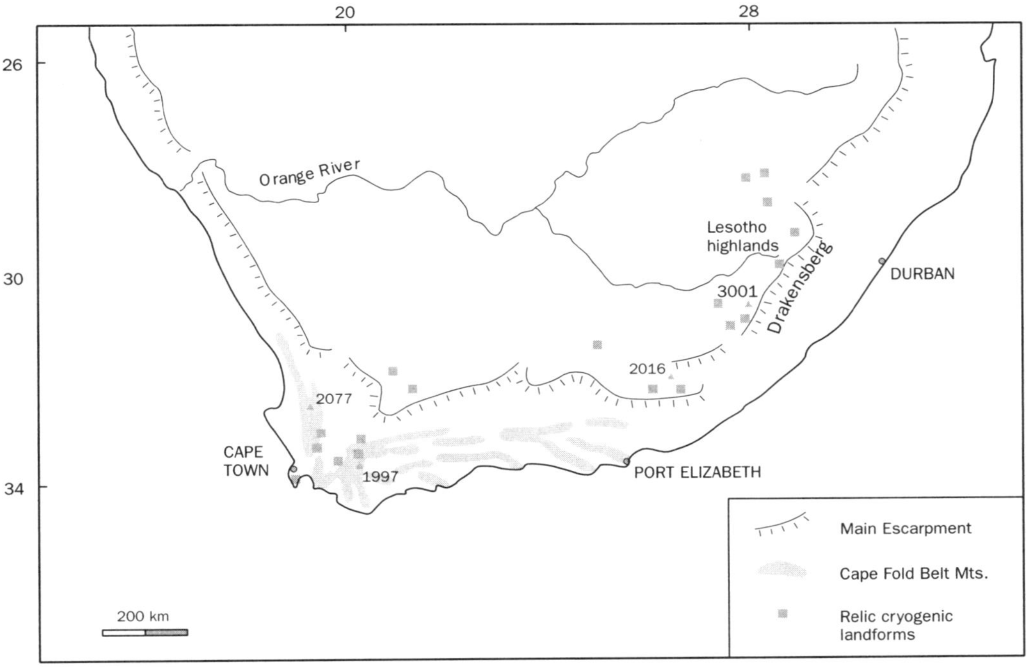

Southern African Quaternary geomorphology

Southern Africa has experienced at least seven glacial events in the past, the most significant being the Dwyka Glaciation (300 million years ago) of which tillite remnants (a rock formed from glacial till) are particularly conspicuous in the region (see Chapter 11). It is virtually impossible, however, to envisage geomorphological evolution over such large time scales. Consequently, much of the geomorphological evidence for climatic change has focused on the Quaternary period (the last two million years) with greatest evidence coming from the late Pleistocene and Holocene epochs (the last 40 000 years), both of which fall within the range of carbon dating.

Southern Africa has undergone continuous climate changes throughout the Quaternary, ranging from alternating glacial and interglacial conditions (with a periodicity of about 100 000 years) to shorter-term fluctuations in climate and weather conditions (see Chapter 2; Partridge 1997). These climatic changes have left their imprint on the landscape.

Southern African glacial and periglacial geomorphology

Under current climatic conditions, glacial and periglacial conditions worldwide are restricted to high-latitude and high-altitude regions. Possible evidence for glacial and periglacial conditions during the Quaternary has been found in the mountainous regions of southern Africa. A substantial proportion of geomorphological research in the region, particularly over the past two decades, has focused on past and present glacial and periglacial features (Boelhouwers 1995). Although few problems are encountered in defining a 'glacial' region, the definition of a 'periglacial' region is more problematic. The term 'periglacial' was originally used to denote the non-glaciated but frost-affected regions adjacent to the Pleistocene ice sheets. This definition was later refined by means of a combination of temperature and precipitation conditions; for example, periglacial activity occurs where mean annual temperatures are below 0°C and mean annual precipitation is below 1 500 mm, regardless of the proximity to glacial activity (see Thorn 1992). According to these and other suggestions as to what constitutes a periglacial environment, no region in southern Africa is truly periglacial in nature today. Studies on current periglacial activities related to frost action are restricted to the Western Cape, the Eastern Cape Drakensberg and the High Drakensberg/Lesotho highlands, suggesting the existence of current marginal periglacial conditions in the higher mountain areas of southern Africa (Lewis 1996; Boelhouwers 1998a; Grab 1998).

Current (marginal) periglacial activity

Active periglacial features are recorded in the Lesotho mountains and to a lesser extent the adjacent escarpment, and in the Hex River Mountains of the Western Cape. In the Lesotho highlands these features include thufur (frost mounds), sorted patterned ground caused by soil freezing and thawing, and non-sorted patterned ground (associated with circular plant growth and needle-ice-raked ground) above 3 000 m (Boelhouwers 1991; Grab 1998). Sorted patterned ground is also found in the Eastern Cape Drakensberg at altitudes above 2 900 m (Lewis 1996).

Other features include stepped microrelief, which occurs both above and below the escarpment in the form of cattle steps and terracettes, but these may not all be attributed to frost action. Stone- and turf-banked lobes and steps occur at high altitudes in the Drakensberg. Since the nature of their formation is unclear, however, reservations as to their periglacial origin can be expressed (Boelhouwers 1991). Active stone- and turf-banked lobes are reported in the Hex River Mountains, where soil frost activity sometimes gives rise to micropatterned ground (Boelhouwers 1998a). Detailed investigations of frost-related activity in the Western Cape mountains are provided by Boelhouwers (1998a) and for the high regions of the Drakensberg by Grab (1998). In general, the current periglacial activity in southern Africa can be described as limited, with the region exhibiting only marginal periglacial activity at high altitudes. The presence of associated features does, however, provide the opportunity for process studies on active features. The marginal nature of periglacial activity and the susceptibility of landscape features to climatic change facilitate studies of landscape response to environmental change. Previous landforms, for example those resulting from an intensification of frost-related processes under a cold climate, may remain in the landscape as relics and indicators of past climatic conditions.

Relic glacial and periglacial landforms in southern Africa

Numerous relic (or fossil) periglacial and glacial features formed under previous depressed-temperature conditions are proposed to exist in the mountainous areas of southern Africa (see Plates 35 and 36, colour section), including the Lesotho highlands, the eastern Free State, the KwaZulu-Natal and Eastern Cape Drakensberg, the Amatola Mountains, and the mountains of the Western Cape (see Figure 13.1). Relevant features include relic erosional and depositional landforms associated with cold-climate geomorphic processes. The types of relic periglacial and glacial features found in the region, with brief descriptions and locations, are listed in Table 13.1. A bibliography on current and relic periglacial activity in southern Africa has been compiled by Boelhouwers (1998b), and further descriptions of periglacial processes are given by French (1997).

An altitudinal distribution of periglacial features, where specific features are associated with specific altitudinal (and hence temperature) zones, has long been proposed for southern Africa (see Marker 1995). The intention behind an altitudinal zonation of relic features is to establish temperature and precipitation conditions and their altitudinal and latitudinal distribution across the region during the period of their formation (the Quaternary), and thus to contribute to palaeoenvironmental reconstruction. Marker (1995) correlates fossil Pleistocene periglacial and glacial evidence to altitude and latitude (see Figure 13.2). Her findings show minimum altitudes for relic cryogenic features to have an inverse relationship to latitude, thus establishing minimum altitudes for cryogenic activity (and hence a snowline) for southern Africa.

This has important palaeoenvironmental implications. However, the rate of decrease in the minimum altitude of cryogenic activity from about 2 000 m at 28.5° S to sea level at around 34° S is quite remarkable, given the lower rates of altitude–latitude temperature depressions for Colorado, New Mexico, and Australia (Marker 1992). Alternatively, it is possible that the altitudinal distribution of cryogenic features in southern Africa is predominantly dictated by lithological, altitudinal, and relief distribution. Given that the subcontinent has experienced different phases and extremes of cold-temperature conditions and precipitation in the Pleistocene epoch (Partridge 1997), exact correlation of cryogenic activity to landforms at specific time periods is difficult, and constructing any general model will prove problematic.

Process–form (and therefore climate–form) relationships are difficult to establish even under current conditions. It is not surprising that the exact nature of formation of the supposed relic periglacial and glacial features and their palaeoenvir-

FIGURE 13.2 Altitudinal distributional of relic cryogenic landforms in southern Africa

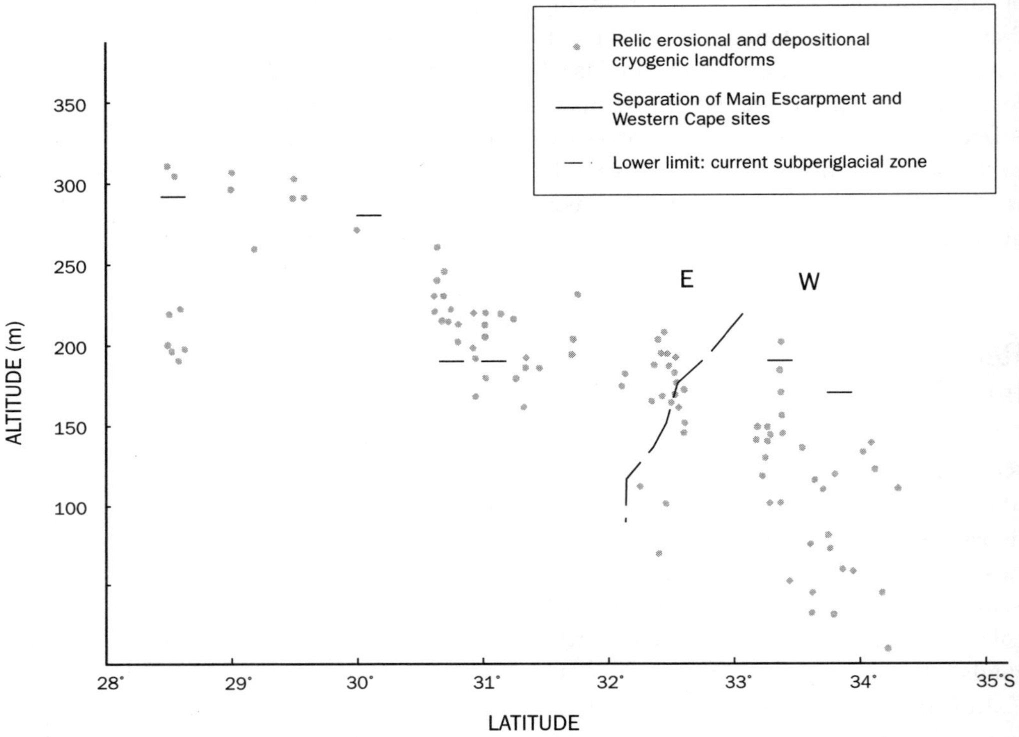

SOURCE: Marker 1995.

onmental implications in southern Africa have raised many debates. Of particular interest is the debate surrounding Quaternary glaciation and marginal glaciation of the mountainous areas. Marker (1991) proposes that there has been marginal glaciation of the Lesotho highlands under arid conditions, while Hall (1994) and Grab (1996) present evidence that suggests localized (niche) glaciation in the vicinity of the Drakensberg escarpment. In contrast to the proposed glacial evidence from the Lesotho highlands, Grab and Hall (1996) suggest possible formation of the Lesotho/Drakensberg hollows by 'bog cirque' development under wet conditions, while Sumner (1995) criticizes evidence presented for escarpment niche glaciation on the basis of the lack of field data and possible alternative modes of formation. Similarly, Pleistocene glacial evidence for the Eastern Cape Drakensberg (Lewis 1996) has been recently criticized, and further investigations and alternative hypotheses require consideration. The presence of as yet undated blockfields and blockstreams in valleys in Lesotho indicates that glaciation could not have postdated their formation. The geomorphic activity of mountainous areas provides a further complication, as it is likely that evidence for any glacial activity may well have been removed. It appears unlikely that a generally accepted model of glaciation in southern Africa in the Quaternary will be reached at this stage.

The evidence presented for previous periglacial activity has not escaped scrutiny. Valley asymmetry has been attributed to aspect contrasts and hence varying processes on slopes with different aspects under periglacial conditions (for example Meiklejohn 1994). Contrasting arguments suggest non-periglacial activity as contributing factors to the asymmetrical development of valleys. The periglacial geomorphology of the eastern Free State has raised similar comment based on the possibility of alternative non-periglacial explanations (Le Roux 1990). In the Eastern Cape Drakensberg, proposed relic rock glaciers generate considerable debate, while the existence of protalus ramparts and other possible ramparts in southern Africa has been questioned by Shakesby (1997). The attribution of angular material in sediments to freeze–thaw weathering processes, and hence specific climatic conditions, has also been criticized on the basis of a generally poor overall understanding of the weathering process (weathering is discussed below).

The above debate can be highlighted further by the fact that most proxy data for the Quaternary indicate a 5°C to 10°C decrease in temperature for southern Africa during the Last Glacial Maximum (LGM) of 18 000 BP (Partridge 1997). It is also known that the environment during this period was more arid than at present and precipitation in the Drakensberg and Lesotho was approximately 70 per cent of current values. Apart from the fact that an 18°C to 24°C drop in temperature would be required for glaciers to have existed at the LGM, glacial environments require substantial amounts of precipitation in the form of snow (which changes over time to glacial ice). Given the existing estimate by most (interdisciplinary) research of a 5°C to 10°C drop in temperature and a drier climate, it is unlikely that glacial conditions could have existed. The indication, therefore, is that conditions around the LGM were more likely periglacial than glacial. A dryer environment, moreover, could have resulted in a relatively inactive periglacial environment. It is thus possible that the most active landscape-forming processes occurred during the warmer interglacial periods under moist conditions, rather than during colder and dryer periods.

Additional problems can be considered. Firstly, more than one process may lead to the same apparent landform or feature. Angular fragments moving in a debris flow may striate (scratch) bedrock surfaces as fragments at the base of a glacier might, while fluvially incised debris deposits may have the same appearance as glacial moraines. Secondly, the duration and size of a geomorphic event and the intensity and duration of subsequent modifying processes will impact on the lifetime of the landform such that impressions of previous climatic conditions can be obliterated.

This discussion does not attempt to provide solutions to the above problems, but serves to highlight contemporary debates in which geomorphological evidence has been used for palaeoenvironmental reconstruction. The interpretation of relic features clearly requires an understanding both of the mechanism of formation of the relic feature and of the effects of subsequent processes operating in these mountain environments, including those at the present day. While international literature can focus on process studies in active periglacial and glacial environments, current southern African geomorphic process studies are limited to slope process investigations where periglacial activity, if present at all, is localized and seasonal. This is nevertheless an important contribution towards understanding the overall environment, and provides insight into geomorphological response to environmental change. Since slope form and slope processes are integrated components of landscape development, both slope development and slope-surface processes will now be considered further.

Slope development and processes

Slope development

The earliest studies concerning landscape evolution emphasized macro-scale features and processes. Modern geomorphological studies have focused more on the slope scale and smaller. The best-known of the older models are those of Davis and King, dating from the late nineteenth century and the 1920s respectively. An excellent comparative discussion of their models, as well as that of Penck, is provided by Thorn (1988).

Davision geomorphology, as it became known, attempted to simplify the changing landscape into stages within a 'geographical cycle' (Thorn 1988). The ultimate landscape that resulted from the Davision stages is almost without relief and is known as a peneplain. Davis's progressive stages of 'youth', 'maturity', and 'old age' are often still used in many school textbooks, despite the fact that the science of geomorphology has progressed beyond these earlier models.

Lester King is recognized to be the last internationally important scientist who considered geomorphology on a global scale. In the southern African context, King is extremely important, as his studies were largely based on work in this subcontinent; in fact, he is often credited with creating an awareness of southern African geomorphology. King's model included uplift, cycles of erosion, and the identification of associated landforms on a global scale (see King 1963). An essential component of his model was the notion of parallel scarp retreat, and, when observing the landscape of the South African interior, it is not difficult to see the origin of his thoughts.

Some more recent models of landscape development have been formulated which link processes to slope and soil development (Dalrymple et al. 1968). Other models illustrating slope form and development rather than regional development appear less problematic, attempting as they do to incorporate both lithological or structural control and geomorphic processes (see Moon 1988 for a review). Scale extrapolation, from the individual moving soil particle to landscape morphology, is a problem inherent in any landscape development model. In many respects, the regional approaches adopted by Davis and King have been abandoned and are now considered to be relevant by only a few researchers (see Thorn 1988). In order to describe landscape change, geomorphology has progressed towards a more quantitative, dynamic, and process-orientated approach focusing on processes operating at the slope scale. Nevertheless, some of these earliest models of landscape evolution have shaped the field of geomorphology.

Slope processes

Mass movement

Considering the diversity of mass-movement phenomena in southern Africa, surprisingly few data are available on either current processes or previous mass-movement events (see Sumner and Meiklejohn 1998). In the Drakensberg, supposed current and relic mass-wasting events include major relic landsliding and rotational slumping incorporating bedrock, active and inactive surficial (translational) landsliding on slopes and riverbanks, stepped micro-relief in the form of terracettes, and the widespread occurrence of soil creep in the form of terracettes (see Table 13.2).

The type and distribution of mass-movement events in the Drakensberg appear to be regulated by the underlying lithology. Larger landsliding events and rockfalls, although evident in the basalt, are primarily associated with the underly-

TABLE 13.2 Classification and characteristics of some mass-movement phenomena found in the Drakensberg

MASS MOVEMENT TYPE	MATERIAL IN MOTION	MOISTURE CONTENT	NATURE AND RATE OF MOVEMENT
Creep	Soil	Low	Slow plastic downslope deformation of soil
Debris flow	Mixture of fine and coarse slope debris	High	Very rapid downslope movement with intergranular shear distributed evenly through the mass
Translational landslide	Bedrock, slope debris, or soil	Low to moderate	Slow to extremely rapid slide with a well-defined shear (failure) plane approximately parallel to the ground surface
Rotational landslide/ slumping	Bedrock, slope debris, or soil	Low to moderate	Slow to extremely rapid rotational movement along a well-defined concave shear plane
Rockfall	Detached bedrock blocks	Low	Free fall of blocks from free faces

ing Clarens Formation sandstones and the transition to the lower Elliot Formation shales. Surficial mass movements, primarily translational landslides and debris flows incorporating soils and slope debris, predominate in the shales. Observations show that few large landslide events incorporating bedrock occur under current conditions, and it is apparent that surficial events are the dominant mass-movement processes under present climatic conditions (see Plate 37). Trigger mechanisms for surficial mass movements probably include increased interflow at the bedrock–regolith interface, soil saturation during high-rainfall events, and reactivation of relic landslide events. The larger relic mass-movement events may have been associated with fluvial incision and slope steepening during the Pliocene uplift. No data are available on event frequency–magnitude relationships for the region, and therefore the overall significance of mass movements to landscape development, both past and present, is not known.

Frost creep caused by freezing and thawing of the ground and possible solifluction activity (slow flow of saturated slope material over an impermeable layer) is found in the marginal periglacial region of the Eastern Cape Drakensberg and the Lesotho highlands (Lewis 1996; Boelhouwers 1995). Turf- and stone-banked steps, possibly evidence of solifluction, have been identified in the Lesotho highlands above 3 000 m. These may be inactive under current conditions (Boelhouwers 1991). In general, mass movements will contribute to landscape degradation and assist soil erosion by disaggregating soil and accelerating surface-wash processes on disturbed surfaces.

Soil erosion

Southern Africa as a whole is susceptible to soil erosion, and the mountainous regions are particularly vulnerable to degradation, as illustrated by high erosivity indices and erodibility (susceptibility) potential (see Chapter 14). Both surface and subsurface erosion are prevalent in southern Africa. Surface erosion processes include rainsplash erosion caused by the impact of falling raindrops on the soil surface. Where rainfall intensity exceeds the infiltration capacity of the soil, runoff is initiated. Runoff has the potential to entrain and transport soil particles, and, where wash is combined with rainsplash disaggregation, erosion is particularly effective. If wash processes are concentrated, soil erosion leads to the development of rills, which can ultimately form gullies. Soil erosion assists in the reworking of slopes and surficial features such as mass-movement events. As elsewhere, soil erosion in the mountainous regions of southern Africa is enhanced by human activity. This includes grazing, for example in the Lesotho highlands, and management practices such as veld-burning programmes, as well as the use of footpaths (see Plate 38) and vehicle tracks in wilderness areas.

The development of soil pipes by subsurface erosion is also prevalent in the Drakensberg. Beckedahl (1998) describes subsurface erosion features associated with dispersive soil types and non-dispersive soil scree slopes in the foothills of the Drakensberg. Evidence for the onset of piping is not obvious in the landscape, but the collapse of a pipe can lead to immediate gully development (see Figure 13.3). Further explanations of soil erosion pro-

FIGURE 13.3 Gulley development by surface and subsurface soil erosion

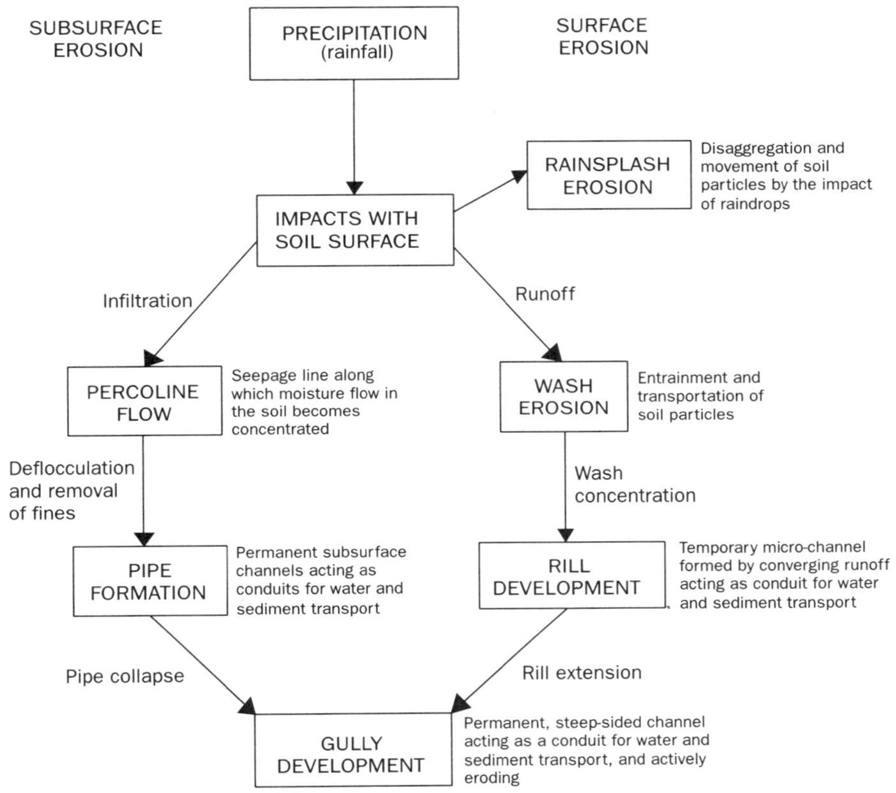

cesses and features can be found in general texts, and details of erosion features found in southern Africa are provided by Beckedahl et al. (1988).

Where moisture is present and temperatures are sufficiently cold for ground freezing to occur, small needle-ice protrusions may grow near the soil surface, lifting the surface layer of soil particles. This needle-ice growth results in soil and vegetation disruption (turf exfoliation), particularly in the Lesotho highlands and down to altitudes of 2 300 m in the Drakensberg in winter (Boelhouwers 1991). Sites where vegetation is scarce, such as footpaths or the banks of streams, are particularly prone to needle-ice disruption. Needle-ice growth is thought to enhance the development of terracettes, particularly in the Lesotho highlands. While soil disrupted by needle-ice may either move downslope under gravity or be eroded by wind, it is likely that wash processes during the wetter summer periods are primarily responsible for the erosion of needle-ice-disrupted soil.

Variables interacting to control the rate and type of soil erosion processes include climatic conditions (erosivity) and erodibility factors such as soil properties, vegetation, geology, and topography (see Figure 13.4). In addition, the human element plays a significant role in modifying the erodibility of soil. Changes in land management and land-use practices have the potential to influence soil erodibility, thus altering the type and rate of the soil erosion process. A further discussion of human-induced soil degradation is provided in Chapter 14. Similarly, changes in climate can influence temperature, precipitation, and vegetation characteristics. During a colder period, for example, the erosive influence of needle-ice may be enhanced and have a greater influence at lower altitudes. Wind erosion, while not considered a major factor at high altitude at present, can be accelerated during dryer phases. Evidence for these former intensified processes might not be maintained in the landscape, due to the domination of successive processes.

Much of the material comprising relic glacial and periglacial features, as well as the material within which current periglacial activity, mass movement, and soil erosion processes take place, originates from the weathering of bedrock. Palaeoenvironmental inferences in southern Africa and worldwide have been drawn

FIGURE 13.4 Variables acting on the soil erosion process

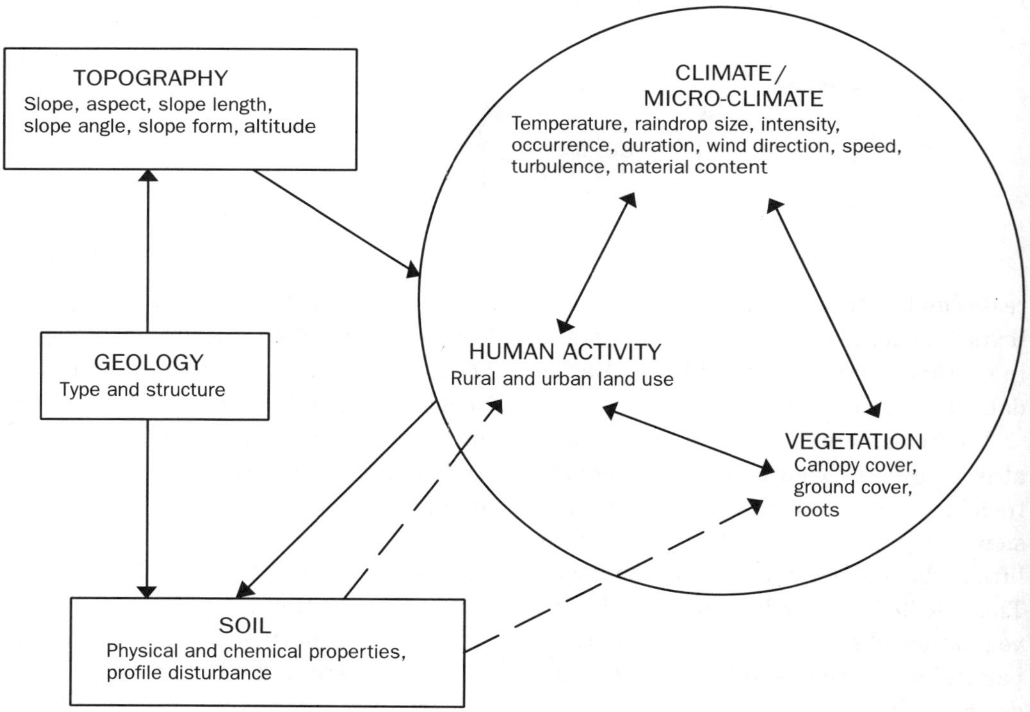

SOURCE: Beckedahl et al. 1988.

directly from the products of weathered bedrock. Weathering is, therefore, a fundamental geomorphic process and is particularly important in the context of palaeoenvironmental reconstruction and the understanding of landscape evolution. (Bibliographies of studies in southern Africa on weathering and slope processes (Sumner and Meiklejohn 1998) and soil erosion (Watson 1998) are available.)

Weathering

Despite the importance of weathering in landscape processes, there is a scarcity of published material on weathering processes active in southern Africa. In most cases, weathering is considered to be one component among other geomorphic processes. It is only relatively recently that research has been conducted into specific weathering processes by a small number of researchers (see Sumner and Meiklejohn 1998). Very few of these studies (for example Meiklejohn 1997) actually monitor field conditions and attempt to infer processes. Recent process studies have concentrated on biological and cryogenic (freeze–thaw) weathering, as well as the deterioration of Drakensberg basalts and indigenous rock art due to weathering processes. It is apparent that Hall's (1988) contention, that considerable research on southern African weathering is still required, still holds.

Notwithstanding the scarcity of weathering studies in southern Africa and the lack of field data worldwide, observations of weathered material have been utilized locally as an indication of palaeoenvironmental conditions. Angular clasts contained within deposits, as well as subangular to angular material constituting screes and blockfields/streams, have been considered the product of cold-climate weathering, notably the freeze–thaw process. Since the international literature indicates shortcomings in linking weathering products to field and laboratory observations, this approach has serious limitations. In order to highlight these problems, both the types of and controls on weathering processes are presented, with a view to highlighting the complex interaction of processes and hence the difficulties in relying on field observations.

Weathering processes and controls on weathering

Weathering is defined as the *in situ* physical (that is, mechanical), chemical, or biological alteration of rock through a variety of processes (see Table 13.3). As weathering takes place *in situ*, no movement of material occurs, and it is therefore apparent that weathering is a contributor to, but cannot result in, landform development (Hall 1988). Landforms arise from the reworking of weathered material by processes such as soil erosion and mass movement.

Chemical weathering involves chemical alteration of rock minerals, while physical weathering is the mechanical alteration of rock where no chemical change takes place. Without knowledge of the specific processes that take place at a small scale, it is not possible to determine how a landscape has been formed. Therefore it is imperative that detailed information regarding this primary alteration of bed-

TABLE 13.3 Rock weathering processes commonly referred to in geomorphic literature

PHYSICAL (MECHANICAL) WEATHERING	CHEMICAL AND BIOCHEMICAL WEATHERING
Dilatation	
Thermal shock	Solution
Thermal fatigue	Hydration
Cryogenic weathering	Acid–base interactions
Salt crystallization	Oxidation and reduction
Hydration of salts	Chelation
Expansion and contraction of salts	Hydrolysis
Hydration	Carbonation
Slaking	
Growth of biological organisms	

rock is understood. Table 13.3 shows the weathering processes that are commonly referred to in the relevant literature. In practice, individual weathering processes are impossible to separate, but for simplification they are normally discussed as separate entities. Similarly, while biological weathering may be classified as mechanical or chemical, texts consider it as a separate issue.

Apart from some chemical processes, rock weathering does not take place in a constant environment. Changes in the environment, even on a very small scale, are required before any breakdown is possible. While the following discussion often refers to specific processes, it will be shown later that rock weathering is a series of interdependent processes with numerous variables interacting to control the rate and type of weathering process. These controls include the properties of the bedrock, climatic and micro-climatic conditions, the length of time available for weathering, and the removal of the weathered product (see Figure 13.5).

Bedrock properties determine the susceptibility to weathering processes and the ease with which weathered products can be removed. Rock mineralogy dictates the chemical composition, albedo, thermal properties, and moisture absorption capacities of a rock. The chemical composition determines the susceptibility of a rock to chemical alteration. Minerals with a lower albedo (darker-coloured minerals) absorb incoming solar radiation to a greater extent than do light-coloured minerals, and will therefore be more prone to thermal weathering processes.

Rock structure is important since discontinuities and other places of inherent weakness are the areas that will be most susceptible to breakdown. Bedding planes (the interface between layers of a sedimentary rock) and joints are avenues for moisture and gaseous intrusion, making them locations for dynamic changes and thus enhanced weathering and weathered product removal. Furthermore, structure influences rock strength and hence its resistance to certain physical weather-

FIGURE 13.5 Variables acting on the weathering process

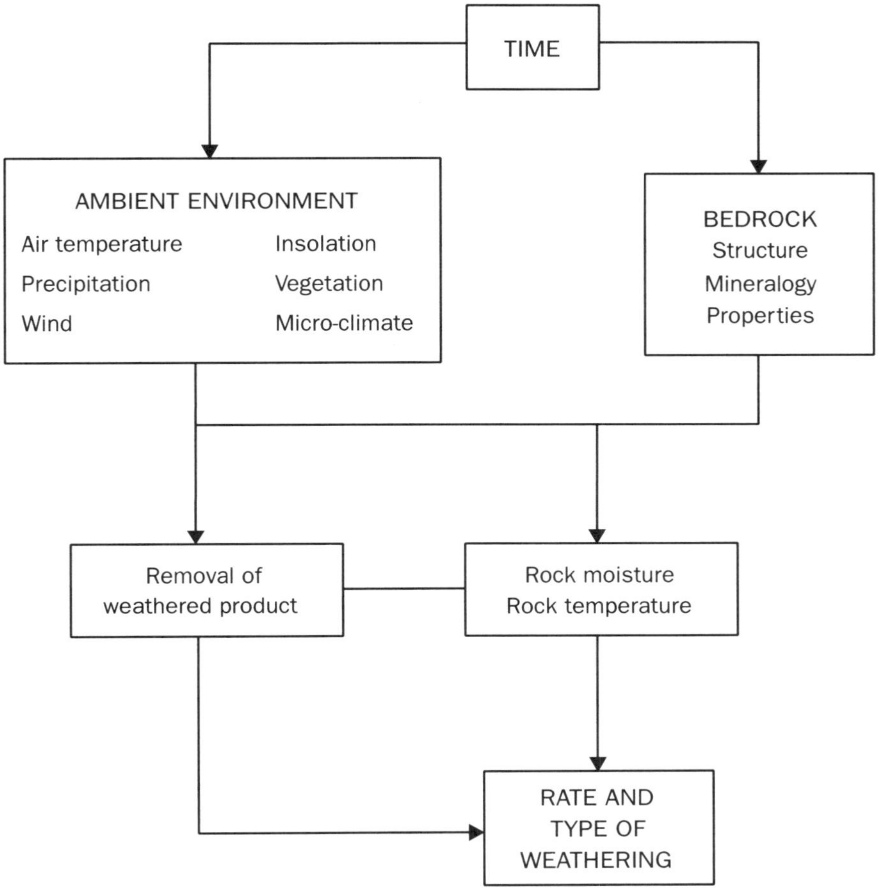

ing processes. Important rock properties include porosity, micro-porosity, permeability, saturation co-efficient, and water absorption capacity. These factors influence moisture movement and the ability of a rock to absorb moisture. This is especially the case with porosity and micro-porosity (micro-porosity being the percentage of rock pores with a diameter less than 1 micron (1 μm = 1 mm^{-3})). Moisture absorption and movement increase with increasing porosity and decreasing micro-porosity.

General weathering models in the past have focused on climatic conditions as the determinants of the rate and type of weathering processes. Weinert (1965) zoned southern Africa according to climatic conditions, noting that chemical weathering will predominate in the eastern regions and physical in the west. Of greater importance in considering weathering processes, however, are the rock moisture and temperature conditions where the weathering is taking place: the so-called weathering front (Pope et al. 1995). The ambient conditions (micro-environment), although influenced by climate and vegetation, have a greater influence on the temperature and moisture conditions at the weathering front. General climate is thus not necessarily an indication of prevalent weathering processes or intensity. The rate and type of processes will vary from site to site over short distances in the landscape. It is still generally accepted, however, that the most favourable conditions for rock weathering are moist, warm climates. On the other hand, cold dry climates are said to be less favourable for rock weathering. This relatively unfavourable weathering environment is possibly the reason for angular material predominating in these areas; the relatively inactive environment may preserve angular fragments.

Given the perceived importance of thermal and cryogenic weathering processes, temperature is obviously an extremely important weathering control. While air temperature may have an influence on rock temperature, insolation has the primary influence. Care should be taken that the actual temperature of the rock is considered, as air temperature cannot be used as a surrogate for rock temperature. Absolute rock temperatures are important in determining rates of chemical reaction (rates of chemical reactions increase with temperature), and variations in rock temperature result in a variety of weathering processes, which are discussed below. While moisture availability is normally associated with precipitation, the relationship is not always direct. The rates of slope-surface drainage and infiltration influence moisture availability, and groundwater quantities may not be directly related to precipitation. Snow may sublimate (evaporate directly from ice crystals) rather than melt, and therefore will not supply moisture directly for weathering processes. The actual rock moisture is of greater importance than the general precipitation conditions. A further influence is wind, which may cause desiccation of rock surfaces and alter rock-surface temperatures.

Through time, bedrock properties are altered by weathering processes, and climatic change and variability influence rates and types of weathering processes. The rate of removal of weathered products is an important factor, since weathering residues may offer protection from insolation (and hence thermal changes) while maintaining higher moisture contents for longer periods. Weathering processes at a site are expected to change through time under different climatic conditions, but may also change through time even under constant climatic conditions due to the accumulation of weathered products.

Weathering process interdependence and interrelationships

In developing a hypothesis about rock weathering it is necessary to consider that no single process can be isolated from the others. Rock breakdown is the result of a complex interdependence and interrelationship between mechanical, chemical, and biological mechanisms.

The major controls on weathering are rock temperature and rock moisture (see Figure 13.5). Rock temperature has an important influence on all weathering mechanisms. For example, it affects the rate of chemical reactions, chemical equilibria, as well as evaporation and condensation cycles (Ollier 1984; Yatsu 1988). The rock moisture regime may be influenced by temporal scales of variation ranging from seasonal changes to more rapid fluctuations occurring within periods of a few minutes. A major influence on the moisture regime is atmospheric humidity, which controls weathering processes near the rock surface. Weathering processes likely to be active at this interface include salt crystallization, hydration/dehydration of minerals and precipitates, solution processes, and hydrolysis (see Figure 13.6).

Attempts to ascertain active weathering processes in the field are difficult and time-consuming. In a long-term study on the deterioration of rock art by weathering of the Clarens Formation in the Drakensberg, it was found that weathering of the sandstone is a complex set of interactions, rather than unrelated individual processes (Meiklejohn 1997). While the rock moisture regime was identified as the major influence on weathering, other controls (for example the thermal regime, differences in mineralogy, and the original sedimentary environment) cannot be ignored. The rock moisture regime itself is influenced by rock temperature, atmospheric moisture, and groundwater moving along rock discontinuities. To emphasize the interdependence and interrelationship between weathering mechanisms from this case study, it is best to attempt to highlight the effect of changing rock moisture conditions.

An increase in rock moisture content due to precipitation, condensation from decreasing rock temperature, or increasing atmospheric moisture will cause hydration of rock minerals, clay minerals, and precipitated salts, solution of rock minerals and precipitates, hydrolysis, and chemical alteration (see Figure 13.6). Alternatively, a decrease in rock moisture content, which may be the result of increasing rock temperature, decreasing atmospheric moisture, or a decrease in the amount of moisture moving along rock discontinuities, will cause dehydration of clay minerals, rock minerals, and precipitated salts, and precipitation (crystallization) of solutes.

The above moisture-controlled weathering processes and other processes related to changes in temperature, such as thermal fatigue, enlarge existing pores and cause granular disintegration. This increases the rock porosity and widens bedding planes. The enlargement of pores will cause a decrease in rock micro-porosity. Lower micro-porosity, higher porosity, and wider bedding planes result in easier moisture infiltration, which in turn causes increased weathering. A further effect of moisture is its ability to remove weathering products. Apart from those taking place along bedding planes, these processes will be most active at the rock surface and within the first few micrometers below it.

This is the very environment in which Bushman paintings are located, and weathering processes therefore cause the deterioration of rock art (see Plate 39). While the above study was generalized with respect to weathering processes in the Clarens Formation, it was found that microclimatic and other environmental variables such as rock moisture, rock

FIGURE 13.6 Effects of changing moisture and temperature conditions on weathering processes

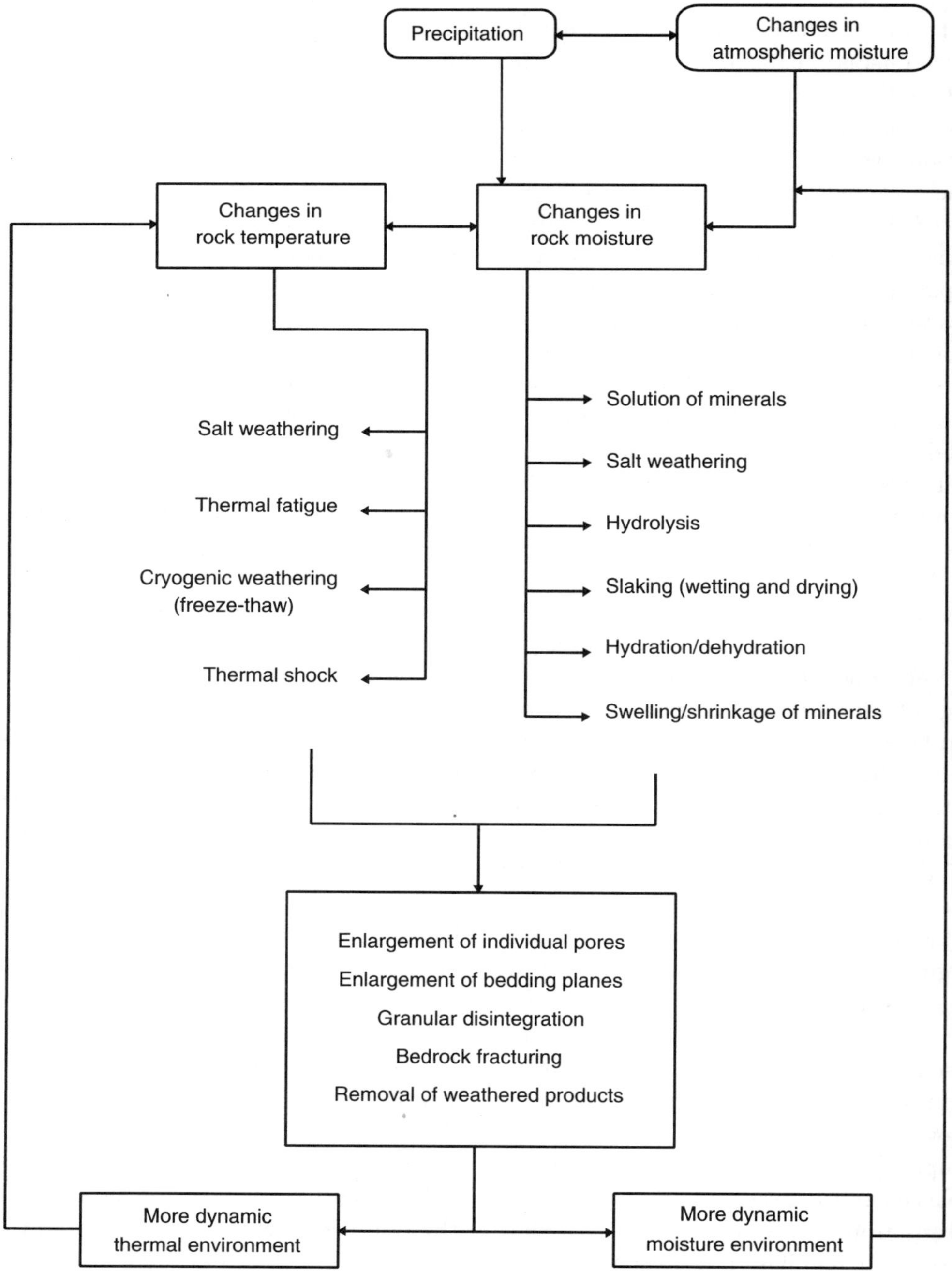

structure, and rock chemistry not only differ from site to site, but also from one location to another within a single shelter. Adding to the difficulty of interpreting weathering, the processes are not only site-specific, but also depend on the precise situation of a rock surface.

It follows that the isolation of processes based on field data is difficult, while an approach based only on field observations is virtually impossible. In addition to the complex integration of processes, it can be seen that weathering, such as that of the Clarens Formation sandstones, can be an 'accelerating' process and that the intensity of processes can alter with time as the physical properties of the material change. To further highlight these process complexities within the context of using weathering products as palaeoenvironmental indicators, the freeze–thaw weathering process will now be considered further.

Weathered products as palaeoenvironmental indicators

Angular rock fragments attributed to cryogenic weathering (freeze–thaw) have been used in southern Africa to argue for previous and contemporary cold environmental conditions (Hall 1991). While the interrelationship of freeze–thaw with other weathering processes is as yet not entirely resolved, neither is the actual mechanism of freeze–thaw clearly understood. Various theories have emerged which attempt to explain the mechanism based on the freezing of moisture *outside* an intact rock and moisture *within* the rock (see Bland and Rolls 1998 for further discussion). These theories include:

- the volumetric expansion of water as it freezes to form ice;
- the freezing of water within a closed system;
- the crystallization pressure of ice;
- capillary frost damage;
- ordered (polar) water;
- hydraulic pressure; and
- the growth of salt crystals during depressed temperatures.

While a number of these processes may take place within a rock during depressed temperatures, the volumetric expansion upon freezing of water external to the rock is the most cited freeze–thaw process. There is a 9 per cent volumetric expansion when water in cracks freezes to form ice, and the resultant stresses are theoretically sufficient to cause the rock to break into angular fragments. This basic theory is problematic, however. The idea of water freezing being solely able to cause angular debris is not plausible, as water can only enter pre-existing cracks and, in any event, the angularity of products will be caused by the bedrock structure (see Plate 40). The possibility of moisture entering cracks and loosening blocks or extending cracks should also be considered. Again a problem exists, in that when water freezes the expansion is likely to occur in the direction of least resistance, in other words towards open space and not towards the crack apex. Crack propagation must then be unlikely.

A further illustration of misconceptions relating to freeze–thaw weathering is seen in many texts in which photographs and diagrams of open cracks are used to illustrate the activity of this weathering mechanism. Any water in these cracks is likely to flow out the side of the cracks before freezing. Further problems exist in the literature concerning the mechanism of freeze–thaw weathering. Air temperatures are often used as a surrogate for rock temperatures, and 0°C is used as the freezing point for water. Measured rock temperature often shows little relationship to air temperature, however. For example, field data from Alexander Island, Antarctica, showed the air temperature to be −2°C, while that for rock was 31°C (Meiklejohn 1994). In southern Africa, air temperatures in the vicinity of the Clarens Formation of the Drakensberg were measured as low as −10°C; the minimum rock temperature recorded simultaneously was +4°C. The freezing point of water is also not necessarily 0°C; both dissolved salts and the ambient pressure will influence the freezing point of liquids.

In many instances, freeze–thaw weathering may not be possible, simply because of the absence of sufficient quantities of moisture. It is unlikely that cryogenic weathering is active in most areas of the continental Antarctica, which is regarded as the driest place on earth. In a similar way, with the exception of the Western Cape mountains, freeze–thaw weathering is unlikely in the highlands of southern Africa; in summer it is too hot, and in winter, apart from rock temperatures not being low enough, there is insufficient moisture.

There is thus little clarity regarding freeze–thaw weathering and its association with landform development in cold environments. Furthermore, the interaction of cryogenic weathering and other weathering processes is poorly understood (Hall 1992, 1995; Thorn 1992). The presence of moisture required for freeze–thaw implies that a range of chemical weathering processes may take place even where air temperatures are low but insolation, and hence rock temperature, is high. Changing temperatures, a requirement for freeze–thaw weathering, and the associated moisture fluctuations give rise to a range of mechanical weathering processes that must operate simultaneously to freeze–thaw. The isolation of one weathering system, freeze–thaw, or a single mechanism of freeze–thaw, is thus virtually impossible in the field and has also proved difficult in the laboratory. This problem is exacerbated by the scarcity of field data on rock moisture and temperature conditions in southern Africa and from regions beyond the subcontinent where cryogenic processes appear active.

The uncertainty with regard to actual processes that cause the weathering of bedrock, and hence the difficulty in understanding landscape evolution, extends beyond the freeze–thaw weathering process. A further example is thermal shattering. Extremely rapid changes in temperature have been noted to result in the shattering of rocks, caused by the inability of the rock to adjust to rapid expansion and contraction. The result is angular fragments from breaks across existing discontinuities (see Plate 40). Laboratory experiments have shown that a rock temperature change of 2°C or more per minute may be sufficient to cause cracks along grain boundaries (Yatsu 1988). Despite such rapid temperature changes being recorded in the field in southern Africa, there is no evidence supporting this weathering mechanism in the Drakensberg mountains (Meiklejohn 1997).

The future of weathering studies in South Africa

It is apparent that there is a need for a thorough understanding of rock weathering processes and their role in the formation of the southern African landscape. Many of the misconceptions about rock weathering result from a scarcity of field data. Furthermore, it is apparent that most laboratory simulations do not use field-based data as a basis for simulation. In this regard, the current research emphasis should be on measuring actual environmental conditions and replicating these under laboratory conditions to determine active weathering mechanisms. With sufficient data regarding active weathering processes and their responses to changing environments, it may be possible to use weathering for palaeoenvironmental research. Within the limits of our current understanding, however, it is suggested that potential inferences from weathered material are highly problematic and should be treated with extreme caution.

Conclusion

By its very definition, geomorphology involves the study of landscape changes through time. Certain environments are particularly appropriate to research on evolving landscapes, no more so than mountain environments. The emphasis of this chapter has been on landscapes in mountain environments, where landscape evolution has been interpreted using relic features attributed to processes operative under previously different climatic conditions. Considering our knowledge of current processes and the problems associated with interpreting processes from relic landforms, the understanding of geomorphic landscape evolution is problematic. Consequently, predicting future landscape response to environmental change is proving difficult.

Other important geomorphological areas in southern Africa exist which can contribute to the geomorphic understanding of our changing environment. This is particularly true for coastal and karst environments. Coastal areas respond dramatically to both environmental and human-induced changes. Karst environments, where limestone dominates the local lithology, are particularly prone to subsidence through the solution of bedrock. It is thus logical to assume that changing environmental conditions could result in these areas becoming more geomorphologically active and susceptible to subsidence. The gold-mining areas of the western parts of Gauteng, eastern North West province, and Centurion in Gauteng are areas affected by this. Investigations into karst environments have also proved particularly useful for palaeoenvironmental reconstruction (Marker 1998).

The changing environment and the impact of humans on it will manifest themselves in the processes that are active in the landscape. A better understanding of these processes will in turn make possible an understanding of how the landscape has responded to changes in the past and allow predictions to be made as to how processes, and therefore landscapes, will respond to environmental changes in the future.

14 Soil resources: distribution, utilization, and degradation

MICHIEL C. LAKER

Soil is one of the basic natural resources on which people depend for their livelihood. About 99 per cent of our food is produced on land, with plant products comprising 93 per cent of the human diet (Borlaug 1997). Consequently, the human carrying capacity of a country or region, in terms of food security, is estimated in terms of the availability of arable land, that is, land which is suitable for successful, economical crop production. Soil is one component of 'land'; others include climate, topography, and vegetation; these are discussed in other chapters.

Soil is a renewable resource in the sense that it can be used repeatedly for growing crops. If it is degraded by misuse beyond the point at which it can recover or be reclaimed to its original state, it becomes a non-renewable resource, because soil formation is an extremely slow process. The 'resilience' or recovery potential of a soil is therefore a very important characteristic.

Soils are very complex systems. The scientific study of soil consequently covers various facets, such as soil physics, soil chemistry, soil fertility, soil biology, pedology (soil formation and classification), soil mineralogy, and so on. Readers are advised to consult a good introductory soil science handbook such as Brady and Weil (1996). Unfortunately no general soils text is available for southern or South Africa. For definitions and explanations of soils terms used in this chapter, the reader is referred to the excellent bilingual (English/Afrikaans) *Glossary of Soil Science,* published by the Soil Science Society of South Africa (Van der Watt and Van Rooyen 1995).

Functions of soils

Soil has various functions. These include the following:

1 Soil serves as a medium for plant growth. It stores and supplies the water and nutrients which plants need. It also supplies the oxygen required for root respiration, without which roots cannot absorb water and nutrients.

2 Soil serves as a regulating system for water supply, acting as a sponge which retains and gradually releases water into the hydrological cycle.
3 Through its ability to adsorb and/or precipitate various chemical constituents, soil purifies water and decontaminates the environment.
4 Soil is an important engineering medium. Almost every house, road, and other structure is built on (or partly in) soil. Soil also provides construction material in the form of earth fill, bricks, and clay for the construction of earth dams.

The ability of a soil to perform the above functions depends on its composition and physical characteristics. These will be reviewed below.

Composition and properties of soil

Physical properties

Soil is defined as the loose outer layer of the lithosphere consisting of material that has undergone changes under the influence of living organisms. The latter distinguishes soil from weathered and/or transported geological materials. Geogenic processes produce the parent materials from which soils are subsequently formed by the process of *pedogenesis*. Soil consists of three phases: a solid phase, a liquid phase (soil water), and a gaseous phase (soil air). In an ideal soil the solid phase occupies about 50 per cent and the liquid and gaseous phases each about 25 per cent of the total soil volume.

The solid phase of soils consists predominantly (usually more than 95 per cent) of inorganic particles of different sizes and varying chemical composition. Only particles less than 2 mm in diameter are considered to be soil or 'fine earth'. Coarser materials are classified as gravel, stones, and so on. Soil particles are divided into different particle size groups according to their diameters: sand (particles which are between 2 and 0.05 mm in diameter), silt (0.05 to 0.002 mm), and clay (less than 0.002 mm). Soil texture is defined as the relative proportions by weight of different particle size groups in the soil. Soil textural classes can be charted on a triangular texture diagram and are given names (for example 'sandy clay loam') for easy communication (see Figure 14.1).

Soil texture is a very important soil property since it determines a large number of other properties which affect the suitability of the soil for different uses. Coarse-textured (sandy) soils are generally freely drained, well-aerated, and favourable for root penetration, but have low water-storage capacities and are infertile. Conversely, fine-textured (clayey) soils are generally poorly drained, poorly aerated, and unfavourable for root penetration, but have large water-storage capacities and are fertile. An exception is the highly weathered, humid tropical clays, which are well-drained, well-aerated, and extremely infertile. Medium-textured soils combine the favourable conditions of both sandy and clayey soils. The above statements are very crude generalizations, however, which may be far from the truth in a specific situation.

FIGURE 14.1 Soil texture diagram

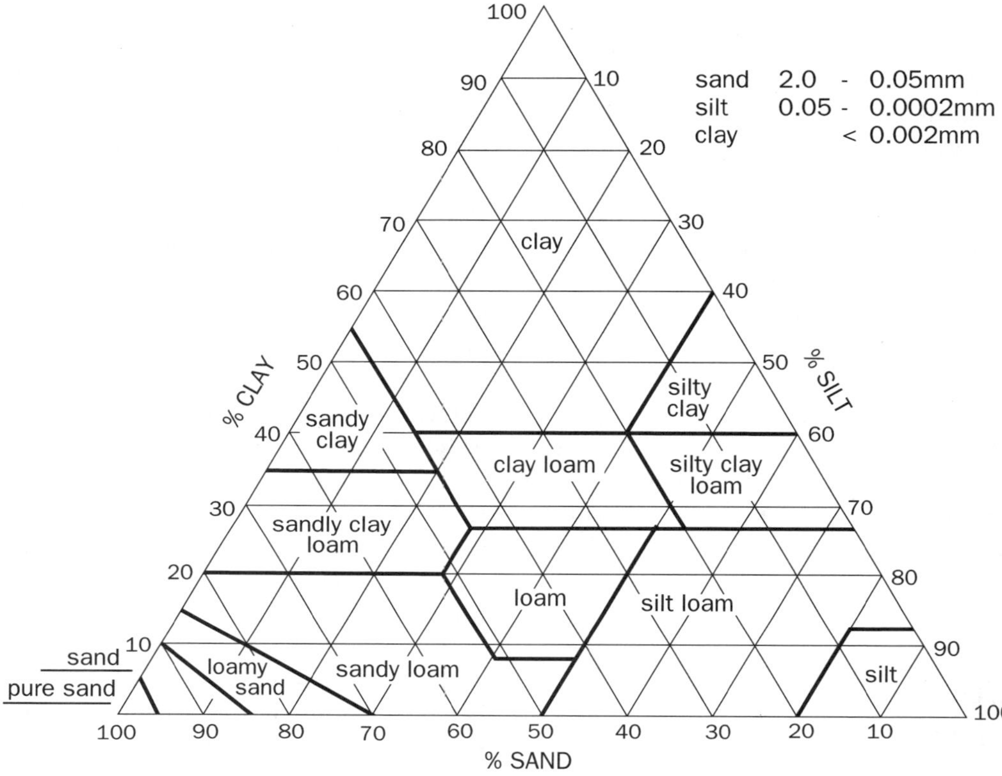

Sand consists almost totally of quartz grains. Even red sands, like those of the Kalahari and the Namib, consist of quartz grains covered by thin iron oxide coatings. Likewise, silt consists mainly of quartz, but also contains significant quantities of other primary minerals (minerals formed during the crystallization of igneous rocks and inherited from the parent material). The clay fractions of soils consist mainly of secondary alumino-silicate clay minerals (minerals formed from the weathering products of primary minerals or other secondary minerals). Although there is a wide variety of these alumino-silicate clay minerals, the clay fractions of soils are dominated by smectites (mainly montmorillonite), hydrous micas (illite), and kaolinite. The clay fractions of highly weathered, humid tropical soils are often dominated by sesquioxides (amorphous oxides and hydroxides mainly of iron and aluminium). The mineralogical composition of the clay fraction of a specific soil is determined by the factors of soil formation (see below) and has an important influence on various soil properties, such as soil fertility, swelling, and so on.

In most soils the mineral particles do not occur as single grains, but are grouped

Clay minerals can be of two types: layer silicate clays, and iron and aluminium oxide clays. **Layer silcrete clays** have a crystalline layered structure, consisting of sheets of silicon, aluminium, magnesium, and/or iron atoms surrounded and held together by oxygen and hydroxy groups. Kaolinite is a 1:1 clay, with one silica sheet and one aluminium sheet. 2:1 clays have one aluminium sheet sandwiched between two silica sheets. Examples are montmorillonite and illite. **Smectites** are a group of 2:1 silicate clays with significant swelling and shrinkage properties; for example montmorillonite.

Iron and aluminium oxide clays develop in highly weathered soils of the tropics and subtropics. Examples are gibbsite and goethite. They lack the layered structure of silicate clays and are not as sticky or plastic.

together in larger units. This grouping of particles into aggregates (*peds*) is called soil structure. Structure is described according to type (for example blocky, granular, columnar), degree (weak, moderate, strong), and size (fine, medium, coarse) (see Brady and Weil 1996: 111–13). Soil structure is determined by various factors, such as texture, clay mineralogy, cation composition, and climate. Together with texture it determines a number of soil characteristics such as permeability to water, aeration, and root penetrability. The degree of structural stability against dispersion or disaggregation by water or implements is an important factor, since it determines the resistance of the soil to erosion, compaction, and/or crusting (surface sealing).

Although organic matter constitutes only a small part (usually less than 5 per cent) of the solid phase of the soil, it has a number of major roles. Firstly, in most soils, decomposed organic matter (humus) is the only agent which stabilizes soil structure. Secondly, humus has a much higher cation exchange capacity than any other soil material. Consequently it makes a large contribution to the soil's ability to store plant-available nutrients and to protect these against leaching from the soil, thus promoting soil fertility. Plant nutrients are also released from organic matter during its decomposition.

Soil water and air are found in the soil pores (the spaces between the soil particles and peds). A dense soil (or soil layer) with low total porosity is, therefore, unfavourable for plant growth. In addi-

Cations are positively charged ions such as Na^{+}, Ca^{++}, and Mg^{++}, the ions of sodium, calcium, and magnesium respectively. Cations are held in the soil by the negative charges which develop on the surface of colloidal particles. These cations can be exchanged for other cations held in the soil solution. The **cation exchange capacity** (CEC) is the sum total of the exchangeable cations that a soil can absorb. It is highly dependent on the relative amounts of the different colloids in the soil. The CEC is high for humus and 2:1 clays, it is much lower for 1:1 clays and iron–aluminium clays. The CEC is a good measure of potential soil fertility.

tion to total porosity, the pore size distribution is also important. Soil pores are subdivided into *macropores* and *micropores*. Drainage of excess water, gas exchange (aeration), and root growth occur in macropores, whereas plant-available soil water is stored (retained against drainage under the force of gravity) in the micropores. A favourable soil thus must have both a high total porosity and a favourable ratio between macropores and micropores. A waterlogged soil is one in which the macropores are filled with water because some limiting layer has prevented drainage from them. The main feature of a waterlogged soil is inadequate aeration.

Soil chemistry

The chemical and mineralogical properties of different soils vary widely. These properties have great impacts on the quality of soils since they control factors such as the susceptibility of the soil to erosion and other forms of degradation, swell–shrink phenomena, and soil fertility.

Both the amount of clay in the soil and its mineralogical composition are of particular importance. In most soils the clay fraction consists almost exclusively of secondary alumino-silicate clay minerals, of which montmorillonite, illite, and kaolinite are most important. The clay mineralogy of a soil is mainly determined by the degree and type of chemical weathering that has occurred. Where weathering has not reached advanced stages, parent material also has a significant influence on clay mineralogy. In slight to moderately weathered or leached environments, 2:1 minerals dominate, with montmorillonite the most widespread. This situation is found in semi-arid to sub-humid areas and/or young landscapes. In some semi-arid to sub-humid areas, where it has been inherited from sedimentary rock parent material, illite is the dominant clay mineral. For example, illite is found in soils developed from Beaufort mudstones and shales in the semi-arid to sub-humid parts of the Eastern Cape and south-eastern Free State.

In the more highly weathered and leached soils of warm, humid climates, kaolinite dominates the clay fraction, whereas in humid subtropical and tropical regions, sesquioxides (iron and aluminium oxides) become a significant component. In highly weathered humid tropical areas, such as the Amazon basin, central Africa, South-East Asia, and the highlands of Madagascar, sesquioxides constitute a very large proportion of the clay fraction.

The colloidal fraction of a soil – that is, clay and humus – is the chemically active part of the soil. An important factor in this regard is the charges that develop on colloid surfaces. These are predominantly negative charges, to which cations (positively charged ions such as sodium (Na^{+}), calcium (Ca^{++}), or magnesium (Mg^{++})) are adsorbed. The number of negative charges in a soil is termed the *cation exchange*

Colloids are very finely divided particles with a large external surface per unit mass of material. Soil colloid surfaces carry either negative or positive charges, with negative charges predominating. The most important soil colloids are clays and humus. They are the seat of most chemical, physical, and biological properties of soils.

capacity (CEC) of the soil. There are large differences between the cation exchange capacities of different soil colloids. The CECs of colloids decrease in the order humus >> smectites > illite > kaolinite (>> denotes much greater than). The CEC of a soil is determined, therefore, by the clay content, the type of clay mineral, and the humus content. The latter is especially important in sandy soils. In strongly acid (low pH) soils, some positive charges develop on kaolinite and many on sesquioxides. This has important implications for soil fertility, as will be explained below.

Dispersion of clay colloids is an important physico-chemical reaction in soils. It causes the disaggregation of the soil into single particles, leading to crusting and/or compaction. Crusting leads to increased runoff and erosion. Also, small individual particles are more easily transported. The dispersivity of a soil, and therefore its inherent susceptibility to crusting and water erosion, is largely a function of its clay mineralogy and cation composition. Of the dominant clay minerals, smectites are the most susceptible to dispersion and kaolinite the least. Illitic soils are often very susceptible to dispersion and water erosion.

The clay mineralogy affects the swelling and shrinkage potential of a soil, which in turn affects soil structure. Soils with high clay contents in which the clay fraction is dominated by swelling clays have strong swell–shrink characteristics, swelling upon wetting and shrinking when drying out. Large cracks are formed in the soil when it is dry. Smectites have strong swell–shrink properties, whereas illite and kaolinite are non-swelling clays. Swelling clay soils cause severe cracking and damage to structures (buildings, roads, and so on) constructed on them. Damage to underground structures, such as pipelines and sewerage systems, is even more severe. Reparation causes only temporary relief, since each wet–dry cycle causes new damage.

The soil structure can also be affected by soil chemistry. Sodium is a highly dispersive cation. Normally an exchangeable sodium percentage (ESP) of 15 per cent is considered to be the threshold above which problems occur. In southern Africa and Australia, sodium-induced dispersion, crusting, and erosion start at much lower ESPs (in some soils as low as 2 or 3 per cent). Magnesium is another problem cation, and high magnesium:calcium ratios (usually above 1.7:1) are often the cause of dispersion and enhanced erosion in many parts of South Africa and elsewhere in the world, such as Australia and Russia. Calcium is a flocculating cation, thus counteracting dispersion, but it cannot stabilize soil aggregates against disintegration. Humus stabilizes soil aggregates, and in highly weathered soils free iron oxides also act as a stabilizing agent.

Exchangeable sodium percentage is the extent to which the absorption complex of a soil (measured as its CEC) is occupied by sodium.

Flocculation is the process in which clay and other soil particles aggregate or clump together.

An important aspect of soil chemistry is the soil pH, a measure of the acidity or alkalinity of the soil solution. The pH value is a measure of the concentration of hydrogen (H^+) ions relative to hydroxyl (OH^-) ions and in theory can range from 1 to 14. In soils the pH normally ranges from

3 to 9. Acidic soils (pH < 7) result from the leaching of calcium and other cations from the exchange complex and their replacement by hydrogen. Such soils are characterized by relatively high concentrations of aluminium and manganese ions and a low concentration of calcium ions. They are therefore said to have a low base status. The most favourable pH for plant growth is slightly acid, in the range 6.5 to 7.0. This is the optimal range for the availability of other nutrients to plants. A pH in this range is also optimal for the activity of soil biota. Acidic soils are associated with very restricted faunal activity, leading to the accumulation of a surface layer of mor humus which remains discrete from the underlying mineral soil. Excessive calcium and a high base status (alkaline soils, pH > 7.0) can reduce the availability of some nutrients. The solubility of soil minerals is affected by pH, most becoming more soluble and hence more readily removed or leached in acid conditions. The solubility of iron and aluminium increases significantly at a pH below 5.

> **Base status** is the extent to which the absorption complex of a soil is saturated with exchangeable cations other than hydrogen and aluminium. It is measured as the base saturation percentage. A high base saturation percentage indicates a tendency towards alkalinity (a high pH).

Soil fertility is strongly dependent upon the clay and humus contents of a soil and the dominant type of clay mineral. This is because most essential plant nutrients are in the form of cations in the soil and the above three factors determine the cation exchange capacity of the soil. Cations which are adsorbed to the negative charges on clay colloids and humus are available to plants, but are protected against leaching from the soil. For this reason, medium-textured to clayey soils in semi-arid to sub-humid areas, with smectites (montmorillonite) or illite as dominant clay minerals, are very fertile soils. It should be kept in mind, however, that soils rich in these clay minerals, especially smectites, have unfavourable physical conditions which severely limit their cropping potential.

In contrast to the former, the highly weathered, intensely leached clay soils of humid subtropical and tropical regions are extremely infertile, and it is very difficult and extremely expensive to improve or maintain soil fertility in these soils. The sesquioxides have no negative charges, and under the strongly acid conditions kaolinite has only a few negative charges and even some positive charges. The positive charges repel cations, which are then leached from the soil. In addition, extremely strong fixation of phosphorus, a key plant nutrient, occurs in such strongly acid, iron-oxide-rich soils.

Sandy soils are relatively infertile because of their low CECs. In high-rainfall areas, such as western Zambia, eastern Angola, and the Congos, soils are extremely infertile as a result of excessive leaching. Fertilizers applied under these conditions are similarly subject to excessive leaching losses. In semi-arid to sub-humid areas, neutral-pH red and yellow sandy soils are relatively favourable because, for example, they retain phosphorus (applied as fertilizer) in plant-available forms.

In semi-arid to arid regions, soils are often alkaline, causing macro-elements such as phosphorus and trace-element nutrients such as iron, zinc, copper, and phosphorus to become unavailable to plants. Trace-element deficiencies are aggravated in soils containing free lime. Deficiencies of these nutrients can be serious in irrigated crops grown in such areas. Alkalinity can also cause serious mineral-nutrient deficiencies in livestock under extensive grazing. Namibia's cattle industry was almost destroyed until it was discovered that copper deficiencies in the diet of the cattle, induced by high soil pH, was the cause of the problem. Several human diseases can also be attributed to mineral-element deficiencies in the soils on which crops are produced.

Soils as natural bodies

Soils are three-dimensional natural bodies. The upper limit is at the interface between the soil and atmosphere. The lower limit grades into non-soil regolith. Laterally, a soil body grades into other soil bodies with different properties and characteristics or into non-soil bodies such as open water or hard rock.

A vertical section through a soil is called a *soil profile*. A soil profile consists of different layers, called *soil horizons*, which have resulted from pedogenesis. Alluvial and colluvial stratifications or cross-bedding due to aeolian (wind) action are not pedogenetic horizons.

In order to simplify categorization, a number of 'master horizons', each designated with a standard horizon symbol and conveying a certain general horizon concept, have been defined. Seven horizons have been designated: an organic horizon, the O horizon; four organic horizons, A, E, B, and G, which comprise the *solum*, or true soil; and two lower horizons which grade into the parent material, the C and R horizons.

O An 'organic' horizon occurring on top of the mineral soil. A high organic-matter content dominates the horizon's properties and characteristics. Typical examples are peat and the layer of leaf litter on a forest floor. Soils with O horizons are scarce in South Africa.

A The surface mineral layer (topsoil) of normal, uneroded soils. It typically has a higher organic-matter content, and is consequently darker, than the underlying horizons. It usually has undergone some eluviation (removal) of materials such as clay and bases. It therefore tends to have a coarser texture and lower base content than the B horizon. In parts of South Africa, for example the Eastern Cape and the Karoo, soils with bleached (light-coloured) A horizons are found, making their identification as A horizons difficult. Where sheet erosion has removed the A horizon, the horizon which is at the surface (the E, B, or G horizon) must retain its conceptual designation and must not be made an A horizon simply because it is at the surface.

E A strongly eluviated horizon occurring directly under an A horizon. It has suffered significant removal of clay, bases, iron oxides, and organic matter. It is, therefore, not only relatively coarse-textured and infertile, but also very light grey to almost white in colour. It is defined and recognized in the field by these colours. Although E horizons occur widely, most soils do not have such horizons.

B The typical subsoil horizon. It is normally an illuvial horizon, that is, an horizon enriched with material such as clay, bases, and/or iron oxides brought into it from the overlying A or E horizons. It typically has a finer texture, higher base content, and/or redder or yellower colour than the overlying A or E horizon. It has a lower organic-matter content than the A horizon. Because of the higher clay content, it often has a coarser and more strongly developed structure than the overlying A horizon, but its structure is usually less stable because of the lower organic-matter content.

G A special sticky, plastic, clayey subsoil horizon showing the typical blue, green, dull yellow, or grey mottled colours of permanently waterlogged conditions, such as that found in wetlands.

C Unconsolidated weathered or partly weathered geologic material. It can consist of alluvial, colluvial, or aeolian drift (transported material), or *in situ* weathered rock (saprolite).

R Essentially unweathered solid rock.

Not all soils have all types of master horizons. A young alluvial soil, for example, has only an A/C horizon sequence (an A horizon over stratified alluvium). Transitional horizons, for example A/B, B/A, or B/C, are sometimes recognized in a profile. It may be necessary to subdivide a master horizon. For example a B horizon may be subdivided into B1, B2, B3, and so on.

The master horizon designations given here are the latest ones used in South Africa and in the major international soil classification systems. Readers are advised to compare the discussions on master horizons in the latest South African soil classification book (Soil Classification Working Group 1991) with those in the previous one (MacVicar et al. 1977) if they seek comparisons with the previous master horizon designations.

Soil-forming processes

The soil found at a specific site is the product of a number of soil-forming processes taking place under the influence of a certain combination of soil-forming factors. Four groups of soil-forming processes are recognized: additions, transformations, translocations (within the profile), and losses. Processes belonging to all four groups occur in all soils, but the specific processes and their relative rates and intensities differ widely. The result is the large range of vastly different soils. The soil-forming processes and a few typical examples of each are illustrated in Figure 14.2.

Factors of soil formation

Five factors of soil formation are recognized, namely climate, parent material, topography, living organisms, and time. Although there is sometimes a tendency to view these as independent factors, the soil at a specific site is the result of an interaction between the combination of soil-forming factors operating at that site.

FIGURE 14.2 Soil-formiing processes

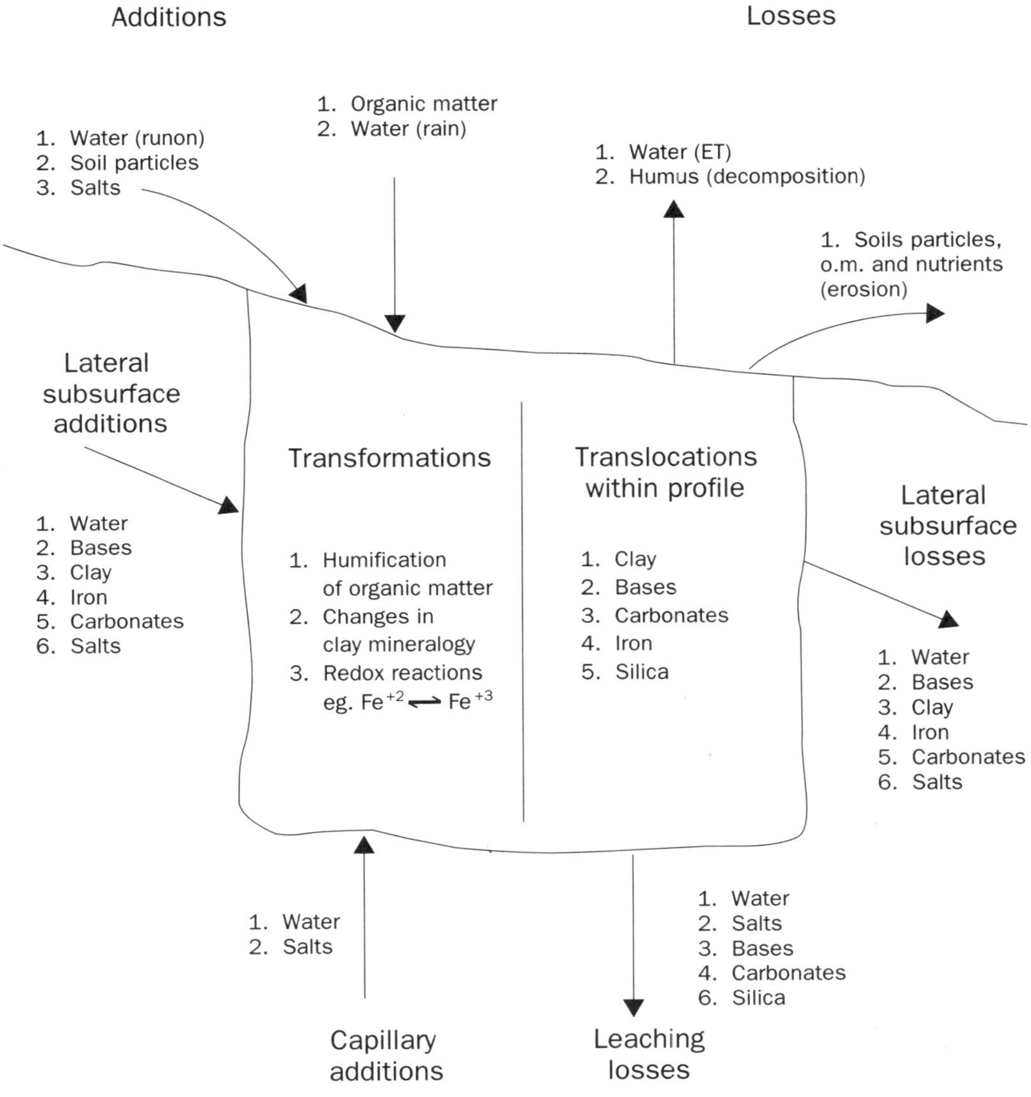

Climate

Rainfall and temperature are the dominant climatic factors affecting soil formation. Water provides the medium in which chemical reactions (transformations) can proceed, as well as the translocation of dissolved or suspended substances and their leaching from the soil. The higher the rainfall, or more correctly the efficiency of the rainfall, the more advanced pedogenesis will be under otherwise identical conditions. Under conditions of high rainfall, increased temperatures promote pedogenesis, since the rates of chemical reactions double with every 10°C increase in temperature. This combination is the main reason for the deep, highly weathered, and extremely infertile soils found in the humid tropics and subtropics.

In arid and semi-arid areas, poorly developed, shallow soils dominate, except in areas with deep aeolian or alluvial deposits. The soils have high base contents and often contain free lime or even lime pans. In extreme arid conditions salt plains ('playas') and/or soils with gypsic horizons (enriched with calcium sulphate) or 'dorbanks' (dor = dry, bank = pan) are found. 'Desert pavement', a hard surface layer containing calcium carbonate, gypsum, or other binding material, is also widespread in arid areas. In semi-arid and arid areas, high temperatures retard pedogenesis because of increased evopotranspiration, thus rendering the low rainfall even less efficient.

In cold regions the low temperatures inhibit pedogenesis, causing poorly developed, shallow soils to dominate.

Parent material

Parent material has a major influence on the properties and qualities of the soils which develop from it. The large land masses at mid to high latitudes in the northern hemisphere are dominated by large areas covered by young, deep, transported materials resulting from the geologically recent ice covering (Brady and Weil 1996). These include glacial till (material deposited beneath a glacier), associated marine and lacustrine (lake) deposits, and, most importantly, the vast areas of loess – deep deposits of fine (mainly silt-sized) aeolian material redistributed from the tills. Soils developed from loess are very fertile, high-potential soils. Loess soils are widespread in the prairies of North America and the Russian steppes. A significant loess area also covers the pampas of Argentina.

Central Africa is dominated by a vast area of sandy aeolian material, often referred to as Kalahari sand, stretching from the Orange to the Congo River. Large areas of sand are found also in North Africa and the Middle East. Parent materials in the rest of the world consist mainly of *in situ* weathered material or local colluvium derived from a variety of igneous, sedimentary, and metamorphic rocks, giving rise to a variety of soils (and often complex soil patterns) at national or regional level.

Topography

Topography is an important soil-forming factor, especially in rolling landscapes on old land surfaces such as those found in East and southern Africa. Repetitive soil patterns known as *toposequences* or *catenas* develop in such landscapes. The general pattern is drier and/or shallower soils on the crests of hills, with soil depth and wetness increasing downslope.

Living organisms

There is a close relationship between natural vegetation and soils. This is clear from the descriptions of the suborders of 'zonal' soils in old soil classification systems. Soil/vegetation associations are somewhat 'chicken and egg' relationships. The vegetation type at a site is determined by the climate and certain inherent properties of the soil (which are themselves products of the climate, parent material, and topography). The vegetation, in turn, may sometimes cause characteristic modifications to the soil. A good example is the widespread podzol/coniferous forest association. A podzol is a soil which develops under acid leaching conditions and is characterized by an acidic mor A horizon and an iron- and humus-enriched B horizon. A combination of sandy parent material and a cool, moist climate is the primary prerequisite for the development of a coniferous forest cover. The acid organic matter produced by the conifers then becomes the driving force for the development of the characteristic podzol profile.

Soil fauna, such as termites, earthworms, and microbes, also play important roles. Furthermore, human activities lead to the development of various 'man-made' or anthropogenic soils.

Time

Soil formation is a slow process, and consequently time is an important soil-forming factor. The middle- to high-latitude regions of the large land masses of the northern hemisphere were covered by ice until quite recent geological times. These areas therefore have chronologically young soil landscapes. In contrast, most tropical and subtropical areas are characterized by very old landscapes, resulting in completely different soilscapes from the former. The high weathering rates under hot, humid conditions, combined with the old age of the landscapes, produce the extremely highly weathered and highly leached soils of the tropics and subtropics. Exceptions are the Andisols, young soils developed on recent volcanic lavas in East Africa, Central and South America, and South-East Asia.

Time as a soil-forming factor is particularly noticeable on the terraces of different ages found along rivers and on coastal dunes.

Taxonomic soil classification

Taxonomic soil classification started towards the end of the nineteenth century in Russia, when Dokuchaev and his school for the first time recognized soils as natural bodies reflecting the influence of soil-forming factors. Before World War II, taxonomic soil classification systems were essentially genetic classifications, based on factors of soil formation rather than on soil properties. The belief in the dominance of climate as a soil-forming factor was reflected in the concept that soils which showed strong climatic associations were called 'normal' or 'zonal' soils. Soils reflecting the dominance of some other soil-forming factor were termed 'intrazonal' soils. Soils which could be found randomly in any zone, for example young alluvial soils, were called 'azonal' soils. It became evident, however, that the zonality concept was a gross oversimplification (MacVicar 1978). Moreover, so-called zonal soils are often found outside their normal climatic zones as a result of the overriding influence of some other factor, such as parent material, thus becoming 'azonal' soils. The zonality concept may help us to understand the diversity of global soils, but it is inadequate as a basis for soil classification.

A major breakthrough came with the publication of the so-called *7th Approximation* (Soil Survey Staff 1960), commissioned by the United States Department of Agriculture (USDA). Two of the main principles embodied in it were (a) that soil properties, not genetic factors, had to be used in soil classification, and (b) that well-defined 'diagnostic horizons' (not to be confused with master horizons) formed the basis for the system. The 'final' version of *Soil Taxonomy*, intended to include all the soils in the world, was published in 1975 (Soil Survey Staff 1975). Updated versions are continuously published as *Keys to Soil Taxonomy*, the latest available version at the time of writing this text being the seventh edition (Soil Survey Staff 1996). Brady and Weil (1996) give an excellent overview of *Soil Taxonomy* and the global distribution of its different soil orders.

Meanwhile, the Food and Agriculture Organization (FAO 1974) prepared a *Soil Map of the World*. Because no generally accepted international classification system existed, they constructed a 'legend' which used internationally well-known soil names and definitions drawn from various national systems. The principle of well-defined diagnostic horizons was followed, drawing mainly on those developed in *Soil Taxonomy*. Over the years the definitions of diagnostic horizons were changed and new ones, not featured in *Soil Taxonomy*, were defined. Soil names and definitions of major soil groups also changed. The latest version of the FAO legend was published in 1990, together with a description of the distribution of soils and their uses.

The 1990 FAO legend still had major deficiencies, causing the International Society of Soil Science to appoint a World Reference Base for Soil Resources (WRB) working group to revise it. Since about 1993 the 'new generation' of WRB members have been making fairly drastic changes to the FAO system. The biggest weakness of the system remained that, like *Soil Taxonomy*, it made good provision for soils of Europe and North America, because the scientists involved came almost exclusively from these areas, but tropical soils were not adequately accommodated and no provision was made for some of the most important soil features found in southern Africa and Australia. Following a workshop by the WRB Working Group South Africa in 1996, provision for the distinctive soils of southern Africa has been made in the WRB classification system, which was published during 1998 in three volumes (ISSS Working Group Reference Base 1998; Deckers et al. 1998; Bridges et al. 1998). At the 1998 Congress of the International Soil Science Society it was decided that the WRB classification system should be used as the standard international reference system.

In 1978 MacVicar reviewed the soil classification for South Africa. At that date the only comprehensive national-scale works were the publications by Van der Merwe (1940, 1962). Van der Eyk et al. (1969) had started the development of a classification system based on the same principles as the USDA's *7th Approximation* of 1960. This was expanded and adapted to become South Africa's first modern-style national soil classification system (MacVicar et al. 1977). As more information became available, especially from the arid regions of the country, the system was fairly drastically revised (Soil Classification Working Group 1991).

As in the USDA's *Soil Taxonomy* and the FAO and WRB systems, the two main features of the 1977 and 1991 South African systems are (a) that soil properties are used for classification, and (b) that rigidly defined diagnostic horizons form the basis of the systems. Unique South African definitions and nomenclature were developed for the diagnostic horizons. Some of these resemble fairly closely the diagnostic horizons of the two major international systems, but others are horizons presently not provided for. Generalized descriptions of the diagnostic horizons and diagnostic materials of the South African system are given in Table 14.1. Note that the full definitions given by the Soil Classification Working Group (1991) must be used when classifying soils.

Both the 1977 and 1991 South African classification systems are two-category systems which use the *soil form* as the higher category. Each soil form is defined by a unique vertical sequence of diagnostic horizons/materials. Soil forms are names after locations where they are commonly found. Typical soil profiles showing the diagnostic horizons of the Kroonstad and Longlands forms are shown in Figure 14.3 by way of example.

The 1977 classification system recognized forty-one soil forms, which were further subdivided into *soil series* as the lowest classification category. Soil series were distinguished in terms of additional properties such as texture. The 1991 classification does not include series, but subdivides some of the 1977 soil forms. Descriptions of the resulting seventy-three soil forms are given by the Soil Classification Working Group (1991).

FIGURE 14.3 Diagnostic horizon sequences for two soil forms

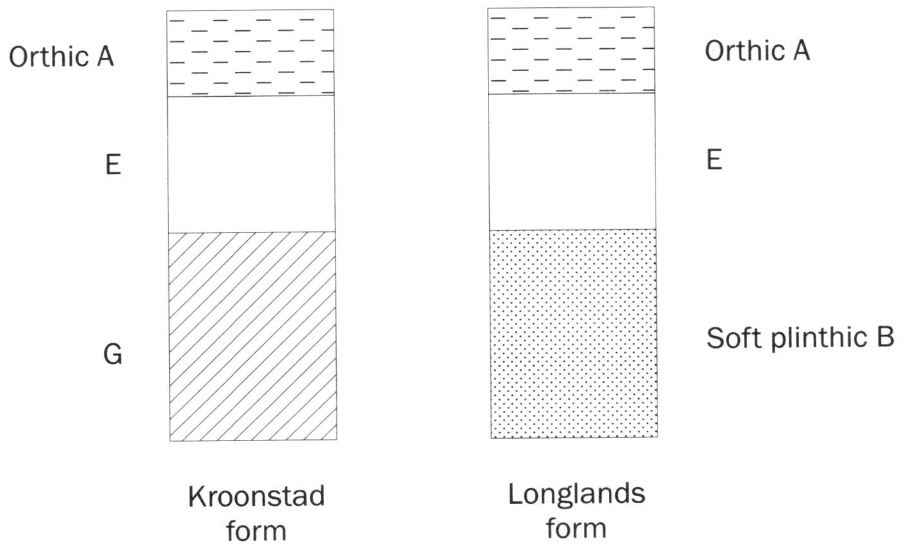

TABLE 14.1 General description of diagnostic horizons and material used for soil form identification in *Soil Classification: A taxonomic system for South Africa*

Diagnostic horizon/material	General description
Surface horizons	
Organic O	Thick organic horizon formed in swamps.
Humic A	Dark, rich in organic matter (humus), with a low exchangeable base cation content. Typically found under dense sourveld grass in high-rainfall areas.
Melanic A	Dark, fertile, with a high exchangeable base content. Has well-developed crumb structure. Typically found under dense sweetveld grass in sub-humid to semi-arid areas on basic igneous rock as parent material.
Vertic A	Topsoil with a high content of swelling clays. Strong swell–shrink properties. Cracks when dry. Identified by cracks and/or slickensides (shiny, polished, grooved planes). Very fertile, but with poor physical conditions. Structures, like houses, subject to severe cracking on soils with these horizons. Root pruning and dwarfing of trees.
Orthic A	All A horizons not qualifying for any of the former.
Subsurface horizons/ materials	
E	Strongly eluviated horizon, i.e., has been/is subject to strong removal of iron, clay, bases, etc. Characterized by a light grey colour in the dry state (some are yellow or red in the moist state). Structureless. Often very hard and dense in dry state. Infertile.
G	Sticky clayey horizon with variegated grey and dull yellow colours. Sometimes also blue-green colours. Indicates near-permanent wetness.
Soft plinthic B	Has abundant high-chroma (bright-coloured) red and/or yellow mottles in a matrix with at least some grey in it. Indicates a zone with fluctuating water-table. Soft and can be cut with a spade in the moist condition.
Hard plinthic B	Indurated layer of iron–manganese concretions. Also known as 'ouklip', petroplinthite, or ferricrete.
Podzol B	Reddish-brown horizon with combined accumulation of iron (and aluminium) and organic matter. Typically found in cool, humid climates on sandy parent material under fynbos vegetation.

Diagnostic horizon/material	General description
Placic pan	Thin, dense layer found in some podzol B horizons. Important in evaluations of podzols for afforestation because placic pans are impenetrable to roots.
Unspecific or unconsolidated material with signs of wetness	Light grey or variegated grey/dull yellow material indicating prolonged wetness, but not qualifying as G or E horizons.
Red apedal B	Uniformly red, without well-developed macro-structure. Uniform red colour indicates a well-drained, well-aerated horizon. Does not contain free lime.
Yellow-brown apedal B	Uniformly red, without well-developed macro-structure. Uniform bright yellow-brown colour indicates a well-drained, well-aerated horizon. Must not be confused with the variegated dull yellow colours of the G horizon, which indicate poor drainage. Does not contain free lime.
Red structured B	Uniformly red, with well-developed 'nutty' blocky structure. Gradual transition from overlying topsoil. Well-aerated, well-drained, and highly resistant to erosion. Associated with basic igneous rocks such as dolerite. Some pedocutanic and prismacutanic B horizons are also red but must not be confused with red structured B horizons because they are unfavourable and highly erodible. Pedocutanic B horizons have non-uniform colours and clear transitions from overlying horizons. Prismacutanic B horizons have prismatic or columnar structure and abrupt transitions from overlying horizon.
Pedocutanic B	Has well-developed blocky structure. Shows clear clay illuviation.
Prismacutanic B	Has prismatic or columnar structure. Abrupt textural and structural transition from overlying horizon. Extremely unstable and highly erodible.
Neocutanic B	Young ('new') subsoil developed in alluvial or colluvial drift. Does not have well-developed structure. Does not contain free lime. Does not qualify as red or yellow-brown apedal B because colour is not diagnostic or is non-uniform.
Lithocutanic B	Young subsoil developed in weathering rock. Characterized by prominent tonguing into partly weathered rock.
Neocarbonate B	Similar to red and yellow-brown apedal and neocutanic B horizon, but containing free lime.

Diagnostic horizon/material	General description
Soft carbonate	A brittle lime pan ('broskalk'). Stores much plant-available water – an important feature for plant survival in the dry areas where lime pans occur.
Hardpan carbonate	A dense, hard lime pan.
Dorbank	'Dry pan'. Reddish-brown hard layer, cemented by silica, found in extreme aridic areas.
Stratified alluvium	Unaltered alluvial or colluvial deposits, still showing the original layering, under an A horizon.
Regic sand	A recent sandy deposit, usually aeolian.

The soils of South Africa

There is no comprehensive up-to-date account of South African soils. The last generalized soil map of South Africa, together with a comprehensive report, was published by Van der Merwe (1962). More detailed reports are available for localized areas, such as the Northern Province and the eastern part of Mpumalanga (Verster 1974), the southern half of the Free State and the Northern Cape north of the Orange River (Eloff 1984), and the Great Karoo, Little Karoo, Bushmanland, and the whole arid area between the western escarpment and the west coast, the latter including Namaqualand and the so-called North West to the south of it (Ellis 1988). A series of excellent 1:250 000 'land-type' maps, accompanied by comprehensive memoirs (reports), have been published during recent years for most of the country by the Institute for Soil, Climate, and Water. Each land type depicts a unique climate–topography–soil relationship. Note that land-type maps should never be used without their accompanying memoirs and are not meant for detailed planning.

The broad soil patterns of South Africa (see Figure 14.4) are mainly determined by climate and geology. The overriding feature is the dominance of poorly developed soils due to the low and inefficient rainfall over most of the country (see Chapter 12). They are described here in terms of their general characteristics. Mention is made of the main soil forms in each area. The reader is referred to the Soil Classification Working Group (1991) for further details of each soil form. The equivalent soil orders as given in the USDA's *Soil Taxonomy* and the WRB/FAO system are also given so that the reader can place South Africa's soils in a global context.

FIGURE 14.4 Generalized soil map of South Africa

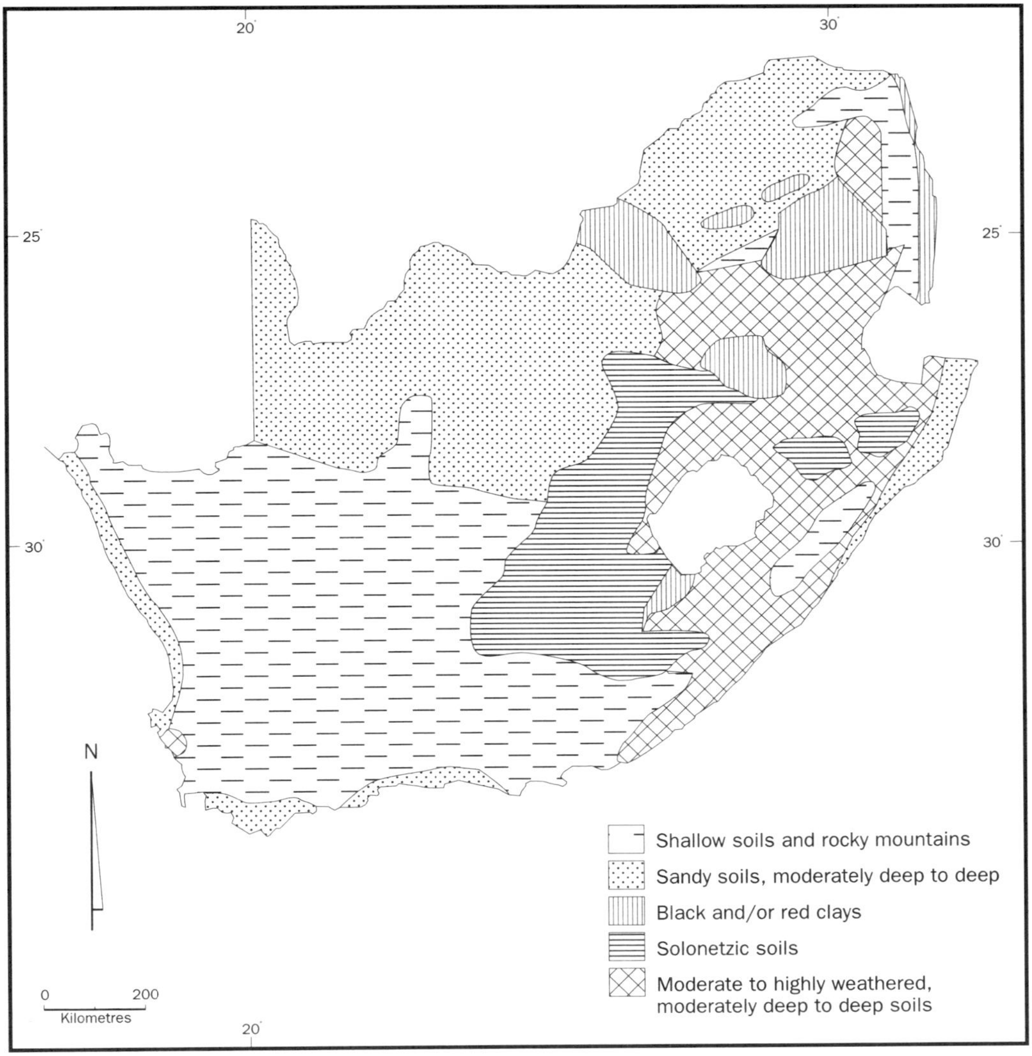

Areas dominated by poorly developed shallow soils

The dry western half of the country, including the Karoo and Little Karoo, is dominated by poorly developed, shallow soils (Lithosols), often less than 30 cm in depth, overlying rock or rock plus lime (Ellis 1988). The dominant soil forms are Mispah and Glenrosa (Soil Classification Working Group 1991). Within this broad area a number of distinct soil assemblages can be recognized.

In the northern and north-western

Karoo and Bushmanland the soils change to predominantly shallow, red, sandy soils on lime, or rock plus lime, probably mainly of the Prieska form (Ellis 1988). Another area of similar soils extends across the Orange River east and north from the Orange–Vaal confluence into the south-western Free State and the Northern Cape. Here significant areas are covered by very shallow soils on hardpan carbonate (lime pans). In the 1977 system these soils used to be classified as the Mispah form, but in the 1991 system most would probably be classified as the Coega form. Glenrosa soils (showing minimal B horizon development) are very scarce here (Eloff 1984). Eloff attributes this to the low/inefficient rainfall and highly resistant parent materials.

In some parts the shallow soils are interspersed with significant areas of moderately deep to deep sandy soils. The red sandy soils on lime correspond to Van der Merwe's 'Kalahari sand on lime'.

A major part of the coastal plateau of the Eastern Cape, adjoining the Karoo, is dominated by shallow soils of the Glenrosa form. This is largely the result of the parent material, an extension of mudstones and shales of the Adelaide Subgroup of the Beaufort sediments into this area. Because of the somewhat higher rainfall, the soils here are slightly deeper than in the Karoo. In the northern part of this soilscape (the southern half of the former Transkei), various types of moderately deep soils with E horizons are relatively abundant due to the high rainfall. A similar mixture of mainly Glenrosa and Cartref soils (the latter having E horizons) is found on the KwaZulu-Natal coastal plateau.

Shallow soils on dorbank occupy more than 80 per cent of the extremely arid area below the western escarpment, including the Knersvlakte and parts of Namaqualand (Ellis 1988). Dorbank is an extremely hard reddish-brown layer cemented by silica. It is identical to the red-brown hardpans of Australia and the duripans of *Soil Taxonomy* and the FAO legend. Less frequent occurrences of dorbank soils are also found at the north-western and eastern edges of the Karoo and in the Little Karoo.

South Africa's main wheat-producing areas, namely the Swartland and the southern Cape Ruens, are dominated by poorly developed soils of the Glenrosa and Swartland forms, with shallow effective depths. This is due to a combination of low rainfall (between 400 and 420 mm per annum throughout both areas), unfavourable geology for the genesis of good soils (Swartland and Bokkeveld shales), and, especially in the Ruens, undulating topography. Significant areas with deep, young soils occur on terraces in parts of the Swartland, however.

Large areas dominated by shallow soils (Glenrosa and Mispah forms) are found in the lowveld of Mpumalanga and the Northern Province and in the southern and west-central areas of the Northern Province. These are mainly associated with relatively low, inefficient rainfall combined with coarse-grained, quartz-rich parent materials (granites, sandstones, and/or quartzites) that weather slowly under such unfavourable moisture supply. A number of other significant areas with similar shallow soils or extensive rocky land are found in the this area. Rocky mountainous areas are generally too small to indicate on Figure 14.4, a notable exception being the Cape Fold Mountains, which cover a large area in the south of the country.

In the USDA's *Soil Taxonomy* most of the poorly developed soils would be classified as Aridisols (because of the aridic soil moisture regime), and some as Entisols or Inceptisols. In the original FAO system most would be Yermosols and some possibly Regosols. In the new WRB/FAO system the shallow soils on rock are Leptosols and those on lime Calcisols, while for the soils with dorbanks a new Durisol major group (from 'duripan') has been created.

Semi-desert aeolian sands and coastal sands

Most of the semi-arid area between the Orange River and the Botswana border (the Kalahari) is covered by deep red and yellow aeolian sand (often deeper than 20 m). The western half of the area is characterized by long, continuous dunes running parallel to each other (Eloff 1984), whereas dunes are mainly absent in its eastern half. The soils are classified as sandy soils of the Hutton (red) and Clovelly (yellow) forms (Eloff 1984), although some of the soils in the drier western area may probably be reclassified as the Namib form of the 1991 system. As stated earlier, this is the southern part of the largest area of sandy soils in the world. Sandy soils are also found in the western half of the Northern Province.

Along the west coast a fairly wide strip of coastal sand is found, with another fairly wide band along the northern KwaZulu-Natal coast. Smaller significant sand patches of this nature are found on the Cape south coast and in the Alexandria–Paterson area near Port Elizabeth.

In *Soil Taxonomy* these soils are classified as Psamments (sandy Entisols), with some of the soils of the eastern Kalahari possibly having undergone enough pedogenesis to be Inceptisols. In the FAO and WRB/FAO systems all deep sandy soils are classified as Arenosols (from the Greek 'arena' = sand).

Solonetzic soils

Solonetzic soils are soils with a structureless, relatively sandy topsoil abruptly (with a sharp transition) overlying a clayey B horizon with prismatic or columnar structure and, usually, a high exchangeable sodium percentage (ESP > 15 per cent) or a high exchangeable sodium-plus-magnesium content in the B horizon. Sometimes there is a thin E horizon between the A and B horizons, giving the so-called solodized Solonetzes. Solonetzic soils are horrible soils from a land-use point of view, as will be discussed later.

Unfortunately, South Africa has a large area dominated by solonetzic soils, sometimes together with other clay soils. The main solonetzic area is in the northern part of the Eastern Cape and the south-eastern Free State. A band of these problem soils stretches through the east-central Free State into the northern and north-eastern Free State. The solonetzes of all these areas are mainly associated with parent materials derived from mudstones and shales of the Tarkastad Subgroup and the Elliott Formation, and to a lesser extent the Adelaide Subgroup, all members of the Beaufort Group of the Karoo geological sequence. Two other significant areas of solonetzes, together with other problem clay soils, occur in northern KwaZulu-Natal. In the area around Estcourt the solonetzes are again associated with mudstones and shales of the Adelaide and Tarkastad Subgroups, and in the other area with mudstones and shales of the Dwyka Formation. These solonetzic soils are mostly classified as Estcourt and Sterkspruit forms, with some classified as the Valsrivier form. The dominance in these specific areas of these parent-material-related problem soils is a major detriment to the agricultural potential of South Africa, since the rainfall in these areas is adequate for dryland cropping.

Solonetzic soils are classified as Natrustalfs in *Soil Taxonomy* if they contain *natric* horizons – that is, if they meet one of the following criteria: either (a) an ESP > 15 per cent in the upper B horizon, or (b) exchangeable (Na + Mg) > exchangeable (Ca + acidity) and an ESP > 15 per cent somewhere lower down in the profile. In the FAO and WRB/FAO systems, soils with natric horizons are classified as Solonetz. Unfortunately, the B horizons of many of the South African soils with typical solonetz morphology and behaviour do not qualify chemically as natric horizons and therefore cannot be classified as Solonetz. Similar classification problems with solonetzic soils are found worldwide.

Clay soils on basic igneous rocks

Various significant areas dominated by black and/or red clay soils, associated with basic igneous rocks as parent materials, occur in the former Transvaal. Some of these are just too small to indicate on Figure 14.4, but are clearly identified on the land-type maps. Most prominent are the black swelling clays on the large dolerite area in the Secunda–Standerton area of central Mpumalanga. Best known are the black swelling clays (with some red clays in between) on the basalt of the Springbok Flats, and similar soils on parent materials of gabbro, norite, and andesite geology which occur in an east–west strip from Pretoria via Rustenburg to the Botswana border. A narrow strip of similar soils on gabbro and norite runs south and north-west of Belfast and Steelpoort and joins up with the area on the Springbok Flats, in the north. Also well known is the strip of such soils on basalt, close to the eastern border of Mpumalanga and the Northern Province, from the Swaziland border through the Kruger National Park to the northern border at the Limpopo River.

'Clay soils' in this context are soils in which the clay content is high throughout the profile, including the A horizon, and in which the clay fraction consists predominantly of smectites, mainly montmorillonite. In addition to the dominant role of parent material in the occurrence of these soils, the areas mentioned all have sub-humid climates with average annual rainfall generally between 600 and 650 mm, evapotranspiration far exceeding rainfall, and definite dry seasons. The highly weathered clay soils of humid areas have not been included in this discussion because of their totally different chemical and physical properties and land-use qualities.

The dominant soils developed on these black and red clays are classified as the Arcadia form because of their vertic A horizons. They are classified as Vertisols in the USDA's *Soil Taxonomy* and the FAO and WRB/FAO systems.

Highly weathered, well-drained soils

A fairly large area of deep, highly weathered, well-drained soils (in some parts with varying quantities of rock and shallow soils) occurs in the southern interior of KwaZulu-Natal, extending in a horseshoe shape into the north-eastern part of the Eastern Cape (the northern part of the former Transkei). Another significant area is found on the eastern escarpment of Mpumalanga, with narrow strips along the rest of the escarpment from the northern border of Lesotho almost to the northern border of the Northern Province. The soils are mainly clayey to loamy and apedal (without structure), with kaolinite dominating the clay fraction. In the KwaZulu-Natal mistbelt and the eastern leg of the 'Transkei horseshoe', soils with humic A horizons (topsoils rich in organic matter, but with a low base status) are abundant.

High rainfall, exceeding evapotranspiration, is the dominant factor in the genesis of these soils. In some areas, notably the eastern leg of the Transkei horseshoe, the presence of Ecca Group sediments as parent material (as opposed to Beaufort Group sediments) is a definite contributing factor.

Most of these soils are dystrophic (highly leached) and mesotrophic (more neutral) soils of the Hutton and Clovelly forms, or represent various soil forms for which a humic A horizon is diagnostic.

In *Soil Taxonomy* most of these soils are classified as Ultisols. In the FAO and WRB/FAO systems most of them will be Ferralsols and some will be Acrisols.

The highveld soil system

A unique South African soil feature is the so-called highveld soil system (MacVicar 1970), which occurs on the highveld, intermingled with other areas of high-potential soil. The highveld system is to be found throughout much of Mpumalanga, the North West province, the north-western and north-eastern Free State, and the north-western part of KwaZulu-Natal. Another quite large area is found in the north-western part of the Northern Province. Smaller significant areas are found in the Eastern Cape. Because of its dispersed nature, the highveld system has not been shown in Figure 14.4.

The distinctive feature of this soil system is the so-called highveld catena or plinthic catena. The characteristic feature of plinthic catenas is the widespread occurrence of soils with 'soft plinthic B' horizons in them. It is, in fact, incorrect to talk about 'the plinthic catena', however, because several variations of plinthic catenas are found in different areas.

The South African soft plinthic B horizon is a special phenomenon. It has abundant high-chroma (bright-coloured) red and/or yellow mottles and/or iron–manganese concretions in a soil matrix that includes at least some grey colours, indicating a soil layer which is subject to a fluctuating water-table. The horizon is soft (it can be cut with a spade). Up to this point the description of the properties and genesis of the soft plinthic B horizon is identical to that of plinthite (*Soil Taxonomy*, FAO) or the plinthic horizon (WRB). To qualify as plinthite, however, it must harden irreversibly upon alternate wetting and drying, which is not the case with the soft plinthic B. The 1998 WRB/FAO system provides for soft plinthite in the form of a 'paraplinthic' lower-level modifier.

A catena is defined as a sequence of soils with similar age and parent material, but which have differences in properties caused by their relative topographical positions on a hillslope. The term *toposequence* is often preferred because it is more descriptive. An idealized example of a plinthic catena is given in Figure 14.5. It is clear why these catenas are typically associated with the eastern highveld, with its rolling topography, relatively high rainfall, and relatively sandy parent materials. Plinthic catenas are not found in the western half of the highveld, as rainfall is too low.

The generally flat western highveld has a relatively deep cover of aeolian sand, which covers dense clay formed from local underlying geological materials. Rain infiltrates the sand very effectively during rainy seasons and accumulates on the dense clay layer, causing the periodic water saturation required for soft plinthite formation. In addition, researchers have shown that a contributing factor is the fact that the clay layer is not flat but undulating, causing 'subsurface catena conditions'.

> The dominant soil forms of the South African classification system found in the plinthic catenas are Hutton, Clovelly, and Avalon, with Bainsvlei or Westleigh significant in some areas. The relative proportions of these soils vary between different areas.

Dolerite as parent material

Intrusions of Jurassic dolerite as dykes or sills into other geological materials are widespread throughout South Africa. The soils developed from dolerite differ greatly from those which developed from the surrounding sedimentary rock, for example. This is most evident on well-drained slopes in sub-humid to humid areas where the dolerite produces red soils that are much more fertile and vastly more stable against erosion than the surrounding soils from other geological materials. Among these are the soils with red structured B horizons (Shortlands form) of the South African classification system, which are classified as Nitisols by WRB/FAO.

Hard plinthite

Hard plinthite, a hard, cemented layer of iron or iron–manganese concretions, occurs widely in South Africa, particularly in the Mpumalanga highveld, KwaZulu-Natal, and the Eastern Cape. Although it is

FIGURE 14.5 Typical example of a highveld catena or plinthic catena

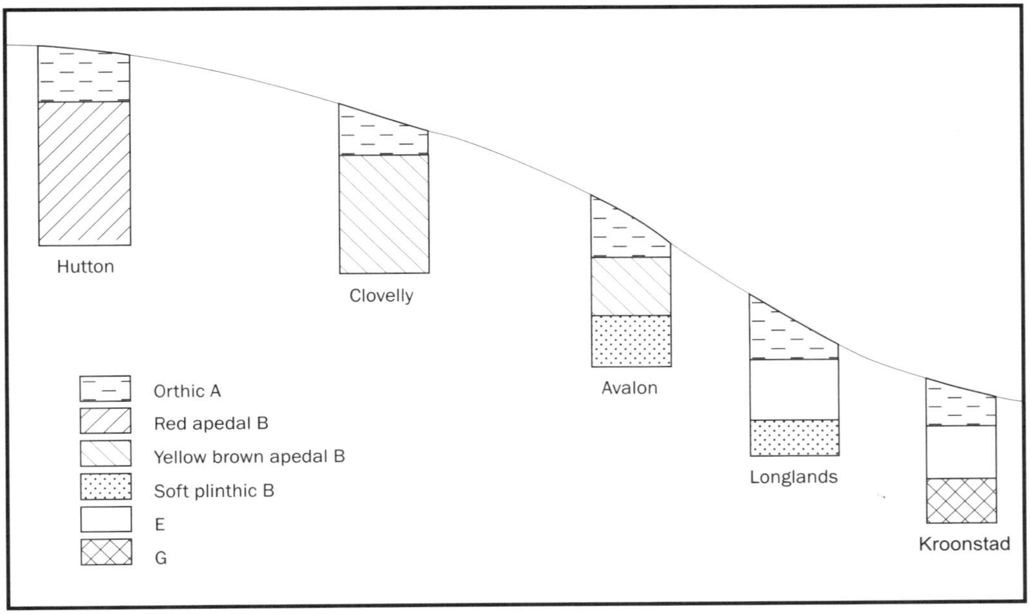

Note: different variations of highveld catenas or plinthic catenas are found

included as a diagnostic horizon in the South African soil classification system, almost all hard plinthites in the country are considered to be palaeo-materials which formed under much wetter conditions than those prevailing at present (see, for example, Verster 1974).

In areas such as the Tyumie and Amatola basins in the Eastern Cape, the hard plinthite is presently breaking up and serving as parent material for new, young soils. This process has also been reported for soils on early-Tertiary surfaces near the Natal coast (MacVicar 1978), as well as by Australian researchers. The soils which develop from these materials have extremely poor physical conditions and are very infertile.

Hard plinthite is popularly known as 'ouklip' (meaning 'old stone') in South Africa. International terms for it include petroplinthite and ferricrete. Africa has the largest areas of petroplinthite in the world (Eswaran et al. 1990). Their extent includes vast areas in the Sahel, under conditions that presently are far too dry for the formation of this material, indicating that it must also be a palaeo-material in that area.

Soils associated with the influence of living organisms

As stated earlier, vegetation types in South Africa are generally determined by soils, and not vice versa. Excellent examples of geology–soil–plant (and animal) relationships are found in the Kruger National Park, geographically so closely spaced that they can be studied in half a day. There are a few cases, however, where soils are strongly influenced by the actions of living organisms. For example, changes in the physical and chemical properties of soils are brought about by termites in the construction of their mounds. The changes depend upon the type of termite. These are 'spot' changes, confined to the immediate location of the mounds, rather than changes to the whole soil landscape.

In the Eastern Cape, giant earthworms, up to seven metres long, create a peculiar soil landscape called 'Debe Hollows', 'Kommetjiesveld' (Afrikaans), or 'Amalinde' (Xhosa) over large contiguous areas. The largest area, covering about 100 000 hectares, is found between Debe Nek and King William's Town (Benians and Van Averbeke 1990).

Podzols are found in the Cape Fold Mountains due to the perfect combination of conditions for the formation of such soils: fynbos vegetation providing acidic organic matter, a cool, moist climate, and sandy parent material (Table Mountain and Witteberg quartzites). In this area of South Africa, fynbos replaces the coniferous forest of northern temperate areas in promoting acidity.

Soils and land use

Land can be utilized in many different ways. Optimal land use is the most beneficial use of each area of land on a sustainable basis without causing unacceptable resource degradation – either on-site or off-site. Each land utilization type (LUT) has specific requirements and tolerances (FAO 1976), while every area of land has specific properties and characteristics which must be translated into qualities in terms of the requirements and tolerances of each possible anticipated LUT. Correct matching of the requirements and tolerances of LUTs with the properties, characteristics, and qualities of each area of land is a prerequisite for optimal land use.

Several environmental factors determine the suitability of an area of land for any specific land use. South Africa's very low agricultural potential is mainly due to its low, inefficient, and unreliable rainfall. The country's poor-quality soils, however, make a major contribution to this situation. The biggest problem in this regard is the large area of non-arable soils, due to extreme erodibility and/or shallow depth, in regions receiving more than 500 mm of rain per annum. An average rainfall above 500 mm per annum is usually considered adequate for rain-fed cropping, although in a province such as the Eastern Cape 600 mm is considered to be the minimum requirement because of the shallow soils and poor seasonal distribution of rain. The occurrence of much steeply sloping land in these higher-rainfall areas aggravates the situation. As a consequence it is estimated that only 13 per cent of South Africa's land is arable (see Figure 14.6), much of this being marginal land for dryland cropping. According to international criteria, a significant proportion of South Africa's 'arable' land would probably be classified as non-arable. The country's low average grain yields are caused mainly by these poor soil and climate resources.

FIGURE 14.6 Potential for rain-fed crop production in South Africa

SOURCE: adapted from Smith 1997.

Only 3 per cent of the country is considered to have high cropping potential. Much of this would not be considered to have high potential if international criteria were applied. A significant proportion of the highest-potential land is in former homelands, where it is severely underutilized due to traditional land tenure systems and agricultural practices. Urban and industrial areas and infrastructure have taken up significant areas of high-potential farmland and the process continues, in some areas, as a result of population movements due to the new political dispensation in the country. In the Mpumalanga highveld the highest-potential areas coincide with the highest concentration of open-cast coal-mining activity.

If the norm of 0.4 ha of arable land per person is accepted, South Africa has a human carrying capacity of 35 million. This is less than the present population, and the high population growth rate is widening the gap rapidly. It is estimated that by about 2005 South Africa will already have to import 80 per cent of its wheat and 40 per cent of its maize requirements in normal years. Ironically, the country's highest-potential maize areas are not well suited for the production of white maize, the preferred choice of consumers. Thus, in the 1996/97 season, for example, there was a shortage of white maize, which is also virtually non-available and expensive on world markets, and a surplus of yellow maize in the country.

South Africa's irrigation potential, too, is restricted. Limited water supply is a major problem in some areas. In other areas where water is more abundant, as in the former Transkei, for example, the localized extent of deep, irrigable soil on the narrow river terraces is the limiting factor. It is estimated that South Africa has an irrigation potential of only 1.5 million hectares. This compares poorly with a country such as Iran, which is the same size as South Africa but has an irrigation potential of 7.5 million hectares. In addition, the transfer of a major portion of the water from the Orange River system to Gauteng via the Lesotho Highlands Water Project poses a threat to irrigation sustainability in the Northern Cape and Eastern Cape provinces.

Africa has similar agricultural resource problems to South Africa. More than 25 per cent of the continent consists of deserts receiving less than 200 mm of rain per annum. The regions with adequate annual rainfall for rain-fed cropping have either unfavourable rainfall distribution patterns or predominantly poor-quality soils that are highly infertile and/or erodible. According to Eswaran et al. (1990), the vast areas of soils with petroplinthite represent some of the more problematic soils of the world and are easily mismanaged. The humid tropical areas are dominated by Ferralsols which have good physical properties, making them less susceptible to erosion than many other soils, but poor chemical properties. Under natural conditions their fertility is maintained by the recycling of nutrients by the deep-rooted trees. Under cultivation their fertility deteriorates very quickly, and massive inputs of fertilizers and lime are required to maintain it.

Fertility management is difficult and expensive because of excessive leaching of most nutrients and strong phosphorus fixation. Fertility is probably the main factor limiting the agricultural productivity of the Ferralsols in the humid tropics. Much less favourable are the Acrisols, Alisols, and Lixisols, which have similar fertility problems to the Ferralsols, but in addition have poor physical conditions and are vulnerable to erosion. On the positive side are the smaller but significant areas of stable, fertile Nitisols found mainly in the highlands of Ethiopia, Kenya, the eastern Democratic Republic of Congo (former Zaire), and Cameroon on basic parent materials. It is no wonder that, during the period from 1970 to 1993, per capita food production in Africa declined by 18 per cent while it increased by 30 per cent in Asia and 16 per cent in South America, the latter two thanks to green revolution technologies which require ideal conditions and/or high inputs.

The situation in South Africa and the African continent stands in stark contrast to that in North America and Europe (including Russia and the Ukraine), with their vast areas of fertile soils. The deep loess parent material in these regions is especially important. It is estimated that about 50 per cent of the contiguous states of the USA (an area the size of the Sahara Desert) is arable, including some of the world's best farmland (Birdwell 1987).

These areas of the USA, which has a population of about 250 million, can produce the food required by about 800 million people. More than half of the deep Chernozems of the former Soviet Union are presently not used for cropping. Chernozems are ranked among the best soils in the world due to their good depth, silt loam textures, and high organic content; they constitute a vast resource for increased grain production in the future.

Human-induced soil degradation

The importance of soil as a resource is often overlooked. The productivity of our land and its ability to grow food and to regulate the water balance are dependent on the soil cover, and degradation of this valuable resource may therefore do more to bring about an environmental crisis than other environmental changes such as climatic change.

In 1990 it was estimated that, worldwide, 1.2 billion hectares of land, an area the size of mainland China and India together, have been seriously degraded by human activities since World War II. In Table 14.2 the extent of human-induced soil degradation is shown by continent and by cause of degradation. These data are from the United Nations Environment Programme (UNEP) project on Global Assessment of Human-Induced Soil Degradation (GLASOD), as summarized by Oldeman (1993). The GLASOD project revealed that, of the total area affected by human-induced soil degradation, 56 per cent suffered water erosion, 28 per cent wind erosion, 12 per cent chemical degradation, and 4 per cent physical degradation. Chemical degradation includes salinization or sodification of irrigated areas, acidification of cultivated land (mainly in high-rainfall areas), reduction in nutrient status (due to both exhaustive cropping and removal of fertile topsoil by erosion), and pollution. Physical degradation includes soil crusting (surface sealing), soil compaction, and waterlogging. All the above-mentioned forms of human-induced soil degradation are problems in South Africa.

Water erosion

Water erosion has been described as South Africa's biggest environmental problem. Expressed per member of the population, the country's soil loss due to human-induced soil erosion is estimated to be twenty times higher than the world average. It is also estimated that South Africa has lost more than 25 per cent of its topsoil since the beginning of the twentieth century.

A basic factor underlying South Africa's water erosion problem is the large proportion of inherently unstable soils which are extremely vulnerable to planning errors and mismanagement. The situation is aggravated by the low and unreliable rainfall and the aggressive type of rain. The high temperatures cause very rapid decomposition of organic matter, thus reducing the main structural stabilizing agent to such low levels that it becomes inefficient as a soil stabilizer. In many areas, steep slopes further aggravate the situation.

TABLE 14.2 Causative factors of soil degradation, expressed in millions of hectares of terrain affected

Continent	Deforest-ation	Overgrazing	Agricultural mismanagement	Over-exploitation	Industrial pollution
Africa	67	243	121	63	+
Asia	298	197	204	46	+
South America	100	68	64	12	–
North and Central America	18	38	91	11	1
Europe	84	50	64	1	21
Australia	12	83	8	–	+
World	579	679	552	133	23

+ degradation due to this cause was reported for the particular continent, but the total area affected is so small that it cannot be expressed in millions of hectares.
– no degradation ascribed to this cause was reported for the particular continent.

SOURCE: Oldeman 1993.

As long as the unstable soils are covered permanently by a dense vegetative cover they are stable and natural erosion is minimal. Grass is especially effective. Grass absorbs the kinetic energy of the raindrops, preventing it from physically disaggregating the surface soil layer. The organic matter from the grass, especially the densely ramified root system, stabilizes the soil, promotes infiltration of water, and reduces surface runoff. The grass cover also reduces the amount and speed of runoff. When the vegetative cover is destroyed by overgrazing or cultivation, the soil becomes unstable due to the lack of organic matter and is also exposed to physical disaggregation by falling raindrops. The results are surface sealing, leading to reduced infiltration, increased runoff, easier removal of disaggregated particles, and thus vastly increased erosion. A comparable example from the Ivory Coast showed that, on a relatively flat slope of only 7 per cent, annual soil loss under maize rose to 90 t/ha and on bare soil to 140 t/ha, compared with only 0.5 t/ha under a dense grass cover (IFAD 1992).

By far the most highly erodible soils

are the solonetzic soils. South Africa's large areas dominated by these soils are therefore particularly vulnerable to erosion. It should be kept in mind that significant areas of these soils are also found on the lower (and flatter!) slopes of landscapes with granite or sandstone as parent materials, for example in the lowveld of Mpumalanga and parts of the Northern Province. All the areas dominated by shallow soils are highly erodible, especially where mudstones and shales of the Beaufort Group are the dominant parent materials. The recovery potential of these erodible soils, especially the solonetzic soils, is extremely low.

Overgrazing is the main human-induced factor causing accelerated water erosion in South Africa. Overgrazing leaves the soil bare and unprotected against the forces of falling and running water (see Figure 14.7), especially when prolonged droughts are followed by torrential rains. This is a problem not only on domestic stock farms, but also in game ranches and game reserves in areas with sandstone and granite as parent materials. The latter is the case in the Waterberg area of the Northern Province and in the lowveld of the Northern Province and Mpumalanga. It is a combined result of a tendency to overstock with game and an unwilling-

FIGURE 14.7 Soil erosion (sheet and donga) due to overgrazing near Alice, Eastern Cape

Note total lack of grass and crusting (surface sealing) of soil

PHOTOGRAPH: J. L. H. Williams

ness (for aesthetic reasons) to fence off sensitive sweetveld areas from sourveld on the hills, leading to overgrazing (and erosion) of the former and under-utilization of the latter.

Injudicious cultivation of inherently unstable soils is another major cause of water erosion in South Africa. This is mainly due to inadequate understanding of the differences in erodibility of the different soils and the consequent use of general criteria or inappropriate erosion models in land-use planning. Ironically, some of the worst-eroded areas in the former Ciskei and Transkei are in cultivated fields that were identified and planned for cultivation by planners during the implementation of 'rehabilitation' or 'betterment' schemes. Figure 14.8, photographed in 1997, gives an example of such a field that was 'planned' in the 1960s and abandoned in the 1970s. See Laker and D'Huyvetter (1988) for a discussion of this.

Water erosion has various serious consequences. It has been estimated that the losses of nitrogen, phosphorus, and potas-

FIGURE 14.8 Eroded cultivated field (middle of photograph), about twenty years after having been abandoned, near Alice, Eastern Cape

Note the contours in the background. The field was identified as suitable for cultivation by land-use planners due to the relatively flat slope. The inherent instability of the soil was not considered. Note the lack of recovery of this originally high potential grassland – now reduced to sparse shrub vegetation

PHOTOGRAPH: J. L. H. Williams

sium from South African soils due to water erosion amounts to R365 million per annum in 1985 equivalent monetary terms (Van der Merwe 1995), causing major reductions in soil fertility. These eroded nutrients often end up in water bodies such as dams, causing their eutrophication and pollution.

Loss of nutrients is not the most serious consequence of soil erosion, however. The most serious problems are the negative impacts on the hydrological cycle and the stability of water supply. Eroded catchments are characterized by excessive runoff, leading to increased frequency and magnitude of flooding. Flash floods become the order rather than the exception. Even moderate, 'normal' rains become flood hazards. The floods are followed by droughts. This is especially the case where upland wetlands, the sponge areas of rivers, are eroded. There is no longer storage and gradual release of water from the soil to the rivers. As a consequence, an induced drought may prevail before the start of an actual drought due to low rainfall (Roux 1990).

Furthermore, water erosion leads to the quick silting up of dams which are supposed to store water for use during droughts. One of the most extreme examples is the Welbedacht Dam in the Caledon River. It was completed in 1973 with a storage capacity of 114 million cubic metres, in order to supply 90 per cent of the water required by the city of Bloemfontein. By 1993 its storage capacity had been reduced to 17 million cubic metres; it had silted up by 85 per cent in twenty years. Water erosion is therefore a major threat to urban populations and industries, which depend on assured, adequate water supplies.

Wind erosion

The extent and impact of wind erosion on a global scale is often underestimated. The infamous 'dust bowl' in the USA during the drought of the 1930s was one of the most extreme cases in crop production areas. A large proportion of Canada's grain-producing areas is also very vulnerable to wind erosion. This is becoming an increasing problem in Germany, too, due to the consolidation of farms into bigger units and the consequent increased sizes of cultivated fields.

Unpublished reports by the Institute for Soil, Climate, and Water indicate that 25 per cent of South Africa is vulnerable to wind erosion. Most prominent are the sandy soils in the arid to semi-arid areas of the Kalahari, the western Free State, and the western part of the North West province. Wind erosion is a major hazard in the maize fields of the latter two areas, due to a combination of sandy soils and strong winds at the stage when the soils are at their driest just before the onset of the next summer rains (August, September, and October). Fields are bare during this period, because they are either left fallow or cultivated for planting the next crop. Sandy soils and strong winds also make the coastal areas of South Africa vulnerable to wind erosion. In the Eastern Cape, attempts at grain production in coastal areas of the former Ciskei and on the red palaeodunes near Paterson became agricultural and environmental disasters due to wind erosion. On the Cape Flats, wind erosion caused widespread damage to the environment until the area was stabilized by the planting of Port Jackson (*Acacia saligna*) trees. Ironically, these are now considered to be causing major environmental problems because of their invasive nature.

Physical soil degradation

In recent decades there has been a sharp increase in the extent and severity of physical soil degradation due to intensification of mechanized agriculture and increased cultivation of marginal land that is more vulnerable to physical degradation. Subsurface soil compaction is a widespread problem. Compacted horizons used to be called 'plough pans', because they were believed to be caused by pressure or smear actions of ploughs. They are now called 'traffic pans', because it has been realized that they are caused mainly by the wheel pressure of heavy tractors and implements. Soil compaction is a widespread severe problem in the cultivated areas of South Africa, because of the inherent susceptibility of the soils to compaction. Particularly vulnerable are the fine, sandy, cultivated soils of the North West province and the Free State, but so are major areas of the Eastern and Northern Cape and some other areas. It has been estimated that approximately 80 per cent of the soils cultivated for maize in South Africa are prone to compaction. Crops become much more vulnerable to drought on compacted soils because roots are restricted to a very shallow topsoil layer and cannot utilize the water stored in the subsoil.

Soil crusting is becoming an increasingly serious problem worldwide. In South Africa it is a very widespread problem, the extent of which has been fully realized only recently. Although it is a problem throughout virtually the whole country, it is particularly widespread and severe in areas such as the Eastern Cape and the Ruens and Swartland of the Western Cape. Crusting leads to poor water infiltration and excessive runoff. This causes soils to remain dry and aggravates erosion. Germination and seedling emergence are seriously inhibited by the dense crust. Crusting makes soils unsuitable for overhead irrigation. Even dripper and micro-jet irrigation cannot be used in many areas because of crusting under these systems. Crusting is a major factor jeopardizing the reclamation of bare patches in overgrazed areas.

Technologies such as zero or minimum tillage are sometimes used to counteract physical soil degradation, but these are not everywhere a success because of other problems related to them. Traffic pans are counteracted successfully by methods such as controlled traffic and ripping, deep cultivation with a tined instrument which breaks up the dense, compacted soil layer.

Chemical soil degradation

Soil acidification is a widespread problem in higher-potential, higher-rainfall areas under intensive cultivation, especially where acidifying nitrogen fertilizers are used. Such fertilizers are essential for achieving top yields. Many of these areas are inherently acidic as a result of the intensive weathering and leaching conditions prevailing under the high-rainfall conditions, the situation being aggravated under intensive cultivation. In South Africa, acidification is particularly a problem in the areas of higher cropping potential in the Mpumalanga highveld, Kwa-

Zulu-Natal, the Eastern and Western Cape, and the eastern Free State. Lime applications to counteract the acidification are inadequate, mainly due to the high transport cost of lime rather than the price of the lime itself.

The salinization and alkalization of irrigated areas are major global forms of soil degradation. They have led to the demise of a number of old civilizations which were built on fertile land along big river basins. The best example is the biblical Mesopotamia, which is now the desert of Iraq along the Tigris and Euphrates Rivers. The Nile escaped this fate of the other big 'old' rivers until the 1950s, when modern technology and international power politics took their toll. For the first time in 5 000 years, production along the Egyptian part of the Nile has declined after the building of the Aswan Dam. On the GLASOD World Map of Human-Induced Soil Degradation, the salinization of this part of the Nile Valley is clear. The Aswan Dam stopped flooding and, consequently, the natural leaching of salts from the soil ceased. Furthermore, relatively saline water leaching from the Sudanese irrigation schemes above the dam now ends up in the dam, instead of going through to the Mediterranean Sea. Modern, intensified irrigation under the hot, desert conditions further accelerates the accumulation of salts. Numerous similar present-day examples are described in various publications. In South Africa it is estimated that approximately 10 per cent of the country's irrigated land has become saline or alkaline.

Soil fertility has declined in many areas due to exhaustive cropping. This trend is fast accelerating in less developed countries, especially in the humid tropics and subtropics, as a result of population pressure. Sustainable shifting cultivation or 'slash and burn' systems, including long bush-fallow periods, used to be practised in these areas. During the long fallow periods, the deep-rooted trees recycled nutrients to the surface layers. Increased population pressure has reduced the fallow periods, and in some cases the bush-fallow systems have been replaced with continuous cropping. Since no or very little fertilizer is applied by the resource-poor farmers, soil fertility has collapsed very quickly in regions such as central Africa. In traditional tribal areas in South Africa, soil fertility has declined as a result of continuous cropping with little or no application of fertilizers.

Soil pollution, with toxic heavy metals, for example, is becoming an increasing problem in highly industrialized areas. According to the GLASOD survey, at present this is predominantly a problem of north-western Europe, where vast quantities of manure are produced under very intensive livestock-production systems on small parcels of land. Unless livestock numbers are reduced drastically, pollution of soil with phosphate and underground water sources with nitrate will create an unsolvable problem. In South Africa, steps are needed to ensure that a similar situation does not, for example, develop in the densely populated Gauteng area through the creation of small, intensive livestock-farming units (such as pig and poultry).

Other forms and sources of degradation

A disconcerting trend in South Africa is the unabating permanent loss of vast areas of its limited high-potential agricultural land to urban development, industry, or mining. Little or nothing is being done to limit this trend. This is in stark contrast to the strict laws of the USA to preserve prime and unique agricultural land for agriculture, despite its vast resources.

Conclusion

Soil is one of the most important natural resources of any country. The distribution of soil types reflects both the physical geography of a region and the potential for human development. Together with water availability, soil is a key factor supporting plant growth and therefore food production. A review of the situation in South Africa points to both the restricted area of high-potential soils and the vulnerability of so many of the country's soils to degradation. A knowledge of the distribution of soil types and their properties is essential for wise land-use planning and for the conservation of soil resources so as to minimize further degradation. It is hoped that this chapter has provided students with an introduction to this important subject.

15 The ecological resource base: biodiversity and conservation

MICHAEL E. MEADOWS

Ecological resources and their significance

The national wealth of any country can be measured in three forms: material, cultural, and biological. Unfortunately, as Wilson (1992) points out, the biological wealth of nations is rarely taken as seriously as its material and cultural counterparts. This is a mistake of monumental proportions, because our very existence is directly dependent on the exploitation of the various ecological resources; we cannot develop our physical and cultural resources without first accessing the ecological systems that provide for our fundamental life-support system. In short, the living world provides us with the food we eat and the air we breathe. Injudicious management of the earth's biological wealth is not merely an environmental inconvenience, but a major global problem that may ultimately result in our extinction.

Ecology is a science that focuses on the linkages between the biological (biotic) components (including ourselves) and the physical (abiotic) components of the earth system. It recognizes that there are complex functional linkages between the biotic and abiotic and identifies the *ecosystem* as the fundamental unit of study: a web which links plants, animals, and their physical environment. Croall and Rankin's (1992) somewhat tongue-in-cheek depiction of the ecosystem (see Figure 15.1) nevertheless contains a serious message: that the role of humans within the ecosystem is all too frequently problematic because of the errant impression that we are separate from it. Viewing the environment as a collection of exploitable resources is inviting trouble if we do not realize the impact that their utilization can have.

Ecosystems provide us with a number of free services, and it is an obvious, albeit ignored, rule that we should attempt to manage and conserve them. In South Africa, as elsewhere, the sustainable exploitation of these services is fundamental to our ability to survive. What are these resources and the services they provide? In order to answer this question

FIGURE 15.1 Croall and Rankin's tongue-in-cheek depiction of human interaction in the ecosystem

The ecosystem is the complex web linking animals, plants, and other life forms in any particular environment, whether it's your windowbox or the whole biosphere. Everything hangs together in the ecosystem – alter one part and you alter the others. Sooner or later.

Humans are only one factor in the ecosystem. But we don't quite see it that way. We set ourselves apart and call all the other factors or species *natural resources* – or simply *nature*.

Human survival depends upon preserving the ecosystem. It's the boundary of existence, the framework of human activity. The ecosystem can do without us, but we can't do without the ecosystem. As long as we live on earth.

Viewing nature as no more than a bundle of resources for human consumption is inviting trouble.

SOURCE: Croall and Rankin 1992.

we need to appreciate that ecosystems have both a particular structure and a range of functions. The basic biological units of an ecosystem are the individual living organisms which belong to its various species. At its simplest level, ecosystem structure is measured by the diversity of organisms, but these organisms are linked to each other, and to their physical habitat, via a number of functioning biological (for example competition, predation, herbivory), chemical (for example photosynthesis, respiration), and physical (for example the hydrological cycle, soil formation and development) processes. Most important among these myriad processes is the phenomenon of *primary productivity*, whereby the energy of solar radiation is converted, in combination with water and carbon dioxide, into carbohydrates by the chlorophyll-containing leaves of green plants. In turn, these carbohydrates are transferred, via the food chain or food web, to other biotic components of the ecosystem, including, of course, ourselves. *Net primary productivity* (measured in tons per hectare per year, or t $ha^{-1} yr^{-1}$) represents the rate at which the process of photosynthesis operates, producing *biomass* (measured in tons per hectare, or t ha^{-1}), which is the quantity of biological material present in an ecosystem at any one time. Other elements or mineral nutrients, extracted mainly by green plants from the physical environment, are recycled in a way that makes them available in appropriate quantities to all the organisms in the ecosystem, again including humans. The various interacting elements of the ecosystem, therefore, are the fundamental resource base that enables the persistence of the human species; we are at once dependent upon and integral to the ecological system.

In this chapter the ecological resources of South Africa are considered in the context of the pattern of various biomes (large-scale ecosystems) that characterize the different and distinctive geographical regions of the country. Because ecosystem boundaries do not coincide with political boundaries, much of this chapter refers to southern Africa in general (incorporating South Africa, Lesotho, Swaziland, Botswana, Zimbabwe, Namibia, and Mozambique). The most important factor which underlies the observed biogeographical patterns is climate, although it must be appreciated that a host of interconnected environmental attributes influence the ecological scenario here. Socio-economic, political, and cultural factors also play their part; indeed there are many ecosystems in southern Africa that so strongly bear the imprint of human activity upon them that they show little resemblance to the 'pristine' environments that must have predated the growth in population and demand for resources that typify the post-colonial era.

Biological diversity, or *biodiversity*, is taken as the ultimate icon of ecosystem identity and status, and therefore it provides the focus for the analysis of ecological resources in this chapter. In short, species richness (the most commonly employed measure of biodiversity) can be considered as an indicator of ecosystem well-being; changes in this parameter, most often a reduction, are indicative of environmental problems. Plant diversity is arguably the most significant component of biodiversity because, firstly, plants are the foundation of the trophic system that links solar radiation with ecosystem function and, secondly, because vegetation is usually the most conspicuous and easily recorded aspect of an ecosystem. For these reasons, this chapter is based on the patterns of plant species diversity in southern Africa and the manner in which different plant species associate to form the various biomes. To begin with, we consider the definition, patterns, and determinants of biodiversity.

Biodiversity in context

> *diversity is a defining characteristic of life. Without diversity life is not conceivable, just like a ball without roundness is not conceivable, or a diamond without hardness is not a diamond.* (Solbrig 1992: 13)

Biodiversity embodies several levels of biological variation, including genetic diversity (variations in genetic characteristics within a species), species diversity or species richness (number of different 'types' of organisms), and ecosystem diversity (range of different types of ecosystems); all these levels of variation are interdependent. Most frequently, biodiversity is represented by species diversity, a kind of inventory of the number of different species within a given geographical area, although even this has several elements, which are shown in Figure 15.2. α-diversity, or point diversity, is the number of species occurring within a particular habitat (say, the number of plant species in a small wetland); β-diversity is between-habitat diversity and takes account of the turnover in species identities along environmental gradients (the number of new species added to an inventory as one moves along, say, an altitudinal gradient); γ-diversity accounts for species turnover within similar habitats but between geographical areas (that is, to what extent species complements are similar in, say, two lowland fynbos areas 100 km apart). Biodiversity in a particular place is an agglomeration of all three measures.

In May 1992 the nations of the world adopted a global Convention on Biological Diversity (Glowka et al. 1994), which has since been signed and ratified by more than 150 countries, including South Africa. This represents a commitment to conserve the earth's biodiversity and to use its biological resources sustainably and equitably. The goals of the convention remain elusive, however, because of political rather than scientific inadequacies. Arguably, the most important recent scientific ideas which assist nations in achieving the goal of the convention are the principles of island biogeography (MacArthur and Wilson 1967) and of minimum viable populations (MVPs). Both of these have resulted in a change of thinking with respect to biological diversity, especially of isolated habitat 'islands' of natural communities in a 'sea' of agricultural or urban landscapes.

The equilibrium theory of island biogeography was developed in the 1960s to account for the observation that species numbers on islands appear to be a function of island area. MacArthur and Wilson's explanation for this pattern centred on the idea of an equilibrium number of species on an island of given size determined by the balance between the rate of newly arriving species (via colonization) and the loss of species through local extinction.

FIGURE 15.2 The three major components of species diversity

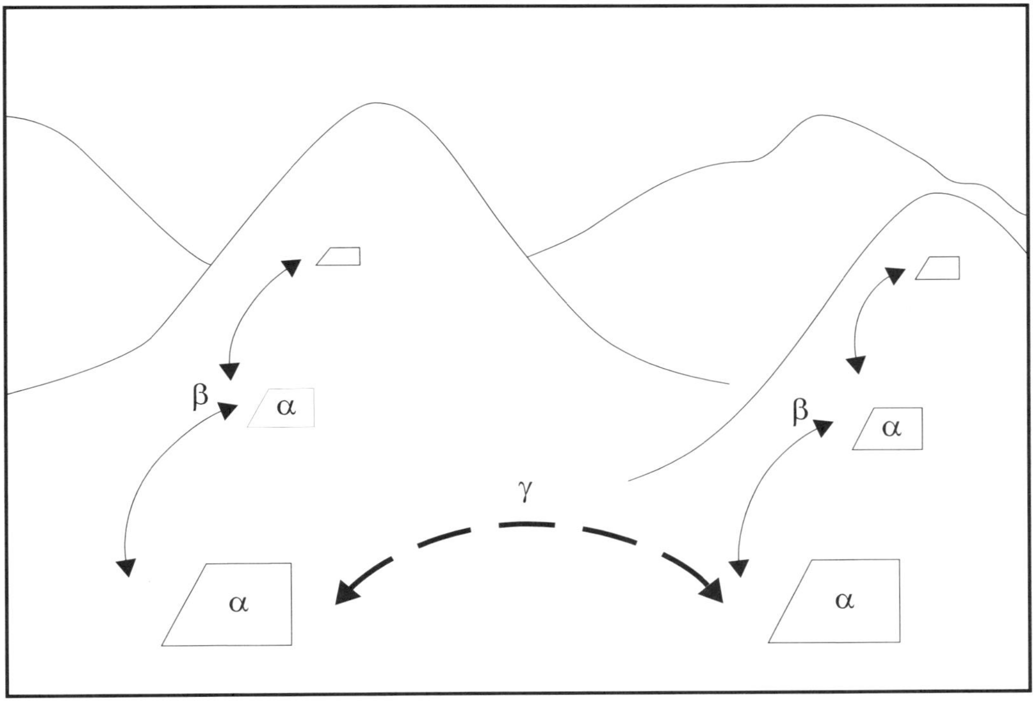

α = species richness within a habitat; β = species turnover between habitats;
γ = species turnover within a habitat between geographical areas

SOURCE: Bond 1985.

Minimum viable population theory focuses on the process of species extinction and the fact that smaller populations are more susceptible to random environmental disturbances and disease.

Sometimes the simplest of questions are impossibly difficult to answer, and so it is with the obvious question: how many different species exist on earth? It is possible to provide a reasonably accurate estimate of the number that have been formally recorded, identified, and classified by systematists, the answer being in the vicinity of 1.5 million (see Figure 15.3). But, as Wilson (1992) notes, we do not know the real number of species on the planet to within even an order of magnitude. This is because much of the earth's biodiversity has remained inconspicuous and unrecorded in the canopy of remote tropical rainforests. Recent investigations have revealed mind-numbing statistics about the diversity of tropical forests: 275 species of ants were collected in eight hectares of Peruvian forest, 300 different types of trees in just a single hectare of forest near Iquito, also in Peru, and up to 18 000 species of beetles in one

FIGURE 15.3 Numbers of formally described species of organisms (according to major group)

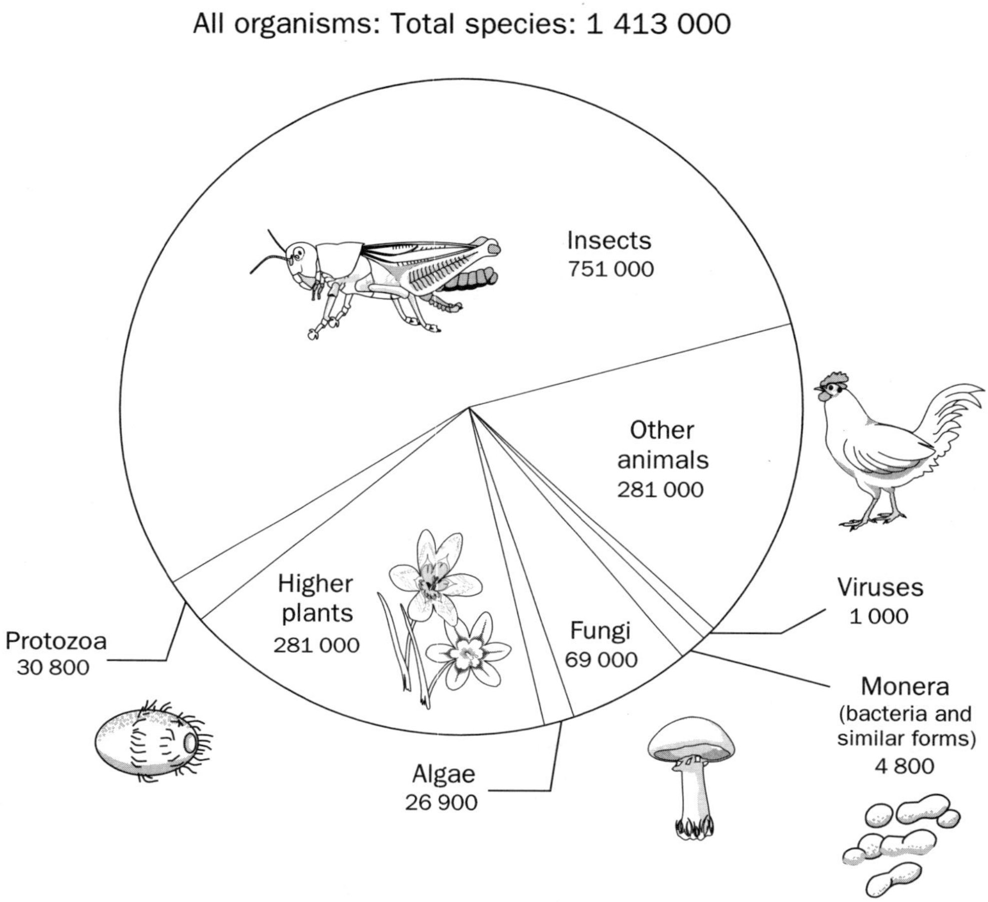

Insects and higher plants dominate the diversity of living organisms known to date, but vast arrays of species remain to be discovered in the bacteria, fungi, and other poorly studied groups. The grand total for all life falls somewhere between 10 and 100 million species

SOURCE: Wilson 1992.

hectare of Panamanian rainforest (Wilson 1992).

Clearly, biodiversity is not evenly distributed across the earth, and there are some places where the concentration of types of organisms is especially great. At the global scale, such biodiversity 'hotspots' (see Figure 15.4) include, of course, many of the tropical rainforest areas, but southern Africa has several localities where species diversity, of flowering plants in particular, is elevated well beyond what may be expected, given its geographical extent. Indeed, southern

FIGURE 15.4 Global biodiversity 'hotspots'

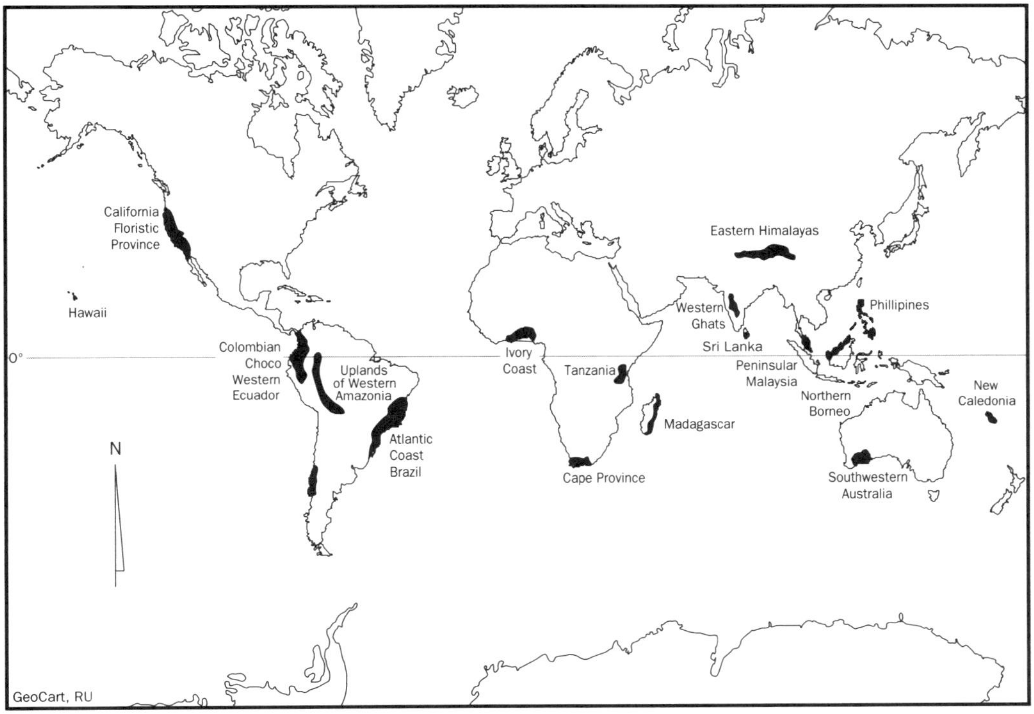

SOURCE: Wilson 1992.

Africa as a whole has a remarkably rich flora, with more than 20 000 species (Arnold and De Wet 1993). Compared to similar-sized regions in Africa, the subcontinent has exceptional levels of plant richness; indeed it is remarkably rich even by comparison with areas of equatorial rainforest (see Table 15.1). Thus, although southern Africa is about the same size as both Algeria and the Democratic Republic of Congo (former Zaire), it has seven times and twice as many plant species, respectively. A further distinctive feature of the southern African flora is its degree of endemism: that is, the extent to which the plant species of the subcontinent are endemic, or restricted, to its geographical area. The levels of endemism are extremely high, since some 80 per cent of its species are found nowhere else; this value is more akin to what might be expected of an isolated oceanic island than a section of a large continent.

This plant diversity is not evenly spread; for example, Namibia occupies more than 30 per cent of the geographical area of southern Africa but contributes only 15 per cent of its species. Cowling and Hilton-Taylor (1994) identify southern Africa's eight centres of plant diversity

TABLE 15.1 Plant-species densities for different regions of the world

Region	No. of species*	Area (10^6 km²)	Species density (per million km²)
Southern Africa	21 000	2.57	8 220
Brazil	40 000	8.46	4 700
Congo (Kinshasa)	11 000	2.35	4 680
India, Pakistan, Bangladesh	20 000	4.89	4 100
Argentina	9 000	2.77	3 250
Australia	25 000	7.71	3 200
Europe	14 000	5.68	2 500
USA	20 000	9.36	2 100
Tropical Africa	30 000	20.00	1 500
Algeria	3 000	2.38	1 300
Sudan	3 000	2.51	1 270

* to the nearest 1 000 species

SOURCE: Cowling et al. 1989; Cowling and Hilton-Taylor 1994.

(see Figure 15.5), which include about 3.5 per cent of the world's flora on just 0.2 per cent of the earth's surface area. The 'hottest' of these hotspots (Myers 1990) is the Cape Floristic Region, where a potent combination of α-, β-, and γ-diversity results in almost 9 000 plant species being squeezed into just 90 000 km².

The factors underlying the remarkable pattern of plant species richness in the subcontinent are not fully understood, although some combination of contemporary environmental heterogeneity and evolutionary history must be at its root. The fact that the region has countless combinations of geological parent material, topography, soils, drainage, climate, fire, and grazing regimes most definitely underlies the impressive biodiversity statistics. For the remainder of this chapter, we turn to an examination of South Africa in particular and its major biomes in order to see how these patterns of biodiversity are given ecological expression.

FIGURE 15.5 Southern African biodiversity hotspots

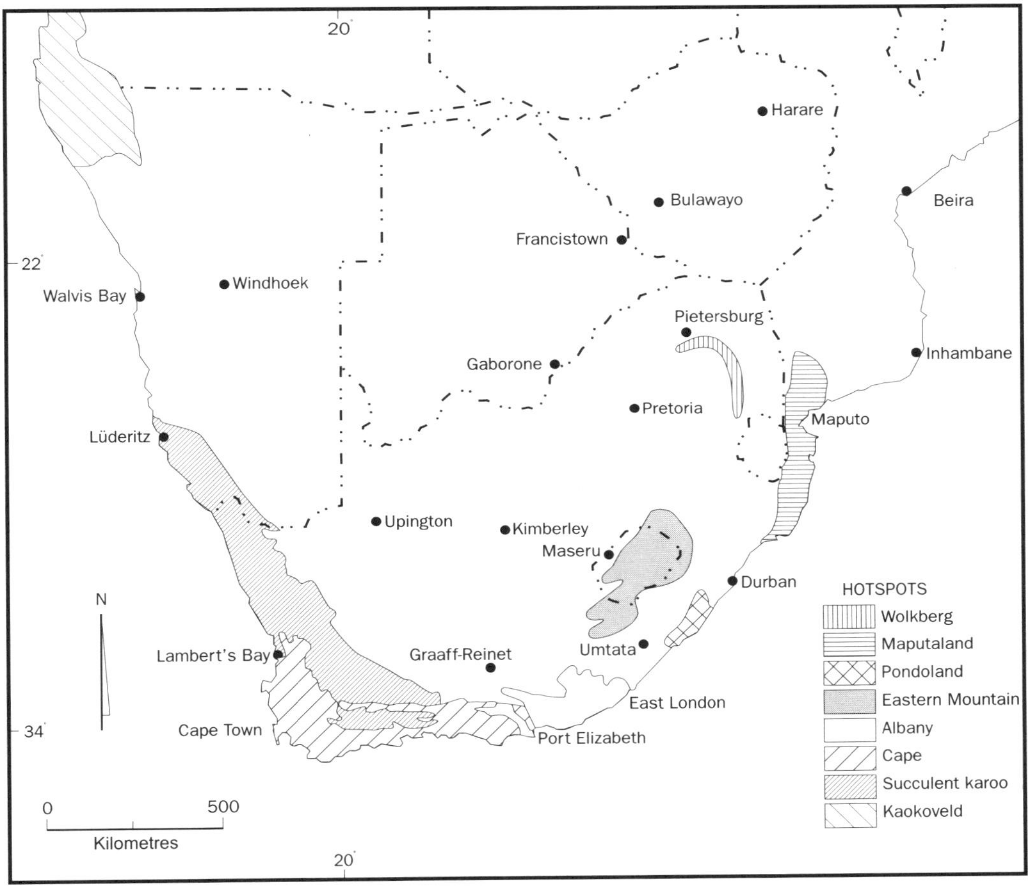

SOURCE: Cowling and Hilton-Taylor 1994.

The major biomes in Southern Africa

As Rutherford (1997) notes, biomes are large-scale ecological units representing major life zones and extending over large natural areas. Conceptually they form patterns discernible at the regional scale, are identified principally by their constituent major vegetation types, and are characterized by a distinctive suite of climatic (primarily) and other physical environmental factors. Rutherford and Westfall (1994) produced the first objective categorization of biomes in southern Africa, a system based on the relative importance of the dominant plant life forms as originally identified by Raunkiaer (1934). The Raunkiaer life-form system is based on the degree of protection afforded new buds in the unfavourable (colder or drier, or both) time of the year (see Figure 15.6). Phanerophytes are mainly trees and large shrubs whose buds are borne on the tips of the branches; chamaephytes are smaller shrubs and herbs with buds borne closer to the ground; hemicryptophytes are plants which die back completely during the unfavourable period and have buds protected by the litter layer; and cryptophytes regenerate from buds which are formed below the ground surface. A fifth category is formed by the therophytes, which are annuals and regenerate only from seed. As can be seen from Figure 15.7, these different physiognomic life forms dominate in different major global biomes, and a similar relative prominence is evident in the seven major biomes identified for southern Africa by Cowling, Richardson, and Pierce (1997) (see Figures 15.8 and 15.9; Table 15.2). The strength of the linkage between these biomes, with their particular vegetation types, and climate, in particular rainfall, becomes obvious when comparing the biome distribution map (Figure 15.9) with that of mean annual precipitation (Plate 32).

FIGURE 15.6 Physiognomic classification of plants (of Raunkiaer 1934) according to the relative position of the perennating buds

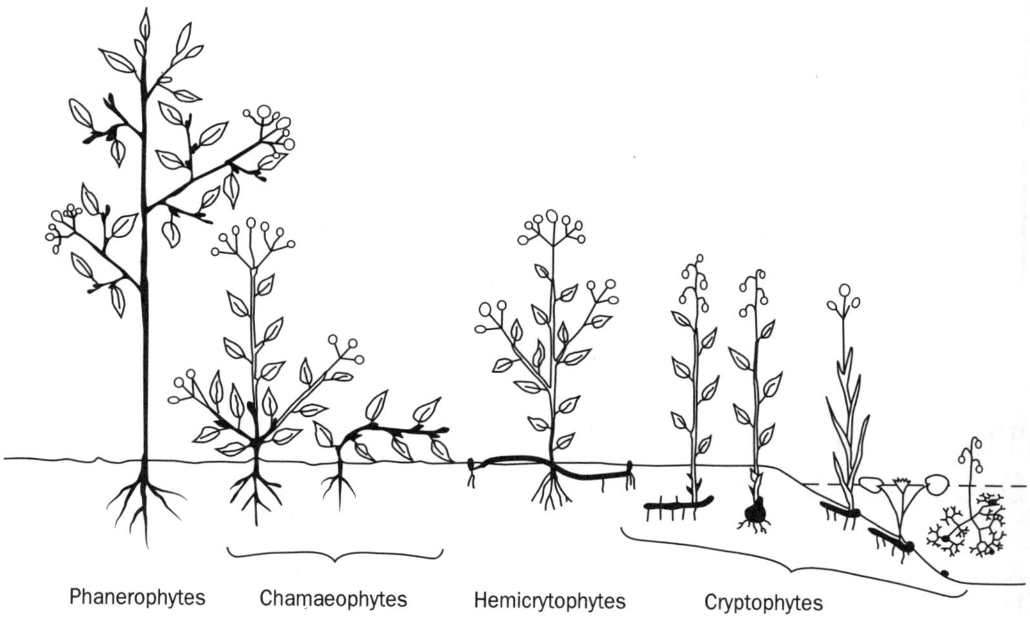

The unshaded parts die back during unfavourable periods. Therophytes (annuals or ephemerals), which persist only as seeds, are not illustrated

SOURCE: Archibold 1995.

FIGURE 15.7 Proportional distribution of different life forms according to the Raunkiaer system

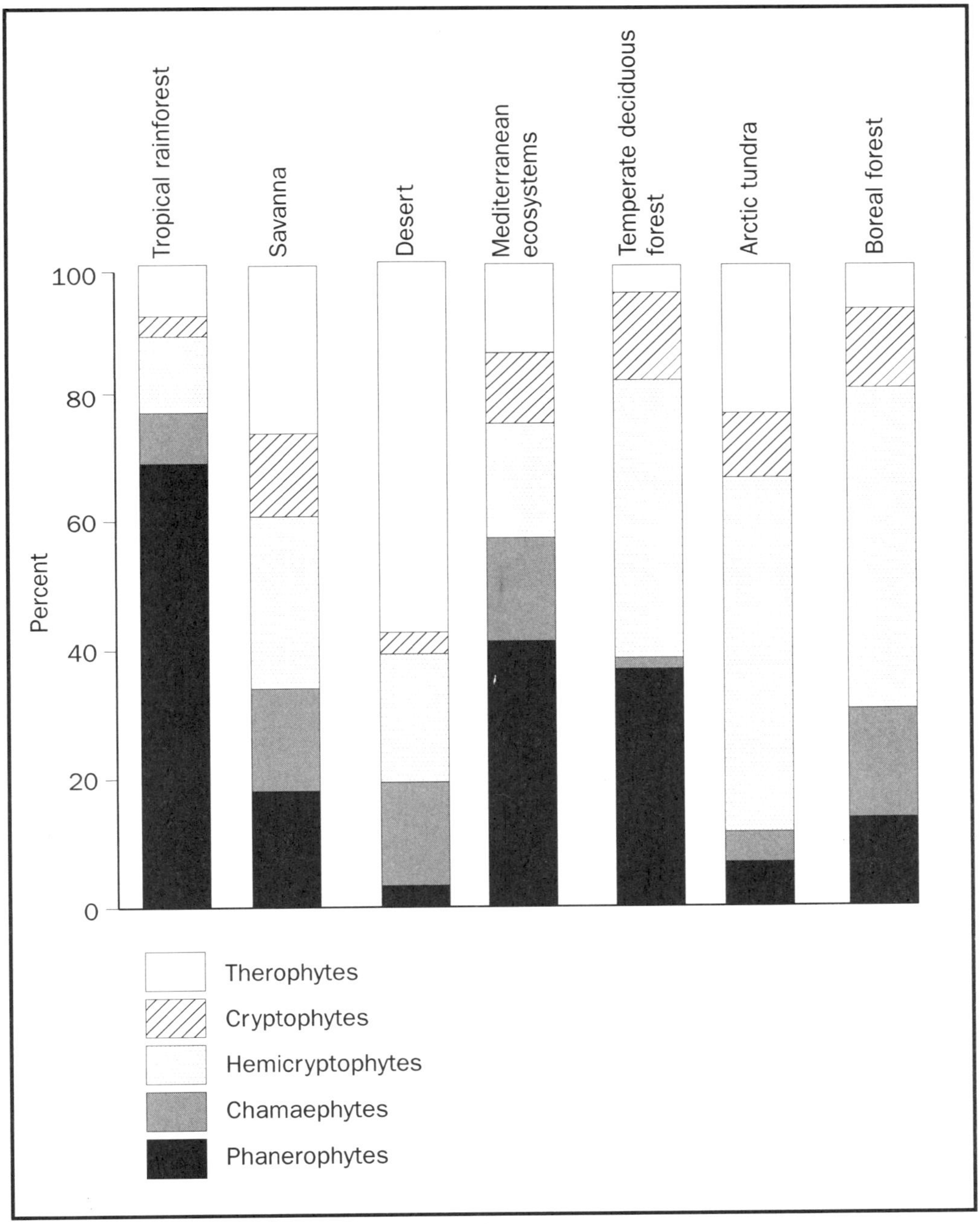

SOURCE: Archibold 1995.

FIGURE 15.8 Dominance, or co-dominance, of four plant life forms according to the Raunkiaer system

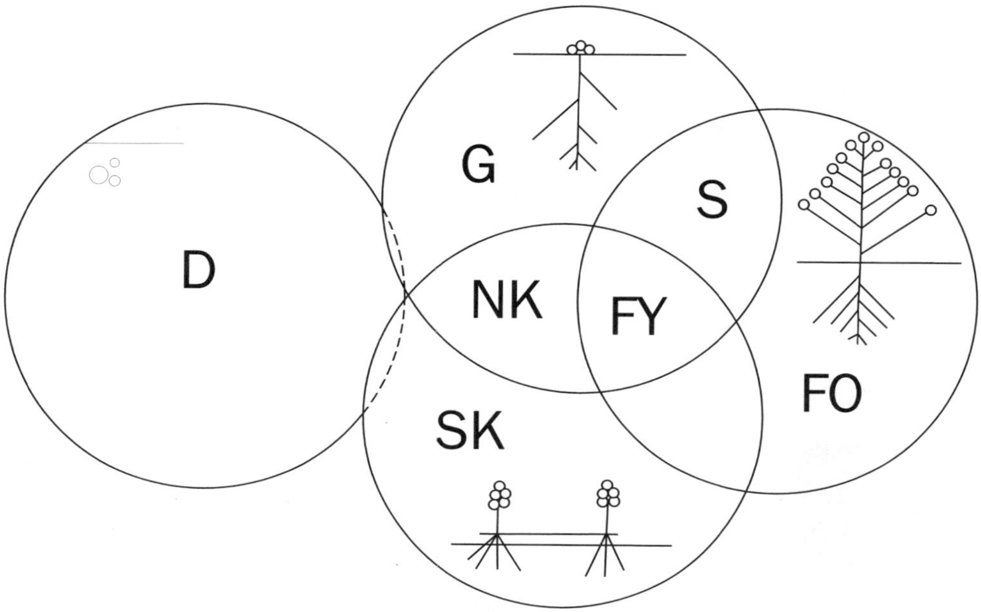

Raunkiaer life forms are represented with phanerophytes on the right, chamaeophytes below, hemicryptophytes above, and therophytes on the left. FO = forest; G = grassland; SK = succulent karoo; D = desert; NK = Nama-karoo; S = savanna; FY = fynbos

SOURCE: Rutherford and Westfall 1994.

Each of the seven biomes is now described in turn, in relation to (a) its fundamental physical environmental determinants, (b) the characteristics of its major plant communities, from biodiversity and plant form to levels of primary productivity, and (c) the degree of human impact and modification. We begin with those biomes associated with subtropical climates.

The savanna biome

The term 'savanna' has been widely applied to plant communities in the tropics and subtropics characterized by various combinations of trees and grasses. They are typified by a continuous ground layer, consisting mainly of grasses (hemicryptophytes) and a discontinuous woody tree or shrub overstorey of phanerophytes

FIGURE 15.9 The seven major biomes of southern Africa

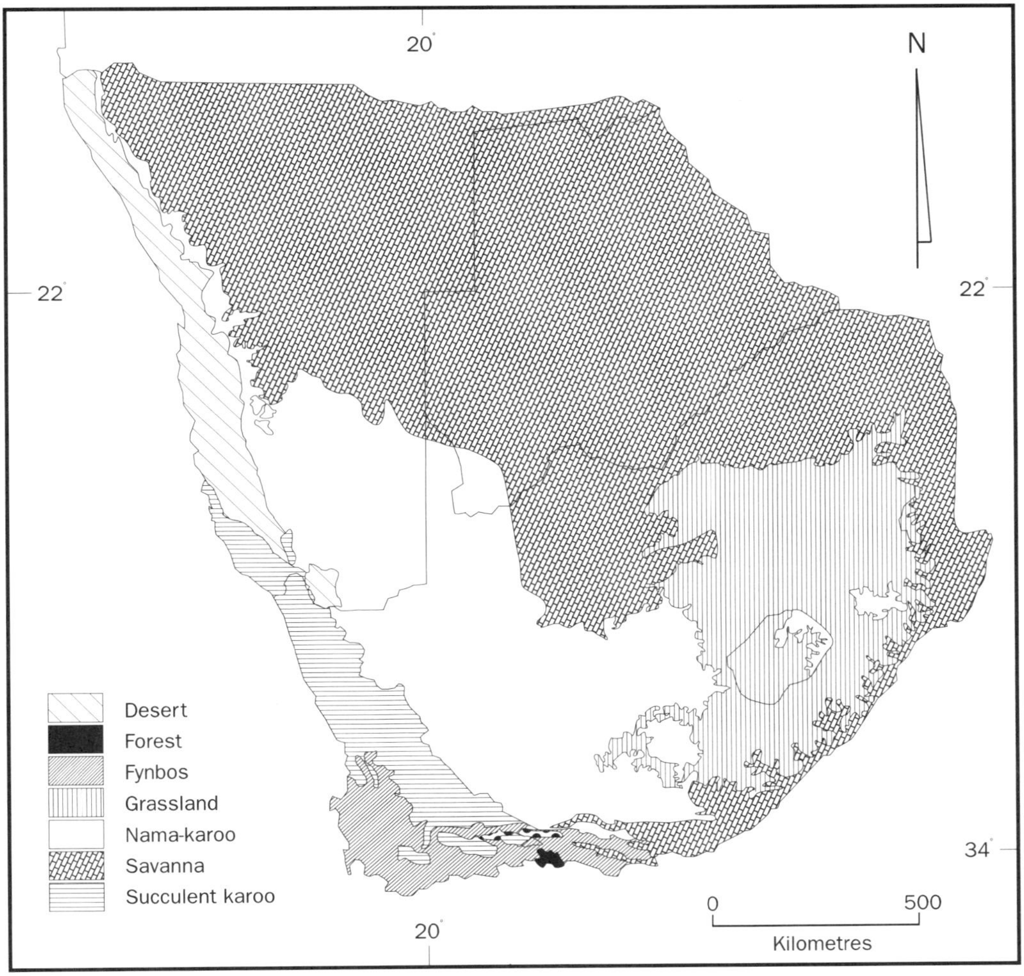

SOURCE: Cowling, Pierce, and Richardson 1997.

(Stock et al. 1997) (see Plate 41, colour section). In practice, there is a wide spectrum of combinations of these two structural elements, although in South Africa the extremes of the continuum are known colloquially as bushveld where the trees and shrubs are most prominent, and grassveld where the grasses are most conspicuous. This biome is easily the most widespread (see Figure 15.9; Table 15.2) in southern Africa as a whole, occupying 54 per cent of the subcontinent (Rutherford 1997), and it makes an important contribution to informal and subsistence economies (Scholes 1997). Indeed, savannas in general have an important global impact through the emission of trace gases from fires, through their incorpora-

TABLE 15.2 Characteristics of the biomes of southern Africa

Biome	Proportion (%)	Area (km^2)	Annual rainfall (mm) (season)	Dominant life forms	Structure
Savanna	52.7	1 435 713	235–900 summer	Hemicryptophytes Phanerophytes	Wooded C4 grasslands
Nama-karoo	22.7	607 235	100–520 summer to all year	Chamaephytes Hemicryptophytes	Dwarf and low open shrublands
Grassland	13.1	349 174	400–2 000 summer	Hemicryptophytes	Grassland; woody plants absent or rare
Desert	4.2	111 147	10–70 summer to aseasonal	Therophytes	Open, ephemeral herbland
Succulent karoo	3.7	100 251	20–290 winter to all year	Chamaephytes	Dwarf and low open succulent shrublands
Fynbos	2.7	71 337	21–3 000 winter to all year	Chamaephytes Phanerophytes Cryptophytes	Evergreen sclerophyllous shrublands
Forest	c. 0.1	568	525–2 000 winter to summer	Phanerophytes	Closed evergreen and semi-deciduous forest

SOURCE: Cowling et al. 1989; Rutherford 1997.

tion of carbon in soils and biomass, and because of their levels of biodiversity.

Climatic conditions are in all cases typified by a strongly developed winter dry period, although this may extend for much of the year at the semi-arid end of the savanna continuum. Mean annual precipitation varies from below 500 mm (in the western portion, associated with much of the Kalahari basin) to above 1 000 mm in parts of the eastern portion, on the coastal platform of Tongaland. Frost may be a factor for more than 100 days per year in some of the higher-lying parts. In all cases, dry-season fire, at variable return intervals, is an important ecological factor. Soils are often weakly developed and shallow, although there is great variation depending on locality. Arenosols in the Kalahari basin, for example, are deep and sandy, whereas Latosols occur on clay-rich substrates. Savanna soils developed under higher-rainfall (mesic) conditions are more dystrophic, or nutrient-poor, and are frequently associated with laterite (a hardpan enriched by sesquioxides of iron and aluminium); those at the drier (xeric) end of the moisture gradient are eutrophic (nutrient-rich) and often develop pedogenic calcrete (a lime-enriched hardpan) within the profile. (See Chapter 14 for a discussion of soils.)

The climatic and pedological factors combine to produce a distinctive pattern in the distribution of the various types of savanna. The main distinction is based on the relative abundance of grasses and trees, and between broad- and fine-leafed savannas (Scholes 1997); the underlying determining factor is nutrient availability, which is strongly influenced by rainfall. Eutrophic, arid savannas are characterized by fine-leafed and often thorny trees and shrubs, typified by various *Acacia* and *Terminalia* species, although there are broader-leafed communities too, for example mopane woodland dominated by *Colophospermum mopane*. The dystrophic savannas are exemplified by *Burkea* and *Brachystegia* (miombo) species. In all cases, however, the trees are predominantly deciduous and pyrophytic (that is, fire-adapted insofar as they are not killed by fire, but subsequently resprout). Throughout this biome, fire is a significant ecological determinant, but it is most frequent where the grass fuel load is highest, as in the mesic savanna (Bond 1997). Fire is also used extensively as a management tool in many other regions, particularly to control bush encroachment.

Primary productivity is variable, again largely dependent upon rainfall, and ranges from 1 t $ha^{-1}yr^{-1}$ in the drier regions, to more than 7 t $ha^{-1}yr^{-1}$ in the mesic parts. Biomass varies along a similar gradient, from 5 to perhaps 60 t ha^{-1}. The level of production is capable of supporting an impressive diversity of grazing herbivores and associated predators, and it is here that the 'big five' of lion, buffalo, leopard, rhinoceros, and elephant are found. These, too, vary in their occurrence due to nutrient availability, tree to grass ratios, and, of course, the degree of human disturbance. The biome is rich in plant species and there are almost 6 000 associated species in southern Africa as a whole (Gibbs Russell 1987), making it second in total richness only to the fynbos. Having said this, the species density is lower than in the fynbos, forest, grassland, and succulent karoo biomes because, while α-diversity is relatively high, there is low turnover and β- and γ-diversity values are generally low (Scholes 1997).

In terms of human impact, the savannas are heavily utilized for informal and subsistence agriculture and they supply substantial numbers of people in southern Africa with protein, in the form of meat from grazing mammals and insects, carbohydrates, and various types of fuel and building materials. Predominant land uses include pastoralism and, where the soils are more nutrient-rich and precipitation is adequate, dryland agriculture. The major environmental threat is fuelwood collection under increasing population pressure (Scholes 1997).

The Nama-karoo biome

Much of the subcontinent is arid to semi-arid, and the ecological response is one of reduced biomass and productivity under severe moisture constraints. By far the most extensive biome of the drier regions of South Africa is the so-called Nama-karoo (see Figure 15.9; Table 15.2), a name which is derived from its distribution in Namaland (southern Namibia) and, of course, the more familiar Karoo of central South Africa. The distribution of the Nama-karoo thus encompasses much of the summer rainfall semi-arid interior of the Western Cape, Eastern Cape, Northern Cape, and southern Namibia. The biome is located north-east of the Cape Fold Mountains on the central plateau, a distinctive geological and geomorphological region characterized by undulating plains lying between 1 000 and 1 400 m interspersed with typical 'kopjies' (Rutherford and Westfall 1994). Soils are generally weakly developed and rich in soluble salts such as calcium, a factor which, combined with the strong degree of aridity, creates demanding conditions for plant growth. Mean annual precipitation, distributed mainly during the summer months, varies from just 100 mm in the west to perhaps 500 mm in the eastern part of the biome. An additional problem is regular frost in the winter months, which at higher altitudes may occur on 180 days of the year (Rutherford and Westfall 1994).

The life forms of the biome have developed specialized adaptations in order to deal with these seemingly harsh environmental conditions. The vegetation is dominated by dwarf chamaephyte shrubs, typified by numerous members of the Asteraceae family, for example *Pentzia incana*, and several species of the genus *Eriocephalus*, with a variable cover of hemicryptophyte grasses, for example *Stipagrostis* spp. (species). Essentially there are two parts to the biome, one in the east with relatively high grass cover, and a second, the vast majority of the area in the west, which is typically a dwarf shrubland developed under a more xeric regime. If grazing pressures are moderate and rainfall is around 500 mm, annual productivity may approach 3 t $ha^{-1} yr^{-1}$ and biomass may achieve 7 t ha^{-1} under an almost complete grass cover. More typically, however, productivity and biomass values are considerably lower, vegetation cover is such that dwarf shrubs are widely spaced, and grass cover is ephemeral. Plant adaptations to the xeric conditions mainly take the form of drought tolerance, rather than avoidance (see Plate 42). The shrubs, for example, are perennials and cope with the dry season by losing their leaves, although leaf and stem succulents also occur and geophytes and annuals may comprise up to half of the species

(Palmer and Hoffman 1997). Perhaps as a result of the significant environmental constraints, plant species diversity is relatively low by comparison with other biomes.

The Nama-karoo biome has been modified rather than transformed by human impact since pre-colonial times. The predominant land use is small livestock production, a system which dates to settlement by colonial farmers in the nineteenth century. The variable grass cover has been the subject of considerable controversy: Acocks (1988) argues that the grass cover has been severely depleted due to increased levels of human disturbance (induced via grazing stock), whereas others (see Dean et al. 1995 for discussion) prefer to view the proportion of grass to shrub cover mainly as a function of changing rainfall inputs. It is likely, in fact, that both variable rainfall and grazing regimes are responsible for the observed vegetation patterns. The large herds of springbok, *Antidorcas marsupialis*, which occupied the Karoo before the advent of sheep farming, are now gone, and the contemporary grazing management favoured by the Department of Agriculture is quite different in practice and effect from that of wild ungulates (cloven-hoofed game) (Palmer and Hoffman 1997). The desertification debate has influenced management of the biome insofar as the officially sanctioned approach is aimed at reducing stocking levels and minimizing the negative impacts of grazing, although it is an approach that Hoffman (1988) has criticized for potentially reducing the abundance of the more palatable species and accelerating, rather than preventing, land degradation. Livestock farming in the Karoo is certainly a difficult enterprise: added to the variability of rainfall and unsuitability of substrate and soil is the problem of large outbreaks of economically damaging insect pests, in particular the Karoo caterpillar, *Loxostege frustalis*, and the brown locust, *Locustana pandalina*, which periodically outcompete the domestic stock in terms of grazing (Rutherford and Westfall 1994).

The succulent karoo biome

Arguably the most unusual of South Africa's biomes, the succulent karoo (see Figure 15.9; Table 15.2) shares aridity as a common environmental factor with the Nama-karoo to the north and east, but the seasonality of the precipitation is characterized by a distinct winter (as opposed to summer) maximum and frosts are less common. The rather surprising result of what is, in many cases, extreme aridity is a proliferation of plant diversity second to none for semi-arid vegetation globally. In the just over 100 000 km^2 of this biome, there are more than 5 000 plant species, many of which are leaf and stem succulent dwarf shrubs (Milton et al. 1997). Succulent karoo is distributed on the coastal plains and intermontane valleys lying to the west of the country, and rainfall in these areas is extremely low (20–290 mm per annum), resulting in a very restricted growing season. Typical soils, shallow and rich in soluble salts, offer further ecological restrictions. Notwithstanding these conditions, it is in the realm of plant species diversity that these dwarf shrublands reveal their most distinctive characteristic. Milton et al. (1997) note that there may be upwards of 100 different species in a 100 m x 100 m plot, more than twice the richness of the equivalent communities in North America and certainly competing with the richest communities in South Africa.

The plant life of the biome shows strong adaptations to the aridity and seasonality of rainfall. These adaptations range from morphological to phenological and physiological. Common morphologies among the chamaephytes include both evergreen and deciduous leaf succulence, especially in members of the Mesembryanthemaceae family. Some succulents have markedly reduced in size, resembling stones or even dung, for example the remarkably long-lived members of the *Lithops* genus (so-called stone plants) (Milton et al. 1997). Phenological adaptations are those related to the timing of growth and flowering. Growth periods are designed to maximize access to the seasonal availability of moisture and the occurrence of growth-suitable temperatures. There is also marked reproductive seasonality in many species, a feature which is especially well known in the spring displays of mass flowering, mainly among annuals of the Asteraceae family. Although such displays are an increasingly popular tourist attraction, sadly they are not necessarily a very healthy environmental signal, and the more spectacular shows are often associated with the most degraded land (Milton et al. 1997). Physiological adaptation is exemplified by the group of succulents known as the CAM plants, which have developed drought tolerance by way of a highly specialized photosynthetic pathway system known as crassulacean acid metabolism (CAM). These CAM plants, associated particularly with the Mesembryanthemaceae and Crassulaceae families, are able to keep their stomata closed during the day and hence gain the competitive advantage of simultaneously photosynthesizing and reducing moisture loss through transpiration. Consequent upon the obvious environmental constraints, productivity and biomass values are among the lowest in South Africa, and would reach 1 $ha^{-1} yr^{-1}$ and 2 t ha^{-1}, respectively, only in the more mesic parts of the biome (Rutherford and Westfall 1994).

In spite of this lack of productivity, the succulent karoo biome has been utilized by people for a very long time. Khoisan hunter-gatherers are recorded throughout the late Pleistocene, and there are records of use by Khoikhoi nomadic herders for some 2 000 years (Smith 1986, cited in Milton et al. 1997). The nineteenth-century trekboers also practised a form of nomadic herding, although they were eventually replaced by settled European pastoralists. Stocking rates have declined markedly throughout the twentieth century, leading to largely untested suggestions in the literature that this has been a function of grazing-induced severe environmental degradation (Milton et al. 1997).

The desert biome

The desert biome is represented by the Namib Desert, which, because of its hyperarid environmental conditions and unique biota, has attracted considerable interest (Jürgens et al. 1997). Rutherford and Westfall (1994) point out that the desert biome does not occur within South Africa per se. Desert landscapes of southern Africa vary from those devoid of any vegetation, for example the Namib sand dunes (see Plate 43), to those where there is a relatively continuous layer of dwarf shrubs.

The Namib forms a broad coastal belt north of Lüderitz in Namibia, and its gently undulating topography is interrupted by occasional inselbergs and by a sand sea with dunes up to 300 metres high. Substrate varies from Tertiary and Quaternary sandstones to considerably older granites and other Basement Complex rocks, although the most distinctive formation is, of course, the mobile sand dunes. Soil type plays an important role, together with climate, in determining the complex pattern of Namib vegetation. Throughout the region, climate is characterized by extremely low precipitation values, ranging from just 13 mm in the west to 70 mm inland, but this is supplemented, especially nearer the coast, by moisture from the frequent fogs. Temperatures may also be extreme, especially at the soil surface, where summer insolation may produce surface temperatures in excess of 50°C (Louw and Seely 1982).

Under such conditions, the vegetation is dominated by therophytes, annuals which generate only from seed, and drought evasion is the major adaptation. Most of these annuals are grasses, although some are succulents; all have short lifespans and appear ephemerally in response to rain. There are also perennials, perhaps the most famous of which is the tap-rooted *Welwitschia mirabilis* (see Plate 44), although, as Rutherford and Westfall (1994) note, this is confined to drainage lines and therefore does not occur within the desert biome *sensu stricto*. Although not typical of the desert itself, occurring in small populations in a specialized habitat, it is indeed a fascinating plant. In effect, *Welwitschia* is a tree with most of its stem buried beneath the ground; it has just two leaves above the surface, which may achieve a length of ten metres and which are able to take in water from fog by channelling the condensed moisture down to the base of their stems. *Welwitschia* is also extremely long-lived, and individuals are thought to survive for more than 1 000 years. The desert fauna is also of great interest. There is an abundant diversity of insects, among the most remarkable of which must be the tenebrionid beetles which collect fog on their backs in the early morning and hence survive in the harshest of environments (Seely 1979).

Biodiversity, biomass, and productivity are all understandably low in this biome, and may indeed be effectively zero on the mobile dunes, although even here there may be intermittent grass cover, especially of the fog-imbibing *Stipagrostis sabulicola* (Louw and Seely 1982). Biomass and productivity increase with moisture availability, but are unlikely anywhere to exceed 1 t ha^{-1} and 0.35 t ha^{-1} yr^{-1} (Rutherford and Westfall 1994). Levels of human disturbance have been relatively low, mainly because these productivity levels are unsuited to sustainable livestock exploitation and the absence of surface water precludes irrigation. There is, however, an increasing focus on tourism in the area, and plans for conservation of the biome will have to take this into account.

The grassland biome

The grasslands of South Africa are east-centrally located and occupy a substantial part of the inland plateau of the country (see Figure 15.9; Table 15.2). Their distinguishing feature is, by definition, the uniformity of vegetation structure, insofar as grasses form the dominant layer. The boundaries between the grasslands and neighbouring biomes appear to be determined mainly by climatic factors, in particular the water balance (Ellery et al. 1992) (see Figure 15.10). On the whole, these landscapes are gently undulating, with substrates dominated by the Karoo rocks on which are developed freely draining acidic red and yellow Latosols and black clay soils. Rainfall is strongly summer-seasonal, with mean annual totals varying from 400 to 2 000 mm in the Drakensberg (Rutherford and Westfall 1994). The other significant climatic factor is the occurrence of frost, which is an added problem because it occurs in the season of lowest moisture availability; at higher altitudes there may be frost on more than 180 days per year (Schulze 1965). The vegetation formation is sometimes referred to as temperate grassland so as to distinguish it from grassed areas to the north associated with the savannas but, as can be seen from Figure 15.9, grasslands extend into the subtropics of

FIGURE 15.10 Schematic representation of the determinants of the boundaries between grassland and adjacent biomes

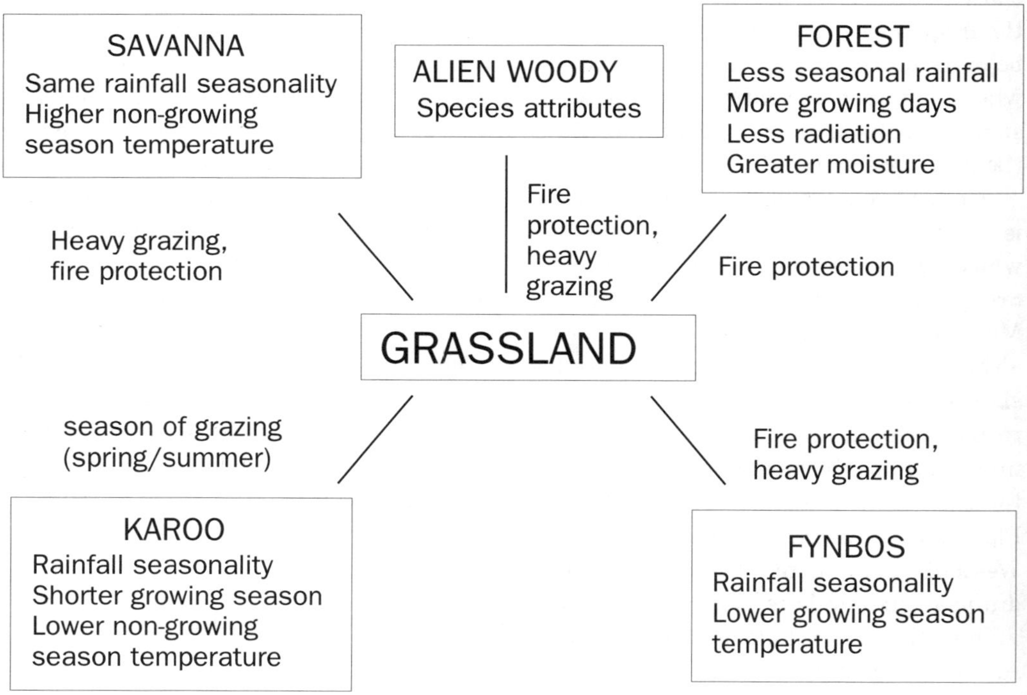

SOURCE: O'Connor and Bredenkamp 1997.

South Africa. Dry-season fires are an important ecological factor; indeed, protection from fire may cause major changes in the character of the grassland and result in marked bush encroachment. Fire exclusion can result in some of the commoner grass species in this biome, such as *Themeda triandra* and *Heteropogon contortus*, becoming moribund; clearly fire is an important component of the ecosystem dynamics of grasslands (Bond 1997).

Hemicryptophyte grasses completely dominate the vegetation, the cover of which is dependent on precipitation, grazing, and fire. The grasses survive regular burning (even as frequently as annual burning) by resprouting from underground rhizomes (plant stems acting as roots), and many are also resistant to grazing because they produce tannins in the epidermal (near-surface) cells of leaf blades. The east–west rainfall gradient in South Africa influences soil nutrient status markedly in the grassland biome, so there is a distinction between so-called sourveld grasslands, developed on oligotrophic soils in the higher-rainfall areas in the east, and sweetveld, developed on more eutrophic soils of the drier west. The sweet grasses, for example *Panicum coloratum*, usually have a lower fibre content and are more palatable to livestock than the sour grasses such as *Aristida junciformis*. The 625 mm annual isohyet is often said to mark the boundary between the two grassland types (Rutherford and Westfall 1994). (An isohyet is a line on a map joining places of equal rainfall amount.) This observation supports the contention that rainfall is the primary environmental factor determining productivity; O'Connor and Bredenkamp (1997) provide data suggesting that primary production varies in parallel with the rainfall gradient. Rutherford and Westfall (1994) suggest maximum productivity values in the wetter areas of around 6 $t\,ha^{-1}\,yr^{-1}$. With the exception of the eastern mountain 'hotspot' (Cowling and Hilton-Taylor 1994), plant diversity in this biome is not regarded as very high.

The degree of human intervention in the grassland biome has been substantial. Not only is it used extensively for agriculture, especially in the form of grain production, but it is also the locale of a significant proportion of the urban population of South Africa. The result has been a major transformation of the character of grassland; the expansion of urban areas and associated industry, together with the intensification and expansion of agriculture, have had major impacts on the biome.

The forest biome

Given the generally arid to semi-arid nature of the South African climate, it is hardly surprising that high-biomass forest is restricted in its distribution (see Figure 15.9, Table 15.2). Rutherford and Westfall's (1994) data include only the largest continuously forested area in the country, around Knysna, although there are substantial fragments in the higher-rainfall areas of the east coast. The fact that forests have a patchy distribution led Acocks (1988) to conclude that the forests are the climatic climax of a larger area of South Africa and were distributed more widely in the past, but that they have been decimated by human activity. This argument is strongly refuted by the palaeoecological evidence of Meadows and Linder

(1993). Notwithstanding the origin of the fragmented distribution, the fact remains that South Africa is not well endowed with indigenous forest. Therefore, although it is one of the most geographically restricted biomes in the country, forest is regarded as biogeographically significant and worthy of considerable conservation effort.

South African forest occurs under a variety of climatic regimes, depending on location. The afromontane forests (forests normally associated with mountains in tropical and equatorial Africa) around Knysna, for example, which are floristically similar to formations occurring throughout the montane region of the African continent, thrive under a temperate climate and all-year rainfall regime. The coastal forests of KwaZulu-Natal, on the other hand, are associated with subtropical conditions in a summer rainfall climate. Relatively free moisture availability, however, characterizes all South African forests: mean annual rainfall varies from a minimum of 500 mm to almost 2 000 mm in the Drakensberg. Frost is absent, as usually is fire, which probably determines the physical boundaries of many individual forests (Midgley et al. 1997). Soils and geomorphology vary according to local conditions; parent material may be dunes in the case of the sand forests of KwaZulu-Natal, or Table Mountain Group sandstones in the Knysna region. Under the relatively favourable climatic circumstances which prevail, the predominant plant physiognomy is phanerophytic, the trees forming a continuous canopy, with an understorey of smaller phanerophytes and chamaephytes (Rutherford and Westfall 1994). Structurally, forests are complex and stratified, with ferns, lianas, and epiphytes (non-parasitic plants that grow on the branches of trees) more abundant here than in any other biome.

Plant species contributing to these formations depend on location. The afromontane element is dominated by, for example, yellowwoods, *Podocarpus* spp., ironwood, *Olea europaea capensis*, and other evergreen tree species (see Plate 45). The subtropical coastal forests, which, structurally at least, are very similar (Rutherford and Westfall 1994), differ floristically; typical dominants include milkwood, *Sideroxylon inerme*, and *Mimusopes caffra*. There is relatively little known about most of the species which constitute South African forests in terms of adaptations and reproductive biology. Fleshy-fruited tree species, with seeds dispersed by birds and mammals, appear to dominate in the Indian Ocean forests, and at least one species, *Balanites maughamii*, has date-like fruits which, like the better-known bushveld *marula*, are dispersed by elephants. This suggests the importance of the fauna in such forests, which, despite their restricted geographical area, are home to more than a quarter of the southern African mammal fauna (Rautenbach 1978, cited in Rutherford and Westfall 1994).

Indeed, the forests are relatively productive systems, with standing crop biomass values which may exceed 230 t ha^{-1} and productivity varying between 6 and 15 t $ha^{-1} yr^{-1}$. In terms of biodiversity, however, they are not especially rich in plant species. Mean plant-species richness in 20 m x 20 m plots varies from 28 in the coastal forests of the Indian Ocean to 15 in the afromontane forests (Everard et al. 1995), values which pale by comparison with, say, the tropical forests of central Africa or South America. Nevertheless, human activity in the biome has a significant potential impact on biodiversity as a whole, and transformation and fragmentation have been progressive. The indigenous forests of South Africa represent a rare resource which is threatened by a range of processes, not least of which is direct exploitation. The Knysna forests have been harvested for timber for a long time, although afforestation with exotics has reduced the pressure in recent years (Midgley et al. 1997) and a sustainable-yield tree-harvesting system is now practised in the region.

The fynbos biome

The fynbos of the Western Cape is probably now South Africa's most widely recognized biome (see Plate 46). Not only has fynbos recently been the subject of considerable scientific effort (Cowling 1992), but it has attracted growing international tourism. Cape Town's Kirstenbosch National Botanical Garden is one of South Africa's premier tourist destinations, along with with the Cape of Good Hope Nature Reserve and Table Mountain itself. Fynbos is a biome worthy of accolades, for it has a unique and species-rich flora which has been accorded the status of floristic kingdom – one of only six globally (see Good 1974).

Situated in the south-western part of the country (see Figure 15.9, Table 15.2), fynbos falls under a mainly winter rainfall regime and the climate in most parts may be described as Mediterranean-type. This description, however, masks a high degree of variability, both in terms of annual amount and seasonality of precipitation. As Cowling, Richardson, and Mustart (1997) note, the predominance of winter rainfall declines towards the east, and the annual totals vary between as little as 210 mm and as much as 3 000 mm. The occurrence of frost is mainly altitude-dependent, but fire, mainly in late summer, is a most important and ubiquitous ecological factor to which the fynbos plant and animal communities have clearly adapted. The topography of the biome is dominated by the Cape Fold Belt, which forms an arc of high relief, inland of which is the Karoo basin and seawards of which is a coastal plain of varying width. Over and above climate, topography and parent material are very important determinants of particular plant communities within the biome: fynbos is a generic term and is in reality a complex mosaic of distinctive plant communities. For example, the shallow, acidic, infertile, and sandy soils developed on the Table Mountain Group sandstones support a markedly different suite of plant species than the nutrient-rich and clayey soils found on Malmesbury shales. Arguably, the type of fynbos which is associated with the sandstone and quartzite mountains of the region represents the biogeographical symbol of this biome, although strictly speaking there is a range of plant communities falling under the umbrella of the term fynbos.

Fynbos is a shrubland co-dominated by evergreen, sclerophyllous (hard-leafed) phanerophytes, chamaephytes, and hemicryptophytes. Vegetation cover depends mainly on rainfall and may be open in drier areas or form a closed canopy in the more mesic areas. Nowhere does plant height exceed three metres. The hemicryptophytes are dominated by members of the Restionaceae family (Cape reeds), while the phanerophyte and chamaephyte shrubs are characterized by members of the Proteaceae, Ericaceae, and Asteraceae families. To cite any plant taxon as dominant, however, is to underplay the prolific plant diversity evident in this biome, for the profusion of different species here is second to none in the subcontinent. Fynbos, at the biome level, represents perhaps one of the hottest of all the biodiversity hotspots (Myers 1990; Cowling and Hilton-Taylor 1994). It is the richest of all the Mediterranean-climate hotspots and has more species than many rainforest areas of similar size; levels of endemism (68 per cent of the plant species are endemic to the biome) are also remarkably high (Cowling and Hilton-Taylor 1994).

Despite this floristic diversity, one of the most striking features of fynbos across its range is its structural uniformity, and almost 90 per cent of the constituent plant species can be classified into just six structural types (see Table 15.3). Possible explanations for this include the prominence of summer drought, fire, and nutrient limitations in the soil, but exactly why strikingly uniform plant physiognomies are found in parallel with such remarkable taxonomic diversity remains something of an unsolved mystery. The interplay of environmental history, contemporary environmental heterogeneity, and patterns of fire must inevitably play their part in determining the high levels of diversity in this biome as a whole. Fire, on a return interval of ten to forty years, certainly helps to maintain the plant diversity, for it is a variable that operates stochastically (at random) to prevent competitive exclusion and facilitate the coexistence of species with different life histories (Cowling, Richardson, and Mustart 1997). Plant diversity has certainly received much more attention than that of the fauna, although diversity of invertebrates is notable (Jarvis 1979).

Because of the scarcity of trees, fynbos has lower biomass values than equivalent communities in other continents (Cowling, Richardson, and Mustart 1997). Biomass normally varies between 6 and 25 t ha^{-1} for mature (pre-fire) communities (Rutherford and Westfall 1994), but may reach 51 t ha^{-1} (Cowling, Richardson, and Mustart 1997). Productivity values range up to 4 t $ha^{-1} yr^{-1}$, again a fire-dependent value, reaching a maximum a few years after fire. The study of fynbos plants has revealed fascinating insights into the means by which they have adapted to their environment. Two phenomena in particular have attracted attention, namely the high proportion of reseeders in the shrub component (see Table 15.3 and Le Maitre and Midgley 1992) and the importance of myrmecochory (seed dispersal by ants) in proteoid shrubs (Bond and Slingsby 1983). Both characteristics are related to the timing and frequency of fire. Reseeding, involving death of the adult plant after fire, and fire stimulation of both seed-set and seed germination, is strongly dependent on the return period of fire, which must clearly exceed the minimum time taken for the new generation to accumulate an adequate seed bank and so ensure post-fire germination success. A significant proportion of the proteoids (plant species with similar habits and structure to Proteaceae) are serotinous, meaning that they retain the seed bank on the adult plant until scorched by fire (Bond 1985) and, in so doing, protect the seed bank from predation by mammals. Specialized seed dispersal conveys a symbiotic advantage to the indigenous ant species, which consume the nutrient-rich elaiosome (a protein-rich, fleshy part of the seeds), and to the participating proteoceous shrubs whose seeds are buried by the dispersing agents. Late-summer burning maximizes the chances of such buried seeds germinating during the succeeding autumn and winter wet period, and minimizes their predation by small mammals.

The degree of human impact on this biome has been very marked and the subject of much debate, especially with regard to the effects of urbanization and the invasion of fynbos communities by alien trees (MacDonald 1984; Richardson et al. 1997; Meadows 1998) (see Plate 47). Invasion by exotic trees is widely regarded as one of the most serious forms of land degradation in the region (Richardson et al. 1997). The twelve most widespread invasives are all trees, many of which hail from Australia; they have all spread within a relatively short period across large areas of the south-western Cape. Richardson et al. (1997) conclude that certain invasive trees, especially *Hakea sericea* (silky hakea) and *Pinus pinaster* (pinaster pine), have caused major modifications to the montane communities in the biome. The silky hakea, for example, covers 4 800 km^2, representing 14 per cent of the area of mountain fynbos, and 19 per cent of the quarter-degree grid squares of the fynbos biome as a whole contain dense stands of pinaster pine (Richardson et al. 1997). The situation in the lowlands is at least as severe, although the successful invading species in this case tend to be *Acacia saligna* and *Acacia cyclops*. The introduction of exotic fauna has also produced degradation problems; *Iridomyrmex humilis*, the so-called Argentine ant (which actually hails from Brazil), has proved especially problematic. This ant fails to bury the seeds of myrmecochorous *Protea* species and hence renders them susceptible to granivory (the eating of seeds), with consequences for species richness and plant community structure.

Another pressure on this biome is urbanization. Expansion of Greater Cape Town has physically consumed large areas of otherwise species-rich vegetation, as well as causing fragmentation of the remaining fynbos areas. McDowell et al. (1991), in their survey of the Cape Flats, show that these processes have accounted for very substantial proportions (up to 57 per cent) of lowland fynbos in the region.

The conservation of ecological resources and biodiversity

The long history of human occupation of southern Africa, added to the population growth and technological developments of recent decades, has had enormous impacts on the biogeography of the subcontinent. The effects have been widespread and diverse. The introduction of domesticated varieties of grasses, for example, has virtually transformed the entire grassland biome; domesticated herbivore species may have had an equivalent effect in the savanna and Nama-karoo biomes. It has often been the physical space occupied by the biological diversity of the region that has been exploited, rather than the biodiversity per se. In other words, it is habitat destruction, rather than its direct utiliza-

TABLE 15.3 Structural and reproductive characteristics of major growth forms in the fynbos biome

Growth form	Structural characteristics	No. of spp.	Dominant families	Age to maturity (years)
Proteoids	Mid–tall (1.5–5 m) shrubs with nanomesophyllous, isobilerateral leaves	120	Proteaceae	1–5
Ericoids	Dwarf–mid (0.5–2 m) shrubs with leptophyllous, rolled leaves	*c.* 3 000	Ericaceae, Asteraceae, Fabaceae, Rutaceae, Protaceae	1–4
Restoids	Dwarf–mid evergreen	*c.* 550	Restionaceae, Cyperaceae, Poaceae	1–2
Geophytes	Evergreen and deciduous herbs with below-ground storage organs	*c.* 1 400	Iridaceae, Liliaceae, Orchidaceae, Amaryllidaceae	1
Fire emphemerals	Forbs and dwarf–low (0.5–1 m) shrubs	*c.* 1 000	Asteraceae, Scrophulariaceae, Fabaceae, Poaaceae, Mesembryanthemaceae	0.5–2
Obligate resprouters	Low–tall shrubs with nano-mesophyllous dorsiventral leaves	102	Anacardiaceae, Celastraceae, Ebenaceae	1–5

SOURCE: Cowling, Richardson, and Mustart 1997.

lifespan (years)	Reseeder/ resprouter (%)	Pollinators	Storage dispersal	Seed size	Seed dispersal	Germination cues
>15	>80	Insects/ birds Rodents Wind	Canopy Soil	Large (>5 mg)	Wind Ants	Cold Fluctuating temperature
>6	>60	Insects Winds Birds	Soil Canopy	Fine (<1 mg) – large	Passive Ants Wind	Fluctuating temperature Heat Cold Smoke
>25	<20	Wind	Soil	small (1 mg) – large	Ants Passive Wind	Largely unknown Smoke
<15	0	Insects Birds	Soil No storage	Fine – large	Wind Ants	Cold Fleshy seeds are non-dormant
<8	>80	Insects Wind	Soil	Fine – small	Wind Passive	Cold Smoke Unknown
>>10	0	Insects	No storage	Large	Birds	Mainly fleshy and non-dormant

tion, which accounts for most of the modification of biogeographical resources. In this sense, reduction in biodiversity mirrors other forms of environmental degradation, such as soil erosion.

Nevertheless, there have been concerted attempts to conserve the biogeographical heritage that is South Africa's biodiversity. South Africa was, in fact, one of the first countries in Africa to proclaim the 'conservation estate', albeit for reasons which initially had nothing to do with the preservation of biodiversity itself, but which were more of a response to the perceived loss of game animal species (Rebelo 1997). In total, South Africa now has 5.1 per cent of its area listed in the national Register of Protected Areas, in which the overt goal of land-use management is conservation of biological diversity. The geographical distribution of this diversity is, however, far from even, and this means that a number of biome types are severely under-represented in the conservation estate (see Table 15.4). This unbalanced situation becomes even more apparent at the plant community scale. For example, Acocks' (1988) 'Central Upper Karoo' has no formally conserved areas at all, and a large number of communities fall well below the national average of 5 per cent (Rebelo 1997). Some plant communities also have exceptionally high concentrations of threatened species. Rebelo (1994) has determined the minimum conservation requirements for southern Africa in general. Using a data set on the distribution of plant species, he established a reserve network that would place at least one population of each plant species in the subcontinent within the geographical boundaries of at least one nature reserve.

TABLE 15.4 Current conservation status of the biomes of South Africa

Biome	Conserved area (x 10^6 ha)	Conserved area (%)
Fynbos	4.835	14.89
Succulent karoo	15.800	2.36
Nama-karoo	24.529	0.45
Desert (Namibia)	7.840	85.00
Grassland	28.929	1.38
Savanna	37.633	9.96
Forest	7.681	9.36

SOURCE: Rebelo 1997.

The exercise, conducted on the basis of a grid size of 25 km x 25 km, results in a comparison of the contemporary reserve network with a 'wish list' of desired conservation areas. The results suggest some important conservation priorities and identifies succulent karoo, grassland, and lowland fynbos plant communities as currently markedly under-represented in the reserve network. Siegfried (1989) has examined the effectiveness of the conservation areas of southern Africa in maintaining populations of animal taxa and reaches the somewhat surprising conclusion that 92 per cent of reptilian, 97 per cent of avian, and 93 per cent of mammalian species are represented in the nature reserves. There is no guarantee, however, that contemporary 'presence' is the equivalent of an assured future for any of these taxa.

The 'best-case scenario' approach of Rebelo (1994) suggests a reserve network incorporating just 12 per cent of South Africa, but this should not release us from an obligation to attempt to maintain high levels of biodiversity in the non-reserve areas. Since, even if the network were fully established, a further 88 per cent of South Africa would fall outside reserve boundaries, it makes sound biogeographical sense to manage those areas also for the essential (and free) ecological resources they provide. Only time will tell if South Africans in the future will look back on the conservation efforts around the turn of the millennium and view them as a success. Surely the very richness of the biogeography of the country, as manifested at several different levels, should act as an incentive in itself to promote the conservation ethic? If not, then perhaps the boom in ecotourism and the millions of dollars that South Africa can earn on the basis of its biodiversity would be more persuasive. The nation's biological diversity has a long history of exploitation and a number of rural communities depend on its 'productive value' (McNeely et al. 1990) for their subsistence. For example, the citizens of Botswana on average consume some 90 kg per year of 'game', harvested wild from the veld; in so doing, biodiversity makes a significant contribution to the protein requirements of rural people. Clearly our subcontinent is blessed with a prolific ecological and biodiversity resource; will our dependence upon it be realized before its very life-giving properties are extinguished?

16 Geography of drainage basins: hydrology, geomorphology, and ecosystem management

KATE ROWNTREE

Rivers provide a vital resource to humanity, they constitute an important group of ecosystems, and they are one of the main agents of landscape evolution. These several functions take place within a *drainage basin* or *river catchment*, the land surface which contributes water and other materials to the river channel and which in turn is shaped by the action of the river and hillslope processes. This chapter will explore the hydrological and geomorphological processes which determine river structure and function. It will emphasize the interrelationships between the river channel and its drainage basin, and will examine the separate but interconnected, and sometimes conflicting, roles that the river plays as a resource, an ecosystem, and an agent of landscape evolution.

Hydrology is the scientific study of the distribution and quality of water within the atmosphere and on the earth's surface, particularly in relation to the replenishment and depletion of the water resources on the land areas of the earth.

Development of the drainage basin concept

The hydrological importance of river basins has been recognized for over 300 years. Hydrology as a scientific study dates back to the end of the seventeenth century, when it was shown that the flow of the Seine could be maintained by the balance of rainfall over its catchment area minus the level of evaporation. The geomorphological significance of drainage basins was not recognized until the middle of the nineteenth century when the 'fluvialist' theory became generally accepted – that is, the theory that landscapes are formed over geological time scales by the continued work of water. The early development of this theory is exemplified by Davis's cycle of fluvial erosion, published in 1889.

Dunne and Leopold (1978: 495) define the drainage basin as 'the area of land that drains water, sediment and dissolved materials to a common outlet at some point along a stream channel'. Ward (1975) notes the strong interaction between the hydrological processes which shape the channel network in the long term and the geomorphological structure of that channel network within its catchment which determines the streamflow response; thus he gives the drainage basin geomorphological credence. Drainage basins are complex interconnected systems which consist not only of the channel network, but also of the hillslopes which contribute water and other materials to the channel. The components of the fluvial system are illustrated in Figure 16.1.

Drainage basins can also be considered from the perspective of landscape ecology, in which the functions of stream and riparian-zone ecosystems are closely controlled by hydrological and geomorphological processes. Ecologists began to give serious attention to the links between stream habitat and drainage basin processes in the early 1980s, and interdisciplinary studies linking hydrology, geomorphology, and stream ecology have resulted in a sig-

The **riparian zone** is the landscape zone which is in immediate contact with the river channel and whose physical features and plant and animal communities are determined by the flow of the river. The riparian zone includes the floodplain.

FIGURE 16.1 Components of the fluvial system

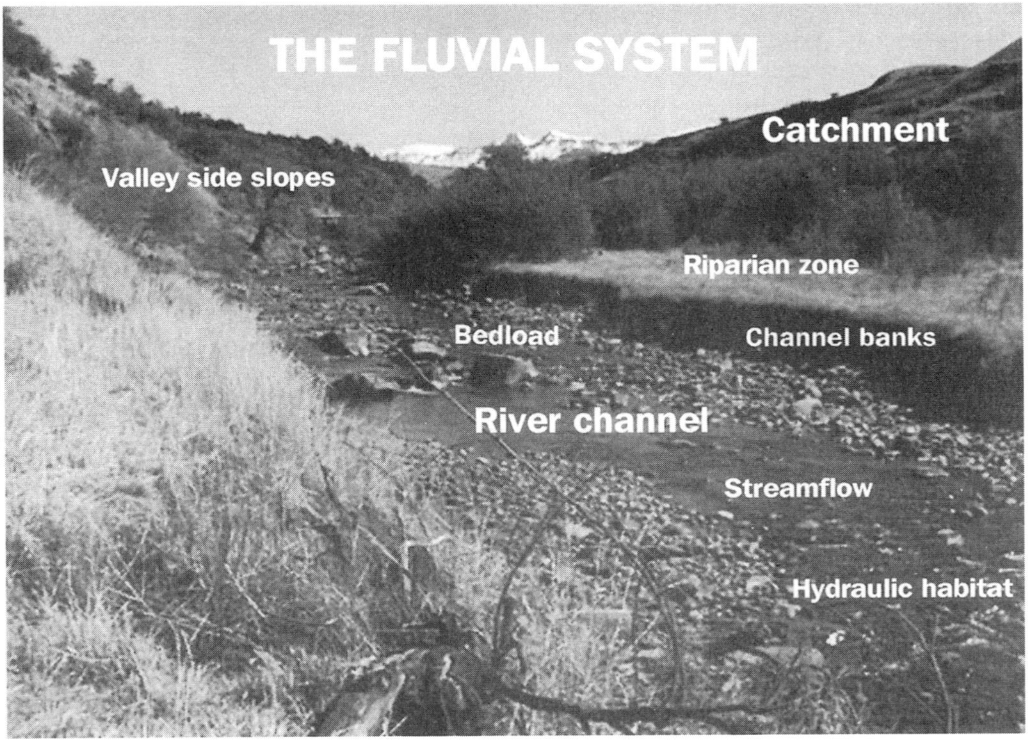

nificant research field. The international recognition of the importance of these links can be seen from recent books from Australia (Gordon et al. 1992), Britain (Newson 1994), and South Africa (Davies and Day 1998), all of which explore the relationship between river ecology, hydrology, geomorphology, and river management. The recognition by South African ecologists of the need to understand both the hydrology and geomorphology of local rivers has provided the catalyst for research in fluvial geomorphology during the 1990s.

Southern African drainage basins

The major rivers of Africa are shown in Plate 48 (colour section). The southern African region contains four major drainage basins: in order of drainage area, the Zambezi, the Orange, the Okavango, and the Limpopo (see Table 16.1). As much as 32 per cent of the African continent is made up of internal drainage systems (Walling 1996). In southern Africa, the

TABLE 16.1 Drainage basin characteristics and hydrology of the major river basins of southern Africa compared to the rest of the world

River	Drainage area ($\times 10^6$ km^2)	Mean annual flow volume (km^3/yr.)	Mean annual sediment yield ($\times 10^6$ tonnes/yr.)
Orange	1.02	11	17
Limpopo	0.41	5	33
Zambezi	1.20	223	20
Okavango	0.73	data not available	data not available
Nile	2.96	30	13
Zaire	3.82	1 250	43
Danube	0.81	206	67
Mississippi	3.27	580	210
Amazon	6.13	6 300	498
Ganges	1.48	971	1 670

SOURCE: Van der Leeden et al. 1990.

Okavango River flows south-east from the Angola highlands to the Okavango basin and Makgadikgadi pans. The so-called Okavango Delta is renowned as a wetland and is the basis of the tourist industry in Botswana.

The rivers of southern Africa reflect the tectonic history of the region since the break-up of Gondwana some 180 million years ago. (The geological history as it relates to landscape evolution is described in detail in Chapter 11.) A long period of stability following the break-up of Gondwana allowed denudation processes to level the landscape. Continental separation produced the marginal escarpment which separates the interior plateau from the coastal lowlands. Significant uplift in the Miocene and Pliocene epochs rejuvenated the river systems. As a result, the interior of the continent is characterized by a plateau with large tectonically induced basins. Rivers arising in the interior flow through these basins before dropping steeply to the coastal margins, often over spectacular waterfalls and through deep gorges. Best known are the Victoria Falls on the Zambezi River. On the other side of the continent the Orange River passes over the Augrabies Falls (see Plate 28).

The rivers of the interior have exhibited major changes in direction over the course of time. (The history of the palaeodrainage of the Orange River is described in Chapter 2.) Considerable research has been carried out by South African geologists and geomorphologists into the pattern of palaeo-river terraces within the Vaal and Orange drainage basins. According to Moon and Partridge (1993), drainage adjustments were in part related to climatic change, but primarily to tectonic activity.

The drainage systems of northern hemisphere temperate areas have been strongly affected by the glacial epochs during the Quaternary period. Not only did glaciation shape the landscape, but it also left a legacy of glacial deposits which are now being reworked by the rivers. There has been no such legacy in southern Africa, where the glacial periods were marked by increased aridity throughout much of the area, while pluvial periods (higher rainfall) marked the interglacials (see Chapter 2). The increased aridity of the glacial periods had two important effects. One was an increase in the area occupied by sand seas, and the second was the development of colluvium – hillslope deposits resulting from slope wash and mass movements. Goudie (1996) notes that colluvium mantles the lower hillslopes of about 20 per cent of the land area south of the Zambezi and is often the site of intense donga formation (gullying). Colluvium thus acts as an important source of sediment for modern river systems.

Over the last 100 to 200 years, increasing population numbers and resource development have transformed the landscape of southern Africa, and in so doing have markedly changed hydrological and geomorphological processes. Examples of contrasting land use in two adjacent catchments in KwaZulu-Natal can be seen in Plate 49. These anthropogenic landscape transformations came later than they did in many countries of the northern temperate latitudes; the time for the landscape to achieve an equilibrium with the new conditions has therefore been more limited. The legacy of change in fluvial systems is manifested in gully erosion, for example, which is widespread throughout the region (see Chapters 13 and 14).

The drainage basin as a hydrological system

The drainage basin water balance

From a hydrological perspective, the drainage basin is a spatially defined system which converts precipitation into streamflow. At its simplest level it can be considered as a black box for which the inputs and outputs can be defined and, hopefully, measured. Measurement of hydrological variables is not considered here, but the reader is referred to the extensive discussions by authors such as Gregory and Walling (1973), Newson (1994), and Ward (1975). Regional patterns of runoff are determined by the balance of precipitation and evaporation over the drainage basin. In calculating the water balance, runoff is given as millimetres per unit area, so as to make the units compatible with those for precipitation and evaporation. It is calculated as the total volume of runoff measured at the drainage basin outlet divided by the drainage basin area.

The regional precipitation and potential evaporation for Africa are shown in Plates 50a and b. Runoff is closely related to the ratio of precipitation to potential evaporation, as given in Plate 50c. Note that *potential evaporation* is determined by the potential of the atmosphere to take up moisture, whereas *actual evaporation* depends also on the available moisture supply which, throughout southern Africa, is often extremely limited. The general dryness of the area relative to equatorial Africa is evident from the maps in Plate 50. It is also evident that southern Africa is wetter than the equivalent latitude to the north, due to the greater oceanic influence. There is a clear gradient of decreasing moisture availability from east to west, with the exception of a narrow strip along the south coast. Runoff coefficients for the area (the ratio of runoff to precipitation) are generally between 0.1 and 0.2, which are low on a world scale. This results in a mean annual runoff of between 10 and 300 mm per annum, compared to 500 and 4 000 mm in the high-rainfall areas of the equatorial belt, where runoff coefficients lie between 0.4 and 0.6. Further discussion on the hydrology of the region is given by Walling (1996).

A detailed account of the regional water balance for South Africa as it effects the distribution of water resources is given by the Department of Water Affairs and Forestry (DWAF) (1986). The mean annual rainfall over much of the country is low (averaging 497 mm), while the potential for evaporation is exceedingly high. Little water is left for runoff. For South Africa as a whole the runoff coefficient is only 10 per cent, meaning that 90 per cent of rainfall is lost through evapotranspiration, the combined loss of water by evaporation from open water surfaces and from the soil and groundwater through transpiration by plants. Controls of evapotranspiration are therefore of paramount importance. The distribution of runoff in relation to major drainage basins for South Africa is shown in Plate 51. The source areas for runoff are seen to be along the eastern escarpment and the Cape Fold Mountains, coincident with high rainfall and lower evaporation potential. Much of the country has very low runoff rates. As a result there is a concentration of streams draining eastwards from the escarpment and in the Western

Cape. These rivers tend to be short and steep. Although the catchment of the Orange River covers 48.5 per cent of the land area of South Africa, it contributes only 22.5 per cent of the total surface runoff. Conversely, the rivers draining the eastern plateau slopes, covering 13 per cent of the country, account for around 43 per cent of total runoff (Department of Water Affairs 1986). These rivers have been targeted as important national water resources, and the pressure to develop them through storage reservoirs and interbasin transfer schemes to less watered areas is considerable (Davies and Day 1998).

Regional patterns of runoff are controlled by the climatic factors of precipitation and potential evapotranspiration, but other factors become important locally. These include the amount of surface storage in the catchment from which water can be evaporated, and the vegetation cover. Important open water storage components include lakes and reservoirs, wetlands and swamps, the river channel, and water held in interception storage on the leaves of vegetation. Water can also be lost from the soil or groundwater if it is accessible to plant roots. Phreatophytes are deep-rooted plants which tap groundwater. Evapotranspiration rates vary with the type of vegetation, its biomass, and its rooting depth. A comparison of the ratio of potential evapotranspiration to *pan evaporation* (the most common direct measure of evaporation potential) for a range of natural veld types and some common irrigated crops is shown in Figure 16.2. Natural veld shows a clear seasonality, with a reduced evaporation demand during the winter season. Irrigated crops show a wide range of water demand, both in terms of seasonality and total demand. Although maize has a very high demand at the height of its growing season, the growing period is short. This is in contrast to sugar cane, which maintains a high demand throughout the year. Sugar cane thus puts great pressure on available water resources in the areas where it is grown (see Plate 18).

Another major user of water is commercial forestry (see Plate 49). Much research effort has been devoted worldwide to estimating water use by pines, eucalyptus, and other forest types in comparison to natural vegetation. Although Newson (1994) cautions that is difficult to derive generalized figures for water use by forest plantations, there have been several attempts to synthesize international literature and to provide common guidelines such as those provided by Bosch and Hewlett (1982). Data for a selection of South African catchments are compared to those for Plynlimon in the Welsh mountains (UK) in Table 16.2. As a percentage of the natural runoff, the increased water loss in South Africa is much greater than that found in the Welsh catchments. It can be seen that, even in the higher-rainfall areas of South Africa, forests can reduce runoff by nearly 50 per cent of that from areas of natural vegetation. Forestry in South Africa is controlled by the Department of Water Affairs and Forestry by means of a permit system. Particularly strict controls are placed on commercial forestry in areas which have been proclaimed as Mountain Catchments in terms of the Mountain Catchments Act of 1970 (see Chapter 19).

A related problem in South Africa is the high use of water by invasive alien vegetation such as various species of *Acacia*, *Hakea*, and *Lantana camera*. A prelimi-

FIGURE 16.2 Seasonal distribution of water use by selected land covers in South Africa

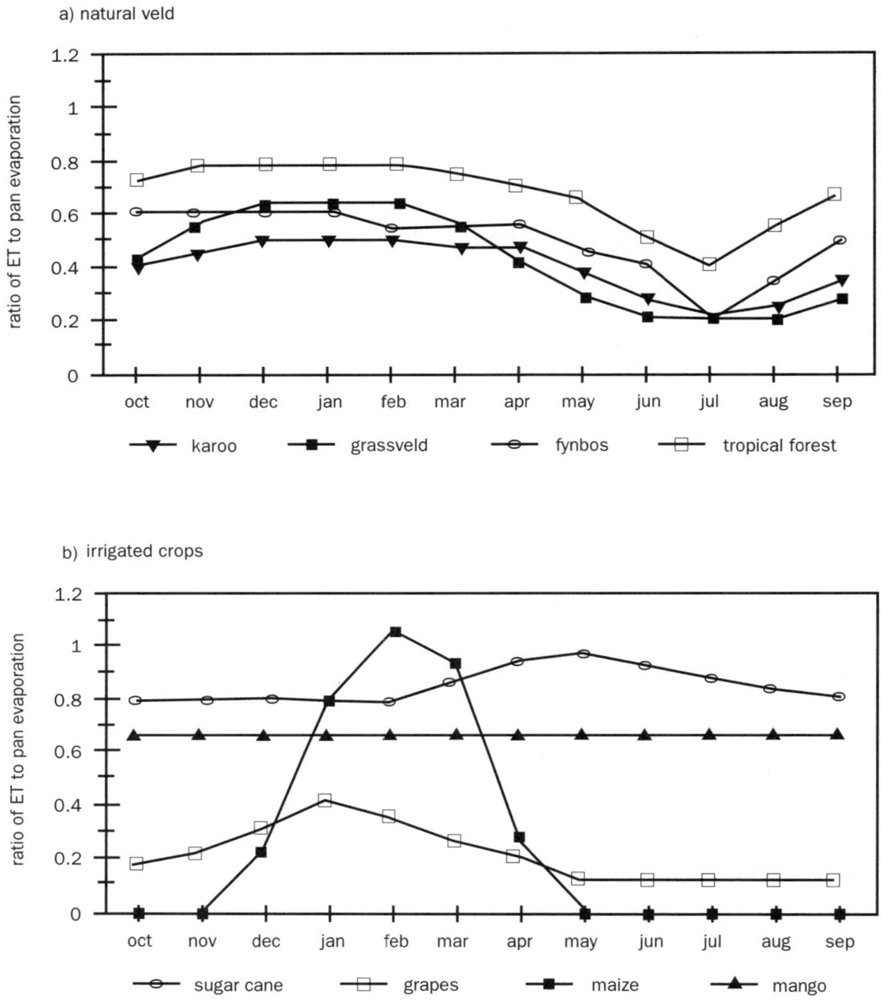

SOURCE: based on data from Midgley et al. 1994.

nary assessment of the problem has been undertaken by Versveld et al. (1998). They estimate that 3.72 per cent of the land area of South Africa has been invaded by alien plants, resulting in an annual water loss of some 1 900 m^3 per hectare of invaded land. This is approximately double that estimated for plantation forestry, due in part to the fact that many invasives are to be found in riparian areas where they have free access to water. The extent of the invaded area varies over the country:

TABLE 16.2 Water use by commercial forestry

Welsh catchment experiments

	WYE: GRASSLAND	SEVERN: FOREST
Mean annual precipitation (MAP) (mm)	2 348	2 213
Streamflow (mm)	1 944	1 364
Evapotranspiration (mm)	405	849
% loss	17	38

South African catchment experiments

RESEARCH AREA	NATURAL VEGETATION	MAP (mm)	NATURAL RUNOFF (mm)	REDUCTION DUE TO AFFORESTATION (mm)
Western Cape	Fynbos	1 300–1 800	530–1 000	170–313
Natal Drakensberg	Grassland	1 400	750	360

it is highest in the Western Cape (16.81 per cent), followed by Mpumalanga (13.28 per cent), KwaZulu-Natal (3.88 per cent), and the Eastern Cape (3.37 per cent). The seriousness of the problem has led the Department of Water Affairs and Forestry to support a Working for Water Programme, a nationwide alien clearance programme designed to increase water yields while at the same time providing employment (see Plate 52).

Hillslope processes

The flow in the river channel depends primarily on hillslope processes, which determine the rate and quantity of runoff. A full review of hillslope hydrology can be found in textbooks such as Ward (1975), Dunne and Leopold (1978), and Newson (1994). Research from the 1970s showed that the pathway water takes across a hillslope depends on a number of different factors, including the slope and soil characteristics, the vegetation cover, and the precipitation characteristics. These different pathways and the variables controlling them are summarized in Figure 16.3a. The different types of flow have important implications not only for the hydrological response of a drainage basin to a rainfall event, but also for the movement of sediment and dissolved material, as can be seen in Figure 16.3b.

Figure 16.3a shows how hillslope runoff is controlled by the soil profile, the slope profile, the vegetation cover, and precipitation characteristics. Two distinct climatic regimes can be recognized: semi-humid and arid. In semi-arid areas a sparse vegetation cover leaves the soil surface open to the impact of raindrops so that soils tend to be compacted and infiltration rates are low. During heavy storms the excess water unable to infiltrate into the soil runs rapidly off the surface as *infiltration-excess overland flow*, causing a flood wave to pass down the channel. The infiltrating water replenishes the soil moisture or percolates down to the groundwater zone, if present. In the dry period between storms, much of the soil moisture is lost to evaporation so that the soil profile rarely becomes saturated. Southern Africa is dominated by semi-arid landscapes in which the mean annual rainfall is below 400 mm. Much of the region's runoff should therefore be determined by the above group of processes. It should be noted that, because of the lack of vegetation cover, infiltration-excess overland flow is associated with soil erosion by surface wash.

In humid areas the more frequent and prolonged rainfall events have two important consequences. Firstly, the soil is better able to support a good vegetation cover, which protects the soil surface and promotes surface infiltration. Secondly, the soil profile is more likely to become saturated, especially where subsurface flows from the adjacent hillslopes become concentrated in hollows and in footslope areas. It is these saturated areas, often adjacent to river channels, that contribute *saturation overland flow* during and immediately after a storm event. Overland flow moves rapidly over the surface into the channels, causing a sharp rise in flow levels. If the area contributing overland flow is large enough due to widespread saturation, a flood may result. Meanwhile, in freely drained soil profiles, water percolates down to the groundwater zone, from where it moves slowly into the river channels, thus contributing to baseflow during dry periods.

Because of the good vegetation cover, saturation overland flow is not associated with high surface-wash rates. Saturation of the soil profile, however, may lead to mass movements or tunnel erosion, as described in Chapter 13. A common soil condition is that the soil profile is characterized by a permeable surface horizon over a less permeable subsoil (for example the duplex soils described in Chapter 14). This encourages the build-up of saturation within the soil profile and lateral movement of water down the slope as *throughflow*. Throughflow can exploit lateral openings in the soil due to cracking or animal activity, causing tunnel erosion or piping. Pipes create a subsurface drainage network which conveys water rapidly to the river channels, thus contributing to stormflow. The collapse of large underground pipes is a common form of gully erosion.

FIGURE 16.3 Hillslope processes in relation to their major controls

a) Hydrological processes

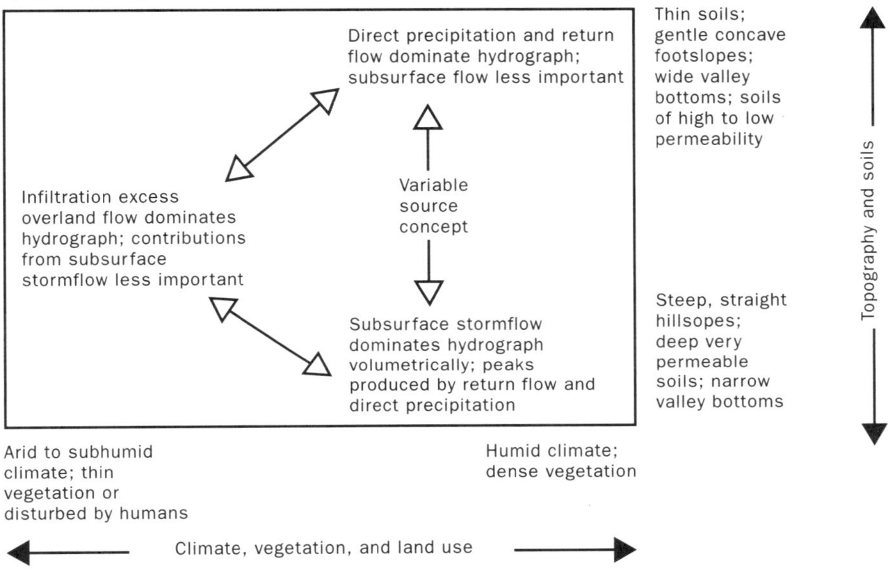

b) Geomorphological processes

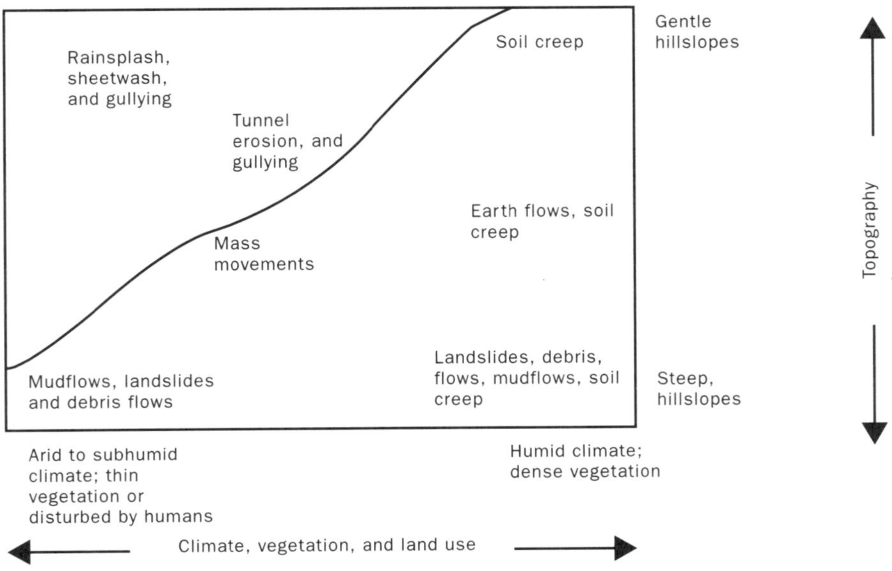

SOURCE: adapted from Dunne and Leopold 1978.

Humid-area runoff processes, as described above, depend on the maintenance of a good vegetation cover. Where the vegetation cover is disturbed, infiltration is severely restricted and the contributing area for infiltration-excess overland flow increases significantly. This may happen, for example, in agricultural areas where the soil surface becomes impacted by vehicular traffic, livestock trampling, or along footpaths, or in urban areas where the vegetation cover is replaced by impermeable tarmac and paving. The hydrological response thus becomes more like that in semi-arid areas, except that rainfalls are heavier and more frequent. The storm runoff component increases at the expense of the baseflow, and the flow regime becomes more variable. Such changes to hillslope hydrology are believed to be widespread in the more humid areas of South Africa, especially where high rural population density is combined with poorly structured soils (see Chapter 14 and Plate 49). The hydrological effects of urbanization in Gauteng province are significant, because Gauteng lies in the headwater area of important drainage basins such as the Olifants River. Increased surface runoff is linked to increased pollution of a river which ultimately flows through the Kruger National Park.

Streamflow

The pattern of floods and baseflow through time constitutes the flow regime of a river. There are three time scales over which streamflow patterns can be considered, the short term (measured in days or weeks), the seasonal pattern (measured in months), and the inter-annual pattern (measured in years) (see Figure 16.4).

The short-term response depends on the frequency and duration of individual storm events and their relationship to the hillslope processes described above. Floods are the response to heavy or prolonged precipitation events, while flows between storm events depend on the baseflow contribution from soil moisture and groundwater. The distinction between stormflow and baseflow is an important one. Stormflow contributes to flooding, which can be a major hazard, while baseflow represents the main water resource that can be abstracted from the river. Stormflow can be utilized efficiently only if it can be stored, for example in a reservoir. Because the main pathway taken by stormflow is over the ground surface, the water picks up much loose material, soil, organic debris, pesticides, fertilizers, and so on. The quality of stormflow therefore tends to be poor. Baseflow, which passes slowly through the soil or rock profile, carries little particulate material, but has time to pick up dissolved material which has been produced by weathering. Baseflow therefore often has a high load of dissolved salts.

The flood hydrology of southern Africa is discussed in detail by Alexander (1990). Table 16.3 gives details of a selection of floods in South Africa. While localized floods over the interior are often the result of intense convective activity, possibly associated with a trough of low pressure bringing in moist tropical air, floods along the south coast are often associated with upper-air cut-off lows (see Chapter 12). Tropical cyclones infrequently affect the eastern half of the country. Ten cyclones have been experienced since 1956, of which

FIGURE 16.4 Streamflow patterns in the Mkomazi River, KwaZulu-Natal

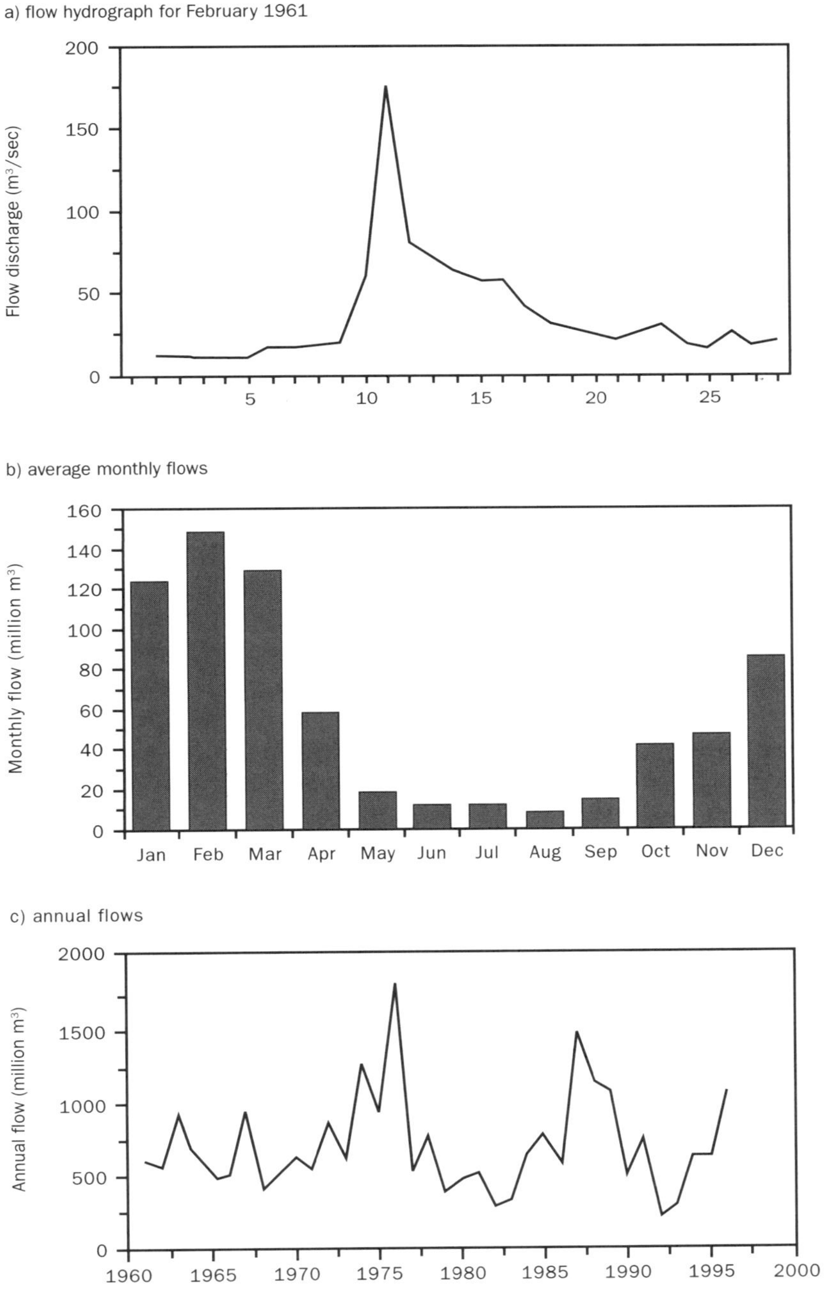

TABLE 16.3 Examples of extreme floods in South Africa

Flood event	Date	Meteorological event	Local peak rainfall	Catchment rainfall (mm)	Catchment/ storm area (km^2)	Peakflow ($m^3 sec^{-1}$)
Nelspruit	Nov. 1975	Convective storm enhanced by cloud seeding	101 mm in 16 min.	data not available	50	no flood produced because of small storm area
Pretoria (Roodepoort Dam)	Jan. 1978	trough of low pressure	280 mm in 8 hr.	110	684	1 510
South-western Cape (Laingsburg)	Jan. 1981	upper-air cut-off low	288 mm in 3 days	142	4 001	5 740
Demoina (Pongola River)	Jan./ Feb. 1984	tropical cyclone	840 mm in 2 days	570	7 831	13 000

SOURCE: Alexander 1990.

Demoina was by far the largest in terms of rainfall. While floods are a fairly regular occurrence along the eastern seaboard from KwaZulu-Natal to the Western Cape, floods in the more arid interior are less common. In 1987/98 very severe flooding occurred in KwaZulu-Natal, the Free State, and the Northern Cape due to widespread rains well above the monthly average in August and September and again in February. The heavy rainfall in KwaZulu-Natal in September caused widespread flooding which resulted in 287 flood-related deaths. Flooding was delayed until February in the interior, by which time the earlier rains had filled dams and natural pans, raised the water-table and increased soil moisture – prerequisites for flooding in this more arid area.

In addition to the extreme floods such as those detailed in Table 16.3, a large number of smaller, less noteworthy events occur more frequently. It is these floods which are the most important for regular channel maintenance and ecosystem functioning. Figure 16.5a shows the flood frequency curves for two Eastern Cape rivers, the Kowie and the Buffalo. The recurrence interval is a measure of the average time interval between floods of a given size or greater, but, as can be seen from Figure 16.5b, large floods are often grouped in consecutive years. The Kowie is situated in a more arid area, as is reflected in the very low values for 'floods' occurring with a frequency greater than once in three years. Noticeable in the flood frequency curves of many South African rivers are the two flood populations: the smaller, more frequent floods, generally absent altogether from the more arid catchments such as the Kowie, and the much larger floods at frequencies greater than once in five years. Flood variability in southern Africa is considered to be high (Walling 1996). The flow volume of the 100-year flood is some eight times greater than that of the mean annual flood. In Europe the equivalent figure is two times, and in North America four times.

Most areas of the world display some degree of seasonality in their precipitation and river regimes. In temperate latitudes there is often a winter maximum due to a higher storm frequency coupled with a low evaporation rate which prevents drying out of the soil. Snowmelt contributes to floods in early spring. In southern Africa, rainfall tends to have a marked seasonality, which is mirrored by streamflow. Most rivers have high summer flows and low winter flows (see Figure 16.4). Only in the Western Cape does a winter rainfall maximum reverse the pattern. High evaporation levels, maintained even during the winter dry season of the Western Cape, rapidly deplete soil moisture, reduce dry-season flow, and heighten the seasonal contrast. High potential evaporation rates even in the winter season result in a reduction in dry-season baseflows throughout the country relative to those in the wet season, and many tributaries may therefore be seasonal.

Walling (1996) notes the extremely high inter-annual variability of southern African precipitation compared to that of the rest of the world. Only Australia shows similar variability. The coefficient of variation (standard deviation/mean) of mean annual runoff is just below 0.8, compared to 0.7 for Australia and between 0.25 and 0.4 for the rest of the world. One particular tendency of the southern African climate is the cyclical nature of rainfall. Over much of the eastern region an

FIGURE 16.5 Annual flood variability

a) flood frequency curves for the Kowie and Buffalo Rivers

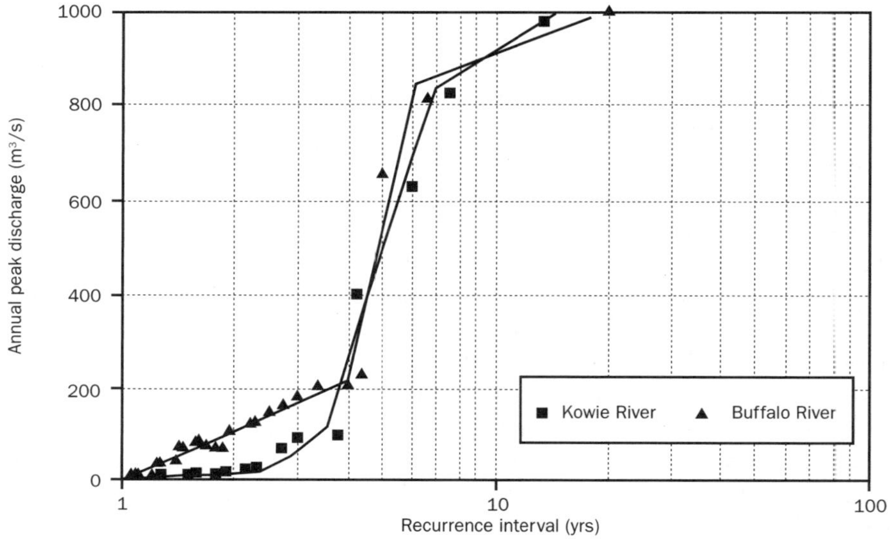

b) flood series, Sabie River

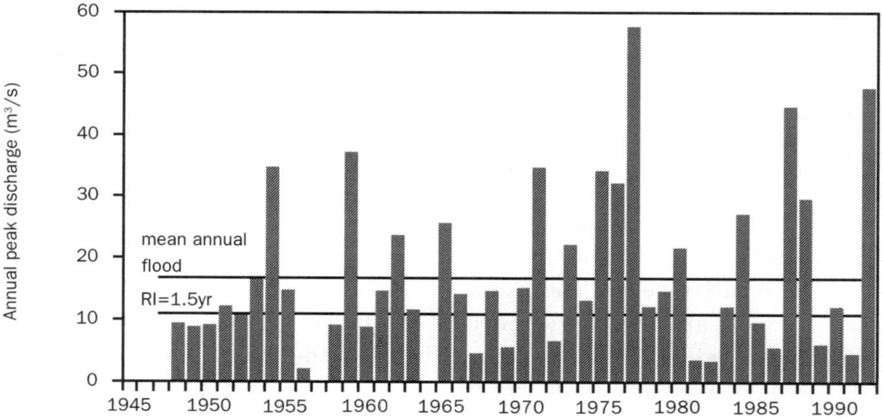

eighteen-year cycle has been noted, as explained in Chapter 12, a cycle which is translated into the flow regime.

An important consequence of this variability is the large storage volume required if a constant developed water supply is to be maintained. The only sizeable natural lake in the region is Lake Malawi, the third largest African lake. The building of dams has made up for the lack of natural storage, however. The African continent has the highest proportion of its runoff impounded in reservoirs (Walling 1996). Lake Kariba on the Zambezi has the largest volume of all anthropogenic African lakes (160 km^3), and Cahora Bassa, also on the Zambezi, is the fourth largest with a volume of 66 km^3 (Adams 1996). Lake Kariba, built in the 1960s, covers an area of 5 250 km^2, and its construction caused the displacement of some 57 000 people. In South Africa alone there are 452 large dams (Davies and Day 1998).

The spatial distribution of streamflow down the channel network deserves attention. Normally, streamflow increases in proportion to catchment area, as flow is augmented by direct hillslope runoff and tributary inputs. This is not always the case, however, especially in semi-arid or arid areas, where losses from the channel may exceed inputs. As the water travels downstream into drier areas, it may move from the channel into the dry river banks or into floodplain storage, where it is subject to evapotranspiration. Further losses include direct evaporation from the channel. In ephemeral streams, which only flow periodically in response to heavy rainfall, flood water must first fill up the storage in the initially dry channel sediments. These various losses are known as *transmission losses*; they are an important component of the hydrology of many South African rivers. The Orange River, for example, suffers a 44 per cent loss of its flow volume as it crosses the dry interior. Likewise, the Zambezi loses 30 per cent of its flow (Walling 1996). Studies of the Sabie River (Mpumalanga) by Birkhead et al. (1997) have shown that the indigenous riparian vegetation in the Kruger National Park accounts for a streamflow reduction of between 1.51 $m^3 s^{-1}$ and 0.25 $m^3 s^{-1}$ for February and August respectively, equivalent to between 5 and 40 per cent of the monthly flow discharge. The highest percentage loss occurs in the winter dry season. Invasive alien riparian vegetation also has a high water use. Experiments in the Zwartkops catchment near Port Elizabeth in the Eastern Cape (Rowntree and Beyers 1999) have demonstrated that Black Wattle in the riparian zone causes an average water loss of 15 litres per hectare per day.

Sediment loads of southern African rivers

The sediment load carried by rivers is largely dependent on contributions from the river catchment. It therefore varies regionally with catchment conditions and is sensitive to catchment management. Much sediment enters the channel with surface runoff, so that any factor which promotes surface runoff will also increase sediment yields. Climate and vegetation are the overriding factors here. In humid climates, despite the high rainfall, surface runoff is limited in extent as long as a dense vegetation cover enhances infiltration processes. Where overland flow does occur, the binding effect of vegetation protects the soil from erosion. As the climate becomes drier, so the vegetation becomes sparser and for a given storm intensity the chance of overland flow increases. No longer is there a protective layer of vegetation, and erosion therefore increases. Hence, the highest sediment yields under natural conditions have been associated with semi-arid climates. As aridity increases further, the frequency of storm events decreases to the point where long-term erosion rates become insignificant.

Land use and land management have a major impact on erosion rates and sediment yields. Erosion increases wherever the vegetation cover is disturbed or the soils are compacted by trampling. In the absence of effective conservation measures, sediment yields increase in proportion to the extent of cultivation or livestock densities in the catchment. Other cases of degradation of vegetation are firewood collection and the myriad of footpaths that are associated with dense rural settlements. Forestry tracks and the many unmetalled roads are a common sediment source area.

Although erosion rates may be high on the catchment slopes, this is not always reflected in the amount of sediment reaching the channel. Unless there is an efficient drainage network to transport the sediment from the slopes, it will often be deposited in footslope areas where it may remain for centuries. The ratio between the soil loss from the slopes and the sediment yield of the drainage basin as measured in the channel itself is known as the *sediment delivery ratio*.

Despite the high erodibility of many of the soils of southern Africa and the extensive areas of degraded land, the sediment loads of the region's rivers are not especially high (see Table 16.1). Walling (1996) reports that measured erosion rates are between one and two orders of magnitude higher than the sediment yield for the same drainage basin, indicating a low sediment delivery ratio. This disparity can be explained by invoking large-scale redistribution and storage of sediment within the system. The discrepancy may, however, be caused by two problems of measurement. Firstly, erosion studies may have been concentrated in those areas most obviously prone to erosion, whereas sediment yield measurements apply to a wider area. Secondly, significant volumes of sediment moved by the larger floods may be unaccounted for.

There is a general lack of data on sediment loads of the region's rivers. The best estimates of long-term yield come from surveys of sediment accumulated in dams. Estimates of sediment yield for South Africa fall within the range of 50 to 1 000 t $km^{-2} yr^{-1}$ (Department of Water Affairs 1986). The regional distribution of sediment yield is shown in Plate 53. High yields are indicated for the catchments of the upper Orange and the lower Vaal, and parts of the Karoo including the Sundays and Great Fish Rivers. Low yields are indicated for the Cape Fold Mountains of the Western Cape, reflecting the resistant nature of the bedrock.

The dissolved loads of southern African rivers are variable and reflect the underlying geology and climate (Walling 1996). Rivers draining the strongly weathered basement rocks of the interior generally have low dissolved loads, whereas rivers draining the sedimentary rocks of the periphery of the continent have higher dissolved loads. Concentrations also increase as the climate becomes more arid. Both the Zambezi and the Orange have relatively high concentrations of dissolved salts (113 and 140 mg per litre, respectively), due to the semi-arid to arid climates of much of their drainage basins. The total dissolved load carried by the Orange River, however, is only about 3 per cent of its suspended load. Bed load accounts for less than 10 per cent of the particulate load carried by the major rivers.

Sediment load
Particulate load is sediment load made up of solid particles.
Dissolved load is sediment load made up of dissolved material.
Bed load is particulate load carried in contact with the bed, normally the coarser load including coarse sand, gravel, cobbles, and boulders.
Suspended load is material carried in suspension, which may become bed load under low velocities. It is normally fine material such as medium to fine sand, silt, and clay.
Wash load: very fine material (fine silt and clay) that is washed into the river from the hillslopes and is carried downstream as suspended load.
It is deposited only under very low velocities, for example in backwaters, lakes, or on the floodplain.

The drainage basin as a geomorphological system

South African geomorphology has been dominated by the landform school, particularly by macro-scale studies of landscape evolution since the break-up of Gondwana (Moon and Partridge 1993). The interpretation of river terraces and the reconstruction of palaeodrainage systems may have contributed to the interpretation of environmental change, but they have done little to increase our understanding of geomorphological processes. Moon and Partridge note a few studies on fluvial deposition, the effects of flood events, and regional sediment yield. There have also been a few ad hoc studies on drainage morphology and long-profile development. It is only since the late 1980s that significant research into fluvial processes and smaller-scale channel features has been carried out. Two main schools of fluvial geomorphology have been established in South Africa, at the University of the Witwatersrand by the Centre for Water in the Environment (CWE), and at Rhodes University by the Geography Department. Both centres have worked closely with ecologists, their research being driven by the need to manage river systems on a sustainable basis. An up-to-date bibliography on South African fluvial geomorphology has been prepared by Dollar (1998).

The work of the CWE has been focused on the rivers of the Kruger National Park, and has followed both a descriptive/classificatory approach and a more process-orientated approach, studying trends in channel change over the last fifty years. Preliminary models of the Sabie drainage basin point towards the role that sediment inputs and changing flow regimes play in determining the distribution of channel-form features.

Geomorphologists at Rhodes University have based their work on a geographically wider field area. A recently completed project has aimed to develop a hierarchical classification system for South African drainage basins and their river systems (Rowntree and Wadeson 1999). A standard terminology was provided for the description of streams at various scales; it is hoped that this terminology will be adopted by all river scientists, including ecologists. Although the study encompassed all landscape scales within a drainage basin, a focus of the research was the link between smaller-scale channel features and stream habitat.

There is an increasing demand for South African geomorphologists to apply their expertise to river management. Researchers are beginning to tackle the fundamental geomorphological problems which will help us answer such questions as: what will be the long-term effects of impoundments (dams and reservoirs) on channel morphology, how do these vary with the type of river system, and how can we manage the flow regime to reduce these impacts? What is the effect of an interbasin transfer scheme on channel morphology? What effect does riparian vegetation have on channel processes, and what happens following clearance?

Our knowledge of the fluvial geomorphology of South African rivers is still rudimentary and localized. The remainder of this section will sketch what we understand to be the main characteristics of our river systems and will point to some of the important questions that are waiting to be answered. The reader is referred to basic texts such as Dunne and Leopold (1978), Gordon et al. (1992), Newson (1994), and Selby (1985) for a general discussion of the underlying geomorphological theory. Rowntree and Wadeson (1999) provide a review of geomorphological processes and channel forms as they relate to South African conditions.

South African fluvial systems

Fluvial processes are driven by two main groups of factors: those determining the supply of sediment to the channel and those determining the capacity for sediment transport or erosion of the channel bed. The sediment supply is largely determined by catchment factors which control rates of hillslope erosion, as discussed in the previous section, and the potential for sediment storage at different points in the system. Sediment transport is directly related to *stream power*, the product of discharge and channel gradient. Channel gradient is determined by the long-term development of the river profile; discharge is a function of climate and catchment characteristics. Also important at the local scale is the resistance to erosion attributed on the one hand to the bedrock or host sediments through which the channel cuts and, on the other, to more dynamic factors such as bank vegetation.

Sediment supply factors

Factors determining sediment supply have been outlined in the previous section on hydrological processes. The distribution of sediment input is not uniform along the length of channel. Because of the high rainfall in the headwater catchments, the vegetation cover is generally good, comprising either natural montane forest or high-altitude grassland. Surface erosion is therefore limited, but mass-movement processes bring mixed material into the channel. Debris flows, for example, are prevalent in the fynbos areas of the Western Cape. Lower down the catchment, human populations increase, and the natural vegetation cover decreases in response to the lower rainfall (see Plate 49). This, combined with the steep rejuvenated slopes, gives rise to a significant supply of sediment along the length of the profile.

The river long profile

The river long profile reflects regional geological events and long-term fluvial action. Uplift and tectonic warping provide the template upon which the profile develops. Over geological time the profile becomes adjusted (or *graded*) to transport the sediment that becomes available to the river channel. A theoretical graded profile developed on homogenous bedrock is concave in shape, like the long profile of the Berg River in the Western Cape (see Figure 16.6a) The steep headwaters are in equilibrium with coarse materials being transported by relatively low flows in low-order streams, whereas the lower-gradient lowland areas are in equilibrium with the transport of finer materials by increasing flows in high-order streams.

This classic long profile may be disrupted by a number of features, including local outcrops of more resistant rock and rejuvenation due to tectonic uplift or a fall in sea level. In South Africa, widespread rejuvenation occurred due to uplift and tilting during both the Miocene and Pliocene epochs (see Chapter 11). The axis of uplift ran more or less parallel to the east and south coast and reached a maximum of 800 m in the Natal Midlands in the middle Pliocene, about four million years ago. As a result, the rivers draining the Great Escarpment in the Eastern Cape, KwaZulu-Natal, and Mpumalanga have steepened long profiles (see Figure 16.6b). This has disrupted the classic bed-material sequence of boulder, cobble, gravel, and sand; sand-bed channels in the lower reaches may be replaced by bedrock, boulder, and cobble. The extent of uplift decreases towards the west, becoming negligible in the Western Cape.

The flow regime and sediment transport

The discharge which over time transports the most sediment through the channel and is responsible for maintaining channel form is known as the *dominant discharge*. Floods are the flow components of most interest to the geomorphologist, as it is high discharges which have sufficient energy to erode and transport channel sediments. Although the largest floods have the greatest transport capacity, they occur infrequently, and it is therefore the intermediate floods that have been found to be the most effective in the long term. Dominant discharge has often been equated with *bankfull discharge*, the discharge which is just sufficient to fill the channel to the top of the banks without overflowing onto the floodplain. In humid areas in the northern hemisphere, bankfull discharge has been found to occur every one or two years.

Floods are less frequent in semi-arid areas than in humid areas, and significant flooding may be absent for several years in succession. When heavy rains do occur, the catchment response is rapid and floods can be severe. The dominant discharge in semi-arid areas therefore relates to lower-frequency floods. For example, dominant discharge has been related to floods with frequencies of between three and ten years in semi-arid areas of Australia (Pickup and Warner 1976).

FIGURE 16.6 Long profiles of South African rivers

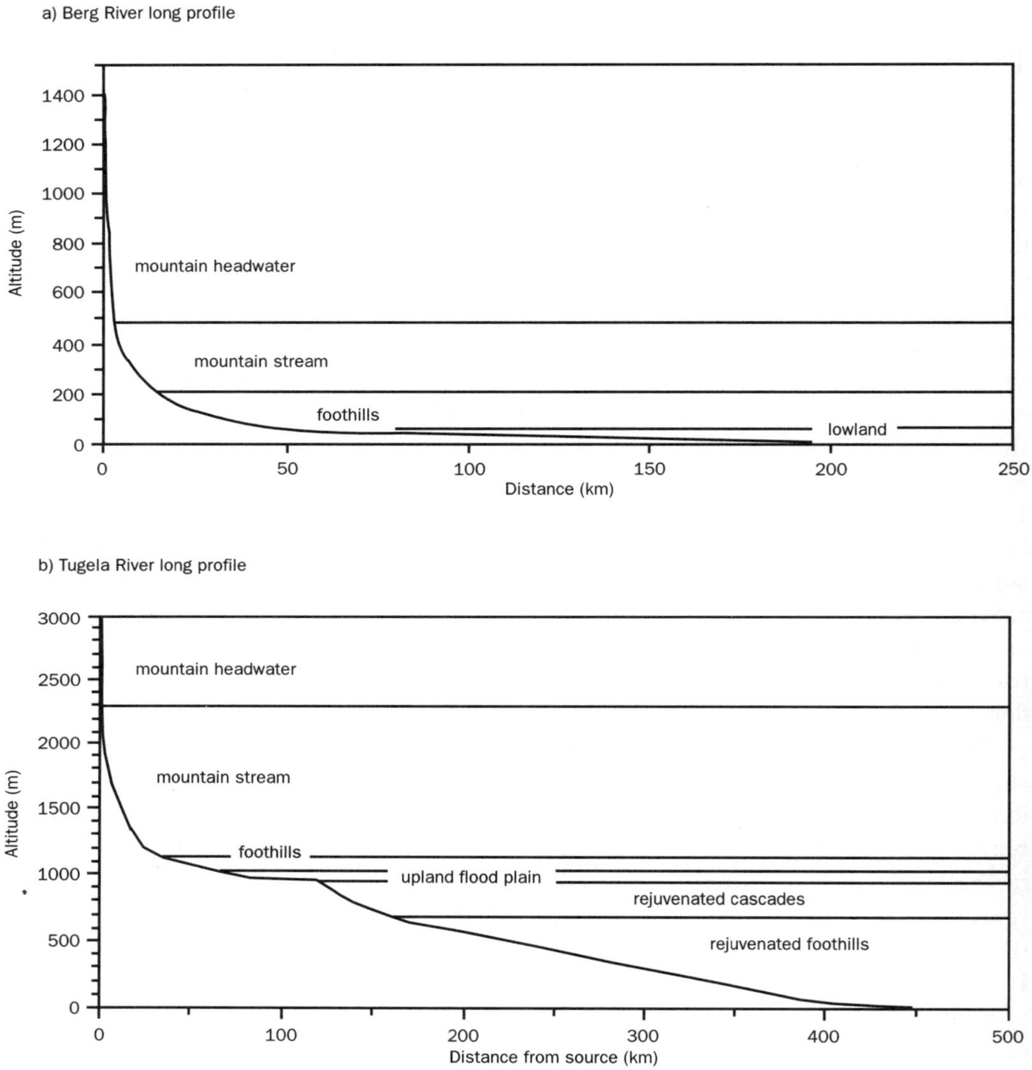

Rivers in southern Africa vary from humid, perennial rivers to semi-arid, ephemeral rivers. All display a high flood variability. No definitive research on dominant discharge has been completed, but it can be expected that the rivers would follow trends found elsewhere. A complication in defining dominant discharge in South African rivers is the definition of bankfull discharge. Many South African rivers, especially those along the eastern seaboard which have been affected by rejuvenation, have complex cross-sections, as illustrated in Figure 16.7. The active channel (Figure 16.7a) is often formed within a larger macro-channel bordered by high terraces (Figure 16.7b). There is no clear floodplain, but instead a narrow flood bench may be formed between the active channel banks and those of the macro-channel. This bench is inundated by more frequent floods and probably is equivalent to the conventional bankfull level. Only extreme floods such as those produced by Demoina or the Laingsburg flood overtop the macro-channel.

The catchments of many South African rivers are characterized by high-rainfall

FIGURE 16.7 Morphological features and processes of the active and macro-channels, Mkomazi River

a) Active-channel features and processes

b) Macro-channel features and processes

headwater areas and low-rainfall, semi-arid lowlands. Except under extreme flood conditions, most of the runoff is generated in the upper catchment, so that the downstream increase in discharge is less marked in South African rivers. Transmission losses into the banks and direct evaporation can further reduce flows downstream.

While flow discharge is a measure of the volume of water available to perform geomorphological work (erosion and sediment transport), the ability of the water to transport sediment depends on a number of factors, not the least of which is the stream velocity. Velocity, in turn, depends on the channel gradient (velocity increases on steeper slopes) and the channel resistance. Bedrock obstructions, boulders, and cobbles all offer considerable resistance to flow; as these features are associated with steeper gradients, the two factors of gradient and resistance tend to balance out so that the average flow velocity is similar in mountain, foothill, and lowland reaches. This is true for both flood and baseflow conditions. The local variability of velocities across and down the channel is, however, very different at different locations along the profile. Mountain streams are characterized by short sections of rapid flow, as water plunges over rock steps and through boulder chutes, separated by sections of very low velocities in bedrock pools. In lowland sand-bed rivers the channels are more homogenous, and this is reflected in their more uniform flow velocities. Smooth flows in lowland channels, undisturbed by roughness elements projecting through the flow, contrast with the white water of cascades and rapids which dominate our perception of mountain streams.

River channels provide the habitat for organic organisms. *Hydraulic biotopes* describe the physical habitat provided by the combined effect of flow and the bed-material assemblage (Rowntree and Wadeson 1999). Hydraulic biotopes are small patches of relatively uniform velocity, flow depth, and bed-material characteristics, and are defined in terms of their surface-flow characteristics (flow type) and substrate (bed material). Examples are rippled flow over coarse gravel, characteristic of the upstream area of many pools, and undular standing waves (unbroken, stationary surface waves) over cobble, characteristic of many riffle areas. The assemblage of hydraulic biotopes in any channel section will tend to change with the flow discharge, reflecting for example the increased depth and velocity experienced at higher discharges. The hydraulic biotope provides a useful scale to link ecology to channel geomorphology. It is presently the focus of ongoing interdisciplinary research.

Channel types and channel adjustment

River channels can be classified into two broad types: *bedrock channels* and *alluvial channels*. In bedrock channels the energy of the stream during flood events is sufficient to transport all available loose material, whether this comes in from the side slopes, the banks, or the bed of the stream itself. Such conditions occur where channel slopes are steep, the bedrock underlying the channel is resistant to weathering, and/or there is a limited input of sediment from the valley slopes. In bedrock channels the geology of the channel bed and its resistance to erosion are the main determinants of channel form. Such channels have been termed controlled channels, as their form is determined by bedrock controls rather than by flow.

Alluvial channels, in contrast, are formed within the sediment which is transported by the river. Both the bed and the banks of the river are composed of sediment which is in temporary storage within the system. The channel form is the result of the balance between the available sediment and the transport capacity of the flow. Channel form can be considered to be in dynamic equilibrium with the flow and sediment regimes. A third channel type is the *mixed channel*, in which alluvial and bedrock sections alternate. Mixed channels are common in South Africa, especially in the rejuvenated sections of the lower profile. Long pools separated by bedrock rapids are common features of mixed channels (Heritage et al. 1997; Van Niekerk et al. 1995).

Riparian vegetation

An important component of the fluvial system is the riparian zone, normally consisting of a floodplain or flood bench, whose characteristics are determined by frequent flood inundation and the deposition of flood sediments. The riparian zone represents an important habitat for phreatic vegetation, which depends on groundwater, and other related plant communities. In addition, temporary pools in the riparian zone provide breeding ground for certain fish species, as well as favoured habitat for other marsh-loving species.

Riparian-zone processes are of significance to the geomorphologist who recognizes a complex interaction between channel morphology, riparian vegetation, and channel processes. Channel processes control vegetation growth by providing the morphological features upon which plants become established, replenish soil materials, and control the frequency of inundation (Van Coller et al. 1999). Vegetation, in turn, provides channel resistance and increases bank stability. Woody vegetation can also promote instability by creating near-bank turbulence or bank collapse under the weight of the vegetation. Blocking of channels by woody debris can result in overbank flows and avulsion, the formation of new channels across the neck of a meander. This process has been observed in the Bell River in the Eastern Cape (Dollar and Rowntree 1995). Various effects of vegetation on channel processes and channel morphology have been reviewed by Rowntree (1991)

The riparian zones of South African rivers are often vegetated with dense stands of woody shrubs which take advantage of the higher moisture levels in these areas. The extensive distribution of alien vegetation in riparian areas has already been noted. Although few studies have been carried out on the geomorphological effects of riparian vegetation in South Africa, they are undoubtedly significant. Where this vegetation has been removed, as in areas of dense rural settlement or floodplain cultivation, the river banks show signs of marked erosion, enhanced by the large number of footpaths and livestock tracks providing access to the river. One of the challenges of river restoration in South Africa is to rehabilitate the riparian vegetation in these areas.

Water resource developments

The widespread occurrence of large dams is a notable feature of South African drainage basins. Dams and interbasin transfer schemes have provided the main infrastructure through which the country's water resources have been managed. On a smaller scale, the cumulative effect of farm dams used to supply water for irrigation and for livestock watering is also significant (Davies and Day 1998).

Dams have a number of effects on both the hydrology and geomorphology of the downstream river (Petts 1979). The large surface area increases evaporation, storage reduces both the frequency and magnitude of floods, and downstream releases to augment baseflows have resulted in an average baseflow increase of nearly 40 per cent for African rivers (Walling 1996). Another effect is that sediment becomes trapped in the reservoir. The suspended sediment load of the Zambezi below the Cahora Bassa Dam is half that measured prior to its construction. The change in the magnitude and frequency of floods and the reduced sediment load have a number of geomorphological effects which vary with the magnitude of the change in the flow regime, the location of the dam relative to sediment sources, and the channel type. Immediately below the dam wall, the reduced sediment load is replenished by erosion of the bed and banks. Further downstream, the reduced flood magnitudes result in insufficient stream power to transport the sediment introduced by tributaries, and aggradation (the build-up of sediment due to deposition) therefore takes place (Petts 1979; Sherrard and Erskine 1991).

The geomorphological effects of impoundments have not been well documented for South Africa. Field observations have shown significant aggradation in the Tugela below the Spioenkop Dam, and loss of sediment in the Goedetrouw Dam on the Mhlatuze River (KwaZulu-Natal) has resulted in a braided river being transformed to a single-thread channel. Tributary bars have been noted downstream of the Sandile Dam on the Keiskamma River and the Bridle Drift Dam on the Buffalo River, both in the Eastern Cape (Rowntree and Dollar 1996).

An important engineering function of a dam is to control floods and protect downstream areas. Although floods are often perceived as a hazard and a 'wasted' resource, they play an important role in maintaining the structure and function of the channel through sediment transport processes. Ecological processes, too, are dependent on floods to scour out fine materials from gravels used by fish for spawning, to provide ecological cues for different life-cycle stages, to flood temporary pans on floodplains used for fish breeding, to water vegetation on the floodplain, and so on. River managers are slowly beginning to accept the importance of floods in maintaining sustainable river ecosystems, and flood releases for river maintenance are being built into reservoir operation rules. South African ecologists have developed what has become known as the Building Block Methodology for determining the ecological reserve of rivers for which impoundments are planned. The three building blocks are the seasonal low flows, freshes, and floods. Low flows are equivalent to the baseflow and provide aquatic habitat. Freshes are small pulses of increased flow which serve to flush out water of poor quality from pools and to winnow fine silts from the channel bed, as well as providing fish spawning cues. Floods have both a geomorphological function, as discussed above, and replenish the riparian zone with water and silt. Further discussions of the method are given by King and Louw (1998) and Rowntree and Wadeson (1998).

Ecology and river management

The National Water Act of 1998 includes a far-reaching departure from the previous Act of 1956, namely the recognition of the Reserve as the basis of water rights (see Chapter 19). The Reserve consists of two parts: the *ecological reserve* and the *basic human needs reserve*. The ecological reserve relates to the quantity and quality of water necessary to protect the sustainable functioning of aquatic ecosystems. Once defined, the Reserve has priority and cannot be allocated to development uses. This is a fundamental change in legal thinking in a country where the environment formerly had no legal protection under the Water Act.

The new law will provide effective protection to the environment only if scientists can give the Reserve a legally defensible definition. For a given length of river it will be necessary to specify and justify the volume, timing, and quality of the flow of water required to maintain ecological functioning. Considerable progress has already been made in recommending instream flow requirements below reservoirs using the Building Block Methodology discussed above. This is now standard practice for any reservoir feasibility study associated with the Department of Water Affairs and Forestry. The challenge now facing hydrologists, geomorphologists, and ecologists is to develop an equivalent cost-effective method that can be applied to every 'significant water course' as required by the Act.

The links between drainage basin processes and the river ecosystem have been highlighted at a number of points in this chapter. As noted, the continued concern for the ecological state of the country's rivers has given the impetus for hydrology to move away from the engineering syndrome of resource estimation, which has characterized South African hydrology for much of this century, towards recognition of the need to get into the river and research the streamflow processes themselves. The importance of understanding contemporary geomorphological process–form relationships has also been recognized. It is interesting to note that Gregory and Walling asserted in 1973 that information from the process school is increasingly of most direct relevance to physical geography and is important for the management of river systems. Likewise, Ward (1975) urged the geographer to apply his or her skills to understand the natural systems and processes of the landscape and to see the drainage basin as a complex interacting ecosystem. It is hoped that this chapter has gone some way towards bringing together the hydrological and geomorphological processes which determine the structure and function of drainage basins and their ecosystems in southern and South Africa.

SECTION IV

People and the environment

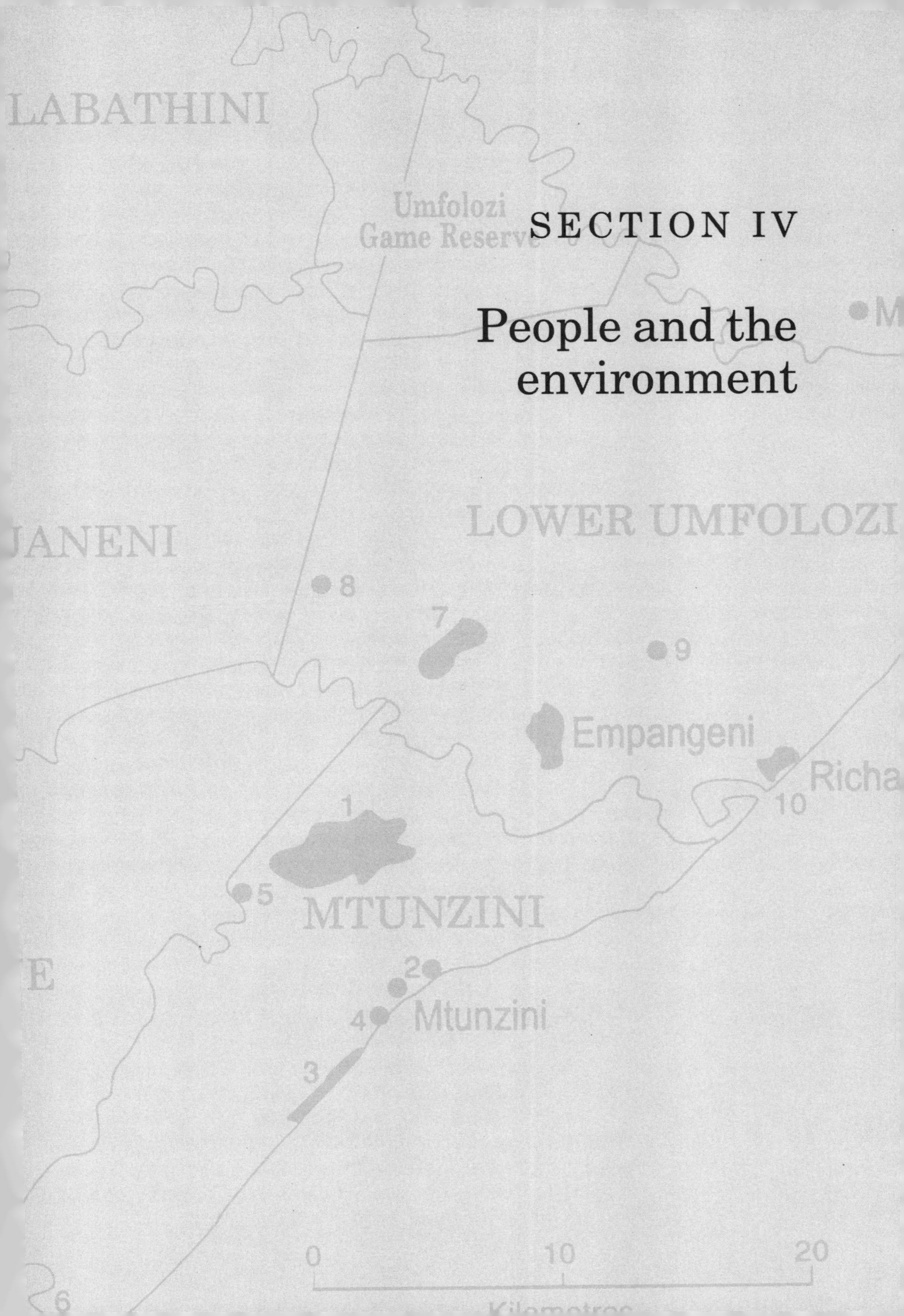

17 Environmental resources and development

PETER OMARA-OJUNGU

The word 'environment' strictly refers to the external conditions or surroundings of a phenomenon. In this chapter, 'environment' is defined as an area or land unit in which the biophysical system interacts with the socio-economic system to produce potential resources and to support a range of existing resource practices for both human and other life forms.

The biophysical system includes the atmosphere (weather and climate), the lithosphere (landforms), the hydrosphere (water in oceans, lakes, rivers, and marshlands), the pedosphere (soils and soil processes), and the biosphere (flora, fauna, and people). The interactions between these five spheres result in a range of environmental conditions and resource potential. At the macro scale these conditions vary from one landform type to another, from one climatic region to another, and from biome to biome. At the micro scale the biophysical interactions create several habitats/biotopes with varying biodiversities. The socio-economic system interacts with the biophysical system, especially at the interface between the lower atmosphere and the ground surface, to determine further the volume, quality, and distribution of resources from a potential and an active point of view. The socio-economic system includes human communities and their numbers, goals, and strategies; institutional frameworks (policies, structures, programmes, and legislation); culture, attitudes, and perceptions; and the economy, level of technology, and position in the global development continuum (from developed to less developed). These attributes of the socio-economic system empower humans to determine the overall nature of the environment.

Resources and resource classification

Resources as a concept have been variously defined by many authors. Zimmerman (1951: 15) presented a classic statement that 'resources are not, they become'. In

1989, Mitchell translated this philosophical viewpoint into an anthropocentric position by stating that an object is at first a neutral phenomenon until human value is attached to it, and then the object becomes a resource. Both the philosophical and anthropocentric views of resources have been further clarified by Omara-Ojungu (1992), who demonstrates that availability for human use, not mere physical presence, is the most important attribute of resources. Differences in value systems, aspirations, and levels of technology will often mean that a neutral phenomenon to a community at one time may become a resource to that community at some other time. This statement underpins the need for resource conservation. Thus a resource may be defined as a functional relationship between people and their biophysical environment from which they obtain goods and services in order to satisfy their needs and aspirations. But this definition is limited to the present generation:

> [A]s we continue to care for the future generations, it must be accepted that all things or resources have value even if they offer no immediate utility to people. Other ecosystem components are also resources because of the functional roles they play in ecosystem stability. (Omara-Ojungu 1992: 2)

There are several ways in which natural resources may be classified. In this chapter only the commonly used bases are delimited; others are discussed by Mitchell (1989), Omara-Ojungu (1992), and O'Riordan (1971).

Resources may be classified as either tangible or intangible resources in terms of their physical parameters. *Tangible resources* can be physically touched and their quantity can be measured; examples are minerals, timber, water, fish, livestock, and land. Conversely, *intangible resources* may be perceived but their quantities are not usually or easily measured; for example landscape beauty, pleasure, and the souls of ancestors. The ability to quantify resources is important for environmental auditing and accountability. Knowing the quantity of resources can help in monitoring and controlling against excessive extraction and environmental change or in establishing a price mechanism so that the values of different resources can be costed and compared.

Another method of classification is based on the relative dynamics or regenerative ability of resources. In this classification, resources are either non-renewable or renewable. *Non-renewable resources* do not regenerate themselves after extraction, or have regeneration rates which are ecologically and economically insignificant. Non-renewable resources are often regarded as finite resources, because their quantities are exhaustible over time. Minerals such as gold, copper, diamonds, rock salt, and oil are resources that do not regenerate themselves.

Renewable resources such as forests, fish, animals, and other plant biomass, solar radiation, wind, and tides are infinite resources because they are all apparently inexhaustible. Without human interference, plant and animal biomass regenerates itself by replacing the old and dead with the young, thus guaranteeing its sustainability. Energy resources from wind, tides, and solar radiation are apparently inexhaustible because their source is extraterrestrial and therefore unaffected by human activities. Solar energy, for example, originates from the stock of matter in the sun, while tidal energy derives from the gravitational force related to the interaction between the earth and the moon. The renewability of biomass resources depends on their initial mass in relation to the rate of extraction. Biomass resources will regenerate themselves only if the rate of extraction is kept below the threshold population size of forest, fish, or livestock, or within their respective carrying capacities.

Soil fertility levels, the ozone layer, underground aquifers, environmental conversion of wastes, and the regulation of soil and hydrological conditions by forests are all examples of renewable physical stock resources. Through physical and biological processes these resources can regenerate themselves within relatively short periods of time, but, like other resources, they are potentially depletable.

Resources may also be classified on a functional basis as *common property resources*, over which communities have collective responsibility regarding rights of use and management for sustainability, for example the atmosphere, the oceans, and communal grazing land, and *symbolic resources*, whose value is subjective and personal, such as the grave of a relative, historical monuments, and, more recently, the new South African national flag.

These classifications have important implications for resource development and management. For example, perceptions on the renewability or non-renewability of resources, or their function as symbols or common property, may affect controls over resource extraction and management approaches.

Environmental processes and development: a conceptual framework

One of the reasons why environmental deterioration has persistently undermined socio-economic development is that the behaviour of the natural environment has often not been clearly conceptualized or understood. When students or readers are plunged directly into a problem such as water quality and the corresponding solutions, for example water pollution abatement, they gain only a sectoral perspective which cannot be applied globally to other resource sectors. Even under pragmatic approaches, applied science must be informed by the theory or theories underlying any proposed action.

This section of the chapter will attempt to provide a general understanding of the behaviour of the natural environment in order to explain the environmental changes that may occur along the continuum of resource development. It is now a well-established practice among ecologists and environmental scientists to conceptualize the natural environmental as a *system*. Russwurm and Summerville (1974) have argued that a systems approach allows one to study the whole, move from the whole to the parts, and return to the whole again. A system is a set of objects with organized and interdependent interactions. The components of a system are often hierarchically arranged, with objects and elements in many subsystems acting individually and in combination to affect the functioning of the whole system.

The natural environment is a system

composed of five integrated components or subsystems:

- the atmosphere (weather and climate);
- the lithosphere (landforms);
- the hydrosphere (water and hydrology);
- the pedosphere (soils); and
- the biosphere (biota).

Each of the five components can be further subdivided into its constituent elements. For example, the biotic subsystem consists of plants and animals, the latter including humans and micro-organisms. The atmospheric subsystem has elements such as air masses, solar radiation, clouds, and precipitation.

At a much larger scale, the natural environment may be comprehended as consisting of only two major components – the *abiotic* and *biotic* subsystems. The abiotic, often called the physical environment, provides the setting, physical support, and sustenance for the biotic environment of plants and animals. Soil provides the link between the abiotic and biotic subsystems and demonstrates their interdependence. Hence, soil incorporates almost all the features of both the abiotic and biotic worlds (gases, water, nutrients, and micro-organisms).

Integration and interdependence among the components and elements of a system take place due to the flows of energy and materials. Energy and materials flow between environmental subsystems, causing *intersystem exchanges* of heat, moisture, nutrients, and particulates. Within each subsystem there are also *intrasystem exchanges* of energy and materials. Because the flows of energy and materials often represent the actual exchange of resources, the quality of any environment (in terms of resource potential) will therefore be determined by both intersystem and intrasystem exchanges.

In the context of systems theory, intersystem and intrasystem exchanges take place in the form of inputs, throughputs, and outputs (see Figure 17.1). The ratio of input to output represents the process by which environmental systems either maintain themselves or undergo change, and the condition of environmental resources depends on their reaction to the input/output ratio. Figure 17.2 illustrates the primary inputs, outputs, and interlinking processes in the natural environmental system. Outputs are the product of a set of interactions between inputs of energy (solar radiation) and materials (nutrients and water) and the natural

FIGURE 17.1 System interaction

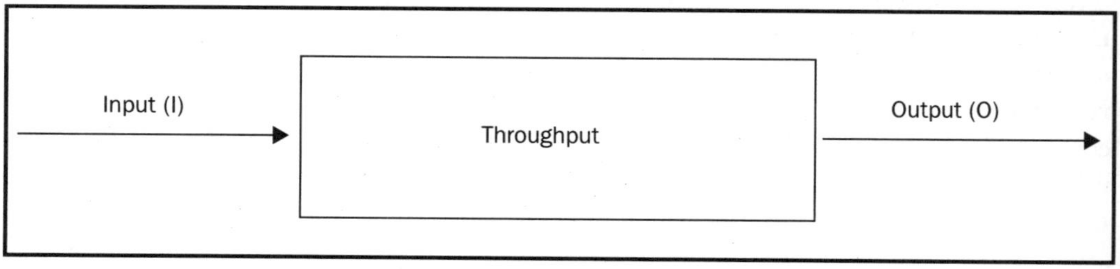

FIGURE 17.2 The natural environmental system

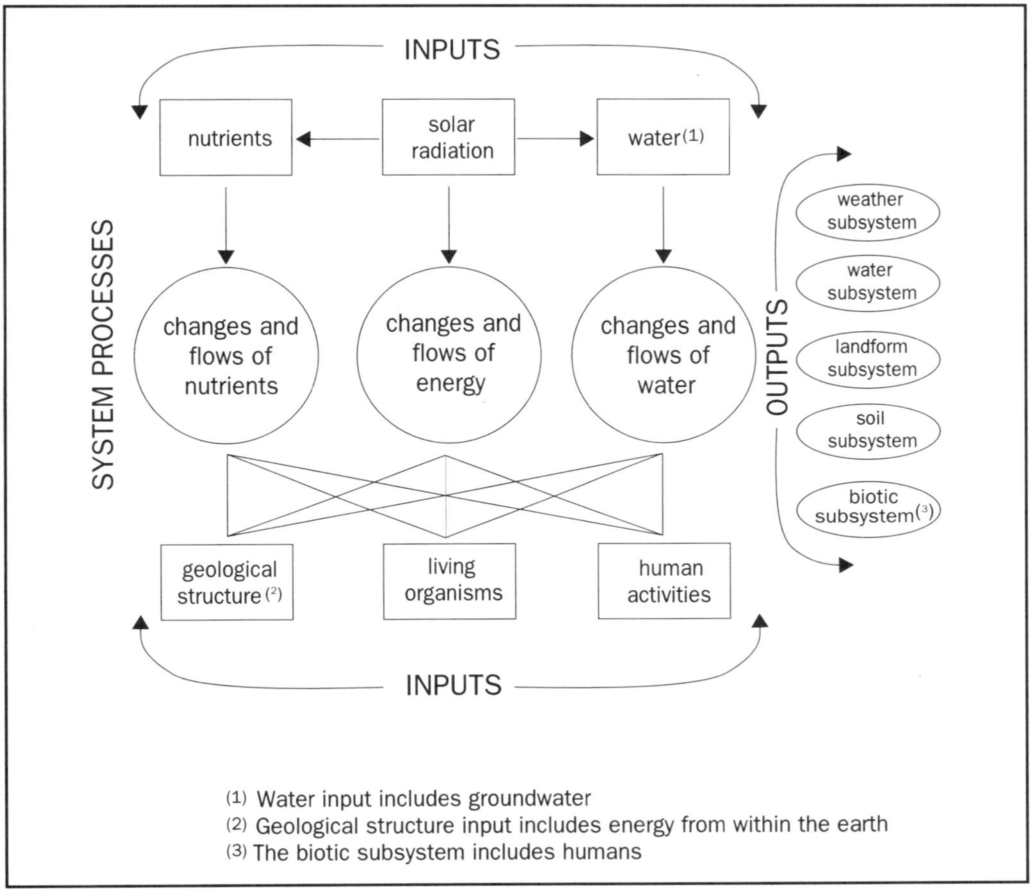

SOURCE: adapted from Russwurm and Sommerville 1974.

system processes. Natural system processes will transform, transport, and use flows of energy and materials to provide a given range of outputs from each of the five subsystems of the natural environment. For example, sediment load into a river subsystem is determined by the process of erosion, which in turn depends on gravitational energy and the kinetic and heat energy of wind, water, and ice. Erosional activities themselves depend on the supply (or input) of weathered bedrock. The subsequent deposition of transported material, in turn, provides inputs to the soil subsystem. The interactions presented in Figure 17.2 imply that inputs and outputs exist at varying hierarchical levels, depending on the scale of analysis. The ratio of input to output prompts an ecosystem to assume three types of trajectory over time: a steady-state trajectory, an upward trajectory, or a downward trajectory (Russwurm and Sommerville 1974).

The steady-state trajectory

The steady-state trajectory is the normal trend in the natural environment or in an environment little affected by human activities. The steady state occurs, for example, in a mature ecosystem such as the climax forests in certain climatic regions. Although the natural environment sometimes behaves like a closed system, it is generally regarded as an open system into which there is constant input, throughput, and output of energy and materials. As it continues to function naturally on its own or under human pressure, the open environmental system will tend towards a steady-state condition. A steady state is an equilibrium condition whereby the ability to do work is balanced by the amount of work to be done. Thus a point will be reached at which there is a balance between input (I) and output (O) of energy and materials; that is, I = O. This implies a balance between extraction and regeneration in the case of renewable biomass resources. In terms of landscape dynamics, the steady state may refer to a balance between the rates of erosive destruction and constructive sediment deposition. For renewable resource development such a steady state may in theory be attained by limiting resource extraction within the carrying capacity of an environment. In mature ecosystems in the natural landscape, death and birth rates are brought to a balance, thus ensuring stability in morphology and biodiversity over long time periods. The actual steady-state level depends on the dynamic changes in the ratio of input to output of energy and materials.

The anthropogenic climax in many farming communities in South Africa and elsewhere may be considered a type of steady-state trajectory over a specified time scale. The anthropogenic climax is a mixed community of indigenous vegetation, imported vegetation, and crop species that sustain themselves in the landscape in spite of relatively long periods of human activity. In South Africa a combination of gum trees, pine, orchards, crop fields, and remnants of indigenous vegetation species is common in the farming landscape. The proportion of species in each area (species diversity) depends on the climate of the region and the degree of human impact. Figure 17.3 shows that the steady-state trajectory is a dynamic equilibrium in which changes that occur are represented by deviations below and above the mean line. Environmental systems in a steady-state trajectory have losses of resources (output) that are replaced by inputs of the same type or magnitude or inputs of similar kinds of energy and matter; in other words, extraction/loss equals regeneration/replacement.

Ideally, mature forests such as the humid evergreen forest of the tropics are characterized by high biodiversity, stability, and equilibrium. The diversity of habitats/biotopes and species, known as biodiversity, is at its maximum when each habitat or organism is different from all the others within the same community. In contrast, biodiversity is lowest when each organism in the community belongs to the same species and occupies the same type

FIGURE 17.3 Dynamic equilibrium

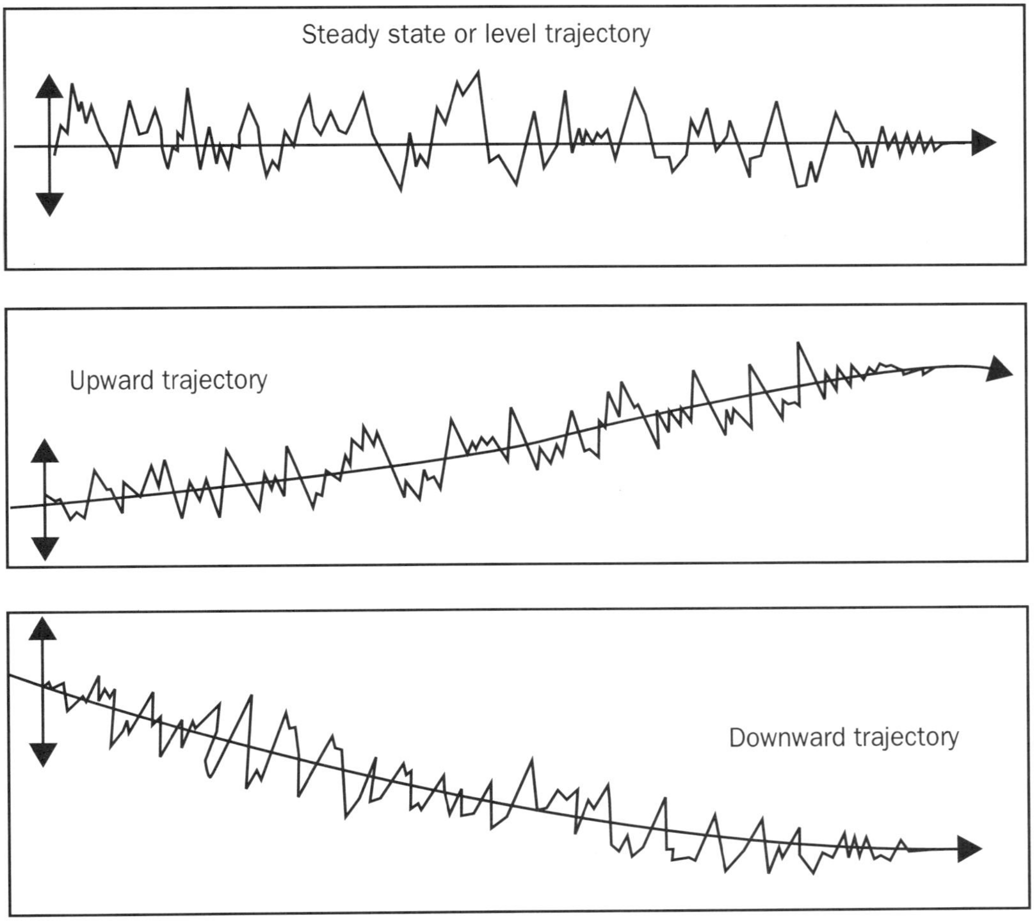

of niche. It is generally accepted that the higher the biodiversity, the more stable the ecosystem or resource base. This is because:

- an ecosystem with a high biodiversity possesses a variety of ways to counteract the effects of changing inputs;
- risks to ecosystem stability are so widely spread that the critical mass for negative impacts may not be reached easily; and
- when one element is eliminated, the chances are high that another element with similar energy or matter may assume the functions of the lost element.

Stability through high biodiversity means that the resource base is diverse and predictable, and can therefore be managed to meet the needs of the present and future generations. For these reasons, modern resource management requires sufficient knowledge of the natural processes that bring about acceptable harmony between environmental resource potential and human demands for natural resources, that is, steady-state trajectory processes.

The downward trajectory

Besides leading to a steady state, the ratio of input to output can also cause either the accumulation or the depletion of energy and matter over time in an ecosystem. Figure 17.2 shows that inputs may derive from natural sources (solar radiation, nutrients, water, geological structure, and living organisms) or from human sources. A downward trajectory, like an upward trajectory, occurs as resources undergo processes of development.

The downward trajectory occurs when the critical level of output of energy and matter exceeds the level of inputs into the ecosystem; that is, O > I. Over time this will lead to resource depletion as the trajectory tends towards a lower level of equilibrium. In various ecosystems a downward trajectory is associated with such trends as:

- overgrazing of palatable livestock feeds;
- overcultivation leading to loss of nutrients;
- overfishing depleting fish stocks; and
- deforestation of timber.

Such a downward trend in resource potential has serious resource development and management implications. Figure 17.4 represents a hypothetical but possible case of a trend towards desertification in a village hitherto surrounded by thick humid forest. Due to excessive deforestation and constant cultivation, the ecosystem declines from forest to bushland, then to grassland, and finally to bare ground surface or sand. Human communities will be forced to adapt by changing their resource needs and practices because of the constant decline in the level of energy and matter from the forest ecosystem to the bare ground conditions. Such adaptation, however, may not keep pace with the rate of decline in the range and quality of resources. For instance, a change from sedentary cultivation to livestock grazing may require a complete overhaul of cultural fabrics and the abandonment of attitudes and technology appropriate for the previous forest ecosystem. In Nepal, such transitional tension has led to large-scale abandonment of farmlands and migration of able-bodied labour to cities and lowland environments (Eckholm 1975).

The upward trajectory

The upward trajectory occurs when the volume or rate of input supplied into the ecosystem exceeds the volume or rate of output leaving the ecosystem. This will cause accumulation of energy and materials, thus leading to growth to a new (higher) level of equilibrium.

Lake eutrophication is a good example of an upward trajectory. Eutrophication occurs where there is a constant discharge of matter such as domestic wastes, nitrates, or phosphates from agrochemicals into a lake at a rate far greater than the lake system can absorb and convert/degrade. In time the mix of chemical

FIGURE 17.4 Ecosystem decline through excessive deforestation and constant cultivation

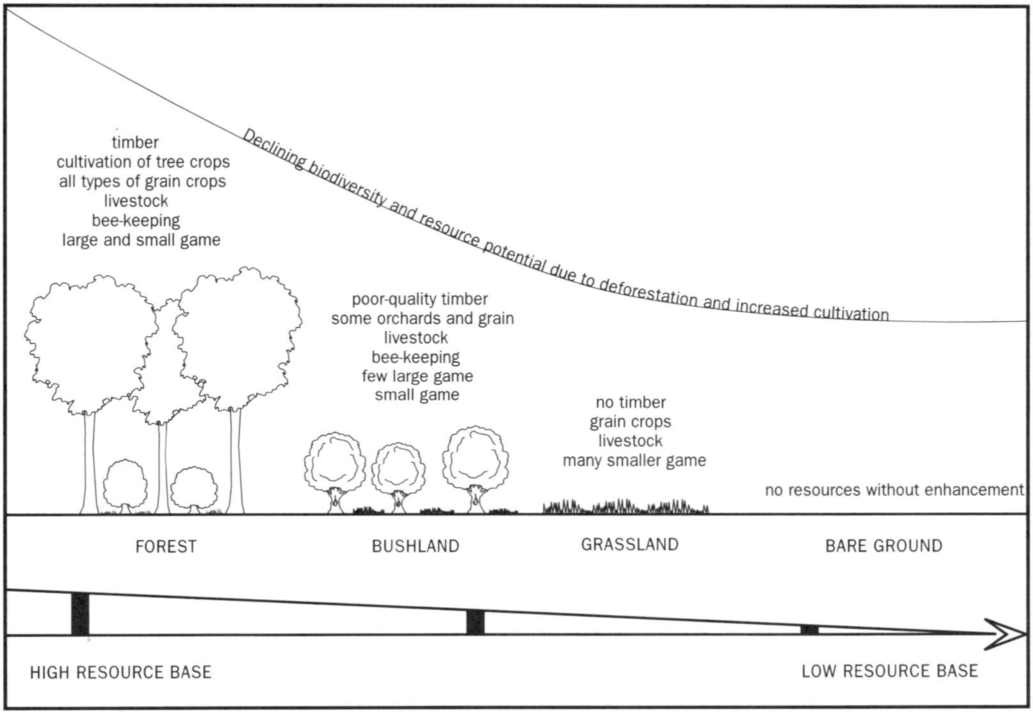

nutrients will promote algal bloom, increased biological oxygen demand, reduced diversity of fish species, and continued accumulation of sediments and nutrients. Due to sediment and nutrient build-up, water hyacinths and rooted plants will begin to colonize the ecosystem until the lake eventually assumes a dryland condition. Eutrophication, therefore, will alter the ecological, economic, and social roles of the lake, and communities whose livelihoods were dependent for years on fishing, water transportation, and water sport will be forced to adapt to new requirements for continued living.

Resources and development

Natural resources without doubt form the foundation of economic and social development. In order to promote economic development by raising living standards, providing infrastructure, employment, education, recreational amenities, health care, and opportunities for self-determination and earning foreign exchange, governments have to rely on either increasing production from agriculture, forestry, fishery, wildlife, and minerals, or promoting manufacturing, trading, services, and tourism. There is, therefore, an obvious link between the natural resource endowment of a nation and its level of socio-economic development. South Africa, for example, exhibits a strong link between resources and development. Because of large deposits of various minerals (gold, diamonds, coal, mica, silver, iron, copper, lead, and tin), South Africa has developed into one of the world's leading countries in terms of mining technology. The proceeds from the mines have had a profound impact on the development of infrastructure, the provision of social services, and the conservation of the quality of the environment itself. The country has a much more limited potential in forestry, water, and agriculture, but wildlife and ecotourism resources have been developed into lucrative commercial ventures.

The link between natural resources, sometimes called *earth capital*, and economic development is further confirmed in Miller's statement that

> environmentalists and a small but growing number of economists and business leaders argue that economic systems depend on resources and services provided by the sun and by earth's basic components and processes, and that a healthy economy is ultimately dependent on a healthy ecosphere. (Miller 1996: 672)

Development in the context of environmental science

Development can be taken to mean the following:

- a new stage of development;
- new products and technological inventions;
- economic growth;
- the advancement of social justice;
- improvements in the value of resources or products; and
- more effective social, political, or business organizations.

Crucial to this chapter, however, are notions of growth, the quality of growth, and equity in welfare systems.

Growth

Growth in economic terms is an increase in the capacity of the national, regional, or local economy to provide goods and services to meet the needs and aspirations of the people. Chapter 5 has shown how the desire for growth has been at the heart of attempts to modernize the economies of the developing world from as long ago as the late 1950s. Such growth, according to Miller (1996), is usually accomplished by maximizing the flow, or throughput, of matter and energy by means of either serving a growing population of consumers or increasing consumption per capita, or both. However, promoting energy and matter consumption patterns in a growing population may lead to deleterious consequences and raises the question of the quality of growth itself – does growth necessarily lead to development? Such concerns are well expressed in Paul Hawken's observation:

> We have reached a point where the value we do add to our economy is now being outweighed by the value we are removing, not only from

> future generations in terms of diminished resources, but from ourselves in terms of unlivable cities, deadening jobs, deteriorating health, and rising crime. In biological terms we have become a parasite and are devouring our host. (Hawken 1995: 74)

The gloomy picture depicted in Hawken's statement typifies environmental conditions in a downward equilibrium trajectory. Although, through technological development, we may be able to avoid some of the negative consequences of growth mentioned by Hawken, the greatest danger to resources and eventually development is the overwhelming preoccupation with growth by national government at the expense of issues such as equity, social justice, and individual freedom for self-determination. In probing into the various perceptions and views about growth, the South African government's Reconstruction and Development Programme policy document comments that,

> where that growth occurs, how sustainable it is, how it is distributed, the degree to which it contributes to building long-term productive capacity and human resource development, and what impact it has on the environment, are the crucial questions when considering reconstruction and development. The RDP integrates growth, development, reconstruction and redistribution into a unified programme. (African National Congress 1994: 4)

Because economic development (especially development based on growth) always consumes both non-renewable and renewable resources, growth will reach a point where the costs generated become more than the worth derived and therefore where the system has become unsustainable. A new way of realignment of the relationship between growth and resource capacity therefore must be sought.

Development efficiency

Another way of addressing development is to focus more on efficiency. In the context of natural resource management this involves increasing the efficiency of resource use so as to produce the same or greater output of goods and services with smaller inputs and throughputs of natural resources and manufactured and human capital. This approach may sustain resources for a while longer, especially if population and consumption patterns are also regulated. But there are limits to the efficiency approach, particularly if development is based on non-renewable or depletable renewable resources. These limits make it difficult to sustain the level of efficiency prescribed by policy in the long run.

Ecologically sustainable development

This scenario is often more favoured by environmental scientists. In this approach, development is regarded as a way of improving human livelihoods by realizing human potential without degrading the life-support capacity of the earth systems. This involves restricting resource extraction and human demands within the natural carrying capacity of an area, region, or nation. To achieve this end, great attention has to be paid to such issues as:

- the wisdom of decisions;
- education;
- resource sharing;
- co-operation;
- environmental protection; and
- restoration and enhancement of the environment.

Whatever scenario for development is chosen, there will always be tension between such conflicting goals as economic gain, ecological feasibility, technological adaptability, and cultural norms. The most glaring tension occurs during assessment or measures of development. Three measures (Gross National Product, quality of life, and environmental degradation indicators) are discussed below.

Measures of development

Gross National Product

Traditionally, economists have measured economic growth by the annual increase in a country's *Gross National Product* (GNP). GNP represents the monetary value of all goods and services produced by the national economy for final use in a particular year. GNP measurement therefore includes the volume of natural resources extracted to meet the needs and aspirations of a nation's population and for export. Real GNP may be derived by adjusting for inflation and may also be used to calculate real GNP per capita (the real GNP divided by the total population) to compute the relationship between economic growth and population expansion.

GNP as a measure of economic growth has been severely criticized over the years on a number of grounds. Firstly, it is misleading because it gives a national picture of wealth distribution, thus hiding inequalities between the wealth belonging to the few and the poverty widespread among the many poor individuals. Secondly, GNP provides a poor measure of the quality of life, environmental quality, and economic health of a nation:

- The negative impacts of producing goods and services through, for example, resource extraction are often not incorporated or are kept hidden.
- Ecological degradation may intensify while GNP shows increasing growth; for example, excessive deforestation leading to high sales of timber, charcoal, and forest products will boost GNP while resulting in intensifying environmental degradation.
- GNP does not always include the effects of new innovations aimed at improving the quality of services to society or at curbing environmental degradation. Thus GNP measures generally omit the costs or gains associated with non-market use of labour, for example subsistence production, mother care, volunteer work, and promotion of environmental awareness.

Because of the preoccupation with market-valued growth, the fixation on increasing GNP has often been labelled 'growth without development'.

Environmental indicators

Increasing numbers of economists have begun to appreciate the need to replace GNP with other indicators that can more reliably measure environmental health and human welfare. Both environmental scientists and an increasing number of economists prefer to replace or supplement GNP with environmental indicators. Such indicators often include measures of the quality of life for humans and of harmful effects on the environment.

There are specific techniques for environmental indicator determination. For example, the annual change in a nation's quality of life can be measured by an indicator developed by William Nordhaus and James Tobin in 1972 (Miller 1996), which they called *net economic welfare* (NEW). In this procedure a price is calculated for each type of environmental degradation (for example pollution), and these values are then subtracted from the GNP to derive the net economic welfare. David Pearce (1993) estimated that pollution and natural resource degradation subtract 1–5 per cent from the GNP of most developed countries and 5–15 per cent from less developed countries. Another indicator called *net national product* (NNP) was developed by Robert Repetto (1992) of the World Resources Institute. This indicator includes the depletion of natural resources as a factor in the GNP.

The United Nations Development Programme developed a *human development index* (HDI) to estimate average quality of life in various countries. The HDI is the aggregate of life expectancy at birth, literacy rates, and real GNP per capita. Miller (1996) notes that in 1992 Canada had the highest HDI, followed by Switzerland, Japan, Sweden, and Norway.

Another, more comprehensive indicator called the *index of sustainable economic welfare* (ISEW) was developed by Dally et al. (1994). This human welfare indicator adjusts per capita GNP for inequalities in income distribution, depletion of nonrenewable resources, loss of wetlands, loss of farmland due to erosion and urbanization, the cost of air and water pollution, and estimates of long-term environmental damage due to ozone depletion and global warming (Miller 1996)

Doubts remain for ecologists and environmentalists, however, as these measures are still associated with economists and the notion that a calculated monetary value is the basic indicator of development.

Sustainable development

The most widely accepted definition of sustainable development is the one put forward in the Brundtland report of the World Commission on Environment and Development (WCED). In the report, entitled *Our Common Future*, the Brundtland Commission defined sustainable development as 'development that meets the needs of the present without compromising the ability of future generations to meet their own needs' (World Commission on Environment and Development 1987: 43). This definition is based on several conceptualizations, including:

- intragenerational and intergenerational equity;
- the non-decline in resource conditions and consumption over time;
- management of resources for sustained yields of goods and opportunities;
- the stability and resilience of ecosystem conditions over time; and
- the full compensation of future generations by the present generation. The compensation level must be similar in value to whatever resource is lost or impaired.

The definition and principles listed above manage to bring economists and environmental scientists (especially ecologists) together because of the promise of shared stability in economic and ecosystem conditions. The requirements for sustainable development have further resulted in a growing number of economists focusing on human needs in preference to human wants. In theory at least, one would argue that satisfaction of needs requires less or fewer resources than satisfaction of wants.

Several factors have resulted in sustainable development being seen as the overriding principle or goal of socio-economic development:

- the excessive scale of economic activity militating directly against ecosystem carrying capacities;
- the increasing failure of ecosystem capacity to support life and services;
- the threat of species extinction, declining biodiversity, and destabilization of ecological relationships;
- the need to regulate and at times restrict resource extraction within the regenerative capacity of renewable natural resources;
- the failure of the market and other interventions in relation to pricing and property rights;
- persistent poverty among a rapidly increasing number of people; and
- existing throughput-increasing technology consumes too much of the earth's resources, especially those which are non-renewable.

These issues prompted the Brundtland Commission to identify seven crucial objectives for environmental and development policies:

1. Reviving growth.
2. Changing the quality of growth in order to emphasize development rather than growth.
3. Meeting essential needs for jobs, food, energy, water, and sanitation.
4. Ensuring a sustainable level of population.
5. Conserving and enhancing the resource base.
6. Reorientating technology and managing risk.
7. Merging environment and economics in decision-making.

Later we will see that the Brundtland Commission has clearly influenced certain South African policies.

How can sustainable development be achieved?

Various authors and countries have used sustainable development as a principle, concept, strategy, or goal. For example, the South African Department of Environmental Affairs and Tourism (Van Riet et al. 1997: 60) has used sustainable development as both a concept and a methodology to develop an index, an electronic atlas, and hard-copy maps showing spatial variations in environmentally sustainable development across South Africa. The atlas can be used interactively, through overlays, to examine human population, environmental resource sensitivity indices, environmental management indicators, environmental resource conservation, and management requirements. The areas of highest potential for environmentally sustainable development are broadly identified as parts of Mpumalanga, northern KwaZulu-Natal, and the northern part of the Eastern Cape, since these are the areas with a high environmental resource base and low population densities.

Some authors, for example Dovers and Handmer (1992), have been concerned about the urgent need to transform sustainable development from a concept to action. They have identified the following key issues which need to be addressed in order to achieve sustainable development:

1 The paradox of technology.
2 Humility or arrogance in the face of technology.
3 Intergenerational and intra-generational equity.
4 Economic growth versus ecological limits.
5 Reconciliation of individual and collective interests.
6 The balance between democracy and purposeful action.
7 Different styles of resilience.
8 The role of optimization.

Dovers and Handmer argue that these eight issues form the start of an agenda for anyone contemplating a sustainable development strategy.

In the following sections, aspects of policies for sustainable development from South Africa and Manitoba in Canada are examined.

Sustainable development in South Africa

Sustainable development was chosen as the route for South Africa to follow by the first post-apartheid administration. Indeed, in 1994, the government of the Republic of South Africa launched the Reconstruction and Development Programme (RDP) as the flagship policy for economic growth and reconstruction, reconciliation, and harmony between the peoples of the country. The policy sought to provide a fresh beginning in human–environment relationships, and it contains features which are directly consistent with the major recommendations of the Brundtland report on sustainable development.

The RDP promoted five major policy programmes which are interlinked and aimed at the following achievements:

- meeting basic needs;
- developing human resources;
- building the economy;
- democratizing the state and society; and
- implementing the RDP.

According to the RDP policy,

> the RDP links reconstruction and development in a process that will lead to growth in all parts of the economy, greater equity through redistribution, and sustainability. The RDP is committed to a programme of sustainable development which addresses the needs of our people without compromising the interests of future generations. (African National Congress 1994: 9)

The RDP intended to meet basic needs by creating opportunities for all South Africans to develop to their full potential. Furthermore, the policy proposed production and household income increases by means of job creation, productivity and efficiency increases, improving conditions of employment, and creating opportunities for all to sustain themselves through productive activity. Living conditions would improve through better access to basic physical and social services, health care, and education and training for urban and rural communities. Finally a social security system and other safety nets would be established to protect the poor, the disabled, the elderly, and other vulnerable groups.

These goals were to be achieved by implementing various public-sector-driven programmes. Various line ministries were therefore responsible for public works, land redistribution and restitution, housing and service delivery, water and sanitation, energy and electrification, telecommunications, transport, and environmental management programmes. Sustainable resource use is the backbone of environmental management, and the RDP policy pledged to ensure that all South Africans, present and future, have the right to a decent quality of life. The equitable access to natural resources, safe and healthy living and working environments, and a participatory decision-making process were all intended to empower communities to manage their natural environments.

Whether these very laudable goals can be achieved, however, is questionable in a country that became increasingly dependent on the neo-liberal globalization development model with the adoption of the Growth, Employment, and Redistribution (GEAR) strategy in 1996 (discussed in Chapter 5). GEAR is above all else an economic growth model with the usual negative environmental consequences described earlier in this chapter. Furthermore, sustainable development is also dependent upon the effectiveness of new environmental legislation, described in Chapter 19.

Sustainable development in Manitoba, Canada

The two boxes below give a set of principles and guidelines for sustainable development that are derived from the experience of Manitoba, Canada (Mitchell 1997). The application of these guidelines and principles is, however, problematic when a country attempts to launch vigorous economic development programmes while respecting the integrity of the culture of its people and its physical landscape. There is frequently a contradiction between the need for economic development on the one hand, and social, cultural, and environmental sustainability on the other.

Concerns about sustainable development

Sustainable development is such a broad concept that it runs the risk of becoming something for everyone and being used as a justification for any action, whether for economic growth or environmental protection. But vagueness and ambiguity (Mitchell 1997) may also allow a degree of flexibility, which would permit perspectives or programmes to be plugged in or plucked out according to changing perspectives and priorities over time and space. Sustainable development is also rooted in the concept that development is capitalistic and may therefore be rejected on ideological grounds.

Is sustainability possible in the face of ever-rising population and consumption patterns which are themselves dependent

Principles for sustainable development

1. **Integration of environmental and economic decisions.** This is a two-way process that must ensure economic decisions adequately reflect environmental impacts, including human health.
2. **Stewardship.** This requires that we manage the environment and the economy for the benefit of present and future generations. We must recognize that we are caretakers of the environment and economy for the benefit of present and future generations of Manitoba. A balance must be struck between today's decisions and tomorrow's impacts.
3. **Shared responsibility.** All Manitobans acknowledge responsibility for sustaining the environment and economy, with each being accountable for decisions and actions, in a spirit of partnership and open co-operation.
4. **Prevention.** We must anticipate, prevent, or mitigate significant adverse environmental and economic impacts of policy, programmes, and decisions.
5. **Conservation.** This requires that we maintain essential ecological processes, biological diversity, and the life-support systems of our environment; harvest reusable resources on a sustained yield basis; and make wise and efficient use of our renewable and non-renewable resources.
6. **Recycling.** We must endeavour to reduce, reuse, and recover the product of our society.
7. **Enhancement.** We must enhance the long-term productive capability, quality, and capacity of our natural ecosystems.
8. **Rehabilitation and reclamation.** We must endeavour to restore damaged or degraded environments to beneficial use. Rehabilitation and reclamation requires ameliorating damage caused in the past. Future policies, programmes, and developments should take into consideration the need for rehabilitation and reclamation.
9. **Scientific and technological innovation.** Our research, development, testing, and implementation of technologies are essential to further environmental quality, including human health and economic growth.
10. **Global responsibility.** This requires that we think globally when we act locally. Global responsibility requires that we recognize there are no boundaries to our environment, and that there is ecological interdependence among provinces and nations. There is a need to work co-operatively within Canada and internationally to accelerate the merger of environment and economics in decision-making and to develop comprehensive and equitable solutions to problems.

Guidelines for sustainable development

1 **Efficient use of resources.** We shall encourage and support the development and application of systems for proper resource pricing, demand management, and resource allocation together with incentives and disincentives to encourage efficient use of resources and full environmental costing of decisions and developments.

2 **Public participation.** We shall establish appropriate forums which encourage and provide the opportunity for consultation and meaningful participation in decision-making processes by all Manitobans. We shall endeavour to ensure due process, prior notification, and appropriate and timely redress for those affected by policies, programmes, decisions, and developments.

3 **Understanding and respect.** We shall be aware that we share a common physical, social, and economic environment in Manitoba. Understanding and respect for differing social and economic views, values, traditions, and aspirations are necessary for equitable management of these common resources. Consideration must be given to the aspirations, needs, and views of various regions and groups in Manitoba.

4 **Access to adequate information.** We shall encourage and support the improvement and refinement of our environmental and economic information base and promote the opportunity for equal and timely access to information by all Manitobans.

5 **Integrated decision-making and planning.** We shall encourage and support decision-making and planning processes that are open, cross-sectoral, incorporate time horizons relevant to long-term implications, and are efficient and timely.

6 **Substitution.** We shall encourage and promote the development and use of substitutes for scarce resources where they are both environmentally sound and economically viable.

on stocks of resources which are either non-renewable or have potential for depletion? This concern echoes ideas from the laws of thermodynamics. The first law of thermodynamics states that energy can neither be destroyed nor created, but can only be transformed from one state to another. This law is essential for the promotion of conservation on a sustainable basis because it declares that, with the exception of incoming solar energy, we have no stock of resources other than what is available now in the ecosphere. The change of state of resources may be beneficial at times, for example the transformation of coal into heat energy, but the ultimate change will definitely end up in the production of vast amounts of waste. The first law of thermodynamics can be interpreted to mean that we must optimize our use of resources among the present and future generations while at the same time

The first law of thermodynamics: In all physical and chemical changes, energy is neither created nor destroyed, but it may be converted from one form to another.
The second law of thermodynamics: When energy is changed from one form to another, some of the useful energy is always degraded to lower-quality, more dispersed, less useful energy (Miller 1996).

requiring the handling of waste products arising from human production and resource consumption. A change of state of matter and energy also implies that external or unforeseen effects such as pollution may be the inevitable (rather than accidental) consequences of the transformation of matter and energy.

The second law of thermodynamics states that in a closed system, as matter and energy are used, the degree of entropy increases because the amount of concentrated energy and matter declines over time. This raises a fundamental question about the long-term viability of sustainable development before even considering the distinction between non-renewable and renewable stocks of resources. Will human civilization inevitably decline once global stocks of energy and matter reach maximum entropy or become dissipated? Neither the second law nor sustainable development as a concept specifies time frames for the inevitable collapse of the ecosphere or for achieving sustainable development. The issue of time and space scales is important for defining and particularly for predicting sustainable development. Robert Constanza (in Miller 1996: 8) observes that 'when one says a system has achieved sustainability, one does not mean an infinite lifespan, but rather a lifespan that is consistent with the system's expected time and space scale'. But what is that expected time and space scale?

Understanding sustainability requires clear specifications of the goals of sustainable development in relation to the maturity of a given economic and ecological system. Once the full lifespan of a mature economic and ecological system is known, sustainability is achievable within the lifespan of the system. Sustainability should not be associated with an indefinite or infinite time or space scale if something practical is to be gained.

The ultimate dissipation of global energy and matter as predicted by the second law of thermodynamics may never occur during the lifespan of the human species. What is likely to happen is that, through resource substitution and other innovations of both inputs and outputs, the accompanying shifts may postpone the ultimate collapse of the total global system, thus allowing us to achieve relative sustainability in some but not all systems at a time.

On the basis of physical laws, the concept of equity appears untenable or unnecessary except on moral or ethical grounds. The environment for sustainable development is an interactive system guided by flows of energy and matter. Newton's law states that interaction between two phenomena is proportional to the product of their mass and inversely proportional to the square of the distance between them. This law implies that in interactive systems uniformity of mass is not necessary, and that interactions occur depending on the distance between the masses. There is therefore no readily perceivable analogue to equity in interactive physical systems or in Newton's model. Equity is based purely on ethical or moral grounds which can be determined by various ideologies and cultures which may be dominant at different times.

The final concern is about the relationship between the present and future generations. In the strategy for sustainable development it is often argued that, to be fair and to achieve equity, the present generation must pass the same stock of resources existing during its own time to future generations. But how can this happen when sustainable development also allows for resource substitution, which implies the loss of some stock of resources? Economists have argued that the needs of the future generations can still be realized by compensating them with the equivalent, even in monetary terms, of what has been lost by the present generation (Mitchell 1997; O'Riordan 1995; Perman et al. 1996). Two problems arise during such compensation. The first concerns the fairness of substituting resources with the equivalent resource. Some resources such as landscape beauty are so subjective and personal in value that their equivalence is inconceivable. The second concerns the bias involved in determining what is equivalent and beneficial to future generations.

Furthermore, sustainable development analysts have not been specific about the distance in time between the present and future generations. Which generation do we pass resources to? The assumption is perhaps that a generation passes resources to the following generation. But for resources to sustain their inherent quality and value across successive generations, the ecosphere must demonstrate unforeseeable resilience and stability in character and processes.

Ultimately, these concerns leave the impression that sustainable development must be implemented cautiously, because it is more of an ideal yardstick against which present and future resource development and management initiatives must be measured and compared. Sustainability, as suggested, is a state of development relative to the lifespan of a particular economic and ecological system. It should not always be treated in absolute terms.

18 People, recreation, and the environment

LINDISIZWE M. MAGI

The coming of the third millennium has served to focus our attention on the long-term survival of our natural resources and environment. It is in this connection that this chapter seeks to reflect on the relationships between people, outdoor recreation, and the environment. The demand for the use of recreation resources within the natural environment has attracted much attention in recent times because of increases in potentially destabilizing factors such as population growth, higher disposable income, greater leisure time, higher mobility, and improved education and standards of living (Carlson et al. 1979; Torkildsen 1986; Weiskopf 1982). This demand has resulted in the overuse of the natural environment and a need for better environmental management practices.

The history of natural or outdoor recreation resource utilization in South Africa is both short and sporadic. This is in part due to the relatively new and increased demand for recreation on a large scale. As the population grows and socio-economic changes take place, a completely different recreation scene is emerging. What has been observable in South Africa, particularly before the advent of democracy in 1994, is that most of the outdoor or natural recreation resources and facilities were solely used by and reserved for whites. The distribution of these resources is shown in Figure 18.1. This exclusiveness was not achieved and encouraged by discriminatory legislation alone, but was also a factor of the socio-economic inequalities prevalent in the country. These inequalities were and are manifest in such areas as education, employment, health, residence, and social and outdoor recreation facilities. Many writers in this field (Donaldson 1995; Magi 1986; Mahaye 1996; Mwandla 1995) argue that recreation should be recognized as a common human need, the satisfaction of which must be equitable and not specific to limited individuals or social or ethnic groups.

Considering that the geographical and environmental issues associated with this subject are numerous, broad, and far-reaching, the discussion in this chapter will focus on achieving four fundamental objectives:

FIGURE 18.1 Location of major outdoor recreation resources in South Africa

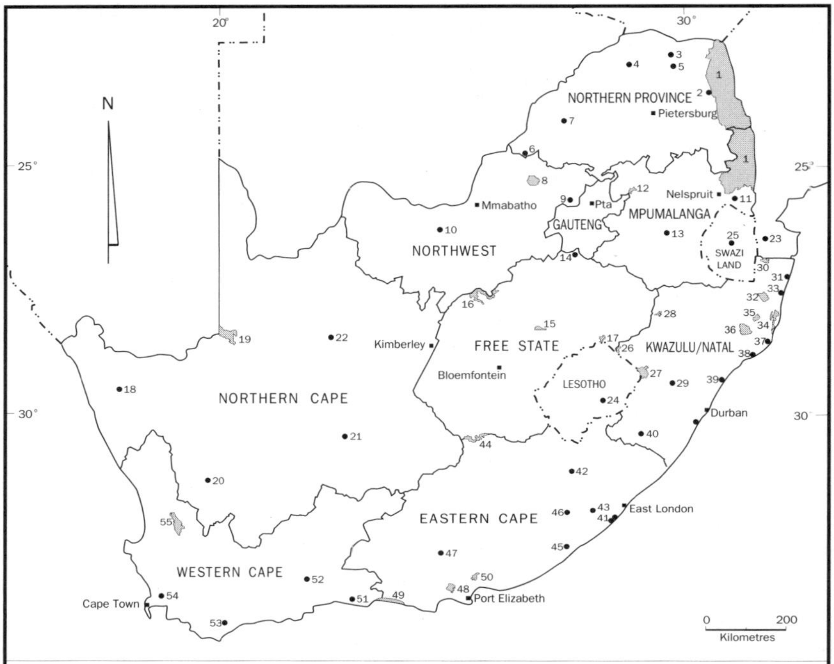

Northern Province
1. Kruger National Park
2. Great Letaba Game Park
3. Messina, Honnet & Njallele Nature Reserve
4. Langjan Nature Reserve
5. Magoebaskloof Hiking Trail
6. Pilansberg Nature Reserve
7. Hans Strydom Dam Nature Reserve

Northwest
8. Rustenburg Hiking Trail
9. Hartebeespoort Dam Nature Reserve
10. Lichtenburg Nature Reserve

Mpumalanga
1. Kruger National Park
11. Malelane Nature Reserve
12. Loskop Dam Nature Reserve
13. Douglas Dam Nature Reserve

Free State
14. Vaal Dam Nature Reserve
15. Erfenis Dam Nature & Willem Pretorius Game Reserve
16. Sandveld Nature Reserve
17. Golden Gate Highlands National Park

Northern Cape
18. Hester Malan Nature Reserve
19. Augrabies Falls
20. Akkerendam Nature Reserve
21. Great Karoo Nature Reserve System
22. Sishen Nature Reserve

Lesotho & Swaziland
23. Lebombo Mountain Range
24. Lesotho-Drakensberg Mountains
25. Mbabane Nature Reserve

KwaZulu/Natal
23. Lebombo Mountain Range
26. Royal Natal National Park
27. Giant's Castle Game Reserve
28. Chelmsford Public Resort Nature Reserve
29. Umgeni Valley Nature Reserve and Howick Falls
30. Ndumu Game Reserve
31. Kosi Bay Nature Reserve
32. Mkuze Game Reserve
33. Sodwana Bay National Park
34. Great St Lucia Wetland Park
35. Hluhluwe Game Reserve
36. Umfolozi Game Reserve
37. Richards Bay Nature Reserve
38. Umlalazi Nature Reserve
39. Durban Natural Beach
40. Coleford Nature Reserve

Eastern Cape
41. Cwebe Nature Reserve & Dwesa Nature Reserve
42. Tsolwana Game Park
43. East London System of Nature Reserves
44. Hendrik Verwoerd Dam Nature Reserve
45. Kowie Nature Reseve
46. Mountain Zebra Hiking Trail
47. Karoo Nature Reserve Trail
48. Groendal Wilderness Area
49. Tsitsikama Coastal National Park
50. Addo Elephant National Park

Western Cape
51. Knysna National Forest Reserve
52. Cango Caves Nature Reserve
53. Bredasdorp Nature Reserve
54. Table Mountain Nature Reserve
55. Cedarberg Wilderness Area

- to clarify the meaning of various concepts important for understanding relationships between people, recreation, and the environment;
- to identify some of the local writers and literary materials used in the study of outdoor recreation and, to a lesser extent, in environmental studies;
- to trace some of the historical accounts describing the cultural principles and values of various societies and regimes as they relate to recreation and environmental affairs; and
- to give some illustratory discussion by citing recreation- and environment-related case studies from South Africa and the USA.

The basis for discussion in this chapter is the premise that, since the history of spatial and socio-economic inequalities in South Africa has favoured whites, African people may well have a negative attitude towards the usage of natural outdoor recreation resources and environments, such as beaches, forests, game reserves, wilderness areas, rivers, lakes, and estuaries. The primary significance of discussing this subject matter is the need to learn more about South Africans' perceptions of the meaning and operation of their environment, whether they be educated or not. For example, the recreationist or environmentalist needs to understand that designating a natural area as 'wilderness' is as much an act of cultural preference as the industrialist bulldozing a hillside for a superhighway or industrial plant. If in the near future significant socio-political changes occur that directly affect the majority of South Africans, then the problem under discussion will, without doubt, become even more important for decision-makers and outdoor recreation and environmental planners.

This chapter is modelled in such a way that it takes a recreation–geographical perspective and at the same time gives two types of information sources. The first is conceptual, concerned with the history of, and research in, the field of outdoor recreation and environmental affairs. The second is derived from case studies which aim to introduce a practical basis into the discussion. It is anticipated that this chapter will contribute to the theoretical fields of outdoor recreation geography and environmental studies, as well as providing additional information about the relationships that exist between people, recreation, and the environment (Adams 1990; Brockman 1979; Hugo et al. 1997; Larkin et al. 1981; Magi 1989b; Pigram 1983; Sarre and Blunden 1995).

Explanation of concepts

It is essential in this area of geography to explain some of the more frequently used concepts. This chapter uses the term 'perception' to describe people's judgements, attitudes, goals, feelings, and beliefs towards outdoor recreation and environmental usage and management. It is important to understand what the terms 'leisure' and 'recreation' mean in the context of this discussion. These terms are closely interrelated and are used interchangeably though not synonymously. The terms at times mean different things to different people, and generate some confusion if used indiscriminately. The simplest distinction is for us to identify leisure with time (specifically with unobligated time), and recreation with activities undertaken voluntarily for pleasure and satisfaction during that time (Pigram 1983). This explanation is not always true; at times recreation is associated with free time in Western societies and obligated time in African societies.

Another important term is 'outdoor recreation', which is used in a general sense to include activities associated with both the natural and cultural environments. As a result, the terms 'outdoor recreation' and 'natural recreation' have often been used synonymously to denote recreation activities which take place out of doors in a natural setting (McCall and McCall 1977). A definition which attempts to give a more specific explanation of outdoor recreation is given by Kraus (1966: 130), who states that '[i]t should consist of those recreational activities that can best be carried on out of doors, and which have in some way a direct relationship or dependence on nature, or that place the participant in direct contact with the elements'. Without doubt, people's recreational activities and the natural environment are the core elements of outdoor recreation.

Also important for our discussion is the term 'sustainable development'. As we saw in the previous chapter, this was defined by the World Commission on Environment and Development (often called the Brundtland Commission) as meeting the needs of the present communities 'without compromising the ability of future generations to meet their own needs' (World Commission on Environment and Development 1987: 8). The concept 'sustainable outdoor recreation', as is the case with 'sustainable environmental development', relates to the use of nature-based facilities or the pursuit of leisure activities in such a manner that the people–environment relationship is kept in ecological balance and is not compromised by overuse and environmental degradation (Cooper et al. 1996; SARDC 1994).

The term 'environment' is widely used and means many things to different people. In this chapter it broadly refers to all external situations which affect people and organisms in their lifetimes. More specifically, it suggests the natural setting within which people participate in activities, one of which is outdoor recreation. The concept of the sustainable environment reflects a belief that the natural environment is a place with finite space and resources, and that continued population growth, production, and consumption will ultimately put severe stress on natural processes (SARDC 1994). These processes are needed for the natural environment to renew itself and thus maintain the resource bases of air, water, and soil that support life (Miller 1988). It cannot be overemphasized that people need to work with nature and not against it in order to prevent environmental overload and resource depletion. As was discussed in Chapter 17, there is a need for people to control population growth as well as to reduce the unnecessary use and waste of natural resources.

Another important term to define is that of 'African', which is used in this chapter synonymously with the term 'black' and refers to the indigenous people of South Africa. The concept 'township' refers to African residential areas adjacent to major towns in South Africa (these are discussed more fully in Chapter 8).

Studies of outdoor recreation and environmental issues

To date, relatively few studies (Kruger et al. 1997; Magi 1992; Preston 1983) have focused attention on outdoor recreation in combination with environmental issues. In considering what has been written on this subject, let us first examine the basic outdoor recreation literature available to geographers. Furthermore, we can look at what is available in studies of environmental affairs, and, finally, we can consider some geographical perspectives on recreation behaviour and the relationship between people, recreation, and the environment.

Behavioural recreation studies

Outdoor recreation research has gone beyond the stage of listing recreation activities and outlining the 'what' of them. In general, there is now a need and a willingness to apply the theories and concepts of other disciplines in order to reveal the 'how' and 'why' of recreation activities and experiences. If recreation studies could move from undiluted theory to practice in understanding the use of the natural environment, this would be useful for the managers of environmental systems.

The most common approach to the study of outdoor recreation and leisure has been to investigate recreation activities. This is the 'easy way out', because activities are concrete acts. They have a specifiable duration, they can be observed, and they represent actual behaviour. Meyersohn (1969) lists three ways in which activities have been studied:

- the amount of time used for them;
- the amount of money spent on them; and
- the amount of interest in them.

These three approaches offer insights into the use of, and demand for, outdoor recreation activities, but usually fail to reveal the inner experience and behaviour of the recreationist. Nor can they reveal the cultural values and beliefs associated with interest in outdoor recreation activities and the natural environment. In Chapter 1 we read that these are the concerns of researchers using the behaviouralist approach.

An example of an international study seeking to establish reasons why people participate in outdoor recreation activities is that conducted by Crandall (1979). He investigated 222 respondents to show that social interaction is an important component of outdoor recreation activities and that individual perceptions are more important in understanding recreation behaviour than are traditional measures such as hours of participation, expenditure on activities, or social class variables.

Crandall's (1979) work is supported by Pigram (1983: 24), who states that perceptions or personal judgements 'are a function of the perceiver's past experience, present values, motivation and need'. The studies cited above, and those of Mercer (1971, 1977) and Neulinger (1981), are theoretically valid and relevant in their emphasis of the behavioural approach in understanding the use and management of outdoor recreation and the environment.

Outdoor recreation and the environment

Only a few years ago, studies on people and natural or outdoor recreation were relatively rare. Since the 1970s there has been an increase of interest in looking at people's aspirations and motivations for participating in outdoor recreation. The study by the Outdoor Recreation Resources Review Commission (ORRRC 1962) is regarded as the first study to tackle the subject in a comprehensive manner.

Outdoor recreation is, by definition, natural-resource-related, and involves recreation activities which need to be managed in a sustainable manner. Recently, attention has been given to the situation in which recreation takes place as a major influence on people, causing them to experience pleasure. There is general agreement in studies of outdoor recreation that all human action should be interpreted with reference to the ecological environment in which most of the recreation activities occur. In support of this approach, many surveys (for example Price 1979) have been undertaken to elicit people's feelings about the recreation experience in natural or outdoor areas.

People and the environment

Environmental issues in South Africa have been dominated and influenced by white middle-class values. Among other things, the aim of this chapter is to draw attention to people–environment relationships, as well as to explore the feelings of Africans towards environmental issues. For our purpose, environmental issues are defined as those relating to the impact of human interventions on nature and natural processes, interventions which pose a threat to human existence in general and outdoor recreation in particular (SARDC 1994).

There are several ways of explaining the relationship between people and the environment. The key paradigm operating in this field, identified by Knill (1991) as the dominant Western environmental approach, sees humans as separate from and superior to nature. It discounts the role of God and sees man as the supreme operator, who thus is superior to all the other constituents of nature.

The dominant Western environmental approach is in conflict with what we can call Africanist ecology. This approach advocates a closer relationship between people and nature, and incorporates principles of *ubuntu*, or African humanism. Its main argument is that humans use and sustain the God-given natural resources which are primarily under the stewardship of the community. People are expected to use natural resources for the benefit of family and community. Approaches similar to Africanist ecology have begun to emerge in recent Western writings, in the form of social ecology, deep ecology, and eco-feminism (Cotgrove and Duff 1980). Space constraints do not permit any extended discussion of these approaches.

The environmental awareness which has developed in South Africa has, for the most part, been taken up by a few voluntary and issue-driven agencies, and, to a large extent, the media. Until recently there has been little official participation by education, health, and political structures. Environmental issues which have generated considerable public outrage and debate in South Africa, and particularly in KwaZulu-Natal, include:

- the use of pesticides;
- river and air pollution by paper-making companies;
- the dumping of chrome chemicals near Durban;
- dune mining in St Lucia;
- forced removals in Maputaland;
- the occupation of Dukuduku Forest;
- deposits of effluent in the sea at Richard's Bay; and
- wildlife poaching in Hluhluwe game reserve.

The effects of these environmental issues are suffered more acutely in African communities than elsewhere, yet it is apparent that Africans have not started addressing these problems.

Over the past two decades there has been an increased interest in environmental issues among South African geographers. Three research agendas geared to encouraging environmental awareness have been identified:

1 Nature experience through communicative approaches.
2 Environmental programme design and planning.
3 Environmental curriculum development (Cotgrove and Duff 1980; Knill 1991).

It seems, however, that fewer studies have paid attention to environmental factors relating to outdoor recreation and African community issues. The Southern African Research and Documentation Centre (SARDC 1994) has made a concerted effort to produce a comprehensive report about the state of the environment in southern Africa.

Recreation studies in South Africa

It has already been suggested that there are few studies of how South Africans, and particularly African people, perceive natural recreation resources. Much of the material reviewed in this chapter has been written by human geographers working in the fields of recreation geography and environmental studies.

A ground-breaking attempt in the field of recreation geography in South Africa was the recreation review undertaken by Steyn (1976). His work, together with the bibliographic inventories produced by the Human Sciences Research Council (1984), indicates that this field includes a variety of contributors such as planners, state and semi-state agencies, social and natural scientists, and university-based research institutes. The organizations listed include the Department of Environmental Planning and Energy, the Department of Sport and Recreation, the Department of Nature and Environmental Conservation, the Natal Town and Regional Planning Commission, the Natal Parks Board, and the Wildlife Society of South Africa. A total of fourteen recreation geographers have worked in this field directly or indirectly under the auspices of some of these organizations, but their contributions remain small in the face of the research in outdoor recreation which still needs to be undertaken.

Over the past three decades, and notably during the 1980s, there has been an increase in outdoor recreation studies, which have converged into five broad areas.

Physical aspects of the recreation system

This type of study places emphasis on the regional occurrence of outdoor recreation resources; examples are Gibbon (1976) and Steyn et al. (1982).

Outdoor recreation behaviour patterns and perceptions

These studies pay attention to South African people's attitudes, perceptions, preferences, and general behaviour towards recreation facilities and activities. Examples of research can be found in the work of Butler-Adam (1981, 1986), Magi (1986, 1989a, 1992), Mahaye (1996), and Mwandla (1995).

Conceptualization of recreation processes

The few studies conducted in this area have attempted to generate more theoretical and conceptual frameworks based on local issues, and they have emphasized recreation in the context of environmentally sustainable development processes. Contributors in this field include Butler-Adam (1981, 1984), Kruger et al. (1997), Lamont (1983), and Magi (1986, 1988, 1989b).

Recreators or recreation users in action

Studies in this category describe the action space of recreators or recreation users; authors include Hugo (1974), Richards (1994), Sutcliffe (1981), and Van der Wal and Steyn (1981).

Outdoor recreation systems and tourism

The studies in this area attempt to integrate outdoor recreation systems with tourism; writers include Ferrario (1981), Fowkes (1992), and Kock (1993).

What is most striking in this summary of local studies is that little attention has been paid to recreation issues relating to Africans. This situation cannot be blamed entirely on apartheid, but also on the scarcity and inactivity of African writers in outdoor recreation matters. It needs to be mentioned, however, that recreation is a modern phenomenon which is completely different, both conceptually and spatially, from what used to exist before the emergence of an industrialized market economy in South Africa. Notwithstanding these varied restraints, it is essential to understand outdoor recreation as interacting with the natural environment within which people act out their life experiences. It should be enjoyed by all, and not limited to individuals or groups in any particular society.

Values and the environment

The most interesting thing about human behaviour is that it relates to the environment by determining or influencing what reality is going to be like. It is generally accepted that different cultures perceive and classify the world in varied ways by using different symbols. These help to give meaning to the world (the way that scientists do this is discussed more fully in Chapter 1). To a large extent, natural resources in South Africa are viewed from Western cultural perspectives (Burger 1981), as well as the functional choices which have come to be known as the dominant Western environmental paradigms (Catton and Dunlap 1978). Resources are cultural creations, however, and therefore should also be seen from an African value perspective (SARDC 1994). This notion is gaining acceptance among social scientists, because they are beginning to realize that even in so-called simple societies the role of cultural values in resource evaluation, use, and management is vital.

Our understanding of current African values and of Africans' interpretation of the natural environment can be delimited in four distinct periods: African indigenous accounts, colonial accounts, apartheid-related accounts, and new South African accounts.

African indigenous accounts

In establishing a link between people, culture, and nature, it is necessary to pose a central question: what bearing do African values and beliefs have on our perception of the natural recreation environment? The relationship between Africans and the natural environment in the pre-colonial period needs discussion. It is worth noting that pre-colonial life in South Africa was recorded by European historians, missionaries, adventurers, and explorers, who brought their bias to bear on indigenous achievements, artefacts, and technologies. As a result, little is known of the ecological strategies and environmental perceptions of African pastoralists. Equally, documentation does suggest that colonization brought with it a serious deterioration in people–land relationships (Kay 1970). This is evident in that a change from traditional foodstuffs to maize monoculture and cattle-keeping, and the laying waste of the natural environment, increased the pressure on the land.

There is without doubt a paucity of accounts written by Africans about their understanding of natural and environmental resources. Many of the cited accounts have been passed on through the oral tradition (Paden and Soja 1970). Philosophical relationships and coexistence between people, the environment, and nature have been recorded over the last eighty years in the creative literature of African scholars, starting with Molema (1920) and more recently including Nyembezi (1954), Vilakazi (1965), Ndebele (1972), Manganyi (1973), Mphahlele (1974), and Bengu (1975). These writers describe the following three cultural beliefs:

1. People and nature are part of an inseparable continuum and are not two independent and opposing realities (Mphahlele 1974; Ruch and Anyanwu 1984).
2. The land in traditional African society is communally owned and there is collective societal responsibility (Vilakazi 1965).
3. African people's family names and eulogia indicate their destiny and are related to the natural and physical environment in which they find themselves (Molema 1920).

Philosophically speaking, the African's respect for the land is a point of reverence of life. This deep respect is usually maintained to the point where the exploitation, sale, and destruction of land are actively discouraged. In contrast, the arrival of Europeans in South Africa introduced a period of buying and selling of land. It is from that period that land came to be regarded as an economic possession and an 'object' of exchange.

In a traditional African situation, no distinction is made between an act or behaviour, on the one hand, and a belief or a system of beliefs on the other. Accordingly, what people do is motivated by what they believe, and what they believe springs from what they do and experience. This is one of the intriguing aspects of African beliefs: the absence of a concrete distinction between the spiritual and the physical. In some African communities in South Africa, for example, aspects of the physical natural environment are seen in a spiritual way as a uniting force and as responsible for making what is alive or constitutive of life itself. To illustrate, a physical feature such as the sea is seen as a constituent of life. People immerse themselves in the sea to rid themselves of ailments and negative forces – so-called bad luck. Furthermore, some African Christian churches baptize their members in the sea to provide a more spiritually based experience than would be the case if they were to use other sources of water.

Colonial accounts

Let us now consider, within the limits of available objective information, early African attitudes to nature and the environment in relation to broad African values. Land, flora, fauna, and recreation are key pointers of environmental awareness emerging from early colonial writings. Land in African communities of those days was, and still is, communally owned. A slight exception to this practice occurred in cases where land was loaned out to new families for subsistence purposes. In contrast to this situation, colonial writers reported that land around St Lucia in colonial Natal was rough and luxuriant (Ludlow 1882) and was sold by Europeans for as little as four shillings an acre (Jenkinson 1882). They considered this environment to be hostile and ready for conquest.

Accounts from early explorers indicated that the flora was characterized by a large variety of tropical trees that grew with rich luxuriance. This kind of environment was used by local indigenous inhabitants but was not, it seems, overused and wasted in any way. Several writers (Jenkinson 1882; Ludlow 1882) reported on the cultural and symbolic significance of plants to most African communities. For example, the annual rite of the first fruits, *ukweshwama*, was performed by community leaders, and only after this was it regarded as permissible to gather the new fruit, herbs, and other edible plants of the season. This ritual had both a spiritual and functional significance for the natural recreation landscape. The few remaining indigenous forests such as Ongoye and Dlinza in KwaZulu-Natal still contain native vegetation of ornamental trees, shrubs, and herbs. The forests were heavily exploited for rare ornamental trees such black stinkwood, yellowwood, and cycads by commercial lumber companies (Huntley 1965).

It is common knowledge that the influx of Europeans to South Africa heralded the introduction of an anti-conservation era encouraged by conflicts over space and the rule of the gun. This resulted in the indiscriminate shooting of animals (Bryant 1929, 1949), thus bringing several species of big game to the verge of extinction. Animal extermination expeditions are known to have been very common, organized from time to time in the name of 'taming nature' and 'civilizing' the African landscape. An example is the infamous slaughter of 1860, led by Prince Alfred, Duke of Edinburgh, in which 600 head of game were killed in just a single day (Comrie-Greig 1984). Such exploitation and waste were inconceivable within traditional African values and norms founded on a sense of people's solidarity with nature, that is, the coexistence of people and nature.

It has been reported that many African communities engaged in organized hunts for subsistence purposes. According to Bryant, however, it is significant to observe that,

> Among the well-ordered and well-disciplined people as were the Zulus, a public hunt was by no means the mere disorderly dispersal of a chasing, slaughtering rabble of savages, as one might suppose. It was a systematically and scientifically arranged ... and devotional exercise. (Bryant 1949: 682)

It is apparent from the above that African people traditionally treated wildlife with respect and prudence, which suggests that there was a healthy relationship between people and the natural environment in those societies. It also seems reasonable to conclude that hunting as a form of outdoor recreation was not widely practised, since hunts were associated with subsistence living.

Colonial writers described outdoor recreation very superficially, and they associated it only with hunting. It is important to note that recreation activities were not categorically separated from what can be regarded as subsistence-orientated activities. In fact, outdoor recreation activities were, and perhaps still are, interwoven into the general fabric of life. This is still apparent in activities that integrate work and non-work situations, obligated time and non-obligated time situations, and which involve more subjective than objective perceptions of the recreation act.

Apartheid-related accounts

With the passage of time and through the processes of industrialization and modernization, blacks seem to have assimilated some European perceptions and perspectives of outdoor recreation. During the apartheid era, however, many outdoor recreation facilities were segregated to prevent social contact between the races. Restrictions were achieved by means of the Separate Amenities Act of 1953, the Group Areas Act of 1955, and the earlier-promulgated Land Act of 1936. Under these laws, many blacks were restricted from using outdoor recreation facilities and participating in environment-related activities. It is noteworthy that there has been a shift in access and control since the advent of the new democratic order in South Africa (this is summarized in Figure 18.2).

FIGURE 18.2 The shift in recreational control: before and after the Separate Amenities Act

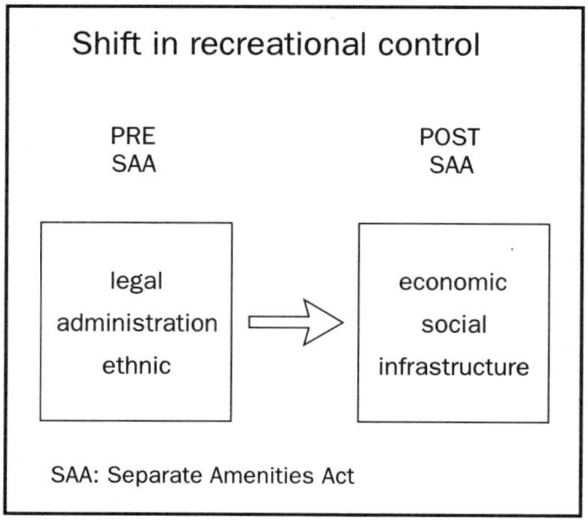

It can be argued that the Western-orientated lifestyle of blacks is a result of the experiences they endured under apartheid (Burger 1981). These experiences include having to:

- comply with the processes of Western urbanization and industrialization (Hugo 1974);
- contend with legalized discrimination and spatial inequalities in everyday living (Magi 1992; Donaldson 1995);
- endure being landless, homeless, unemployed, and kept away from their natural habitat and family (Ramphele 1981); and
- suffer the indignity of being denied entrance into recreation areas and facilities (Butler-Adam 1981; Magi 1992).

Under apartheid, blacks in South Africa were painfully aware of the relatively luxurious and easy way of life their white counterparts led. This perception has provided sufficient motivation for them to aspire to the same type of life. Unfortunately this one-way cross-cultural influence tends to be to the disadvantage of the natural environment.

New South African perspectives

The new democracy in South Africa has introduced different approaches to outdoor recreation activities and environmental management planning. This transformation, which has been promoted by the Reconstruction and Development Programme, as well as by several Green Papers and White Papers associated with recreation, sport, tourism, environmental policy, and environmental impact management systems, is expected to improve recreational lifestyles in South Africa. The broad aims of these policy documents are to empower the previously disadvantaged communities, to improve their physical environment, and to broaden participation in the planning, usage, and management of natural recreation resources and facilities (South Africa 1993, 1994, 1996a, 1996b, 1997).

Furthermore, these policy documents have introduced new concepts, agendas, and thinking to South Africans. Environmental policies are now being guided by principles such as accountability, affordability, equity, inclusivity, and sustainability. They address the inequalities between different communities in terms of access to recreational and environmental opportunities, and they attempt to institute an all-inclusive outdoor recreation and natural environment participation policy. Democratization of participation is now a basic principle in recreation and environmental policy formulations, and the policies have introduced sustainable resource management and environmental impact assessment practices. Finally, there is both a concern for and an awareness of concepts such as biodiversity, conservation, environmental awareness, ecotourism, and environmental rights (Coke 1983; African National Congress 1993; SARDC 1994; Grove 1995).

The introduction of environmental impact assessment (EIA) procedures, for example, is aimed at assisting decision-makers with guidelines for achieving sustainable development of the environment. According to Lengoasa (1997: 4), 'an EIA is a process to gather and evaluate environmental information, so as to provide sufficient supporting arguments to evaluate the overall impact' on the environment. The collected information is meant to assist decision-makers to make value judgements in choosing one development project over another.

Case studies in recreation environment

It is important to understand the existing relationship between people and the environment, particularly the environment which is used for recreation purposes. If the relationship is well understood, then the environment can be used in a sustainable manner – that is, used in such a way that it is not harmed irreversibly. As a result, people could continue to afford to maintain the environment, and it would continue to remain habitable. In order to place this relationship in its context, four case studies covering experiences of outdoor recreation, one in a Western country and three in South Africa, are discussed. The South African case studies were written in the last two of the recreation periods delimited above: the apartheid and post-apartheid periods.

The Washington State case study

This study on outdoor recreation, entitled *Aquatic Resources and Recreation Behaviour* (Cheek and Field 1977), was undertaken between May 1975 and May 1976 in the State of Washington in the USA. The aim of the study was to examine emerging relationships between resource bases and recreation behaviour, mainly using water-based recreation activities and aquatic resources. The resource bases identified were river (such as shown in Plate 54, colour section), lake, ocean, swamp/marsh, forest, mountain, farm, and town/city. There were five outdoor recreation activities: water; nature study and food gathering; hiking, camping, and picnicking; recreation vehicle driving; and sports and games. The questions to be answered concerned the number of outdoor recreation activities per individual resource.

Study design

A random sample of Washington households was identified for investigation, and 7 287 respondents were established for mailed questionnaires and telephone interviews. The sample was divided into three groups:

- those participating as eligible household individuals;

- those who participated alone or with someone else; and
- those who participated on the basis of an activity event (there were 31 649 activity events for 3 048 households).

Findings

The findings of this study were as follows:

- All resource bases, with the exception of swamp/marsh, were the focus of a large number of different outdoor recreation activities.
- The water-related resource bases were found to be more popular than the mountain, farm, and town resource bases.
- Water-based activities constituted 14–30 per cent of all outdoor recreation activities taking place.
- The largest proportion of water recreation activities (30 per cent) occurred at the lakeside; the second most (28 per cent) were found to be at the ocean.
- The recreational vehicle and the hiking, camping, and picnicking groups of activities tended to be equally distributed across three or more resource bases.
- The human community resource base of town/city was not only an important locale for sport and games, but was also important as a locale for the nature study and food-gathering group of activities.
- It was observed that water-related resource bases also received about 28 per cent of the outdoor recreation participation of households not reporting any water activity participation.

What can be inferred from this study is that in a Western country water-based recreation activities are far more important than food-gathering activities. This would generally not be the case in South Africa. The water-based recreation activities are also strongly associated with recreation vehicle driving, sports and games, as well as hiking, camping, and picnicking activities. It would be interesting to note the results of such a study if it were to be replicated in South Africa or a similar developing country.

The Albert Falls public resort study

The recreation case study of the Albert Falls public resort (KwaZulu-Natal) by Butler-Adam (1981) sought to establish behavioural patterns of people using the resource, to reveal who used the facilities (see Plate 55) and what recreators felt about the facilities and how they would like to see them developed.

Study design

The case study was designed to have two major samples, namely 6 761 recreators on-site and 3 970 respondents in the service area around the resort, comprising potential recreators and members of the public. The sample was taken from all population groups, since all were allowed to use the facility.

Findings

After a rigorous computer analysis, the study arrived at the following findings:

- The resort was open to all population groups, yet it was predominantly used by whites (49 per cent) and Indians

(41 per cent), and less so by coloureds (8 per cent) and blacks (1 per cent).
- The image which the resort had earned was that it was predominantly for groups and families who were young and middle class.
- To most whites, Albert Falls was part of an array of recreational opportunities at their disposal, whereas to coloureds, Indians, and blacks it was one of the very few places they could go to that was near at hand.
- All groups were attracted by the relaxation and water-based qualities of the resort. For the less privileged groups the fishing potential was the most important, reflecting their socio-economic conditions.
- On the question of sharing the facility, the general feeling favoured integration of facilities and activities. However, negative separatist notions were expressed by some white and Indian respondents.

In conclusion, Butler-Adam (1981: 22) remarks as follows: 'Yesterday's practice of the segregation of people on the basis of colour has been, and continues to be, replaced by a need for the segregation of activities on the basis of simple human experience'. What can be inferred here is that there is a need for more case studies that seek to reveal the preferences of all South Africans, and perhaps blacks in particular. The use and management of outdoor recreation resources can be understood only if behavioural patterns of people in the context of their environment are known.

The Northern Province study

The next case study (Donaldson 1995) focused on urban blacks in the Northern Province and is a post-apartheid assessment of outdoor recreation resources. The aim of the study was to determine three things: first, the distribution of outdoor recreation resources within the province; second, urban black people's outdoor recreation needs, preferences, and demands for facility-orientated outdoor recreation resources; and, finally, considerations towards forward planning.

Study design

Research data were collected from outdoor recreation administrators, recreation agencies and institutions, and urban dwellers. The sample consisted of six recreation authorities and 355 urban dwellers randomly selected from six townships. The activities investigated included hiking, picnicking, camping, caravanning, and visiting game reserves and nature reserves.

Findings

The findings of the study related to the supply and demand, administration, and planning of outdoor recreation resources:
- Camping and caravanning facilities were found to be under-utilized and over-supplied in the Northern Province.
- Outdoor recreation activities such as camping and caravanning were not popular among African township dwellers, whereas picnicking and visits to game and nature reserves were more popular.

- Outdoor recreation planning in the province was inadequate, as a result of the apartheid-related spatial fragmentation and duplication of services. The post-apartheid integration of urban authorities has been slow.
- The outdoor recreation imbalances need to be addressed through affirmative planning approaches in areas such as education, marketing, public participation, and structural and physical changes.

The study recommends that the imbalances of the past can be overcome by public participation and reorganization of administration. Also important is the adoption of a holistic affirmative planning approach in education, marketing, tourism, and urban outdoor recreation development. From this case study it can be concluded that Africans in the study area have lost touch with environmental issues, and that these issues are influenced by industrial and urban lifestyles.

The environmental concepts and issues study

This case study (Magi 1993) was undertaken in the Lower Umfolozi–Mtunzini districts of KwaZulu-Natal (see Figure 18.3). The aims of the study were to identify African perceptions of environmental issues, to reveal the extent to which environmental concepts were understood, and to indicate the extent to which school and college environmental curricula were seen to address environmental issues.

Study design

Information was collected from 433 randomly selected respondents. This sample was stratified into teachers, students, and the general public. Environmental issues considered for perceptual analysis included the following: food and water resources, energy, environment-related diseases, land degradation, pollution, population increase, nature conservation, pesticides, and toxic and nuclear waste dumping.

Findings

The results of this study reflected what the subjects felt about environmental concepts and issues, rather than what outside observers believed they knew about what such subjects wanted. The findings were as follows:

- The meaning of environmental issues among Africans has been influenced by aesthetic rather than functional choices, as well as by the dominant Western environmental paradigm.
- When asked to respond to the following definition of environmental issues – 'human practices and interventions responsible for negative impacts on the natural and human environments, which pose a threat to human existence' – almost 60 per cent of the public and 40 per cent of students and teachers did not understand the concept of environmental issues.
- In ranking social issues which they considered important, the respondents ranked education, housing, and jobs high, while environmental issues were ranked low.

FIGURE 18.3 Recreational resources in the Lower Umfolozi–Mtunzini districts

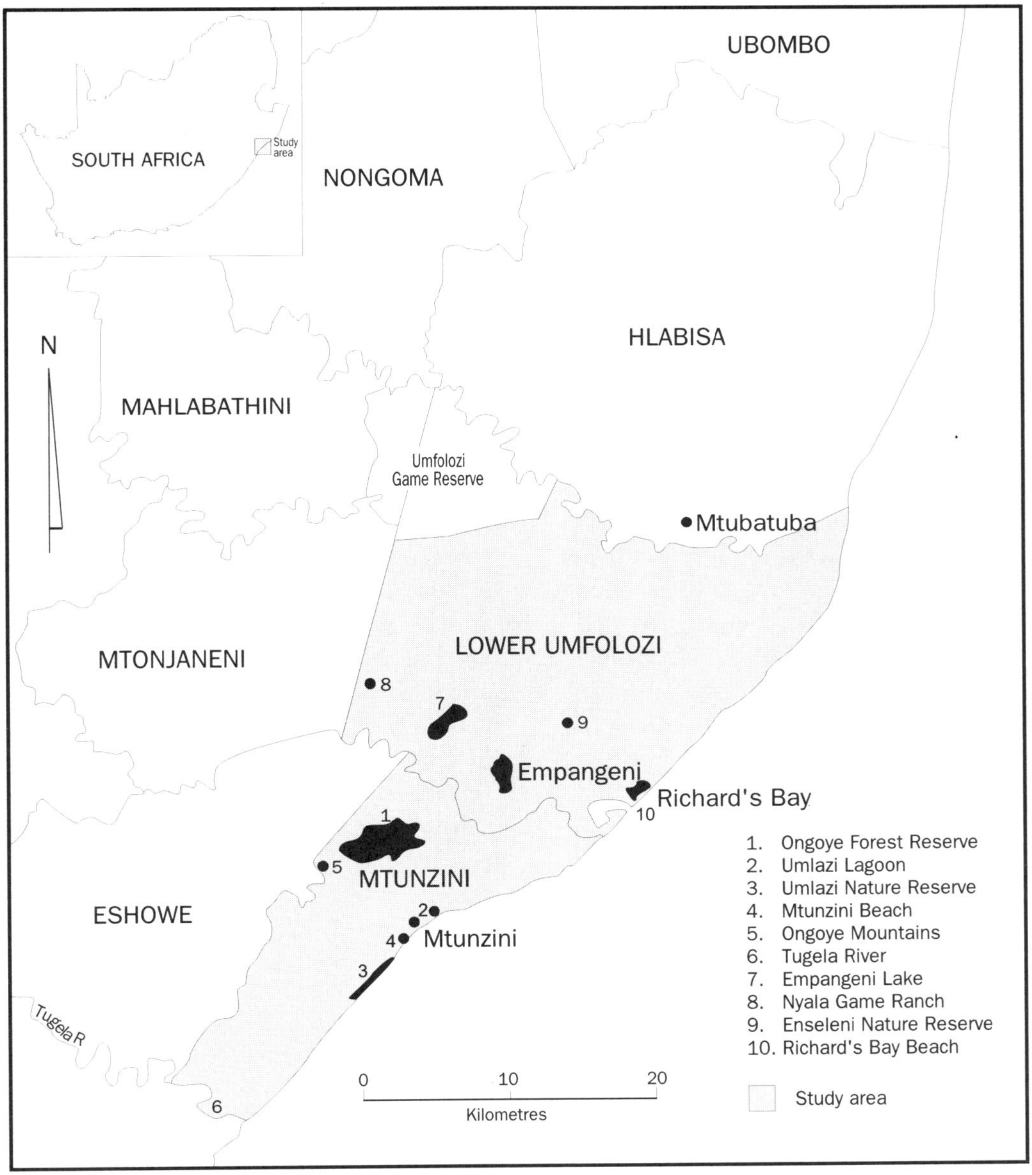

- The public perceived functional issues (food, water, energy, disease, and land) to be more important than the more theoretical and obscure issues (pollution, population, pesticides, and toxic and nuclear waste dumping). This applied particularly to the African community's knowledge of environmental issues.
- A majority of respondents (58 per cent) indicated that they were not sure that they would do anything if it were found that their jobs were responsible for environmental degradation. On the other hand, more teachers were prepared to do something about it.
- Four environmental issues relating to food, water, land use, and population were allocated the highest number of teaching periods (10–23). Respondents also revealed that these issues were 'well taught' at colleges and 'moderately taught' in schools.
- A majority of respondents supported the setting up of environmentally conscious agendas on condition that these improved the state of employment, housing, and education, and eradicated violence, food scarcity, and the problems caused by apartheid.

From this case study it was concluded that Africans in the study area have lost touch with environmental issues, and that these issues are now influenced by industrial and urban lifestyles. These modern lifestyles have tended to ignore the fact that the key to environmental change is the way in which people adapt to the changing environment.

Conclusion

This chapter has attempted to present explanations for various outdoor recreation- and environment-related terms and issues. It has identified some of the writers who have contributed to the field of outdoor recreation and environmental studies, reviewed historical accounts interpreting cultural principles and values, and used case studies to illustrate the connections between values and perceptions in outdoor recreation resource use. Finally, reference has been made to policies recently introduced in South Africa. These policies have started a new debate concerning the role of African values in outdoor recreation and environment studies.

19 Law and environment in South Africa

BRIAN PECKHAM AND
KATE ROWNTREE

Law has been described as the flower of civilization, one of the great achievements of human evolution, in that it represents a conscious attempt to promote an ordered, and acceptable, set of norms for human relationships and for the way in which humans deal with the world around them. Unlike the laws of nature, such as the law of gravity, which describe the way in which the natural world functions, normative laws prescribe human behaviour, setting standards for behaviour in specified circumstances.

Geography encompasses the study of the interrelationship between human activity and the natural environment. While the natural environment places constraints on human activity, human activity impacts on the natural environment through the utilization and degradation of resources. Environmental law is the tool kit through which society regulates or manages its activities so as to minimize negative impacts on the natural environment (giving the environment legal protection) or to promote sustainable resource utilization (thus giving society, the resource users, long-term legal protection).

Thus, if society controls resource utilization through the law, the law is fundamental to the geography of resource utilization. Moreover, the natural environment puts constraints on resource use which, along with social and political systems, have moulded the law in different regions of the world. Thus the geography of a region is both the basis and result of environmental law.

In South Africa there are a number of geographic factors which have affected resource availability and have helped to shape environmental law. For example, the spatial and temporal variability of rainfall and water resources and the propensity for drought are reflected in the National Water Act of 1998 and the related Mountain Catchment Areas Act of 1970. Provincial nature conservation ordinances have been promulgated to protect South Africa's remarkable biodiversity, and the Conservation of Agricultural Resources Act of 1983 has addressed the need for soil conservation which arises from the highly erodible nature of our soils.

The recent political shifts and the new Constitution of 1994 have had major impacts on both the social geography and environmental law of South Africa. Legal access to resources has changed, and universal human rights and environmental rights are both recognized by the Constitution.

As this book is aimed at undergraduate geography students, who are probably without any specific legal background, two preliminary issues need attention. These are, firstly, introducing readers to the concept of law with a view to facilitating a general idea of how law functions in socio-political life, and, secondly, formulating a view of the environment upon which to base a definition of environmental law. Thereafter, a brief outline of the development and general substance of some parts of relevant international and South African law will be presented and evaluated.

The general goal of the chapter is to show the complex nature of environmental regulation in modern states, and to underline the importance of creating relevant, co-ordinated, and effective legal norms, procedures, and institutions, if environmental well-being, sustainability, and ultimate human well-being are to be achieved. For a more in-depth review of many of these issues, the reader is referred to Fuggle and Rabie's (1992) edited volume on *Environmental Management in South Africa*.

Introduction to environmental law

Law

Law has been defined in a number of ways. These include: commands given to people by God for the ordering of society; the dictates of reason for the just ordering of human affairs; the commands of the sovereign; an element of capitalist society's ideological superstructure which is instrumental in preserving capitalist hegemony over the proletariat; and normative rules made, enforced in a prescribed way, and seen as authoritative by the members of society. More recently, law has been seen as encompassing more than rules of behaviour, but as being primarily concerned with establishing and enforcing the fundamental or human rights of individuals.

For the purpose of this chapter, a working definition of law can be given as a rule or set of rules, enforceable by the courts, regulating the relationship between the state and its subjects, and the conduct of subjects toward one another (Hanks 1989); or, that which must be followed and obeyed by citizens subject to sanctions or legal consequences (*Blacks Law Dictionary* 1979). Note that, while it is often seen as comprising only rules, law embodies principles and standards that serve to regulate the actions of humans which affect not only other people, but also other parts of their environment.

Whatever one's view, law is an inescapable fact of modern society, which affects all areas of human existence, and is used in various ways to promote human ends. It manifests itself in two primary forms, namely *municipal law* and *international law*. Municipal law applies only within the boundaries of the country concerned, while international law applies to the relationships of countries with each other.

Some of the various categories into which South African law is often divided in academic and legal practice are outlined in Figure 19.1. It is clear from the diagram that law extends its influence into a myriad aspects of human existence. It also has the potential to influence the way in which humans relate to the natural environment. Thus, in recent times, so-called environmental law has made its appearance.

Environment

The external conditions or surroundings of a person or thing constitute their environment (Hanks 1989). This wide definition requires narrowing down for the purposes of environmental legislation,

FIGURE 19.1 Summary of divisions of South African law

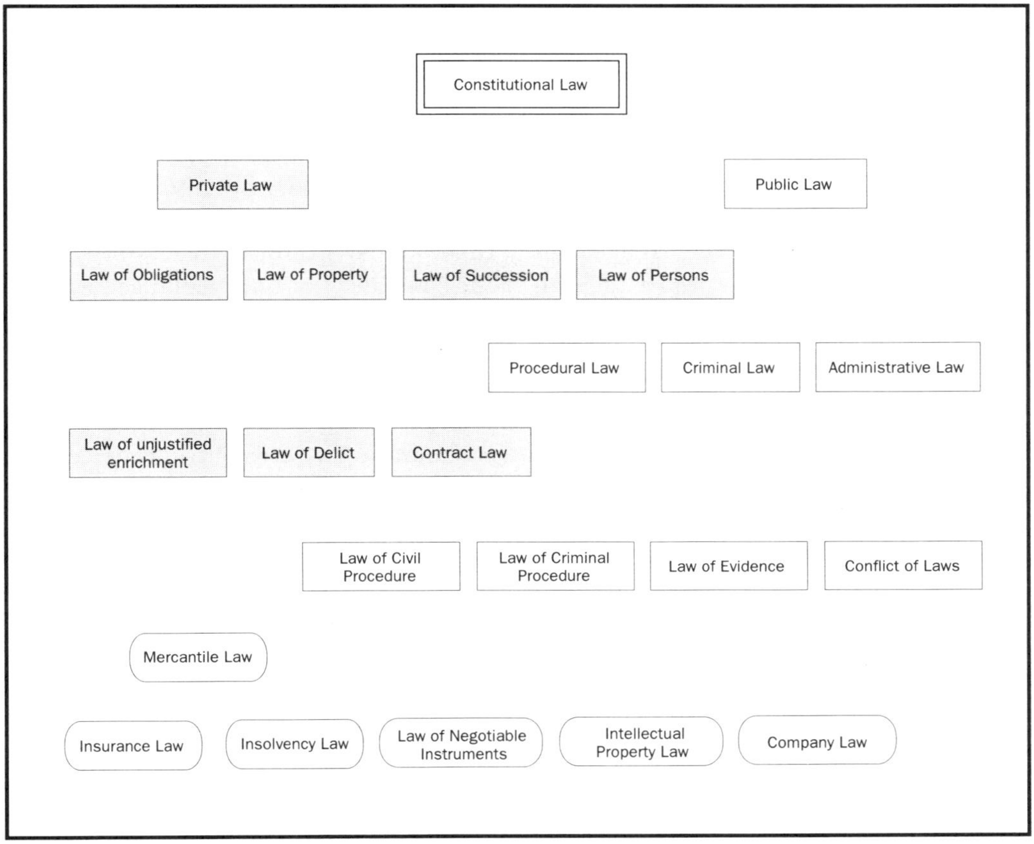

otherwise all law would have to be included in the field of environmental law, because all law ultimately relates to those external conditions or surroundings. South African law does not provide much help in focusing the discussion, as it defines environment as 'the aggregate of surrounding objects, conditions and influences that influence the life and habits of man or any other organism or collection of organisms' (South Africa 1989). This does not take matters much further, except that the reference to 'organisms' offers a tenuous link to the term 'environment' as used in an ecological context: 'the external surroundings in which a plant or animal lives which influence its development and behaviour' (Hanks 1989). Definitions of and legal approaches to the environment are discussed by Rabie (1992). The fact that there is no generally accepted definition of 'the environment' does not seem to have been a stumbling block to either writers or practitioners until quite recently, when the increasingly apparent need for an holistic approach to environmental management has made imperative the proper identification of all relevant areas (Rabie 1996).

Law, environment, and values

Environmental issues have for thousands of years been the subject of human attention, including attempts at regulation in both municipal and international law. During the last half of the twentieth century, law has evolved in response to the increasing attention demanded by environmental issues, but often in an ad hoc, uncoordinated, and ineffectual way. Municipal legal systems provide many examples of this sort of approach to environmental needs, and those of southern African countries – including South Africa – are no exception. There are many reasons for this, some rooted in ethical mores, some in the socio-political realities of the time and place, and others in the realities of the biosphere.

Among the more important global environmental problems which have brought about some of these changes are the following: global warming (the greenhouse effect); deforestation; rapid urbanization and loss of agricultural land; the depletion of stratospheric ozone layers; acid rain; depletion and contamination of water supplies; contamination of the oceans; desertification; overpopulation; extinction of species of plant and living creatures; depletion of non-renewable resources; decreasing agricultural productivity; instability created by resource conflict; and the threat of nuclear war. All these issues present potentially disastrous scenarios for the inhabitants of the earth. All of them similarly cannot be contained within the boundaries of a particular country, but exercise an influence over neighbouring territories as well. They cannot be tackled on an individual and ad hoc basis but require concerted attention by the global community. These realities, coupled with other realities of the twentieth century, have brought about a radically new way of looking at international law and obligations. Geography students will be – or become – acutely aware of the great problems created by many of these factors in the African and particularly southern African context. The increasing concern for the environment in South Africa has been reviewed by Rabie and Fuggle (1992).

Environmental law

Unlike most other areas of law, environmental law cannot yet claim characteristic legal principles as a means of defining its scope and content. This has to be done on the basis of environmental principles which define the field and make possible the identification and grouping together of apposite legal principles and norms. These, however, do not necessarily, or even usually, embody any environmental values per se.

Recognition of a truly environmental norm in a legal system could ground the development of environmentally specific legal principles and enable a more appropriate definition of environmental law's domain. Whether this will ever happen remains to be seen.

International environmental law and organizations

International law

International law can be defined as those rules and norms regulating the conduct of states and other entities such as individuals and international organizations which are recognized as having international personality in their relations with each other (Wallace 1986). It is different from municipal law in that the 'persons' of international law are states and other participants such as internationally recognized non-governmental organizations (NGOs) and organizations which function at a level above that of the ordinary legal person. All states are treated as equally sovereign, even though in practice some are much more powerful than others.

International law is created in two principal ways: either through the evolution of generally accepted practices of states, called *customary international law*, or through agreements entered into between states, usually called *treaties* (sometimes conventions, protocols, and so on).

International customary law represents those principles of international law which reflect long-established values among the international community of states. Custom comes into being through consistent general application in the community of nations. This also provides a way in which new values – in environmental law, examples such as intergenerational equity, common heritage, and sustainable development come to mind – could in due course find their way into customary international law. Although the creation of new customary law depends on general acceptance and adherence, once this has happened it is not possible for a state to claim that the relevant principles do not apply to it. Identifying when some new concept has become part of international customary law is problematic, but the pressing issues raised by world population growth, energy production, pollution, and agricultural demands, as well as issues of environmental security, are likely to give rise to new developments in international customary law within a few generations.

Treaties are currently the most common means of establishing binding international rules. They come into existence between states, or between states and other organizations subject to international law

(Birnie and Boyle 1992). Treaties usually do not come into effect immediately. Often approval by a number of countries is required before they come into force. Moreover, most countries require some sort of approval by their parliament, or other government structures, of the terms imposed by the treaty. In South Africa, for example, the Constitution provides that international agreements bind the Republic only after 'approval by resolution in the National Assembly and the National Council of Provinces unless it is an international agreement of a technical administrative or executive nature or one which does not require ratification or accession' (South Africa 1996: S231). (Note that the prefix S denotes the section of the relevant Act). A further qualification is imposed (S231(4)): international agreements become law in the Republic only when they are enacted into law by national legislation, although self-executing provisions will be regarded as law unless they are inconsistent with the Constitution or existing legislation in South Africa. The Constitution contains a number of other relevant provisions, perhaps the most significant of which states that '[w]hen interpreting any legislation any court must prefer any reasonable interpretation of the legislation that is consistent with international law over any alternative interpretation that is inconsistent with international law' (S233).

Thus, only after the necessary municipal approval has been obtained is a state in a position to ratify a treaty. The treaty will enter into force and become binding only once it has been ratified by that particular country, and a specific number of other countries if such is required. It then becomes binding on all parties. States which have signed the treaty, however, are required not to go against their commitment to the treaty in the period between signature and ratification. It is clear that this process makes the creation of international environmental law a potentially cumbersome process.

Usually treaties would not bind third states which are not party to them, unless of course the provisions of the treaty come to be accepted as generally binding on all countries – acquiring roughly the same status as customary international environmental law. Often countries enter into treaty obligations subject to specific reservations, excluding themselves from commitments in terms of some or other specified section or portion of the convention. The possibility of making reservations, therefore, can be seen as an instrument for encouraging rather wider participation in certain treaties than would otherwise be the case. It also, however, creates weaknesses in treaties aimed at protection of the environment, as not all states necessarily commit themselves to fulfilling all of the obligations envisaged in the treaty. Some treaties prohibit reservations, and any country with reservations would not agree to that particular treaty. It should be noted, however, that the 1969 Vienna Convention on the Law of Treaties brought about an important innovation in international treaty law: the recognition of the concept of '*jus cogens* ... which is the notion that there are certain basic norms of international law which ... are so fundamental ... that no state can opt out of ... observance' (Binnie and Boyle 1992: 14).

The characteristics of international law are in many ways quite different from those of municipal law; for example, it is neither created or imposed by an international legislature nor enforced by a policing system. It relies mainly on consensus and co-operation between nation states for its effectiveness. Where a state is strong enough, it might, if it so chooses, ignore the rules and the system would fail – at least in the short term. Once established, however, these rules impose obligations on states which may not be unilaterally ignored or modified. While the absence of an effective sanction against a state's failure to adhere to international law sometimes creates the impression in the short term that international law is weak and ineffectual, economic and moral sanctions have often proved effective in the long term, even where the use of force has been impossible or undesirable. The principles of international law and their application to the conservation of the environment and prevention of pollution are considered in detail by Devine and Erasmus (1992).

International environmental law

The various environmental problems of the second half of the twentieth century are a major source of international concern and attention. Their prominence makes co-operation between nations absolutely vital. Thus international environmental law has made its appearance, defined as follows by Birnie and Boyle:

> [it] can be assumed to be the aggregate of all the rules and principles aimed at protecting the global environment and controlling activities within national jurisdiction that may affect another state's environment or areas beyond national jurisdiction, it involves questions of sovereignty, jurisdiction, regulation, and state responsibility and liability. It can also involve individuals' responsibility under municipal law and decisions of national courts ... International law sources are the same as those from which all international law emanates ... [as it is] a specialised branch of that general law. (Birnie and Boyle 1992: 9)

Birnie and Boyle (1992) note the following as possible sources of evolving general principles to ground new customary international environmental law: general assembly resolutions; certain treaty provisions which reflect, in due course, the general values accepted by the majority of states (for example declarations of principles adopted at the United Nations – or at other international conferences); and proposals from the International Law Commission. Before they achieve this status, however, they remain soft law (see below) and do not satisfy the traditional view that custom (incorporating *jus cogens*) and treaty are the sources of international law. States cannot be bound unless the rule is a customary one, or unless they have clearly consented to the relevant terms.

Laws sourced in international custom, treaties, and *jus cogens* or established general principles are known as *hard law*. Alongside the hard law is what has come to be known as *soft law*. Soft law includes the policies and conduct of nation states and other environmental international organizations which have come into being mostly during the latter part of the twentieth century. These include the United

Nations and its specialized agencies such as the Food and Agriculture Organization (FAO) and the United Nations Environment Programme (UNEP). They also include many other organizations such as regional fisheries, pollution and trade commissions and organizations, as well as numerable international conferences convened by the UN or other relevant agencies in a global, regional, or local context. Among the ad hoc international conferences mentioned are the United Nations Conference on the Human Environment, Stockholm, 5–16 June 1972, and the United Nations Conference on Environment and Development, Rio de Janeiro, 3–14 June 1992, which produced the Rio Declaration, among others. The Man and the Biosphere Programme was developed at the Biosphere Conference in 1968.

Another organization of environmental relevance is the Organization for Economic Co-operation and Development (OECD), which is made up of countries which were party to the post–World War II Marshall Plan in Europe, with the addition of Australia, New Zealand, and Japan, among others. While not directly relevant to most states in Africa, the programmes of the OECD are useful in providing guidelines for developing countries in connection with environmental issues such as chemical control programmes, pollution control, economic practices affecting environmental protection, waste management policies concerning hazardous waste, water quality control, and environmental monitoring. One of the particularly relevant organizations within the OECD is the ad hoc Group on Environmental Assessment and Development Assistance, which focuses attention on assisting and protecting the environment in developing countries by incorporating environmental standards in projects undertaken in those countries.

The International Union of Conservation of Nature and Natural Resources, along with a number of other NGOs and states, took the initiatives which led to the Convention on International Trade in Endangered Species of Wild Fauna and Flora (CITES). CITES is aimed at controlling international wildlife trade in endangered species, and is binding on South Africa as well as other countries. The Ramsar Convention has its origins in similar organizations: a meeting to discuss the increasing loss of migratory bird species' habitat as a result of activities such as the draining of wetlands and introduction of flood control in particular areas led to the Convention on Wetlands of International Importance, Especially with Respect to Wildfowl Habitat, held at Ramsar in Iran in 1971. The Convention Concerning the Protection of World Cultural and Natural Heritage came into being through the efforts of the United Nations Educational, Scientific, and Educational Organization (UNESCO), as well as other NGOs and several states. A World Heritage Fund was established in terms of the convention to assist developing countries in particular to preserve and protect areas of valuable cultural and natural phenomena. Thus, for example, assistance has been given to the Democratic Republic of Congo to protect the white rhinoceros.

Conferences and conventions such as these have led to the publication of various instruments containing relevant principles, such as codes of practice, declarations, occasional treaties, as well as recommendations and guidelines or standards for application by the participants.

All of these are often regarded as soft law, and need to be taken into account in establishing the exact parameters of customary international environmental law. Many of tomorrow's binding principles are incorporated in today's soft law, which is not necessarily enforceable between all nations. Birnie and Boyle (1992) point out that, in a new and rapidly developing field such as that of international environmental law, soft law is most useful and widely employed in that the field is uncertain and constantly changing. It also has the advantage of providing guidelines without necessitating the authoritative scientific definition of the standards required for proper action on environmental issues. For hard law to come into being, such standards would have to be established and agreed upon.

South Africa is party to and has incurred environmental obligations in terms of numerous treaties and conventions, such as those listed in Table 19.1. These have been selected as examples of the wide range of possible treaty obligations which a country might incur.

A global development which is giving environmental issues even greater impact in international law and strengthening the sort of obligations placed upon nation states is the concept of *human rights*. Historically, most legal systems were structured around two premises. The first was a philosophical commitment to a positivist definition of law which entailed recognizing law on the basis of its emanating from a recognized source, for example a sovereign or parliament. The second was the ability of whatever authorities were involved to ensure compliance with the rules and institutional structures and functions which were created in terms of that positivistic model. While this remains true to some degree of politico-legal arrangements in any particular country, the increasing acceptance of human rights approaches to law and its content has altered the accepted notions of both municipal and international law.

The human rights philosophy negates the idea that individuals can acquire legal rights only where the law necessarily or specifically recognizes such rights, or, in the case of international law, that binding international law obligations can be created only by acceptance either in the sense of customary international law or in the form of treaties agreed to by particular states. This leads to a revision of the idea that international law is not law in the true sense because of the absence of potential sanctioning organizations. Along with environmental realities which know no political or geographical boundaries, international law acquires increasing importance on the basis of the recognition of human rights as grounding a particular set of rights which are due to all individuals and which govern their relationship not only among themselves but also in respect of their governments and other governments.

It can be seen that this approach, if it is generally accepted, will lead to a radical revision in the way in which both municipal and international law function. Many countries have formulated constitutions which include specific recognition of human rights – among them the recognition of environmental rights on the part of persons in those countries. But the logic of the human rights approach is based on such rights being pre-legal in their existence. They do not require specific recognition in a legal system for an individual to

TABLE 19.1 Treaties and conventions with environmental obligations to which South Africa is party

TREATY OR CONVENTION	PLACE AND DATE
The International Convention for the Regulation of Whaling	Washington, 2 December 1946
The Convention on Fishing and Conservation of the Living Resources of the High Seas	Geneva, 29 April 1958
The Antarctic Treaty	Washington, 1 December 1959
The Treaty Banning Nuclear Weapons Tests in the Atmosphere, in Outer Space, and Under Water	Moscow, 5 August 1963
The Agreement Concerning Rivers of Mutual Interest Between Portugal, Mozambique, Swaziland, and the Republic of South Africa	13 October 1964
The International Convention Relating to Intervention on the High Seas in Cases of Oil Pollution Casualties	Brussels, 23 November 1969
The Convention on Wetlands of International Importance, Especially with Respect to Waterfowl Habitat	Ramsar, 2 February 1971
The Convention for the Conservation of Antarctic Seals	London, 1 June 1972
The Convention on the Prevention of Marine Pollution by Dumping of Wastes and Other Matter	London, Mexico City, Moscow, 29 December 1972
The Convention on International Trade in Endangered Species of Wild Fauna and Flora (CITES)	Washington, 3 March 1973
The Vienna Convention for the Protection of the Ozone Layer	Vienna, 22 March 1985
The Convention on Assistance in the Case of a Nuclear Accident or Radiological Emergency	Vienna, 26 September 1986
The Montreal Protocol on Substances that Deplete the Ozone Layer	Montreal, 16 September 1987
The Basel Convention on the Control of Transboundary Movement of Hazardous Wastes and Their Disposal	Basel, 22 March 1989
The Convention on Biological Diversity	Rio de Janeiro, 5 June 1992

have them or for the legal system to enforce them. The approach is controversial, and is not always widely accepted. Bills of rights have been incorporated in the constitutional arrangements of countries as widely diverse as the United States, Canada, various countries in Europe, Brazil, Malaysia, India, and South Africa. These constitutions may or may not contain specific environmental rights, although examples of constitutions containing environmental rights are found in the cases of South Africa and Namibia. The implications of such rights for the general legal system are considered by Boyle and Anderson (1996) and Van der Vyfer (1992), and will be discussed below with reference to South Africa.

Environmental law in South Africa

Historical development

Municipal legal systems have tried to deal with environmental issues in various ways for thousands of years. Rules emanating from municipal legal systems apply only within the boundaries of particular nation states, and quite often different approaches to the same environmental problem are found on opposite sides of a political border. Unfortunately, political borders do not always coincide with geographical or environmental ones, and the result is an uncoordinated and possibly ineffective overall management of a problem.

Most ancient city civilizations had rules which regulated environmental issues such as the disposal of sewage, waste, and other nuisances in settled areas. In ancient Rome there were many rules relating to such problems. Numerous statutes from medieval Europe and England provide examples of how smoke, sewage, and other waste were regulated in the cities. There are numerous examples of rules relating to the use of common natural resources such as rivers, commonages, forests, and wildlife.

South African environmental law has a long and chequered history. Environmental problems have been addressed in legislation, as well as in common law rules, for centuries. The content and focus of the relevant rules and principles have changed with changing values and needs. They have also reflected the ruling political ideologies of the time in varying ways. The early development of South African environmental law is very similar to that outlined above, displaying a focus on the pollution of drinking water and controlling the hunting of game in its earliest forms. The old Roman remedies formed part of South African law from 1652 onwards and were available to persons aggrieved by polluting activities on the part of others. These still form part of South African law (Rabie 1991).

Although an interest in conserving water supplies was evinced in early Cape legislation, a more specifically environmental concern in legislation made its appearance only in the last part of the nineteenth century. More conservation legislation appeared, regulating wildlife protection as well as the exploitation of natural forest areas. Perhaps the earliest of these was the declaration of forest reserve areas under the laws of the then Cape Colony. The forerunners of the Dwesa and Cwebe forest reserves on the Transkei coast appear to have been proclaimed as early as 1888. Other conservation areas were proclaimed during the 1890s in the Transvaal Republic and also in the colony of Natal. Public health legislation controlled activities which might affect the health of inhabitants of the cities and towns. Thus legislation was enacted dealing with pollution. Water conservation, the regulation of the use of the seashore, and the conservation of forests formed a focus of environmental legislation during that period. Rabie (1996) notes that the National Parks Act of 1926 represents perhaps the most important conservation statute of the early twentieth century.

The period around the Second World War marked a watershed in the sort of problems faced by South African society (like much of the rest of the world) and was a time when numerous environmentally focused laws made their appearance.

Developments during the period up to 1970 reflect an increasing global interest in environmental matters, and in South Africa this led to the promulgation of sea fisheries legislation, legislation focused on conserving soil resources and forest resources, as well as the protection of the water resource from pollution and from certain forms of over-exploitation. The coming of the nuclear age saw controls imposed on the use and disposal of radioactive materials, and, at a more local level, the promulgation of provincial ordinances aimed at promoting the conservation of wildlife and various other fauna and flora made their appearance. The legislation mentioned here usually took the form of Acts of Parliament, some of which were later replaced by other parliamentary enactments as the need for environmental protection was perceived to have grown. Since the early seventies a new concern with environmental issues in countries such as the United States and those of Europe has been reflected in legislative developments in South Africa. The rather unsophisticated environmental legislation of earlier times now began to take a more specifically environmental focus. It also began to embody new methods for ensuring that environmental interests were taken into account rather more carefully than they had in the past.

Before considering the way in which legislation developed from the 1970s, it is enlightening to briefly examine the evolution of a particular piece of legislation, namely the Water Act (No. 54 of 1956), and its successor, the National Water Act (No. 36 of 1998). When the Act was first promulgated in 1956 it aimed at replacing earlier irrigation and water quality legislation in view of the increasing need to supply water not only for agricultural purposes but also for urban and industrial uses, and to protect the resource from industrial pollution. An examination of the various amendments to this Act in the years from 1956 will show how the focus of attention shifted first from the regulation of irrigation water to that of providing water for urban and industrial needs, then to the protection of the resource and ensuring its proper management when it came to the construction of large dams and irrigation schemes, and later to its protection in the interest of other users, such as recreational users.

An example of resource protection is the introduction in more recent times of guidelines relating to the quality of water which might be discharged back into the original source. These stipulate that the water should be of a quality suitable for the purposes of the normal use to which that water is put. Under the old legislation, no account was taken of the protection of the environment other than where direct human use was affected, for example by water quality. The environment, however, came to be recognized as a competing user of water in, for example, environmental impact assessments (EIAs) of large dams commissioned by the Department of Water Affairs and Forestry (DWAF).

The increasing importance of environmental values in relation to water have now led to the recognition of the resource itself as requiring specific protection in an environmental sense. Under the new National Water Act of 1998, the environment no longer has the status of a competing user, but is recognized in a narrow sense as a primary interest to be protected, although in a broader sense it is the human right to a sound environment that is the ultimate interest to be protected (see S24 of the Bill of Rights).

The National Water Act of 1998 is a comprehensive piece of legislation that replaces the previous ad hoc legislation. It is based on guiding principles which 'recognise the basic human needs of present and future generations, the need to protect water resources, the need to share some water resources with other countries, the need to promote social and economic development through the use of water and the need to establish suitable institutions in order to achieve the purpose of the Act' (South Africa 1998: Chapter 1). This new Act is fundamental to the management of one of the most important components of our environment: water. One aspect in particular is singled out for special mention: the Reserve.

Under Section 6, the national water strategy must provide for the requirements of the Reserve, which is described as follows:

> The Reserve consists of two parts – the basic human needs reserve and the ecological reserve. The basic human needs reserve provides for the essential needs of individuals served by the water resource in question and includes water for drinking, for food preparation and personal hygiene. The ecological reserve relates to the water required to protect the aquatic ecosystems of the water resource. The Reserve refers to both the quantity and quality of the water in the resource, and will vary depending on the class of the resource. (South Africa 1998: Chapter 3)

Before water resources can be protected by law, they must first be classified in terms of their resource quality objectives. These are determined with respect to the Reserve, the flow characteristics and quality of the water resource, the instream and riparian habitat, the characteristics and distribution of aquatic biota, the regulation or prohibition of instream or land-based activities which may affect the quantity of water in or quality of the water resource, and any other characteristic.

Under Section 36, streamflow reduction activities (including commercial afforestation) will be allowed only if, among other criteria, they do not impinge on the Reserve. Furthermore, the Minister of Water Affairs and Forestry is obliged to establish national monitoring systems, in part to monitor the health of aquatic ecosystems (S137).

It is clear from the new legislation that the environmental considerations are of fundamental importance. The river is no longer seen simply as a conduit carrying water that can be exploited by riparian owners and industry at will, but as a resource in its own right, comprising an ecosystem which extends from the channel itself to the riparian zone. The challenge to ecologists and other environmentalists is to ensure that we have a sufficient understanding of the way in which this dynamic ecosystem functions such that requirements stipulated for the Reserve can be defended in a court of law.

The altered environmental values of the times at which the various amendments to the water legislation were promulgated are quite evident in the phrasing and focus of these amendments. No doubt these sorts of factors are evident in the legislation of other countries, including southern African countries, where the older attitude prevalent in South Africa until the 1970s is quite evident in legislation inherited from the former colonial governments.

The sudden surge in environmental awareness in the early 1970s focused legislative attention on issues not previously provided for. In South Africa a plethora of various types of environmental legislation made its appearance. The scattered collection of legislation reflects various ad hoc solutions to environmental problems as they became evident and required addressing. Examples of this legislation can be seen in Figure 19.2. Although a measure of consistency and focus is evident in the legislation, quite often the efforts of different parts of government have been uncoordinated and possibly even counterproductive to each other's attempts at protecting the environment – witness the effects of the grand apartheid legislation on demographics, of huge, centrally planned water schemes, or of industrial developments upon environmental interests. The limited weight given to environmental interests in the mid twentieth century is quite evident in the manner in which these interests were subordinated to developmental and ideological interests.

Current developments in environmental law in South Africa

Although common law provides some principles useful in an environmental context, it recognizes no fundamental environmental ethic. These principles, therefore, are only incidentally relevant. Most of the environmental law is contained in legislation such as Acts of Parliament, Acts and ordinances of the various provincial legislatures, local authority by-laws, and regulations passed in terms of legislation.

The legislation can be divided into different categories according to whether it is fully and specifically environmentally focused, partially of environmental relevance, or perhaps only incidentally of relevance to the environment. Examples of each category are as follows:

- specifically environmental: the Sea Bird and Seals Protection Act (No. 46 of 1973), the National Parks Act (No. 57 of 1976), the Environment Conservation Act (No. 73 of 1989), and the various provincial nature conservation ordinances promulgated during the seventies and eighties;
- partially environmental: the Sea Fisheries Act (No. 58 of 1973), the

FIGURE 19.2 Examples of environmental law in South Africa, 1997

ACT	Air pollution	Use and protection of fresh-water resources	Land and soil conser-vation	Protection of plants and animals
Animals Protection Act 71 of 1962				●
Provincial Nature Conservation Ordinances		○		●
Game Theft Act 105 of 1991				●
Seabirds and Seals Protection Act 46 of 1973				●
Forest Act 122 of 1984		○		●
Agricultural Pests Act 36 of 1983			○	○
Conservation of Agricultural Resources Act 43 of 1983		○	●	
Water Act 54 of 1956 National Water Act 36 of 1998		●	○	
Mountain Catchment Areas Act 63 of 1970		●	○	○
Lake Areas Development Act of 1975		●		
Atmospheric Pollution Act 45 of 1965	●			
Health Act 63 of 1977	○	○	○	
Road Traffic Act 29 of 1989	○			
Minerals Act 63 of 1977		○		
Fertilizers, Farm Feeds, Agricltural Remedies and Stock Remedies Act 36 of 1947			○	
Hazardous Substances Act 15 of 1973	○	○	○	
Common Law: Law of Nuisance	○	○	○	
Key	●	Act directly related to the resource	○	Act indirectly related to the resource

Conservation of Agricultural Resources Act (No. 43 of 1983), the Nuclear Energy Act (No. 131 of 1993), and the Water Act (No. 63 of 1998);
- incidentally relevant: the Sea Shore Act (No. 21 of 1935), the Income Tax Act (No. 58 of 1962), and the Insolvency Act (No. 34 of 1926).

It should be noted that there is no single piece of dedicated legislation in South Africa which regulates the entire spectrum of what is called environmental law in this chapter. As Rabie (1996) notes, it is unlikely that such a comprehensive clarification is feasible. It is also unlikely that any attempt will be made to formulate such a piece of legislation.

The Environmental Conservation Act of 1989

This relatively recent piece of legislation creates a framework within which most environmental issues in South Africa might be regulated, and within which both new and existing environmental law can be co-ordinated. The substantive regulation is not achieved within the Act, which instead empowers the Minister of Environmental Affairs and Tourism, and in some cases the Minister of Water Affairs and Forestry and provincial or local authorities, to make legally binding policies or regulations on a wide number of issues. For a more detailed review of the Act, see Rabie (1992).

Although it did not have much substance as originally promulgated, and evoked criticism for having missed the opportunity to make an important contribution to environmental practice in South Africa, the Environmental Conservation Act was innovative in that it offered the only definition of environment in South Africa law (S1), introduced the concept of compulsory public consultation on the content of policies and regulations to be published under the Act (S32), was binding on the state (S40), and made possible vastly increased penalties for environmental crimes (S29, S30). It also represented the coming of age of the environmental ministry as a distinct and vital arm of government. Later amendments and the publication of policies and regulations have strengthened the Act further, and have made it a vital instrument to the definition and fulfilment of government's environmental and constitutional obligations.

Thus the Act provides for the formulation of general policy for environmental conservation, as well as policies on specific aspects of the environment, which are binding on all areas of government, including provincial and local government, as well as other specified institutions (S3(1) and S1). The Director General of the Department of Environmental Affairs and Tourism (DEAT) is given powers to ensure such compliance (S3(2)). The Act also provides for the declaration of defined areas as 'protected natural environments' (S16) and as 'special nature reserves' (S18). In the case of privately owned land, protected natural environments can be declared by a provincial administrator in order to promote the preservation of specific ecological processes, natural systems, natural beauty, species of indigenous wildlife, or biotic diversity. Special nature reserves can be declared in the case of land, water, or sea areas owned or exclusively controlled by the state, or private land or water at the owner's request, for the purpose of the protection of the environment in such an area.

The Act creates several useful structures to assist in the investigation of environmental issues, the formulation and co-ordination of policy, and the management of sensitive environments (S4 – Council for the Environment; S14 – Committee for Environmental Co-ordination; S15 – Board of Investigation; and S17 – Management Advisory Committees in respect of protected natural environment).

Sections relating to pollution control (S19–20) and the control of activities which may have a harmful effect on the environment (S21–23) are supplemented by further sections empowering the publication of legally binding regulations regarding waste management, littering (S24), noise, vibration (S24A), and shock (S25), environmental impact reports (S26), limited development areas (S27), and international conventions, treaties, and agreements (S27A).

The government department responsible for administering the Environmental Conservation Act is the Department of Environmental Affairs and Tourism. Since 1994, a new sector of government has been created at the provincial level, and certain powers have been delegated to the provincial legislature. The provincial ministries are not necessarily the same as those at national level, and may combine the functions of two national-level ministries. Table 19.2 lists the provincial-level ministries as they relate to the environment. The reader is encouraged to reflect on the way in which the environment is perceived in different areas of the country, and the extent to which this may facilitate or impede the enactment of environmental legislation. In the Eastern Cape, for example, environment is subsumed under Economic Affairs, Environment, and Tourism, reflecting a recognition of the importance of the environment to the local economy, particularly through its relationship to tourism. On the whole, this will probably have a positive effect in resolving environmental issues, but where there is a real conflict of interest between the economy and the environment, the environment has no independent spokesperson, and economic pressures may well win the day.

The Constitution

The new Constitution (The Constitution of the Republic of South Africa, Act 108 of 1996), has numerous implications for environmental law, some of which are examined in greater detail below. On 27 April 1994 South Africa adopted a model of government which included constitutional sovereignty and marked a departure from the erstwhile Westminster model of parliamentary sovereignty. The exercise of all governmental power is now subject to an entrenched Bill of Rights, which includes an environmental clause. The Bill of Rights applies to all law and all organs of state in South Africa. This includes the common law, which must be interpreted and applied in the light of the Bill of Rights (S8).

There are a number of provisions in the Bill of Rights which are relevant to the environment. The most obvious of these is the environmental right. Under Section 24, everyone has the right (a) to an environment that is not harmful to their health or well-being, and (b) to have the environ-

TABLE 19.2 Environmental departments at the national and provincial level

REGION	DEPARTMENT RESPONSIBLE FOR ENVIRONMENTAL AFFAIRS
National	Department of Environmental Affairs and Tourism
Eastern Cape	Economic Affairs, Environment, and Tourism
Free State	Environmental Affairs and Tourism
Gauteng	Environmental Affairs
KwaZulu-Natal	Traditional and Environmental Affairs
Mpumalanga	Environmental Affairs and Tourism
Northern Cape	Agriculture, Nature Conservation, and Land Reform
Northern Province	Agriculture, Land, and Environment
North West	Tourism, Nature Conservation, and Environmental Affairs
Western Cape	Environmental and Cultural Affairs

ment protected, for the benefit of present and future generations, through reasonable legislative and other measures that (i) prevent pollution and ecological degradation, (ii) promote conservation, and (iii) secure ecologically sustainable development and use of natural resources while promoting justifiable economic and social development. This right does not give the environment itself any legal rights; rather, it gives humans the right to a healthy environment. Just how environmentally friendly the effects of this clause will prove remains to be seen, as there is no general agreement on what constitutes sustainable activity, or, for that matter, environmental well-being.

Many matters relating to the environment fall within the legislative and executive competence of provincial or local governments. This results from the way in which the Constitution regulates the structures of government, and their respective powers and competences as set out in Part B of Schedules 4 and 5 of the Constitution.

Schedule 4 lists the functional areas of concurrent national and provincial legislative competence. Part A refers to the administration of indigenous forests, agriculture, education, environment, health services, nature conservation, pollution control, public works, regional planning and traffic regulation, soil conservation, tourism, and urban and rural development. Part B refers to air pollution, building regulations, local tourism, municipal planning, municipal issues, and water and sanitation services.

Schedule 5 contains a list of functional areas of exclusive provincial legislative competence, which means that provinces may legislate to the exclusion of national legislation in connection with these issues. Part A (provincial planning) relates to recreation and amenities, roads and traffic, while Part B (local government) relates to beaches and amusement facilities, cleansing, control of public nuisance, fencing, local amenities, municipal roads, noise pollution, refuse removal, refuse dumps and solid waste disposal, street lighting, and traffic.

Other rights embedded in the Constitution are also relevant to environmental issues. For example, the property right (S25) will play a significant role in determining the power of government to redistribute land and other resources such as water. Freedom of trade, occupation, and profession (S22), and the right to health care, food, water, and social security (S27), are similarly significant. The right to education includes a right to adult basic education (S29), which will no doubt in due course be argued to extend to a right to education on environmental issues, as this is vital in the long term to the survival of humans. The cultural and religious rights (S30 and 31) may also be relevant. For example, particular forms of cultural discrimination in provincial environmental legislation which prohibit particular forms of hunting may well find themselves challenged on the basis of such rights.

The rights to information and to just administrative action (S32 and 33) offer very powerful tools for environmentalists in the sense that they entitle interested persons to information held by government, as well as information held by persons other than government, and also require lawful, reasonable, and procedurally fair administrative action where this is undertaken by government functionaries. Thus, for example, the application of Sections 2 and 3 of the Environmental Conservation Act of 1989, which requires that decision-making in various government departments must take into account the needs of sustainability and environmental protection, will be strongly shaped by this right to just administrative action.

The enforcement of rights section (S38) of the Bill of Rights solves the *locus standi* problem previously faced by interested environmentalists in obtaining access to the courts. *Locus standi* is a term used to denote a person's legal standing, or legal right to approach a court for a court order on a specific issue. If a person does not have such a right, then a court will not hear their case. In the past, individuals had to prove that they were personally affected before they would have *locus standi* in terms of the common law. It was not possible to act on someone else's behalf. Under the Constitution anyone may act in their own interest, on behalf of other persons, or in the general public interest, in approaching the courts for the enforcement of constitutional rights. Thus, if the issue at hand can be brought under any of the fundamental rights, including the right to just administrative action, the question of *locus standi* is made much simpler.

According to Glazewski,

> the inclusion of an environmental clause in the new South African Constitution is a first step towards developing... an equitable future environmental management system

> as well as an acknowledgment that past environmental injustices have to be redressed ... The immediate and real challenge is to ensure that the constitutional foundation which has been laid is translated into effective and meaningful action. The substantive environmental right complemented by the procedural provisions in the Bill of Rights provide the basis for a much needed 'distinctive principles of South African environmental law' ... [T]he clauses provide a foothold whereby environmental considerations can be given judicial regard by the courts using their review powers as well as ensuring greater public participation in decision making ... The ground work has been laid and some tools have been provided to restore environmental justice in a country which has suffered environmental and political degradation ... [and] it is now up to all South Africans to take them up and ensure that the new South Africa is a sustainable one. (Glazewski 1989: 177, 196)

The environmental right as set out in the Constitution has been taken up in environmental legislation, specifically through the National Environmental Management Act (No. 101d of 1998), passed by Parliament in November 1998. The Act sets out the following rights, applicable to all individuals:

1 The right to a sustainable environment.
2 The right for all stakeholders to be consulted on impact assessments (S2 and Chapter 5).
3 The right to participate in dispute resolution (S29).
4 The right to refuse to do hazardous work (S29).
5 The right to report environmental risks (S31).
6 The right to information necessary to fulfil environmental responsibilities (S31).
7 The right to demand that the environment is taken care of (S28).
8 The right to legal standing to enforce environmental laws (S32 and S33).

This Act thus gives individuals and environmental pressure groups strong legal backing to protect the environment.

Other legislation

As indicated above, there are a large number of laws which are relevant to the way in which people relate to the environment. Examples of these laws as they related to air pollution, use and protection of freshwater resources, conservation of land and soil, and protection of plants and animals in 1997 are given in Figure 19.2. It can be seen from this chart that for any environmental component there is a range of legislation that can be applied, and that one Act may cover a number of different components of the environment.

1998 saw a plethora of environment-related legislation going through Parliament. The National Water Act of 1998 has already been referred to. Other important Acts include the Marine Living Resources Act (No. 18 of 1998), the National Forests Act (No. 71 of 1998), the National Parks Act (No. 83 of 1998), the National Veld and Forest Fire Act (No. 101 of 1998), and the National Heritage Act (No. 139 of 1998).

Details of all recent legislation can be found on the South African government web site: http://www.polity.org.za/govdocs/legislation/. Updates and discussions on legislature and developments of environmental interest can be found on the South Africa Environment Home Page: http://www.oneworld.org/saep/.

Since 1994, all laws must be interpreted and applied in the light of the Constitution, and as supplemented by the Environmental Conservation Act of 1989 and the National Environmental Management Act of 1998. It should have become quite evident by this stage that South Africa's environmental legislation is quite complex and fairly comprehensive, but still requires further reconciliation and co-ordination in content, and in institutional control, enforcement, and application. This is necessarily a dynamic and ongoing process, which will never end, simply because the nature and extent of environmental challenges to be met are ever-changing. So are the value paradigms within which ethics, politics, law, economics, and government operate, which complicates the issues even further.

Conclusion and evaluation

It is evident that environmental law is a highly complex field, with a multitude of different areas of relevance. Rabie concludes that environmental law consists of

> a potpourri of legal norms encountered in a number of conventional fields of law ... [I]t serves an omnibus function, accommodating principles of traditional fields of law, which are united by their common object in serving environmental conservation. (Rabie 1992: 97)

Because of their complex nature and constantly evolving goals, environmental laws quite soon begin to display dissonance and become difficult to apply unless subjected to an ongoing process of review to bring them into line with evolving socio-political, scientific, and environmental realities. This is nicely illustrated by the extremely fluid state of South Africa's environmental law at present, as well as the integral role of geographers and environmental lawyers and policy-makers in each other's disciplines.

References

INTRODUCTION

Association of American Geographers. 1997. *What is Geography?* [Online]. Available: http://www.aag.org/Careers/What_is_geog.html [26 November 1998].

Bird, J. H. 1973. 'Desiderata for a definition; or is geography what geographers do?', *Area* 5: 201–3.

Christopher, A. J. 1982. *South Africa*. London: Longman.

———. 1984. *South Africa: the Impact of Past Geographies*. Cape Town: Juta.

Clarke, J. 1991. *Back to Earth: South Africa's Environmental Challenges*. Halfway House: Southern Books.

Cole, M. M. 1960. *South Africa*. London: Methuen.

Corbett, J. D., and R. F. O'Brien. 1997. *Spatial Characterization Tool – Africa* (v 1.0) [CD-ROM]. Available: Texas A & M University System [May 1998].

Dardis, G. F., and B. Moon (eds.). 1988. *Geomorphological Studies in Southern Africa*. Rotterdam: Balkema.

Department of Water Affairs. 1986. *Management of the Water Resources of the Republic of South Africa*. Cape Town: CTP Book Printers.

Enpat. Environmental Potential Atlas [CD-ROM]. 1998. Available: Department of Environmental Affairs and Tourism [November 1998].

Fair, T. J. D. 1982. *South Africa: Spatial Frameworks for Development*. Cape Town: Juta.

Fuggle, R. F., and M. A. Rabie (eds.). 1983. *Environmental Concerns in South Africa*. Cape Town: Juta.

Gregory, K. J. 1990. *The Lie of the Land: An Encyclopedia of Physical Geography*. Amsterdam: Time Life.

King, L. C. 1963. *South African Scenery* (3rd ed.). Edinburgh: Oliver and Boyd.

Lemon, A. 1976. *Apartheid: a Geography of Separation*. Farnborough: Saxon House.

———. 1987. *Apartheid in Transition*. Aldershot: Gower.

——— (ed.). 1991. *Homes Apart. South Africa's Segregated Cities*. London: Paul Chapman.

——— (ed.). 1995. *The Geography of Change in South Africa*. Chichester: Wiley & Sons.

McCarthy, J. J., and D. Smit. 1984. *South African City: Theory in Analysis and Planning*. Cape Town: Juta.

Meadows, M. 1985. *Biogeography and Ecosystems of South Africa*. Cape Town: Juta.

Moon, B., and G. F. Dardis (eds.). 1988. *The Geomorphology of Southern Africa*. Johannesburg: Southern Books.

National Research Council. 1997. *Rediscovering Geography: New Relevance for Science and Society*. Washington, DC: National Academy Press.

Preston-Whyte, R. A., and P. D. Tyson. 1988. *The Atmosphere and Weather of Southern Africa*. Cape Town: Oxford University Press.

Rogerson, C., and J. J. McCarthy (eds.). 1992. *Geography in a Changing South Africa*. Cape Town: Oxford University Press.

Smith, D. M. (ed.). 1982. *Living under Apartheid: Aspects of Urbanization and Social Change in South Africa*. London: Allen & Unwin.

———. 1985. *Apartheid in South Africa*. Cambridge: Cambridge University Press.

——— (ed.). 1992. *The Apartheid City and Beyond: Urbanization and Social Change in South Africa*. London: Routledge.

Tyson, P. D. 1986. *Climatic Change and Variability in Southern Africa*. Cape Town: Oxford University Press.

Water Research Commission. 1997. *Surface Water Resources of South Africa 1990. Digital Data* (2nd ed.) [CD-ROM]. Available: Water Research Commission [December 1997].

Wellington, H. 1955. *Southern Africa. A Geographical Study (Volume I: Physical Geography; Volume II: Economic and Human Geography)*. Cambridge: Cambridge University Press.

CHAPTER 1

Abler, R. F., J. S. Adams, and P. R. Gould. 1971. *Spatial organisation: the geographer's view of the world*. Englewood Cliffs, NJ: Prentice-Hall.

Agnew, J., and S. Corbridge. 1985. 'The new geopolitics: the dynamics of geopolitical disorder', in R. Johnston and P. Taylor (eds.), *A world in crisis: geographical perspectives*. Oxford: Blackwell.

Agnew, J., D. Livingstone, and A. Rogers. 1996. *Human geography: an essential anthology*. Cambridge, MA: Blackwell.

Allen, J. 1983. 'Property Relations and Landlordism: A Realist Approach', *Environment and Planning D: Society and Space* (6): 339–65.

Appleton, J. 1994. *How I made the world: shaping a view of landscape*. Hull: University of Hull Press.

Bhaskar, R. 1975. *A realist theory of science*. Leeds Books Ltd.

———. 1986. *Scientific realism and human emancipation*. London: Verso.

———. 1989. *Reclaiming Reality*. London: Verso.

Bodman, A. 1992. 'Holes in the fabric: more on the master weavers in human geography', *Transactions of the Institute of British Geographers* 17: 21–37.

Bowlby, S., J. Lewis, L. Mcdowell, and J. Foord. 1989. 'The geography of gender', in R. Peet and N. Thrift (eds.), *New Models in geography: the political economy perspective*, Vol 2: 157–75. London: Unwin Hyman.

Brookfield, H. 1969. 'On the environment as perceived', *Progress in Human Geography 1*. London: Edward Arnold. 51–80.

Buttimer, A. 1976. 'The dynamism of life world', *Annals of the Association of American Geographers* 66 (2): 277–92.

Chalmers, A. F. 1982. *What is this thing called science?* Milton Keynes: Open University Press.

Cloke, P., C. Philo, and D. Sadler. 1991. *Approaching human geography: an introduction to contemporary theoretical debates*. London: Paul Chapman Publishing.

Cox, K. R., and R. G. Golledge (eds.). 1969. *Behavioural problems in geography*. Evanston: Northwestern University Press.

Duncan, J. 1995. 'Landscape geography 1993–1994', *Progress in Human Geography* 19: 414–22.

Duncan, J., and D. Ley. 1982. 'Structural Marxism and human geography: a critical assessment', *Annals of the Association of American Geographers* 72: 30–59.

Entrikin, J. N. 1976. 'Contemporary humanism in geography', *Annals of the Association of American Geographers* 70: 43—50.

Giddens, A. 1982. *The constitution of society: outline of a theory of structuration*. Cambridge: Polity Press.

Gould, P. R., and R. White. 1974. *Mental maps*. Harmondsworth: Penguin.

Gregory, D. 1989. 'Areal differentiation and post-modern geography', in D. Gregory and R. Walford (eds.), *Horizons in Human Geography*. London: Macmillan.

Griggs, R. A. 1996. 'The geopolitics of peace in post-apartheid South Africa: a neo-realist analysis of KwaZulu-Natal', *Geojournal* 39 (4): 387–96.

Guelke, L. 1974. 'An idealist alternative in human geography', *Annals of the Association of American Geographers* 14: 193–202.

———. 1982. *Historical understanding in geography: an idealist approach*. Cambridge: Cambridge University Press.

Harré, R. 1986. *Varieties of realism*. Oxford: Basil Blackwell.

Harris, R. C. 1978. 'The historical mind and the practice of geography', in D. Ley and M. S. Samuels (eds.), *Humanistic geography: problems and prospects*. London: Croom Helm.

Hartshorne, R. 1939. *The nature of geography*. Lancaster: American Association of Geographers.

Harvey, D. 1969. *Explanation in geography*. London: Edward Arnold.

———. 1973. *Social justice and the city*. London: Edward Arnold.

Hay, A. 1985. 'Scientific method in geography', in R. J. Johnston (ed.), *The future of geography*. London: Methuen.

Johnston, R. 1997. *Geography and geographers: Anglo-American human geography since 1945* London: Arnold.

Johnston R., D. Gregory, and D. Smith. 1994. *The dictionary of human geography*. Oxford: Blackwell.

Kuhn, T. S. 1970. *The structure of scientific revolutions* (2nd ed.). Chicago: University of Chicago Press.

Lynch, K. 1960. *The image of the city*. Cambridge, MA: MIT.

McDowell, L. 1986. 'Beyond patriarchy: a class-based explanation of women's subordination', *Antipode* 18: 311–21.

Outhwaite, W. 1987. *New philosophies of social science*. London: MacMillan Education.

Peet, R. 1998. *Modern geographical thought*. Oxford: Blackwell.

Peet, R., and N. Thrift. 1989. *New models in geography: the political economy perspective*, Vol 2. London: Unwin Hyman.

Pickles, J. 1992. 'Texts, hermeneutics, and propaganda', in T. J. Barnes and J. S. Duncan (eds.), *Writing worlds: discourse, text, and metaphor in the representation of landscape*. London: Routledge. 193–230.

Popper, K. 1968. *The logic of scientific discovery*. London: Hutchinson.

———. 1969. *Conjectures and refutations*. London: Routledge and Kegan Paul.

Pratt, A. C. 1995. 'Putting critical realism to work: the practical implications for geographical research', *Progress in Human Geography* 19 (1): 67.

Pred, A. 1967. *Behavior and location*. Lund Studies in Geography, Series B, No. 27.

———. 1973. 'Urbanization, domestic planning problems, and Swedish geographic research', *Progress in Geography* 5: 1–77.

———. 1977. 'The choreography of existence: comments on Hagerstrand's time geography and its usefulness', *Economic Geography* 53: 207–21.

Relph, E. 1981. 'Phenomenology', in M. E. Harvey and B. P. Holly (eds.), *Themes in geographic thought*. London: Croom Helm.

Rose, G. 1993. *Feminism and geography*. Cambridge: Polity Press.

Samuels, M. 1978. 'Existentialism and human geography', in D. Ley and M. Samuels (eds.), *Humanistic geography: prospects and problems*. London: Croom Helm.

Sarre, P. 1987. 'Realism in practice', *Area* 19: 3–10.

Sayer, A. 1984. *Method in social science: a realist approach*. London: Routledge.

———. 1985. 'Realism and Geography', in R. J. Johnston (ed.), *The Future of Geography*. London: Methuen.

———. 1992. *Method in social science: a realist approach* (2nd ed.). London: Routledge.

Schaefer, F. K. 1953. 'Exceptionalism in Geography: A Methological Examination', *Annals of the Association of American Geographers* 43: 226–49.

Tuan, Y-F. 1974a. 'Space and place: humanistic perspective', *Progress in Geography* 6: 211–52.

———. 1974b. *Topophilia*. Englewood Cliffs, NJ: Prentice-Hall.

———. 1977. *Space and place: the perspective of experience*. London: Edward Arnold.

Unwin, T. 1992. *The Place of Geography*. Essex: Longman Scientific and Technical.

Werlen, B. 1993. *Society, Action, and Space: An Alternative Human Geography*. London: Routledge.
White, G. F. 1973. 'Natural hazards research', in R. J. Chorley (ed.), *Directions in Geography*. London: Methuen.
Wolpert, J. 1964. 'The decision process in spatial context', *Annals of the Association of American Geographers* 54: 537–8.

CHAPTER 2

Acocks, J. P. H. 1953. *Veld Types of South Africa*. Memoirs of the Botanical Survey of South Africa, 28. Pretoria: Government Printer.
Adams, W. M. 1996. 'Conservation and development', in W. M. Adams, A. S. Goudie, and A. R. Orme (eds.), *The Physical Geography of Africa*. New York: Oxford University Press. 367–82.
Avery, D. M. 1992. 'Man and/or climate? Environmental degradation and micromammalian community structure in South Africa during the last millennium', *South African Journal of Science* 88: 483–89.
Axelrod, D. I., and P. M. Raven. 1978. 'Late Cretaceous and Tertiary vegetation history of Africa', in M. J. A. Werger (ed.), *Biogeography and Ecology of Southern Africa*. The Hague: A. A. Balkema. 77–130.
Binns, T. 1990. 'Is desertification a myth?', *Geography* 75: 106–13.
Botha, G. A., L. Scott, J. C. Vogel, and V. von Brunn. 1992. 'Palaeosols and Palaeoenvironments during the Late Pleistocene Hypothermal in Northern Natal', *South African Journal of Science* 88: 508–12.
Brink, A. B. A., A. van Schalkwyk, T. C. Partridge, D. C. Midgley, J. M. Ball, and S. J. J. Geldenhuys. 1990. 'The changing impact of urbanization and mining on the geological environment', *South African Journal of Science* 86: 434–40.
Cock, J. 1991. 'Going Green at the Grassroots: The environment as a political issue', in J. Cock and E. Koch (eds.), *Going Green: People, politics and the environment in South Africa*. Cape Town: Oxford University Press. 1–17.
Davies, B., and J. Day. 1998. *Vanishing Waters*. Cape Town: University of Cape Town Press.
Davies, T. A., W. W. Hay, J. R. Southam, and T. R. Worsley. 1977. 'Estimates of Cenozoic oceanic sedimentation rates', *Science* 197: 53–5.
Deacon, J., and N. Lancaster. 1988. *Late Quaternary Palaeoenvironments of Southern Africa*. Oxford: Clarendon Press.
Diab, R. 1990. 'Where has all the ozone gone?', in R. Preston-Whyte (ed.), *Rotating the Cube: Environmental Strategies for the 1990s*, Indicator SA Issue Focus, Department of Geographical Science, University of Natal, Durban: 33–6.
Dingle, R. V., and Q. B. Hendey. 1984. 'Mesozoic and Tertiary Sediment Supply to the Western Cape Basin and Palaeodrainage Systems in South-western Africa', *Marine Geology* 56: 13–26
Dollar, E. S. J. 1998. 'Palaeofluvial geomorphology in Southern Africa: A review', *Progress in Physical Geography* 22 (3): 325–49
Du Pisani, A. L., and T. C. Partridge. 1990. 'Effects of global warming on crop production in South Africa', *South African Journal of Science* 86: 306–10.
Garland, G. G. 1990. 'The soil spoilers', in R. Preston-Whyte (ed.), *Rotating the Cube*, Indicator Project South Africa, Robprint. 69–72.
Goudie, A. S. 1993a. *The human impact on the natural environment*. Oxford: Blackwell.
———. 1993b. *The nature of the environment*. Oxford: Blackwell.
———. 1996. 'Climate: past and present', in W. M. Adams, A. S. Goudie, and A. R. Orme (eds.), *The Physical Geography of Africa*. New York: Oxford University Press. 34–59.
Goudie, A. S., and S. Stokes. 1998. *Quaternary Environmental Change*. Oxford: Oxford University Press.
Hoffman, M. T. 1997. 'Human impacts on vegetation', in R. M. Cowling and D. M. Richardson (eds.), *Vegetation of Southern Africa*. Cambridge: Cambridge University Press.
Howell, N. 1979. *Demography of the Dobe !Kung*. New York: Academic Press.
Illenberger, W. K., and B. T. Verhagen. 1990. 'Environmental history and dating of coastal dunefields', *South African Journal of Science* 86: 311–14.
Johnson, D. L., and L. A. Lewis. 1995. *Land degradation: creation and destruction*. Oxford: Blackwell.
Kates, R. W., and B. L. Turner, and W. C. Clark. 1991. 'The great transformation', in B. L. Turner (ed.), *The Earth as Transformed by Human Action*. Cambridge: Cambridge University Press.
Koch, E. 1991. 'Rainbow Alliances: Community struggles around ecological problems', in J. Cock and E. Koch (eds.), *Going Green: People, politics and the environment in South Africa*. Cape Town: Oxford University Press. 20–32.
Lindesay, J. A. 1990. 'Mechanisms of climatic change: a review', *South African Journal of Science* 86: 340–49.
Martin, A. K. 1987. 'Comparison of Sedimentation Rates in the Natal Valley, south-west Indian Ocean, with Modern Sediment Yields in East Coast Rivers of Southern Africa', *South African Journal of Science* 83: 716–24.
Meadows, M. E. 1996. 'Biogeography', in W. M. Adams, A. S. Goudie, and A. R. Orme (eds.), *The Physical Geography of Africa*. New York: Oxford University Press. 161–72.
Meyer, W. B., and B. Turner (eds.). 1994. *Changes in Land Use and Land Cover: a Global Perspective*. Cambridge: Cambridge University Press.
Moore, P. D., B. Chaloner, and P. Stott. 1996. *Global Environmental Change*. Oxford: Blackwell.
Nicholson, S. E. 1996. 'Environmental changes within the historical record', in W. M. Adams, A. S. Goudie, and A. R. Orme (eds.), *The Physical Geography of Africa*. New York: Oxford University Press. 60–87.
Palmer, A. R., C. G. Hobson, and M. T. Hoffman. 1990. 'Vegetation change in a semi-arid succulent dwarf shrubland in the eastern Cape, South Africa', *South African Journal of Science* 86: 392–5.
Partridge, T. C. 1990. 'Cainozoic Environmental Changes in Southern Africa', *South African Journal of Science* 86: 315–17
———. 1997. 'Cainozoic environmental change in southern Africa', *Progress in Physical Geography* 21 (1): 3–22.
Partridge, T. C., and R. R. Maud. 1987. 'Geomorphic evolution of southern Africa since the Mesozoic', *South African Journal of Geology* 90: 179–208.
Partridge, T. C., D. M. Avery, G. A. Botha, J. S. Brink, J. Deacon, R. S. Herbert, R. R. Maud, A. Scholtz, L. Scott, A. S. Talma, and J. S. Vogel. 1990. 'Late Pleistocene and Holocene climate change in southern Africa', *South African Journal of Science* 86: 302–5.
Prell, W. L., W. H. Hutson, and D. F. Williams. 1979. 'The subtropical convergence and late Quaternary circula-

tion in the southern Indian Ocean', *Marine Micropalaeontology* 4: 225–34.
Schulze, R. E. 1990. 'Climate change and hydrological response in southern Africa: heading towards the future', *South African Journal of Science* 86: 373–81.
———. 1997. 'Impacts of global climate change in a hydrologically vulnerable region: challenges to South African hydrologists', *Progress in Physical Geography* 21 (1): 113–36.
Scott, L. 1982. 'A Late Quaternary pollen record from the Transvaal Bushveld, South Africa', *Quaternary Research* 17: 339–70.
———. 1988. 'The Pretoria saltpan: a unique source of Quaternary palaeoenvironmental information', *South African Journal of Science* 84: 560–62.
Shaw, P. A. 1988. 'Lakes and Pans', in B. P. Moon and G. F. Dardis (eds.), *The Geomorphology of Southern Africa*. Johannesburg: Southern Books. 120–40.
Shaw, P. A., and D. S. G. Thomas. 1988. 'Lake Caprivi: a Late Quaternary Lake below the Zambezi and Mid-Kalahari drainage systems', *Zietschrift für Geomorphology* 32: 329–37.
Stocking, M. A. 1996. 'Soil Erosion', in W. M. Adams, A. S. Goudie, and A. R. Orme (eds.), *The Physical Geography of Africa*. New York: Oxford University Press. 326–41.
Sugden, J. M. 1989. 'Late Quaternary palaeoecology of the central and marginal uplands of the Karoo, South Africa'. Unpublished Ph.D. thesis, University of Cape Town.
Summerfield, M. A. 1996. 'Tectonics, geology and long-term landscape development', in W. M. Adams, A. S. Goudie, and A. R. Orme (eds.), *The Physical Geography of Africa*. New York: Oxford University Press. 1–17.
Talma, A. S., and Vogel, J. C. 1988. 'Southern Cape environmental temperatures since the Last Glacial', in I. A. W. MacDonald and R. J. M. Crawford (eds.), *Long-Term Data Series Relating to southern Africa's Renewable Natural Resources*. South African National Scientific Programmes Report No. 157. CSIR, Pretoria. 395–96.
Tiffen, M., M. Mortimore, and F. Gichuki. 1993. *More People, Less Erosion*. Chichester: Wiley.
Turner, B. L., R. E. Kasperson, B. W. Meyer, K. M. Dow, D. Golding, J. X. Kasperson, R. C. Mitchell, and S. J. Rattick. 1990. 'Two types of global environmental change: definitional and spatial-scale issues in their human dimensions', *Global Environmental Change* 1: 14–22.
Tyson, P. D. 1986. *Climate change and variability in South Africa*. Cape Town: Oxford University Press.
———. 1990. 'Modelling climatic change in southern Africa: a review of available methods', *South African Journal of Science* 86: 318–29.
Tyson, P. D., and J. A. Lindesay. 1992. 'The climate of the last 2000 years in southern Africa', *The Holocene* 2 (3): 271–8.
Zawada, P. K., J. Hattingh, and D. van Bladeren. 1996. *Palaeoflood Hydrological Analysis of of Selected South African Rivers*. Water Research Commission Report No. 509/1/96, Pretoria.

CHAPTER 3

Bickford-Smith, V. 1995. *Ethnic pride and racial prejudice in Victorian Cape Town*. Johannesburg: Witwatersrand University Press.
Christopher, A. J. 1984. *South Africa: The impact of past geographies*. Cape Town: Juta.
———. 1994. *The Atlas of Apartheid*. London: Routledge.
Crush, J. 1987. *The struggle for Swazi labour, 1890–1920*. Montreal: McGill-Queen's University Press.
———. 1994. 'Scripting the Compound: Power and Space in the South African Mining Industry', *Environment and Planning D: Society and Space* 12: 301–24.
Etherington, N. 1989. 'The "Shepstone system" in the Colony of Natal and beyond the borders', in A. Duminy and B. Guest (eds.), *Natal and Zululand from earliest times to 1910: A new history*. Pietermaritzburg: University of Natal Press and Shuter and Shooter.
Giliomee, H. 1979. 'The growth of Afrikaner identity', in A. Duminy and B. Guest (eds.), *Ethnic power mobilized: Can South Africa change?* New Haven and London: Yale University Press.
Guelke, L. 1989. 'Freehold farmers and frontier settlers, 1657–1780', in R. Elphick and H. Giliomee (eds.), *The shaping of South African society, 1652–1840*. Middletown: Wesleyan University Press.
Hart, D. M., and G. Pirie. 1984. 'The sight and soul of Sophiatown'. *Geographical Review* 74 (1): 38–47.
Keegan, T. 1996. *Colonial South Africa and the origins of the racial order*. London: Leicester University Press.
Lester, A. 1996. *From colonization to democracy: A new historical geography of South Africa*. London: I. B. Tauris.
———. 1997. 'The margins of order: Strategies of segregation on the eastern Cape frontier, 1806–*c*. 1850', *Journal of Southern African Studies* 23 (4): 635–53.
———. 1998. '"Otherness" and the frontiers of empire: The eastern Cape Colony, 1806–*c*. 1850', *Journal of Historical Geography* 24 (1): 2–19.
Mabin, A. 1986. 'Labour, capital, class struggle and the origins of residential segregation in Kimberley, 1880–1920', *Journal of Historical Geography* 12 (1): 4–26
Maharaj, B. 1995. 'The local state and residential segregation: Durban and the prelude to the Group Areas Act', *South African Geographical Journal* 77 (1): 33–41.
Marks, S. 1985a. 'Southern Africa, 1867–1886', in R. Oliver and G. N. Sanderson (eds.), *The Cambridge history of Africa*. Cambridge: Cambridge University Press.
———. 1985b. 'Southern and Central Africa, 1886–1910', in R. Oliver and G. N. Sanderson (eds.), *The Cambridge history of Africa*. Cambridge: Cambridge University Press.
Marks, S., and S. Trapido. 1979. 'Lord Milner and the South African State', *History Workshop Journal* 8: 50–80.
———. 1987. 'The politics of race, class and nationalism', in S. Marks and S. Trapido (eds.), *The politics of race, class and nationalism in twentieth century South Africa*. London: Longman.
Maylam, P. 1990. 'The rise and decline of urban apartheid in South Africa', *African Affairs* 89: 57–84.
Newton-King, S. 1994. 'The enemy within', in N. Worden and C. Crais (eds.), *Breaking the chains: Slavery and its legacy in the nineteenth-century Cape Colony*. Johannesburg: Witwatersrand University Press.
Oliver, R., and G. N. Sanderson (eds.). 1985. *The Cambridge History of Africa, Volume 6*. Cambridge: Cambridge University Press.
Parnell, S. 1988. 'Land acquisition and the changing residential face of Johannesburg, 1930–1955', *Area* 20: 307–14.

———. 1993. 'Creating racial privilege: The origins of South African public health and town planning legislation', *Journal of Southern African Studies* 19: 471–88.
Parnell, S., and G. Pirie. 1991. 'Johannesburg', in A. Lemon (ed.), *Homes apart: South Africa's segregated cities*. London, Bloomington, and Cape Town: Paul Chapman, Indiana University Press, and David Philip.
Parsons, N. 1993. *A new history of Southern Africa*. London: Macmillan.
Pirie, G. 1991. 'Kimberley', in A. Lemon (ed.), *Homes apart: South Africa's segregated cities*. London, Bloomington, and Cape Town: Paul Chapman, Indiana University Press, and David Philip.
Posel, D. 1991. *The making of apartheid, 1948–61: Conflict and compromise*. Oxford: Oxford University Press.
Price, R. 1991. *The apartheid state in crisis: Political transformation in South Africa, 1975–1990*. Oxford: Oxford University Press.
Robinson, J. 1996. *The power of apartheid: State, power and space in South African cities*. Oxford: Butterworth Heinemann.
Scully, P. 1994. 'Private and public worlds of emancipation in the rural Western Cape, *c*. 1830–42', in N. Worden and C. Crais (eds.), *Breaking the chains: Slavery and its legacy in the nineteenth-century Cape Colony*. Johannesburg: Witwatersrand University Press.
Shell, R. C. H. 1994. *Children of bondage: A social history of the slave society at the Cape of Good Hope, 1652–1838*. Johannesburg: Witwatersrand University Press.
Swanson, M. 1995. 'The sanitation syndrome: Bubonic plague and urban native policy in the Cape Colony, 1900–09', in W. Beinart and S. Dubow (eds.), *Segregation and apartheid in twentieth century South Africa*. London: Routledge.
Thompson, L. 1990. *A History of South Africa*. New Haven: Yale University Press.
Van Onselen, C. 1982. *Studies in the social and economic history of the Witwatersrand, 1886–1914* (2 vols). Johannesburg: Ravan Press.
Western, J. 1996. *Outcast Cape Town*. Berkeley: University of California Press.

CHAPTER 4

African National Congress. 1994. *The Reconstruction and Development Programme*. Johannesburg: Umanyano.
Central Statistical Services. 1997. *Preliminary Census Report*. Pretoria: Central Statistical Services.
Christopher, A. J. 1984. *The British empire at its zenith*. London: Croom Helm.
Cobbett, W., et al. 1987. 'South Africa's Regional Political Economy: A Critical Analysis of Reform Strategy in the 1980s', in R. Tomlinson and M. Addelson (ed.), *Regional Restructuring under Apartheid: Urban and Regional Policies in Contemporary South Africa*. Johannesburg: Ravan Press.
February, V. 1991. *The Afrikaners of South Africa*. London: Kegan Paul International.
Fox, R. 1995. 'Regional proposals: their constitutional and geographical significance', in A. Lemon (ed.), *The Geography of Change in South Africa*. Chichester: Wiley.
Griffiths, I. L., and D. C. Funnell. 1991. 'The abortive Swazi land deal', *African Affairs* 90 (358): 51–64.
Katzenellenbogen, S. 1996. 'It didn't happen at Berlin: Politics, economics and ignorance in the setting of Africa's colonial boundaries', in Paul Nugent and A. I. Asiwaju (ed.), *African boundaries: barriers, conduits and opportunities*. London: Pinter.
Khosa, M., and Y. Muthien. 1997. 'The expert, the public and the politician: drawing South Africa's new provincial boundaries', *South African Geographical Journal* 79 (1): 1–12.
Lemon, A. 1987. *Apartheid in transition*. Aldershot: Gower.
———. 1991. *Homes Apart: South Africa's Segregated Cities*. London: Paul Chapman.
——— (ed.). 1995. *The Geography of Change in South Africa*. Chichester: Gower.
———. 1996a. 'Lesotho and the new South Africa: the question of incorporation', *Geographical Journal* 162 (3): 263–72.
———. 1996b. 'The new political geography of the local state in South Africa', *Malaysian Journal of Tropical Geography* 27 (2): 35–45.
Lupton, M., and C. Mather. 1997. 'The anti-politics machine: GIS and the reconstruction of the Johannesburg local state', *Political Geography* 16 (7): 565–80.
Muthien, Y., and M. Khosa. 1995. '"The Kingdom, the Volkstaat and the New South Africa": Drawing South Africa's New Regional Boundaries', *Journal of Southern African Studies* 21 (10): 303–22.
Nugent, P., and A. I. Asiwaju. 1996. *African boundaries: barriers, conduits and opportunities*. London: Pinter.
Platzky, L., and C. Walker. 1985. *The Surplus People: Forced Removals in South Africa*. Johannesburg: Ravan Press.
Sidaway, J. D., and R. Gibb. 1998. 'SADC, COMESA, SACU: Contradictory formats for regional integration in southern Africa?', in D. Simon (ed.), *South Africa in Southern Africa: reconfiguring the region*. Oxford: James Currey.
Sidaway, J. D., and D. Simon. 1993. 'Geopolitical transition and state formation: the changing political geographies of Angola, Mozambique and Namibia', *Journal of Southern African Studies* 19 (1): 6–28.
Simon, D. 1991. *Independent Namibia one year on*. Conflict Studies 239, Research Institute for the Study of Conflict and Terrorism, London. Reprinted, 1995, in W. Gutteridge (ed.), *South Africa: from apartheid to national unity*. Aldershot: Dartmouth.
———. 1992. 'Reform in South Africa and modernization of the apartheid city', in D. Drakakis-Smith (ed.), *Urban and Regional Change in Southern Africa*. London: Routledge.
———. 1996a. 'Strategic territory and territorial strategy: the geopolitics of Walvis Bay's reintegration into Namibia', *Political Geography* 15 (2): 193–219.
———. 1996b. 'Restructuring the local state in post-apartheid cities: Namibian experience and lessons for South Africa'. *African Affairs* 95: 51–84.
———. 1996c. 'What's in a map? Regional restructuring and the state in independent Namibia', *Regional Development Dialogue* 17 (2): 1–31.
——— (ed.). 1998. *South Africa in Southern Africa: reconfiguring the region*. Oxford: James Currey.
South African Institute of Race Relations. 1996. *Fast Facts No. 5 / 96*. Johannesburg: SAIRR.
Swilling, M. 1997. 'Building democratic local urban governance in southern Africa', in M. Swilling (ed.), *Governing Africa's cities*. Johannesburg: Witwatersrand University Press.
Wilson, F., and M. Ramphele. 1989. *Uprooting Poverty:*

The South African Challenge. Cape Town: David Philip.
Wood, B. 1991. 'Preventing the vacuum: determinants of the Namibia settlement', *Journal of Southern African Studies* 17 (4): 742–63.

CHAPTER 5

Ahwireng-Obeng, F., and P. J. McGowan. 1998. 'Partner or Hegemon? South Africa in Africa', *Journal of Contemporary African Studies* 16 (1): 5–38.
African National Congress. 1994. *The Reconstruction and Development Programme*. Johannesburg: Umanyano.
Berry, B. J. L., E. C. Conkling, and D. Michael Ray. 1997. *The Global Economy in Transition* (2nd ed.). Upper Saddle River: Prentice-Hall.
Binns, T. (ed.). 1995. *People and Environment in Africa*. Chichester: John Wiley.
Brett, E. A. 1985. *The World Economy Since the War: The Politics of Uneven Development*. Basingstoke: Macmillan.
Carim, X. 1993. *Strategic Perspectives for Southern Africa in the 1990s*. Centre for Southern African Studies: University of the Western Cape, Bellville.
Cassim, F. 1988. 'Growth, crisis and change in the South African economy', in J. Suckling and L. White (eds.), *After Apartheid: Renewal of the South African Economy*. London: James Curry. 1–18.
Development Bank of Southern Africa. 1991. *Economic and Social Memorandum: Region D: 1990*. Halfway House: DBSA.
De Souza, A. R. 1990. *A Geography of the World Economy*. New York: Macmillan.
Economist Intelligence Unit 1998a. *Country Report: First Quarter 1998, Botswana and Lesotho*. London: Economist Intelligence Unit.
———. 1998b. *Country Report: South Africa: 1st Quarter 1998*. London: Economist Intelligence Unit.
Fair, T. J. D. 1982. *South Africa: Spatial Frameworks for Development*. Cape Town: Juta.
Harvey, C., and S. R. Lewis. 1990. *Policy Choice and Development Performance in Botswana*. Basingstoke: Macmillan.
Healey, M. J., and B. W. Ilbery. 1990. *Location and Change: Perspectives on Economic Geography*. Oxford: Oxford University Press.
Hodd, M. 1991. *The Economies of Africa*. Aldershot: Dartmouth.
Judd, D., and M. Parkinson. 1990. *Leadership and Urban Regeneration: Cities in North America and Europe*. Newbury Park: Sage.
Knox, P., and J. Agnew. 1989. *The Geography of the World Economy* (1st ed.). London: Edward Arnold.
———. 1994. *The Geography of the World Economy* (2nd ed.). London: Edward Arnold.
Libby, R. T. 1987. *The Politics of Economic Power in Southern Africa*. Princeton: Princeton University Press.
Lipton, M., and S. Simkins. 1993. *State and Market in Post Apartheid South Africa*. Johannesburg: Witwatersrand University Press.
Louw, C. 1992. 'What hope the Windhoek Treaty?' *Die Suid-Afrikaan* 41: 10–11.
May, J. D., and S. E. Stavrou. 1989. *The Informal Sector: socio-economic dynamics and growth in the greater Durban Metropolitan Region*. Centre for Social and Development Studies, University of Natal, Durban.
Mwanza, A. M. (ed.). 1992. *Structural Adjustment Programmes in SADC*. Harare: SAPES Books.
Nattrass, N., and E. Ardington. 1990. *The Political Economy of South Africa*. Cape Town: Oxford University Press.
Nattrass, N., and J. Seekings. 1996. 'The Challenge Ahead: Unemployment and inequality in South Africa', *SA Labour Bulletin* 20 (1): 66–72.
Nel, E. L. 1996. 'Regional and Local Economic Development Strategies in the Eastern Cape and Guidelines for Future Development'. Unpublished Ph.D. thesis, Rhodes University, Grahamstown.
Pickles, J. 1991. 'Industrial restructuring, peripheral industrialization, and rural development in South Africa', *Antipode* 23 (1): 68–91.
Rogerson, C. M. 1991. 'Beyond Racial Fordism; Restructuring industry in the "new" South Africa', *Tijdschrift voor Economische en Sociale Geografie* 82 (5): 355–66.
Rogerson, C. M., and G. H. Pirie. 1979. 'Apartheid, urbanization, and regional planning in South Africa', in R. A. Obudho and S. El-Shakhs (eds.), *Developing Urban Systems in Africa*. New York: Praeger. 323–4.
Scammell, W. M. 1980. *The International Economy Since 1945*. London: Macmillan.
South Africa. 1996. *Growth, Employment and Redistribution: A Macroeconomic Strategy*. Pretoria: Department of Finance.
———. 1998a. *Board for Regional Industrial Development – Statistics*. Pretoria: Department of Trade and Industry.
———. 1998b. White Paper on Local Government. Pretoria: Ministry of Provincial Affairs and Constitutional Development.
Stutterheim Development Foundation. 1997. *Stutterheim Integrated Local Development*. Stutterheim: Stutterheim Development Foundation.
Sylvester, C. 1991. *Zimbabwe: The Terrain of Contradictory Development*. Boulder: Westview.
Taylor, D. R. F., and F. Mackenzie (eds.). 1992. *Development from Within: Survival in Rural Africa*. London: Routledge.
Todaro, M. P. 1994. *Economic Development* (5th ed.). Burnt Mill: Longman.
Tomlinson, R., and M. Addelson. 1987. *Regional Restructuring Under Apartheid: Urban and Regional Policies in Contemporary South Africa*. Johannesburg: Ravan Press.
Wolpe, H. 1972. 'Capitalism and cheap labour power in South Africa', *Economy and Society* 1 (4): 431–3.
World Bank. 1994. *World Development Report*. Washington, DC: World Bank.

CHAPTER 6

Badenhorst, C. 1992. 'Popular culture and cultural geography', in C. Rogerson and J. McCarthy (eds.), *Geography in a changing South Africa: progress and prospects*. Cape Town: Oxford University Press.
Christopher, A. J. 1976. *Southern Africa*. Folkestone: Dawson.
———. 1994. 'Regionalisation and ethnicity in South Africa, 1990–1994', *Area* 27: 1–11.
Cloke, P., C. Philo, and D. Sadler. 1991. *Approaching human geography: an introduction to contemporary theoretical debates*. London: Paul Chapman Publishing.
Crosby, A. W. 1986. *Ecological imperialism: the biological expansion of Europe, 900–1900*. Cambridge: Cambridge University Press.
Daniel, J. B. McI. 1973. 'A geographical study of pre-

Shakan Zululand', *South African Geographical Journal* 55: 23–31.
De Blij, H. J. 1993. *Human geography: culture, society and space*. New York: Wiley & Sons.
Edney, M. H. 1997. *Mapping an empire: the geographical construction of British India, 1765–1843*. Chicago and London: University of Chicago Press.
Forbes, V. S. 1965. *Pioneer Travellers in South Africa*. Cape Town: Balkema.
Fox, R. C., and D. J. Tipler. 1996. 'Rural and urban population trends in the Eastern Cape Province 1936–1991'. Institute of Social and Economic Research Development Studies Working Paper 68, Rhodes University, Grahamstown.
Goodall, B. 1987. *The Penguin Dictionary of Human Geography*. Harmondsworth: Penguin.
Grove, R. H. 1995. *Green imperialism*. Cambridge: Cambridge University Press.
Harvey, D. 1989. *The condition of postmodernity: an enquiry into the origins of cultural change*. Oxford: Blackwell.
Haswell, R. 1979. 'South African towns on European plans', *Geographical Magazine* 51: 686–94.
Jackson, P. 1989. *Maps of meaning*. London: Unwin Hyman.
———. 1992. 'Social and cultural geography', in A. Rogers, H. Viles, and A. Goudie (eds.), *Students' companion to geography*. Oxford: Blackwell.
Johnston, R. J., D. Gregory, and D. M. Smith (eds.). 1986 and 1994. *Dictionary of human geography*. Oxford: Blackwell.
Johnston, R. J., P. J. Taylor, and M. J. Watts (eds.). 1995. *Geographies of global change: remapping the world in the late twentieth century*. Oxford: Blackwell.
Kobayashi, A. 1993. 'Multiculturalism: representing a Canadian institution', in J. Duncan and D. Ley (eds.), *Place / Culture / Representation*. London: Routledge.
Lammas, R. 1992. 'Voices and visions of place: new terrain for humanistic geography', in C. Rogerson and J. McCarthy (eds.), *Geography in a changing South Africa: progress and prospects*. Cape Town: Oxford University Press.
———. 1993. 'Portraits of place: images of Southern Africa in landscape painting'. Unpublished M.A. thesis, University of the Witwatersrand.
Mitchell, D. 1995. 'There's no such thing as culture: towards a reconceptualization of the idea of culture in geography', *Transactions of the Institute of British Geographers*, N.S. 20: 102–16.
Rubinstein, J. M. 1996. *The cultural landscape: an introduction to human geography*. Englewood Cliffs, NJ: Prentice-Hall.
Silva, P. 1996. *A dictionary of South African English on historical principles*. Cape Town: Oxford University Press.
South Africa. 1955. *Summary of the report of the Commission for the socio-economic development of the Bantu areas within the Union of South Africa*. UG 61/1955. Pretoria: Government Printer.
Van der Merwe, I. J., and L. O. van Niekerk. 1994. *Language in South Africa: distribution and change*. Stellenbosch: University of Stellenbosch.
Williams, R. 1976. *Keywords: a vocabulary of culture and society*. London: Fontana/Croom Helm.
Wilmsen, E., S. Dubow, and M. Sharp. 1993. 'Introduction: Ethnicity, identity and nationalism in South Africa', *Journal of Southern African Studies* 20 (3): 347–53.
Zelinsky, W. 1973. *The cultural geography of the United States*. Englewood Cliffs, NJ: Prentice-Hall.

CHAPTER 7

Ashford, L. S. 1995. 'New perspectives on population: Lessons from Cairo', *Population Bulletin* 50 (1). Washington, DC: Population Reference Bureau.
Barrett, H. R. 1992. *Population Geography*. Harlow: Oliver and Boyd.
Boserup, E. 1965. *The Conditions of Agricultural Growth*. London: Allen & Unwin.
Calitz, J. M. 1996. 'Population of South Africa: Updated estimates, scenarios and projections, 1990–2020'. Development Paper 109, Centre for Policy and Information. Halfway House: Development Bank of South Africa.
Editors Inc. 1997. *SA 97–98: South Africa at a glance*. Craighall: Editors Inc.
Haldenwang, B. B. 1993. 'The demography of Africa'. Occasional Paper 22, Institute for Futures Research, University of Stellenbosch.
———. 1997. 'A socio-demographic profile of the Southern African Development Community region'. Occasional Paper 36, Institute for Futures Research, University of Stellenbosch.
Haldenwang, B. B., and S. C. Boshoff. 1996. 'Forecasts of the South African population 1991–2026'. Occasional Paper 26, Institute for Futures Research, University of Stellenbosch.
Haub, C., and M. Yanagishita. 1996. *World Population Data Sheet 1996*. Washington, DC: Population Reference Bureau.
Kane, H. 1995. 'The hour of departure: forces that create refugees and migrants'. Worldwatch Paper 125. Washington, DC: The Worldwatch Institute.
Meadows, D. H., D. L. Meadows, and J. Randers. 1972. *The Limits to growth: a report for the Club of Rome's project on the predicament of mankind*. New York: Universe Books.
Plane, D. A., and P. A. Rogerson. 1994. *The geographical analysis of population – with applications to planning and business*. New York: John Wiley and Sons.
Ribe, H. 1995. *Malawi – Human resources and poverty: profile and priorities for action*. Washington, DC: World Bank, Southern Africa Department.
Roudi, F., and L. Ashford. 1996. *Men and family planning in Africa*. Washington, DC: Population Reference Bureau.
South Africa 1995. *Human development index (HDI) for the RSA: 1980 and 1991*. Pretoria: Central Statistics.
———. 1996a. The Constitution of the Republic of South Africa. Act 108 of 1996. Pretoria: Government Printer.
———. 1996b. *Marriages and divorces 1995*. Pretoria: Central Statistics.
———. 1996c. *October household survey 1995*. Pretoria: Central Statistics.
———. 1996d. Draft White Paper for a Population Policy. Pretoria: Government Printer.
———. 1997. *Census '96: preliminary estimates of the size of the population of South Africa*. Pretoria: Central Statistics.
Steenkamp, H. A. 1997. *Population estimates for the RSA by magisterial district and province*. Report 243. Pretoria: Bureau of Market Research.
Thiessen, H. 1997. *Measuring the real world: A textbook of applied statistical methods*. New York: John Wiley & Sons.
United Nations Population Fund. 1996. *The state of world population (1996) – Changing places: population,*

development and the urban future. New York: UNFPA.
Van der Merwe, I. J., and L. O. van Niekerk. 1994. *Language in South Africa – Distribution and Change*. Stellenbosch: University of Stellenbosch.
Witherick, M. E. 1990. *Population geography*. London: Longman.
World Bank. 1996. *World resources 1996–97: A guide to the global environment*. New York: Oxford University Press.
Yanagishita, M., and N. Yinger. 1995. *The world's women 1995*. Data Sheet. Washington, DC: Population Reference Bureau.

CHAPTER 8

Berry, B. J. L. 1970. 'The geography of the United States in the year 2000'. *Transactions of the Institute of British Geographers* 51: 21–53.
Carter, H. 1995. *The Study of Urban Geography* (4th ed.). New York: John Wiley.
Christopher, A. J. 1983. 'From Flint to Soweto: reflections on the colonial origins of the apartheid city', *Area* 15 (2): 145–9.
———. 1989. 'Spatial variations in the application of residential segregation in South African cities', *Geoforum* 20 (3): 253–67.
———. 1994. *The Atlas of Apartheid*. Johannesburg: Witwatersrand University Press.
Crankshaw, O., and S. Parnell. 1996. 'Housing policy and the need for an urbanization policy in the new South Africa', *Urban Forum* 7 (2): 232–7.
Davies, R. J. 1981. 'The spatial formation of the South African city'. *Geojournal*, Supplementary Issue 2: 59–72.
Drakakis-Smith, D. 1987. *The third world city*. London: Methuen.
HABITAT. 1996. *An urbanizing world: Global report on human settlements 1996* Oxford: Oxford University Press, for the United Nations Centre for Human Settlements (HABITAT).
Harris, N. (ed.). 1992. *Cities in the 1990s: the challenge for the developing countries*. London: UCL Press.
Hart, G. 1989. 'On grey areas'. *South African Geographical Journal* 71 (2): 81–8.
Hattingh, P. S., and A. C. Horn. 1991. 'Pretoria', in Lemon (ed.), *Homes apart: South Africa's segregated cities*. Cape Town: David Philip.
Hoselitz, B. F. 1957. 'Generative and parasitic cities', *Economic Development and Cultural Change* 3: 278–94.
Lelyveld, J. 1987. *Move your shadow: South Africa, Black and White*. London: Sphere.
Lemon, A. 1977. 'Urban primacy and regional economic development in the third world'. *Proceedings of the Geographical Association of Rhodesia* 10: 3–14.
———. 1982. 'Migrant labour and frontier commuters: reorganizing South Africa's Black labor supply', in D. M. Smith (ed.), *Living under Apartheid*. London: George Allen and Unwin.
——— (ed.). 1991. *Homes apart: South Africa's segregated cities*. Cape Town: David Philip.
——— (ed.). 1995. *The Geography of Change in South Africa*. Chichester: John Wiley.
———. 1996a. 'The new political geography of the local state in South Africa'. *Malaysian Journal of Tropical Geography* 27 (2): 35–45.
———. 1996b. 'Post-apartheid cities: planning for a new society', in R. J. Davies (ed.), *Contemporary City Structuring*. Cape Town: IGU Commission on Urban Development and Society of South African Geographers. 62–73.
———. 1999. 'Urbanization strategy in the new South Africa: the role of secondary cities and small towns', in C. de Wet and R. Fox (eds.), *Transforming Settlement in Southern Africa*. Edinburgh: Edinburgh University Press.
Lipton, M. 1977. *Why the poor stay poor: urban bias in world development*. London: Temple Smith.
Lupton, M., and C. Mather. 1996. '"The anti-politics machine": GIS and the reconstruction of the Johannesburg local state', *Political Geography* 16 (7): 565–80.
Mabin, A. 1989. 'Struggle for the city: urbanisation and political strategies of the South African state', *Social Dynamics* 15 (1): 1–28.
Maylam, P. 1990. 'The rise and decline of urban apartheid in South Africa', *African Affairs* 89 (354): 57–84.
Perlman, J. 1976. *The myths of marginality. urban poverty and politics in Rio de Janeiro*. Los Angeles: University of California Press.
Potter, R. 1992. *Urbanisation in the third world*. Oxford: Oxford University Press.
Robinson, J. 1996. *The Power of apartheid: state, power and space in South African cities*. Oxford: Butterworth Heinemann.
Rogerson, C. M. 1989. 'Rethinking national urban policies: lessons for South Africa', *South African Geographical Journal* 71 (3): 134–41.
Sidaway, J., and M. Power. 1995. 'Sociospatial transformations in the 'postsocialist' periphery: the case of Maputo, Mozambique', *Environment and Planning A* 27: 1463–91.
Simon, D. 1997. 'Urbanization, globalization and economic crisis in Africa', in C. Rakodi (ed.), *The urban challenge in Africa: growth and management of its large cities*. New York: United Nations University Press.
Smith, N. 1979. 'Towards a theory of gentrification: a back to the city movement by capital not people', *Journal of the American Planning Institute* 45: 538–48.
Toynbee, A. 1970. *Cities on the Move*. Oxford: Oxford University Press.
United Nations. 1995. *World Urbanization Prospects: the 1994 Revision*. New York: Population Division.
Western, J. 1981. *Outcast Cape Town*. London: George Allen and Unwin.
Winters, C. 1982. 'Urban Morphogenesis in Francophone Black Africa', *Geographical Review* 72 (2): 139–54.

CHAPTER 9

Andrew, M. 1991. 'A Geographical Study of Agricultural Change Since the 1930s in Shixini Location, Gatyana District, Transkei'. Unpublished M.A. thesis, Rhodes University, Grahamstown.
Bryceson, D. F., and J. Howe. 1996. 'An agrarian continent in transition', in S. Ellis (ed.), *Africa Now*. London: James Currey.
Chambers, R. 1997. *Whose Reality Counts: Putting the Last First*. London: ITB.
Christopher, A. J. 1994. *The Atlas of Apartheid*. London: Routledge.
Dixon, C. 1990. *Rural Development in the Third World*. London: Routledge.
Elliott, J. A. 1994. *An Introduction to Sustainable Development*. London: Routledge.

Enpat. Environmental Potential Atlas. Disk 1 [CD-ROM]. 1998. Available: Department of Environmental Affairs and Tourism [November 1998].

FAO [Food and Agriculture Organization]. 1998a. *The Rural-Urban Factfile* [Online]. Available: http://www.fao.org/news/factfile/ff9811-e.htm [29 December 1998].

———. 1998b. *World Food Summit: Food Security Situation and Issues in the Africa region* [Online]. Available: http://www.fao.org/wfs/resource/english/arc96-4.htm#WORLD1 [29 December 1998].

———. 1998c. *World Map. Food Supply Gap Between Rich and Poor Nations* [Online]. Available: http://www.fao.org/news/1998/981204-e.htm [29 December 1998].

GeographyCal. Teaching and Learning Technology Programme [CD-ROM]. 1997. Available: University of Leicester [October 1997].

Jaetzold, R., and H. Schmidt. 1982. *Farm Management Handbook of Kenya Vol II / A. West Kenya*. Rossdorf: German Agency for Technical Cooperation.

Johnston, R. J. 1984. *City and Society*. London: Hutchinson.

Le Heron, R. B. 1993. *Globalised Agriculture: Political Choice*. Oxford: Pergamon.

Lemon, A. 1976. *Apartheid: A Geography of Separation*. Farnborough: Saxon House.

———. 1987. *Apartheid in Transition*. Aldershot: Gower.

McGladdery, K. 1992. 'Political land division in South Africa in the context of agro-ecological zones'. Unpublished Geography III project, Rhodes University, Grahamstown.

Pickles, J., and D. Weiner. 1991. 'Rural and regional restructuring of apartheid: ideology, development policy and the competition for space'. *Antipode* 23 (1): 2–32.

Potts, D. 1992. 'The changing geography of southern Africa', in G. P. Chapman and K. M. Baker (eds.), *The Changing Geography of Africa and the Middle East*. London: Routledge.

Redclift, M. 1987. *Sustainable Development: Exploring the Contradictions*. London: Methuen.

South Africa. 1995. *Rural Development Strategy of the Government of National Unity*. Pretoria: Government Printer.

———. 1997. White Paper on South African Land Policy. Pretoria: Department of Land Affairs.

Tiffen, M., M. Mortimore, and F. Gichuki. 1995. *More People, Less Erosion. Environmental Recovery in Kenya*. Chichester: John Wiley.

Vance, J. E. 1970. *The Merchant's World. The Geography of Wholesaling*. Englewood Cliffs, NJ: Prentice-Hall.

Van Onselen, C. 1996. *The Seed is Mine: The Life of Kas Maine, a South African Sharecropper, 1894–1985*. Cape Town: David Philip.

Van Zyl, J., J. Kirsten, and H. P. Binswanger (eds.). 1996. *Agricultural Land Reform in South Africa*. Cape Town: Oxford University Press.

Waugh, D. 1995. *Geography. An Integrated Approach*. Walton-on-Thames: Nelson and Sons.

Zinyama, L. M. 1995. 'Sustainability of smallholder food production systems in southern Africa: the case of Zimbabwe', in T. Binns (ed.), *People and Environment in Africa*. Chichester: John Wiley.

CHAPTER 10

Hilling, D. 1996. *Transport in developing countries*. London: Routledge.

Hoyle, B. S., and J. Charlier. 1995. 'Inter-port competition in developing countries: an East African case study', *Journal of Transport Geography* 3 (2): 87–103.

Hoyle, B. S., and R. D. Knowles (eds.). 1992. *Modern transport geography*. London: Belhaven/Wiley.

Iheduru, O. C. 1996. 'Post-apartheid South Africa and its neighbours: a maritime transport perspective', *Journal of Modern African Studies* 34 (1): 1–26.

Khosa, M. M. 1990. 'Changing patterns of 'black' bus subsidies in the apartheid city, 1944–1986', *GeoJournal* 22 (3): 251–9.

———. 1991a. 'Capital accumulation in the black taxi industry', in E. Preston-Whyte and C. Rogerson (eds.), *South Africa's informal economy*. Cape Town: Oxford University Press.

———. 1991b. 'Routes, ranks and rebels: feuding in the taxi revolution', *Journal of Southern African Studies* 18 (1): 232–51.

———. 1992. 'Changing state policy and the black taxi industry in Soweto', in D. M. Smith (ed.), *The apartheid city and beyond*. London: Routledge.

———. 1994. 'Accumulation and labour relations in the black taxi industry', *Transformation* 24: 55–71.

———. 1995. 'Transport and popular struggles in South Africa', *Antipode* 27 (2): 167–88.

Lloyd's List. 1997. Special Report: South Africa. 27 June. London. 6–10.

Mabin, A. 1986. 'Labour, capital, class struggle and the origins of residential segregation in Kimberley, 1880–1920', *Journal of Historical Geography* 12 (1): 4–26.

Mann, M., and S. Segal. 1985. 'Carrying apartheid's burden', *Work in Progress* 38: 19–23.

Mutizwa-Mangiza, N. D. 1993. 'Urban informal transport policy: the case of emergency taxis in Harare', in L. M. Zinyama, D. S. Tevera, and S. D. Cumming (eds.), *Harare: the growth and problems of the city*. Harare: University of Zimbabwe Publications.

Maunder, D. A. C., and T. C. Mbara. 1995. *The initial effects of introducing commuter omnibus services in Harare, Zimbabwe*. Crowthorne, UK: Transport Research Laboratory Report 123.

Mwase, N. 1984. 'Transport bottleneck and Tanzania's economic problems', *Transport Reviews* 4 (2): 121–41.

———. 1987. 'Zambia, the TAZARA and the alternative outlets to the sea', *Transport Reviews* 7 (3): 191–206.

Noah, M. R. 1995. 'Minibus taxi safety issues, operators, patrons and the state in the Eastern Cape Province of South Africa, 1982–1994'. Unpublished Ph.D. thesis, University of London.

Pirie, G. H. 1982. 'The decivilizing rails: railways and underdevelopment in southern Africa' *Tijdschrift voor Economische en Sociale Geografie* 73 (4): 22–8.

———. 1987. 'African township railways and the South African state, 1902–1963', *Journal of Historical Geography* 13 (3): 283–95.

———. 1990. 'Dismantling bus apartheid in South Africa, 1975–1990', *Africa Insight* 20 (2): 111–17.

———. 1993. ' Slaughter by steam: railway subjugation of ox-wagon transport in the eastern Cape and Transkei, 1886–1910', *International Journal of African Historical Studies* 26 (2): 319–43.

Reichardt, M., and D. Duncan. 1990. 'Rail transport and the political economy of Southern Africa, 1965–1980', *Africa Insight* 20 (2): 100–109.

Simon, D. 1996. *Transport and development in the Third World*. London: Routledge.

———. 1998. 'Desert enclave to regional gateway? Walvis Bay's reintegration into Namibia', in D. Simon (ed.),

South Africa in Southern Africa: reconfiguring the region. Oxford: James Currey; Cape Town: David Philip; Cincinatti: Ohio University Press.
South Africa. 1996a. Green Paper on National Transport Policy. (March). Pretoria: Department of Transport.
———. 1996b. White Paper on National Transport Policy. (August). Pretoria: Department of Transport.

CHAPTER 11

Buttrick, D. B., J. L. Van Rooy, and R. Ligthelm. 1993. 'Environmental geological aspects of the dolomites of South Africa', *Journal of African Earth Sciences* 16 (1/2): 53–61.
Callaghan, C. C., P. G. Eriksson, and C. P. Snyman. 1991. 'The sedimentology of the Waterberg Group, South Africa: an overview', *Journal of African Earth Sciences* 13 (1): 121–39.
De Wit, M. J., and L. D. Ashwall (eds.). 1996. *Greenstone belts*. Oxford University Press.
De Wit, M. J., and A. Hynes. 1995. 'The onset of interaction between the hydrosphere and oceanic crust, and the origin of the first continental lithosphere', in M. P. Coward and A. C. Ries (eds.), *Early Precambrian Processes*. London: Geological Society. 1–9.
Eriksson, K. A. 1995. 'Crustal growth, surface processes, and atmospheric evolution on the early Earth', in M. P. Coward and A. C. Ries (eds.), *Early Precambrian Processes*. London: Geological Society. 11–25.
Eriksson, P. G., and B. F. F. Reczko. 1995. 'The sedimentary and tectonic setting of the Transvaal Supergroup floor rocks to the Bushveld Complex', *Journal of African Earth Sciences* 21: 487–504.
Eriksson, P. G., K. C. Condie, H. Tirsgaard, W. U. Mueller, W. Altermann, A. D. Miall, L. B. Aspler, O. Catuneanu, and J. R. Chiarenzelli. 1998. 'Precambrian clastic sedimentation systems', *Sedimentary Geology*, in press.
Hamblin, W. K. 1992. *Earth's dynamic systems* (6th ed.). New York: Macmillan.
McCourt, S. 1995. 'The crustal architecture of the Kaapvaal crustal block South Africa, between 3.5 and 2.0 Ga. A synopsis', *Mineralium Deposita* 30 (2): 89–97.
Partridge, T. C., and R. R. Maud. 1987. 'Geomorphic evolution of southern Africa since the Mesozoic', *South African Journal of Geology* 90 (2): 179–208.
Petters, S. W. 1991. *Regional geology of Africa*. Berlin: Springer.
Pickering, K. T., and L. A. Owen. 1994. *An Introduction to global environmental issues*. London: Routledge.
Schopf, J. W. 1983. *Earth's earliest biosphere: its origin and evolution*. Princeton, NJ: Princeton University Press.
———. 1992. 'The oldest fossils and what they mean', in J. W. Schopf (ed.), *Major events in the history of life*. Boston: Jones and Bartlett. 29–64.
Smith, R. M. H., P. G. Eriksson, and W. J. Botha. 1993. 'A review of the stratigraphy and sedimentary environments of the Karoo-aged basins of Southern Africa', *Journal of African Earth Sciences* 16 (1/2): 143–69.
Snyman, C. P. (ed.). 1996. *Geologie vir Suid-Afrika* (Volumes 1 and 2). Department of Geology, University of Pretoria.
South African Committee for Stratigraphy. 1980. *Stratigraphy of South Africa (Compiler L. E. Kent). Lithostratigraphy of the Republic of South Africa, South West Africa (Namibia) and the Republics of Bophuthatswana, Transkei and Venda*. Handbook 8, Geological Survey of South Africa. Pretoria: Government Printer.
Stanistreet, I. G., and T. S. McCarthy. 1991. 'Changing tectono-sedimentary scenarios relevant to the development of the Late Archaean Witwatersrand Basin', *Journal of African Earth Sciences* 13 (1): 65–81.
Strahler, A. N. 1981. *Physical Geology*. New York: Harper and Row.
Tankard, A. J., M. P. A. Jackson, K. A. Eriksson, D. K. Hobday, D. R. Hunter, and W. E. L. Minter. 1982. *Crustal evolution of Southern Africa*. New York: Springer.
Tarbuck, E. J., and F. K. Lutgens. 1987. *The Earth: an introduction to physical geology* (2nd ed.). Columbus, OH: Merrill.
Thomas, R. J., M. W. von Veh, and S. McCourt. 1993. 'The tectonic evolution of Southern Africa: an overview', *Journal of African Earth Sciences* 16 (1/2): 5–24.
Truswell, J. F. 1977. *The geological evolution of South Africa*. Johannesburg: Purnell.
Van der Westhuizen, W. A., H. De Bruiyn, and P. G. Meintjies. 1991. 'The Ventersdorp Supergroup: an overview', *Journal of African Earth Sciences* 13 (1): 83–105.
Vermaak, C. F. 1997. 'A brief overview of South Africa's mineral industry: world context and changing local circumstances', *Mineralium Deposita* 32 (4): 312–22.
Von Gruenewaldt, G., and R. E. Harmer. 1993. 'Tectonic setting of Proterozoic layered intrusions with special reference to the Bushveld Complex', in K. C. Condie (ed.), *Proterozoic crustal evolution*. Amsterdam: Elsevier. 181–213.

CHAPTER 12

Adams, L. 1993. 'A rural voice, Strategies for drought relief', *Indicator South Africa* 10 (4): 41–6.
Ballard, C. 1986. 'Drought and economic disasters: South Africa in the 1980s', *Journal of Interdisciplinary History* 17: 359–78.
Bruwer, J. J. 1989. 'Drought policy in the Republic of South Africa, Part I', *Drought Network News* 1 (3): 14–16.
Dent, M. C., R. E. Schulze, H. M. M. Wills, and S. D. Lynch. 1987. 'Spatial and temporal analysis of the recent drought in the summer rainfall region of Southern Africa', *Water SA* 13: 37–42.
Freeman, C. 1984. 'Drought and agricultural decline in Bophuthatswana', in South African Research Services (eds.), *South African Review II*. Johannesburg: Ravan Press. 284–89.
Harrison, M. S. J. 1984. 'The annual rainfall cycle over the central interior of South Africa', *South African Geographical Journal* 66: 47–64.
———. 1988. 'Rainfall and precipitable water relationships over the central interior of South Africa', *South African Geographical Journal* 70: 100–111.
Hughes, W. S., and R. C. Balling. 1997. 'Urban influences on South African temperature series', *International Journal of Climatology* 16.
Hulme, M. 1996. 'Recent climate change in the world's drylands', *Geophysical Research Letters* 23: 61–4.
Jallow, S. S. 1995. 'Identification of and response to drought by local communities in Fulladu West District, The Gambia', *Singapore Journal of Tropical Geography* 16: 22–41.
Joubert, A., and B. Hewitson. 1997. 'Simulating present and future climates of southern Africa using General Circulation Models', *Progress in Physical Geography* 21: 51–78

Joubert, A., and S. J. Mason. 1996. 'Droughts over southern Africa in a doubled-CO2 climate', *International Journal of Climatology* 16: 1–8.
Joubert, A., and P. Tyson. 1996. 'Equilibrium and fully coupled GCM simulations of future southern African climates', *Southern African Journal of Science* 92: 471–84
Jury, M. R. 1993. 'A preliminary note on rainfall and vegetational trends in the south-western Cape', *South African Journal of Botany* 59: 265–9.
Lindesay, J. A. 1988. 'South African rainfall, the Southern Oscillation and a Southern Hemisphere semi-annual cycle', *Journal of Climatology* 8: 17–30.
———. 1998. 'Present Climates of Southern Africa', in J. E. Hobbs, J. A. Lindesay, and H. D. Bridgman (eds.), *Climates of the Southern Continents Present, Past and Future*. New York: Wiley.
Lindesay, J. A., and M. R. Jury. 1991. 'Atmospheric circulation controls and characteristics of a flood event in central South Africa', *International Journal of Climatology* 11: 609–27.
Lindesay, J. A., and C. H. Vogel. 1990. 'Historical evidence for Southern Oscillation-southern African rainfall relationships', *International Journal of Climatology* 10: 679–89.
Mapuwose, R. M. 1985. 'Pattern and impact of drought in the SADCC countries', *Proceedings of the Symposium on Drought in Africa*, International development research centre, the Canadian International development Agency and the International Council of Scientific Unions, Ottawa, 12–14 August, 1985. 27–36.
Mason, S. J. 1995. 'Sea-surface temperatures – South African rainfall associations, 1910–1989', *International Journal of Climatology* 15: 119–35.
Mason, S. J., and A. M. Joubert. 1997. 'Simulated changes in extreme rainfall over southern Africa', *International Journal of Climatology* 17: 291–301.
Mason, S. J., and M. R. Jury. 1997. 'Climatic variability and change over southern Africa: a reflection on underlying processes', *Progress in Physical Geography* 21 (1): 23–50.
Muller, M. J., and P. D. Tyson. 1988. 'Winter rainfall over the interior of South Africa during extreme dry years', *South African Geographical Journal* 70: 20–30.
Ngara, T., D. L. McNaughton, and S. Lineham. 1983. 'Seasonal rainfall fluctuations in Zimbabwe', *Zimbabwe Agricultural Journal* 80: 149–50.
Preston-Whyte, R. A., and P. D. Tyson. 1988. *Atmosphere and Weather of Southern Africa*. Cape Town: Oxford University Press.
Rowntree, K. M. 1988. 'Rainfall characteristics, rainfall reliability and the definition of drought: Baringo District, Kenya', *South African Geographical Journal* 71 (2): 74–80.
Spio, K., and L. Nchabeleng. 1997. Workshop on drought mitigation in the Northern Province of South Africa. Proceedings of the Drought Mitigation Workshop, Pietersburg, 11–12 March, 1997, held under the auspices of Disaster Mitigation for Sustainable Livelihoods Unit for Southern African Development Education and Policy Research, University of the Western Cape.
Taljaard, J. J. 1985. 'Cut-off lows in the South African region', South African Weather Bureau Technical Paper 14, SAWB, Pretoria.
———. 1986. 'Change of rainfall distribution and circulation patterns over southern Africa in summer', *Journal of Climatology* 6: 579–92.
Tyson, P. D. 1986. *Climate Change and Variability in southern Africa*. Cape Town: Oxford University Press.
———. 1996. 'South African Climatic Change and Variability', in L. Y. Shacleton, S. J. Lennon, and G. R. Tosen (eds.), *Global Climate Change and South Africa*. Cleveland, Johannesburg: Environmental Scientific Association.
Tyson, P. D., T. G. J. Dyer, and M. N. Mametse. 1975. 'Secular changes in South African rainfall: 1880 to 1972', *Quarterly Journal of the Royal Meteorological Society* 101: 817–33.
Van Zyl, J. 1993. 'The last straw: drought and the economy', *Indicator SA* 10 (4): 47–51.
Vogel, C. H. 1989. 'A documentary-derived chronology for southern Africa, 1820–1900', *Climate Change* 14: 291–306.
———. 1994. 'Human response to the environment', in T. Binns (ed.), *People and the Environment*. London: Wiley. 249–56.
———. 1998. 'Disaster management in South Africa', *South African Journal of Science* 94 (3): 98–100.
Von Storch, H., and F. W. Zwiers. 1988. 'Recurrence analysis of climate sensitivity experiments', *Journal of Climate* 1: 157–71.
Walker, N. D. 1990. 'Links between South African rainfall and temperature variability of the Aghulas and Benguela current systems', *Journal of Geophysical Research* 95: 3297–319.
Walker, N., J. Taunton-Clark, and J. Pugh. 1984. 'Sea temperatures off the South African west coast as indicators of Benguela warm events', *South African Journal of Science* 80: 72–7.
Wilson, J. F. 1865. 'Water supply in the basin of the river Orange or "Gariep South Africa"', *Journal of the Royal Geographical Society* 35: 106–29.

CHAPTER 13

Beckedahl, H. R. 1998. *Subsurface Soil Erosion Phenomena in South Africa*, Ergänzungsheft 290. Germany: Petermanns Geographische Mitteilungen.
Beckedahl, H. R., T. A. S. Bowyer-Bower, G. F. Dardis, and P. M. Hanvey. 1988. 'Geomorphic effects of soil erosion', in B. P. Moon and G. F. Dardis (eds.), *The Geomorphology of Southern Africa*. Johannesburg: Southern. 249–76.
Bland, W., and D. Rolls. 1998. *Weathering: An Introduction to the Scientific Principles*. London: Arnold.
Boelhouwers, J. C. 1991. 'Present-day periglacial activity in the Natal Drakensberg, southern Africa: a short review', *Permafrost and Periglacial Processes* 2: 5–12.
———. 1995. 'Some perspectives on southern African periglacial geomorphology', *South African Journal of Science* 91: 163–65.
———. 1998a. *The present-day frost action environment and its geomorphological significance in the Western Cape mountains, South Africa*. School of Environmental Studies Occasional Publication Series no. 2, University of the Western Cape.
———. 1998b. 'Periglacial Geomorphology in Southern Africa: a Bibliography'. Southern African Association of Geomorphologists Occasional Paper No. 8.
Darymple, J. B., R. J. Blong, and A. J. Conacher. 1968. 'A hypothetical nine unit landsurface model', *Zeitschrift für Geomorfologie* 12: 60–76.
French, H. M. 1997. *The Periglacial Environment* (2nd ed.). London: Longman.
Grab, S. W. 1996. 'Debris deposits in the high Drakensberg, South Africa: possible indicators for plateau,

niche and cirque glaciation', *Zeitschrift für Geomorfologie*, Suppl. Ed. 103: 389–403.
———. 1998. 'Periglacial research in Africa: past, present and future', *Progress in Physical Geography* 22: 375–84.
Grab, S. W., and K. J. Hall. 1996. 'North-facing hollows in the Lesotho/Drakensberg mountains: hypothetical palaeoenvironmental reconstructions?', *South African Journal of Science* 92: 183–4.
Hall, K. J. 1988. 'Weathering', in B. P. Moon and G. F. Dardis (eds.), *The Geomorphology of Southern Africa*. Johannesburg: Southern. 12–29.
———. 1991. 'The allocation of the freeze-thaw weathering mechanism in geocryological studies: a critical comment', *South African Geographical Journal* 73: 10–13.
———. 1992. 'A discussion on the need for greater rigour in southern African cryogenic studies', *South African Geographical Journal* 74: 69–71.
———. 1994. 'Cutbacks in the Natal Escarpment: an hypothesis on their origin', *South African Journal of Science* 90: 263–4.
———. 1995. 'Freeze-thaw weathering: the cold region "panacea"', *Polar Geography and Geology* 19: 79–87.
King, L. C. 1963. *South African Scenery* (3rd ed.). Edinburgh: Oliver and Boyd.
Le Roux, J. S. 1990. 'Discussion of periglacial geomorphology at Golden Gate Highlands National Park, Orange Free State, South Africa', *South African Geographer* 17: 129–31.
Lewis, C. A. 1996. 'Glacial features', in C. A. Lewis (ed.), *The Geomorphology of the Eastern Cape*. Grahamstown: Grocott and Sherry.
Marker, M. E. 1992. 'Periglacial landforms of southern Africa compared with those of Colorado and New Mexico, USA', *South African Geographical Journal* 74: 8–12.
———. 1995. 'Further data for a Pleistocene periglacial gradient in southern Africa', *Transactions of the Royal Society of South Africa* 50: 49–58.
———. 1998. 'Cenozoic climatic change in southern Africa: the evidence from Geomorphology, 1967–1996, A personal view', *South African Geographical Journal* 80: 1–8.
Meadows, M. E. 1988. 'Landforms and Quaternary climatic change', in B. P. Moon and G. F. Dardis (eds.), *The Geomorphology of Southern Africa*. Johannesberg: Southern. 296–316.
Meiklejohn, K. I. 1994. 'Valley asymmetry on south-eastern Alexander Island, Antarctica, and valley forms in the high Drakensberg, southern Africa', *South African Geographical Journal* 76: 68–72.
———. 1997. 'The role of moisture in the weathering of the Clarens Formation in the KwaZulu/Natal Drakensberg: Implications for the deterioration and preservation of indigenous rock art', *South African Geographical Journal* 4: 199–206.
Moon, B. P. 1988. 'Hillslope form and development', in B. P. Moon and G. F. Dardis (eds.), *The Geomorphology of Southen Africa*. Johannesberg: Southern. 57–77.
———. 1990. 'Rock mass strength as a control of slope development: evidence from southern Africa', *South African Geographical Journal* 72: 50–53.
Ollier, C. D. 1984. *Weathering*. New York: Longman.
Partridge, T. C. 1997. 'Cainozoic environmental change in southern Africa, with special emphasis on the last 200 000 years', *Progress in Physical Geography* 21: 3–22.
Pope, G. A., R. I. Dorn, and J. C. Dixon. 1995. 'A new conceptual model for understanding geographical variations in weathering', *Annals of the Association of American Geographers* 85: 38–64.
Shakesby, R. A. 1997. 'Pronival (protalus) ramparts: a review of forms, processes, diagnostic criteria and palaeoenvironmental implications', *Progress in Physical Geography* 21: 394–418.
Sumner, P. D. 1995. 'Cutbacks in the Natal Drakensberg Escarpment: comments on an hypothesis on their origin', *South African Journal of Science* 91: 285–6. (Response by K. Hall, p. 286.)
Sumner, P. D., and K. I. Meiklejohn. 1998. 'Weathering and Slope Processes in Southern Africa: A Bibliography'. Southern African Association of Geomorphologists Occasional Paper No. 6.
Thorn, C. E. 1988. *Introduction to Theoretical Geomorphology*. Boston: Unwin Hyman.
———. 1992. 'Periglacial geomorphology: what? where? when?', in J. C. Dixon and A. D. Abrahams (eds.), *Periglacial Geomorphology*. Chichester: Wiley. 1–30.
Watson, H. K. 1998. 'Erosion, Sediment Yield and Sediment Transport in Southern Africa: A Bibliography'. Southern African Association of Geomorphologists Occasional Paper No. 3.
Weinert, H. H. 1965. 'Climatic factors affecting the weathering of igneous rocks', *Agricultural Meteorology* 2: 27–42.
Yatsu, E. 1988. *The Nature of Weathering: an Introduction*. Tokyo: Sozosha.

CHAPTER 14

Benians, G., and W. van Averbeke. 1990. 'Some chemical, pedological and land-forming effects of giant earthworms (Microchaetus microchaetus) in Ciskei'. *Proceedings of the 16th Congress of the Soil Science Society of South Africa*. Soil Science Society of South Africa, Pretoria. 275–86.
Birdwell, B. T. 1987. 'Some of the World's best farmland is here', in W. Whyte (ed.), *Our American Land, 1987 USDA Yearbook of Agriculture*, Washington, DC: USDA, SCS. 91–5
Borlaug, N. E. 1997. 'Fertilizers and the green revolution: Past contributions and future challenges', *FSSA Journal 1997*: 11–26.
Brady, N. C., and R. R. Weil. 1996. *The nature and properties of soils* (11th ed.). Upper Saddle River: Prentice-Hall.
Bridges, E. M., N. H. Batjes, and F. O. Nachtergaele (eds.). 1998. *World Reference Base for Soil Resources: Atlas*. Leuven: Acco.
Deckers, J. A., F. O. Nachtergaele, and O. C. Spaargaren (eds.). 1998. *World Reference Base for Soil Resources: Introduction*. Leuven: Acco.
Ellis, F. 1988. 'Die gronde van die Karoo' [The soils of the Karoo]. Ph.D. thesis, Department of Soil and Agricultural Water Science, University of Stellenbosch.
Eloff, J. F. 1984. 'Die grondhulpbronne van die Vrystaatstreek' [The soil resources of the Free State Region]. Ph.D. thesis, Department of Soil and Agricultural Water Science, University of Stellenbosch.
Eswaran, H., F. de Coninck, and T. Varghese. 1990. 'Role of plinthite and related forms in soil degradation', *Advances in Soil Science* 11: 109–27.
FAO. 1974. *Soil map of the world. Vol. 1. Legend*. Paris: UNESCO.
———. 1976. 'A framework for land evaluation', *Soils Bulletin* 32. Rome: FAO.

———. 1990. *Soil map of the World. Revised legend.* Rome: FAO.

IFAD. 1992. *Soil and water conservation in sub-Saharan Africa – Towards sustainable production by the rural poor.* Amsterdam: Free University Press.

ISSS Working Group Reference Base. 1998. *World Reference Base for Soil Resources.* FAO.

Laker, M. C., and J. H. H. D'Huyvetter. 1988. 'Slope-erodibility relationships for different soils in a semi-arid region of Ciskei'. *Proceedings of the 5th ISCO Conference* 1. Bangkok: Deptartment of Land Development. 641–53.

MacVicar, C. N. 1970. 'A note on the ecology of soils', *South African Journal of Science* 66: 187–91.

———. 1978. 'Advances in soil classification and genesis in South Africa'. *Proceedings of the 8th National Congress of the Soil Science Society of South Africa.* Technical Communication No. 165. Pretoria: Deptartment of Agricultural Technical Services. 22–40.

MacVicar, C. N., J. M. de Villiers, R. F. Loxton, E. Verster, J. J. N. Lambrechts, F. R. Merryweather, J. le Roux, T. H. van Rooyen, and H. J. von M. Harmse. 1977. *Soil classification: a binomial system for South Africa.* Pretoria: Department of Agricultural Technical Services.

Oldeman, L. R. 1993. *Global extent of soil degradation.* ISRIC Bi-annual Report 1991–92. Wageningen: ISRIC. 19–36.

Roux, P. W. 1990. 'The general condition of the veld in the Republic of South Africa'. *Proceedings of the National Veld Trust Conference.* Pretoria: National Veld Trust. (Conference and pages unnumbered.)

Smith, H. J. 1997. 'Definition and demarcation of high potential land for rainfed annual crop production'. Unpublished M.Inst.Agrar. thesis, University of Pretoria.

Soil Classification Working Group. 1991. *Soil classification: A taxonomic system for South Africa.* Pretoria: Department of Agricultural Development.

Soil Survey Staff. 1960. *Soil classification: A comprehensive system. 7th Approximation.* Washington, DC: USDA, SCS.

———. 1975. *Soil Taxonomy: A basic system of soil classification for making and interpreting soil surveys.* Washington, DC: USDA, SCS.

———. 1996. *Keys to Soil Taxonomy* (7th ed.). Washington, DC: USDA, NRCS.

Van der Eyk, J. J., C. N. MacVicar, and J. M. de Villiers. 1969. 'Soils of the Tugela basin. A study in subtropical Africa', *Natal Town and Regional Planning Reports* Vol. 15. Pietermaritzburg: Natal Town and Regional Planning Commission.

Van der Merwe, A. J. 1995. 'Wise land use: The basis for sustainable growth and development in South Africa'. *Proceedings of the Wise Land Use Symposium.* Pretoria: ISCW. 2–8.

Van der Merwe, C. R. 1940. 'Soil groups and subgroups of South Africa', *Science Bulletin* 231. Chemistry series No. 165. Pretoria: Department of Agriculture and Forestry.

———. 1962. 'Soil groups and subgroups of South Africa', *Science Bull.* 356. Chemistry Series No. 165. Pretoria: Deptartment of Agricultural Technical Services.

Van der Watt, H. v. H., and T. H. van Rooyen. 1995. *A glossary of soil science* (2nd ed.). Pretoria: Soil Science Society of South Africa.

Verster, E. 1974. 'Gronde van die Transvaalstreek' [Soils of the Transvaal region]. D.Sc.Agric. thesis, Department of Soil Science, University of the Orange Free State, Bloemfontein.

CHAPTER 15

Acocks, J. H. 1988. 'Veld types of South Africa', *Memoirs of the Botanical Survey of South Africa* 54: 1–146.

Archibold, O. W. 1995. *Ecology of World Vegetation.* London: Chapman and Hall.

Arnold, T. H., and B. C. de Wet. 1993. 'Plants of southern Africa: names and distribution', *Memoirs of the Botanical Society of South Africa* 62: 1–825.

Bond, W. J. 1985. 'Canopy-stored seed reserves (serotiny) in Cape Proteaceae', *South African Journal of Botany* 51: 181–6.

———. 1989. 'Describing and conserving biotic diversity', in B. J. Huntley (ed.), *Biotic Diversity in Southern Africa: Concepts and Conservation.* Cape Town: Oxford University Press. 2–18.

———. 1997. 'Fire', in R. M. Cowling, D. M. Richardson, and S. M. Pierce (eds.), *Vegetation of Southern Africa.* Cambridge: Cambridge University Press. 21–446.

Bond, W. J., and P. Slingsby. 1983. 'Seed dispersal by ants in shrublands of the Cape Province and its evolutionary implications', *South African Journal of Science* 79: 231–3.

Cowling, R. M. (ed.). 1992. *The Ecology of Fynbos: Nutrients, Fire, Diversity.* Cape Town: Oxford University Press.

Cowling, R. M., and C. Hilton-Taylor. 1994. 'Patterns of plant diversity and endemism in southern Africa: an overview', in B. J. Huntley (ed.), *Botanical Diversity in Southern Africa.* Pretoria: National Botanical Institute.

Cowling, R. M., G. E. Gibbs Russell, M. T. Hoffman, and C. Hilton-Taylor. 1989. 'Patterns of plant species diversity in southern Africa', in B. J. Huntley (ed.), *Biotic Diversity in Southern Africa: Concepts and Conservation.* Cape Town: Oxford University Press. 19–50.

Cowling, R. M., D. M. Richardson, and P. J. Mustart. 1997. 'Fynbos', in R. M. Cowling, D. M. Richardson, and S. M. Pierce (eds.), *Vegetation of Southern Africa.* Cambridge: Cambridge University Press. 99–130.

Cowling, R. M., D. M. Richardson, and S. M. Pierce (eds.). 1997. *Vegetation of Southern Africa.* Cambridge: Cambridge University Press

Croall, S., and W. Rankin. 1992. *Ecology for Beginners.* Cambridge: Icon Books.

Dean, W. R. J., M. T. Hoffman, S. J. Milton, and M. E. Meadows. 1995. 'Desertification in the semi-arid Karoo, South Africa: review and reassessment', *Journal of Arid Environments* 30: 247–64.

Ellery, W. N., M. T. Mentis, and R. J. Scholes. 1992. 'Modelling the location of woody–grassland boundaries', in P. A. Furley, J. Proctor, and J. A. Ratter (eds.), *Forest–Savanna Boundaries.* London: Chapman and Hall. 659–82.

Everard, D. A., J. J. Midgeley, and G. F. van Wyk. 1995. 'Dynamics of some forests in KwaZulu-Natal, South Africa, based on ordinations and size–class distributions', *South African Journal of Botany* 61: 283–92

Gibbs Russell, G. E. 1987. 'Preliminary floristic analysis of the major biomes in southern Africa', *Bothalia* 17: 213–27.

Glowka, L., F. Burhenne-Guilmin, and H. Synge. 1994. *A Guide to the Convention on Biological Diversity.* Gland: IUCN.

Good, R. 1974. *The Geography of the Flowering Plants*

(4th ed.).
Hoffman, M. T. 1988. 'Rationale for Karoo grazing systems: criticisms and research implications', *South African Journal of Science* 84: 556–9.
Jarvis, J. U. M. 1979. 'Zoogeography', *South African National Scientific Programme Report* 40: 82–7.
Jürgens, N., A. Burke, M. K. Seely, and K. M. Jacobsen. 1997. 'Desert', in R. M. Cowling, D. M. Richardson, and S. M. Pierce. *Vegetation of Southern Africa*. Cambridge: Cambridge University Press. 189–214.
Le Maitre, D. C., and J. J. Midgley. 1992. 'Plant reproductive ecology', in R. M. Cowling (ed.), *The Ecology of Fynbos: Nutrients, Fire, Diversity*. Cape Town: Oxford University Press. 135–74.
Louw, G. N., and M. K. Seely. 1982. *Ecology of Desert Organisms*. New York: Longman.
MacArthur, R. H., and E. O. Wilson. 1967. *The Theory of Island Biogeography*. Princeton University Press, Princeton.
Macdonald, I. A. W. 1984. 'Is the fynbos biome especially susceptible to invasion by alien plants? A re-analysis of available data', *South African Journal of Science* 80: 369–77.
McDowell, C. R., A. B. Low, and B. McKenizie. 1991. 'Natural remnants and corridors in Greater Cape Town and their role in threatened plant conservation', in D. A. Saunders and R. J. Hobbs (eds.), *Nature Conservation 2: The Role of Corridors*. Chipping North, Australia: Surrey Beatty and Sons. 27–39.
McNeely, J. A., K. R. Miller, W. V. Reid, R. A. Mittermeier, T. B. and Werner. 1990. *Conserving the World's Biological Diversity*. Gland and Washington: IUCN and World Bank.
Meadows, M. E. 1998. 'The southwestern Cape', in A. J. Conacher and M. Sala (eds.), *Land Degradation in Mediterranean-climate Regions of the World*. London: Wiley. 139–54..
Meadows, M. E., and H. P. Linder. 1993. 'A palaeoecological perspective on the origin of Afromontane grasslands', *Journal of Biogeography* 20: 345–55.
Midgley, J. J., R. M. Cowling, A. H. W. Seydack, and G. F. van Wyk. 1997. 'Forest', in R. M. Cowling, D. M. Richardson, and S. M. Pierce. *Vegetation of Southern Africa*. Cambridge: Cambridge University Press. 278–99.
Milton, S. J., R. I. Yeaton, W. R. J. Dean, and J. H. J. Vlok. 1997. 'Succulent karoo', in R. M. Cowling, D. M. Richardson, and S. M. Pierce. *Vegetation of Southern Africa*. Cambridge: Cambridge University Press. 131–66.
Myers, N. 1990. 'The biodiversity challenge: expanded hot-spots analysis', *The Environmentalist* 8: 1–20.
O'Connor, T. G., and G. J. Bredenkamp. 1997. 'Grassland', in R. M. Cowling, D. M. Richardson, and S. M. Pierce (eds.), *Vegetation of Southern Africa*. Cambridge: Cambridge University Press. 215–57.
Palmer, A. R., and M. T. Hoffman. 1997. 'Nama-karoo', in R. M. Cowling, D. M. Richardson, and S. M. Pierce (eds.), *Vegetation of Southern Africa*. Cambridge: Cambridge University Press. 167–88.
Raunkiaer, C. 1934. *The Life Forms of Plants and Statistical Geography*. Oxford: Oxford University Press.
Rebelo, A. G. 1994. 'Iterative selection procedures: centres of endemism and optimal placement of reserves', in B. J. Huntley (ed.), *Botanical Diversity in Southern Africa*. Pretoria: National Botanical Institute. 321–58.
———. 1997. 'Conservation', in R. M. Cowling, D. M. Richardson, and S. M. Pierce (eds.), *Vegetation of Southern Africa*. Cambridge: Cambridge University Press. 571–90.
Richardson, D. M., I. A. W. Macdonald, J. H. Hoffman, and L. Henderson. 1997. 'Alien plant invasions', in R. M. Cowling, D. M. Richardson, and S. M. Pierce (eds.), *Vegetation of Southern Africa*. Cambridge: Cambridge University Press. 535–70.
Rutherford, M. J. 1997. 'Categorization of biomes', in R. M. Cowling, D. M. Richardson, and S. M. Pierce (eds.), *Vegetation of Southern Africa*. Cambridge: Cambridge University Press. 91–8.
Rutherford, M. J., and R. H. Westfall. 1994. 'Biomes of southern Africa: an objective categorization', *Memoirs of the Botanical Survey of South Africa* 63: 1–94.
Scholes, R. J. 1997. 'Savanna', in R. M. Cowling, D. M. Richardson, and S. M. Pierce (eds.), *Vegetation of Southern Africa*. Cambridge: Cambridge University Press. 258–77.
Schulze, B. R. 1965. *Climate of South Africa, Part 8: General Survey*. Pretoria: Government Printer.
Seely, M. K. 1979. 'Ecology of a living desert: twenty years of research in the Namib', *South African Journal of Science* 75: 298–303.
Siegfried, W. R. 1989. 'Preservation of species in southern African nature reserves', in B. J. Huntley (ed.), *Biotic Diversity in Southern Africa: Concepts and Conservation*. Cape Town: Oxford University Press. 186–201.
Solbrig, O. T. 1992. 'Biodiversity: an introduction', in O. T. Solbrig, H. M. van Emden, and P. G. W. J. van Oordt (eds.), *Biodiversity and Global Change*. Wallingford: CAB International. 13–20.
Stock, W. D., N. Allsopp, F. van der Heyden, and E. T. F. Witkowski. 1997. 'Plant form and function', in R. M. Cowling, D. M. Richardson, and S. M. Pierce (eds.), *Vegetation of Southern Africa*. Cambridge: Cambridge University Press. 376–398.
Wilson, E. O. 1992. *The Diversity of Life*. New York: W. W. Norton.

CHAPTER 16

Adams, W. M. 1996. 'Lakes', in W. M. Adams, A. S. Goudie, and A. R. Orme (eds.), *The Physical geography of Africa*. New York: Oxford University Press. 122–33.
Alexander, W. J. R. 1990. *Flood Hydrology for southern Africa*. Pretoria: The Southern African National Committee on Large Dams.
Birkhead, A. L., B. W. Olbrich, C. S. James, and K. H. Rogers. 1997. 'Developing and integrated approach to predicting the water use of riparian vegetation'. Water Research Commission Report No. 474/1/97.
Bosch, J. M., and J. D. Hewlett. 1982. 'A review of catchment experiments to determine the effect of vegetation changes on water yield and evapotranspiration', *Journal of Hydrology* 55: 3–23.
Corbett, J. D., and R. F. O'Brien 1997. *Spatial Characterization Tool – Africa* (v 1.0) [CD-ROM]. Available: Texas A&M University System [May 1998].
Davies, B., and J. Day. 1998. *Vanishing Waters*. Cape Town: University of Cape Town Press.
Cowling, R. M., and C. Hilton-Taylor. 1994. 'Patterns of plant diversity and endemism in southern Africa: an overview', in B. J. Huntley (ed.), *Botanical Diversity in Southern Africa*. Pretoria: National Botanical Institute.
Department of Water Affairs. 1986. *Management of the*

Water Resources of the Republic of South Africa. Pretoria: DWA.
Dollar, E. S. J. 1998. 'Fluvial geomorphology in southern Africa: a bibliography. Southern African Association of Geomorphologists'. Occasional Paper No. 7.
Dollar. E. S. J., and K. M. Rowntree. 1995. 'Sediment sources, hydroclimatic trends and geomorphic responses in the Bell River, North Eastern Cape, South Africa', *South African Geographical Journal* 77 (1): 21–32
Dunne, T., and L. B. Leopold. 1978. *Water in Environmental Planning*. San Francisco: Freeman.
Gordon, N. D., T. A. McMahon, and B. L. Finlayson. 1992. *Stream Hydrology: an introduction for ecologists*. Chichester: John Wiley.
Goudie, A. S. 1996. 'The geomorphology of the seasonal tropics', in W. M. Adams, A. S. Goudie, and A. R. Orme (eds.), *The Physical Geography of Africa*. New York: Oxford University Press. 148–60.
Gregory, K. J., and D. E. Walling. 1973. *Drainage Basin Form and Process: a geomorphological approach*. London: Edward Arnold.
Heritage, G. L., A. W. van Niekerk, B. P. Moon, L. J. Broadhurst, K. H. Rogers, and C. S. James. 1997. 'The geomorphological response to the changing flow regimes of the Sabie and Letaba River systems'. WRC Report No. 474/1/96.
King, J., and D. Louw. 1998. 'Instream flow requirements for regulated rivers in South Africa using the Building Block Methodology', *Aquatic Ecosystem Health and Management* 1 (2): 109–24.
Midgley, D. C., W. V. Pitman, and B. J. Middleton. 1994. 'Surface Waters of South Africa 1990'. WRC Report No. 298.
Moon, B. P., and T. C. Partridge. 1993. 'Geomorphology in Southern Africa', in H. J. Walker and W. E. Grabau (eds.), *The Evolution of Geomorphology*. Chichester: Wiley. 383–95.
Newson, M. D. 1994. *Hydrology and the River Environment*. Oxford: Oxford University Press.
Petts, G. E. 1979. 'Complex response of river channel morphology subsequent to reservoir construction', *Progress in Physical Geography* 3: 329–62.
Pickup, G., and R. F. Warner. 1976. 'Effects of hydrologic regime on magnitude and frequency of dominant discharge', *Journal of Hydrology* 29: 51–75.
Rowntree, K. M. 1991. 'An assessment of the potential impact of alien invasive vegetation on the geomorphology of river channels in South Africa', *South African Journal of Aquatic Science* 17 (1/2): 28–43.
Rowntree, K. M., and G. J. Beyers. 1999. 'An experimental study of the effect of Acacia mearnsii (black wattle trees) on streamflow in the Sand River, Zwartkops River catchment, Eastern cape'. WRC Report.
Rowntree, K. M., and E. S. J. Dollar. 1996. 'Contemporary fluvial processes and landforms', in C. A. Lewis (ed.), *The Geomorphology of the Eastern Cape*. Grahamstown: Grocott and Sherry.
Rowntree, K. M., and R. A. Wadeson. 1998. 'A geomorphological framework for the assessment of instream flow requirements', *Aquatic Ecosystem Health and Management* 1 (2): 125–41.
———. 1999. 'A hierarchical geomorphological model for the classification of selected South African river systems'. WRC Report No. 497/1/99.
Rowntree, K. M., R. A. Wadeson, and J. O'Keeffe. 1998. 'Geomorphological zonation for ecological river typing', in K. M. Rowntree (ed.), *Proceedings of the Biennial Conference of the Southern African Association of Geomorphologists*, Grahamstown, 28 June–1 July.
Schumm, S. A. 1977. *Fluvial Systems*. New York: John Wiley.
Selby, M. J. 1985. *Earth's Changing Surface: an introduction to geomorphology*. Oxford: Oxford University Press.
Sherrard, J. J., and W. D. Erskine. 1991. 'Complex response of a sand bed stream to upstream impoundment', *Regulated Rivers: Research and Management* 6: 53–70.
Van Coller, A. L., K. H. Rogers, and G. L. Heritage. 1999. 'Linking riparian vegetation types and fluvial geomorphology along the Sabie River within the Kruger National Park, South Africa', *African Journal of Ecology*. In press.
Van der Leeden, F., F. L. Troise, and D. K. Todd. 1990. *The Water Encyclopaedia* (2nd ed.). Michigan: Lewis Publishers.
Van Niekerk, A. W., G. L. Heritage, and B. P. Moon. 1995. 'River classification for management: the geomorphology of the Sabie river in the eastern Transvaal', *South African Geographical Journal* 77 (2): 68–76.
Versveld, D. B., D. C. le Maitre, and R. A. Chapman. 1998. 'Alien invading plants and water resources in South Africa: a preliminary assessment'. WRC Report No. TT 99/98, CSIR No. ENV/S-C 97154.
Walling, D. E. 1996. 'Hydrology and rivers', in W. M. Adams, A. S. Goudie, and A. R. Orme (eds.), *The Physical Geography of Africa*. New York: Oxford University Press. 103–21
Ward, R. C. 1975. *Principles of Hydrology* (2nd ed.). Maidenhead: McGraw-Hill.

CHAPTER 17

African National Congress. 1994. *The Reconstruction and Development Programme*. Johannesburg: Umanyano
Dally, H. E., J. B. Cobb Jr, and C. W. Cobb. 1994. *For the Common Good: Redirecting the Economy Towards Community, the Environment, and Sustainable Future* (2nd ed.). Boston: Beacon Press.
Dovers, S. R., and J. W. Handmer. 1992. 'Uncertainty, sustainability and change', *Global Environmental Change* 2 (4): 262–76.
Eckholm, E. P. 1975. 'The deterioration of mountain environments', *Science* 189: 764–70.
Hawken, P. 1995. *The Ecology of Commerce*. New York: HarperCollins.
Miller, G. T. 1996. *Living in the Environment: Principles, Connections, and Solutions*. New York: Wadsworth.
Mitchell, B. 1989. *Geography and Resource Analysis* (2nd ed.). London: Longman.
———. 1997. *Resource and Environmental Management*. London: Wesley Longman.
Omara-Ojungu, P. H. 1992. *Resource Management in Developing Countries*. London: Longman.
O'Riordan, T. 1971. *Perspectives on Resource Management*. Pion: London.
———. 1995. *Environmental Science for Environmental Management*. Longman: London.
Pearce, D. 1993. *Economic Values and the Natural World*. Cambridge, MA: MIT Press.
Perman, R., Yue Ma, and J. McGilvray. 1996. *Natural Resources and Environmental Economics*. London: Longman.
Repetto, R. 1992. 'Accounting for Environmental Assets', *Scientific American*, June: 94–100.
Russwurm, L. H., and E. Sommerville. 1974. *Man's*

Natural Environment: A Systems Approach. North Scituate, MA: Duxbury Press.
Van Riet, W., P. Claassen, J. van Rensburg, T. van Viegen, and L. du Plessis. 1997. *Environmental Potential Atlas for South Africa*. Pretoria: J. L. van Schalk.
World Commission on Environment and Development. 1987. *Our Common Future*. Oxford: Oxford University Press.
Zimmermann, E. W. 1951. *World Resource and Industries*. New York: Harper and Brothers.

CHAPTER 18

Adams, W. M. 1990. *Green Development: Environment and Sustainability in the Third Word*. London: Routledge.
African National Congress. 1993. *Bill of Rights: Policy Guide*. Marshalltown: ANC Department of Information and Publicity.
Bengu, S. M. E. 1975. *Chasing Gods Not Our Own*. Pietermaritzburg: Shuter and Shooter.
Brockman, C. F. 1979. *Recreational Use of Wild Lands*. New York: McGraw-Hill.
Bryant, A. T. 1929. *Olden Times in Zululand and Natal*. London: Longman, Green, and Co.
———. 1949. *The Zulu People: As They Were Before the White Man Came*. Pietermaritzburg: Shuter and Shooter.
Burger, P. 1981. 'White Man's Game', *Sunday Tribune*, 13 September 1981: 6.
Butler-Adam, J. F. 1981. 'A Recreation Study of Albert Falls Public Resort', *Town and Regional Planning Report – Volume 47*. Pietermaritzburg: Natal Town and Regional Planning Commission.
———. 1986. 'Opportunity Space and Recreation Demand and Preference in KZN'. Institute for Social and Economic Research, Occasional Paper No. 16. University of Durban-Westville.
Carlson, R. E., J. R. MacLean, T. R. Deppe, and J. A. Peterson. 1979. *Recreation and Leisure: The Changing Scene*. Belmont: Wadsworth.
Catton, W. R., and R. E. Dunlap. 1978. 'Environmental Sociology: a New Paradigm', *American Sociologist 13*: 41–9.
Cheek, N. H., and D. R. Field. 1977. 'Aquatic Resources and Recreation Behaviour', *Leisure Sciences* 1 (1): 67–83.
Coke, M. 1983. *Conservation: Recreation Plan for Richard's Bay*. Richard's Bay: Richard's Bay Town Council.
Comrie-Greig, J. 1984. 'The Law of the Jungle', *African Wildlife* 38 (5): 174–83.
Cooper, C., J. Fletcher, D. Gilbert, and S. Wanhill. 1996. *Tourism: Principles and Practice*. London: Longman.
Cotgrove, S., and A. Duff. 1980. 'Environmentalism, Middle-Class Radicalism and Politics', *Sociological Review* 28: 333–51.
Crandall, R. 1979. 'Social Interaction, Affect and Leisure' *Journal of Leisure Research* 11 (13): 165–81.
Donaldson, S. E. 1995. 'The Facility-Orientated Outdoor Recreational Needs and Preferences of Urban Black People in the Northern Province: An Evaluative and Perceptual Study'. Unpublished M.A. thesis, University of the North, Sovenga.
Ferrario, F. F. 1981. *An Evaluation of the Tourist Potential of KwaZulu and Natal*. Durban: KwaZulu Development Corporation.
Fowkes, J. D. 1992. 'An Evaluation of the Contribution of Ecotourism to the Economy of the Republic of Botswana in 1984'. Unpublished M.Com. thesis, University of Stellenbosch.
Gibbon, A. B. 1976. 'Outdoor Recreation Survey of the Port Elizabeth Area'. Unpublished M.A. thesis, Department of Geography, University of Port Elizabeth.
Grove, H. 1995. 'Biodiversity: the essence of life', *Conservation* 10: 2.
Human Sciences Research Council. 1984. *Behavioural Recreation Geography: Bibliography*. Pretoria: HSRC.
Hugo, M. L. 1974. 'An Investigation into the Nature and Magnitude of the Demand for Outdoor Recreation Among the Urban Black Communities of South Africa', *South African Journal of African Affairs* 2: 56–66.
Hugo, M. L., A. T. Viljoen, and J. M. Meeuwis. 1997. *The Ecology of Natural Resource Management: The Quest for Sustainable Living*. Pretoria: Kagiso.
Huntley, B. J. 1965. 'A Preliminary Account of the Ngoye Forest Reserve, Zululand', *Journal of South African Botany* 31: 177–205.
Jenkinson, T. B. 1882. *Amazulu: The Zulus, Their Past History, Manners, Customs and Language*. London: Allen and Co.
Kay, G. 1970. *Rhodesia: A Human Geography*. London: University of London Press.
Knill, G. 1991. 'Towards the Green Paradigm', *South African Geographical Journal* 73 (2): 23–33.
Kock, E. 1993. *Ecotourism: a Tool for Rural Reconstruction in South Africa*. Johannesburg: Group for Environmental Monitoring.
Kraus, R. 1966. *Recreation Today: Programme Planning and Leadership*. New York: Appleton-Century-Croft.
Kruger, F. J., B. W. van Wilgen, A. v. B. Weaver, and T. Greling. 1997. 'Sustainable Development and the Environment: Lessons from the St Lucia Environment Impact Assessment', *South African Journal of Science* 93: 23–33.
Larkin, R. P., G. L. Peters, and C. H. Exline. 1981. *People, Environment, and Place*. Columbus, OH: Merrill.
Lengoasa, J. 1997. 'Implementation of EIA Regulations. Environmental Impact Management', *Quarterly Newsletter, DEAT* 1 (1): 4–5.
Ludlow, W. R. 1882. *Zululand and Cetewayo*. London: Simpkin, Marshall, and Co.
Magi, L. M. 1986. 'Black People's Cognitions of Natural Recreation Resources in the Natal North-Coastal Region'. Unpublished Ph.D thesis, University of Zululand, KwaDlangezwa.
———. 1988. 'What the Concept "Recreation" Means to Blacks'. *Publication Series of the University of Zululand. B. No. 74.*
———. 1989a. 'Cognitized use of natural recreation resources: A black perspective', *Development Southern Africa* 6 (3): 326–39.
———. 1989b. 'Cognition of recreation resources through photographic images', *South African Geographical Journal* 71 (2): 67–73.
———. 1992. 'Outdoor Recreation Perception Changes Since the Repeal of the Separate Amenities Act in South Africa'. Paper presented at the 27th International Geographical Conference of the IGU Conference. Washington, DC, 9–14 August.
———. 1993. 'Dealing in Environmental Concepts and Issues: a Time to Take Stock'. Paper presented at the Quadrennial Conference of The Society for Geography. Port Elizabeth: University of Port Elizabeth, 4–7 July.
Mahaye, A. T. 1996. 'Motorised Outdoor Recreation

Pursuits in the Lower Umfolozi and Hlabisa Coastal Districts of KwaZulu-Natal'. Unpublished M.A. thesis, University of Zululand, KwaDlangezwa.
Manganyi, N. C. 1973. *Being-Black-in-the-World*. Johannesburg: Spro-Cas/Ravan Press.
McCall, J. R., and V. N. McCall. 1977. *Outdoor Recreation: Forest, Park and Wilderness*. Beverley Hills: Benzinger Bruce and Glencoe.
Mercer, D. C. 1971. 'The Role and Perception in the Recreation Experience: A Review and Discussion', *Journal of Leisure Research* 3 (4): 261–76.
———. 1977. *Leisure and Recreation in Australia*. Malvern: Sorrett Publishing.
Meyersohn, R. 1969. 'The Sociology of Leisure in the United States: Introduction and Bibliography, 1945–1965' *Journal of Leisure Research* 1: 53–68.
Miller, G. T. 1988. *Environmental Science: An Introduction*. Belmont, CA: Wadsworth Publishing.
Molema, S. M. 1920. *The Bantu: Past and Present*. Edinburgh: W. Green and Son.
Mphahlele, E. 1974. *The African Image*. London: Faber and Faber.
Mwandla, N. D. 1995. 'Blacks and the Coast: Current Demands and Future Aspirations for Coastal Recreation in the KwaZulu-Natal North Coast'. Unpublished M.A. thesis, University of Zululand, KwaDlangezwa.
Ndebele, N. 1972. 'Black Development', in B. S. Biko (ed.), *Black Viewpoint*. Durban: Spro-Cas Black Community Programmes.
Neulinger, J. 1981. *The Psychology of Leisure* (2nd ed.). Springfield, IL: Charles C. Thomas.
Nyembezi, C. L. 1954. *Zulu Proverbs*. Johannesburg: Witwatersrand University Press.
ORRRC [Outdoor Recreation Resources Review Commission]. 1962. *Summary Report: Outdoor Recreation of America*. Washington, DC: US Government Printing Office.
Paden, J. N., and E. W. Soja. 1970. *The African Experience, II*. London: Heinemann Education Books.
Pigram, J. J. 1983. *Outdoor Recreation and Resource Management*. New York: St Martin's Press.
Preston, M. G. 1983. 'Enhancement of the Awareness of Conservation Issues in Visitors to Three South African Nature Reserves'. Unpublished M.A. thesis, University of Cape Town.
Price, C. 1979. 'Public Preference and the Management of Recreation Congestion', *Regional Studies* 13: 125–39.
Ramphele, M. (ed.). 1981. *Restoring the Land: Environment and Change in a Post-Apartheid South Africa*. London: Panos Institute.
Richards, R. F. 1994. 'Living on Durban's Fringe: A Study of the Leisure Styles of Shack and Peri-Urban Youth'. Unpublished M.A. thesis, University of Natal, Durban.
Ruch, E. A., and K. C. Anyanwu. 1984. *African Philosophy: An Introduction to the Main Philosophical Trends in Contemporary Africa*. Rome: Allen and Co.
SARDC [Southern African Research & Documentation Centre]. 1994. *State of the Environment in Southern Africa*. Johannesburg: Penrose Press.
Sarre, P., and J. Blunden. 1995. *An Overcrowded World? Population, Resources and the Environment*. New York: Oxford University Press.
South Africa. 1993. *Managing South Africa's Environmental Resources: A Possible New Approach*. Pretoria: Government Printer.
———. 1994. *Towards a National Policy for the Use of Environmental Resource Economics in Environmental Impact Management*. Pretoria: Government Printer.
———. 1996a. Green Paper: An Environmental Policy for South Africa. Pretoria: Government Printer.
———. 1996b. White Paper: The Development and Promotion of Tourism in South Africa. Pretoria: Government Printer.
———. 1997. *Environmental Impact Management: EIA Regulations [Draft Guidelines]* Pretoria: Government Printer.
Steyn, J. N. 1976. 'The Geography of Recreation: A Young Research Field', *South African Geographer* 5 (4): 332–41.
Steyn, J. N., P. S. Hattingh, and J. J. Booysen. 1982. 'The Identification of Recreation Pressure Areas and Potential Recreation Areas in Transvaal', *S.A. Journal for Research in Sport, Physical Education and Recreation* 5 (2): 85–101.
Sutcliffe, M. O. 1981. 'A Behavioural Study of Recreation in the Natal Drakensberg'. Natal Town and Regional Planning Report No. 48. Pietermaritzburg: NTRPC.
Torkildsen, G. 1986. *Leisure and Recreation Management*. London: Spon Ltd.
Van der Wal, R. W. E., and J. N. Steyn. 1981. 'Recreation in Metropolitan Areas: Patterns and Problems', *S.A. Journal for Research in Sport, Physical Education and Recreation* 4 (2): 81–96.
Vilakazi, A. 1965. *Zulu Transformations*. Pietermaritzburg: University of Natal Press.
WCED [World Commission on Environment and Development]. 1987. *Our Common Future*. Oxford: Oxford University Press.
Weiskopf, D. C. 1982. *Recreation and Leisure: Improving the Quality of Life*. Boston: Allyn and Bacon.

CHAPTER 19

Birnie, P. W., and A. E. Boyle. 1992. *International Law and Environment*. Oxford: Clarendon Press.
Black's Law Dictionary (3rd. ed.). 1979. St Paul, MN: West Publishing.
Boyle, A., and M. Anderson (eds.). 1996. *Human Rights Approaches to Environmental Protection*. Oxford: Clarendon Press.
Devine, D. J., and M. G. Erasmus. 1992. 'International Environmental Law', in R. F. Fuggle and M. A Rabie (eds.), *Environmental Management in South Africa*. Cape Town: Juta. 155–80.
Fuggle, R. F., and M. A. Rabie (eds.). 1992. *Environmental Management in South Africa*. Cape Town: Juta.
Glazewski, J. 1996. 'Environmental rights and the new South African Constitution', in A. Boyle and M. Anderson (eds.), *Human Rights Approaches to Environmental Protection*. Oxford: Clarendon Press.
Hanks, P. (ed.). 1989. *The Collins Concise Dictionary of the English Language* (2nd ed.). London: Collins.
Rabie, M. A. 1992a. 'Nature and scope of Environmental Law', in R. F. Fuggle and M. A Rabie (eds.), *Environmental Management in South Africa*. Cape Town: Juta. 83–98.
———. 1992b. 'The Environment Conservation Act', in R. F. Fuggle and M. A Rabie (eds.), *Environmental Management in South Africa*. Cape Town: Juta. 99–119.
———. 1996. 'Environmental Conservation', in W. A. Joubert (ed.), *The Law of South Africa*. Vol 9. Durban: Butterworth.
Rabie, M. A., and R. F. Fuggle. 1992. 'The rise of environmental concern', in R. F. Fuggle and M. A Rabie (eds.),

Environmental Management in South Africa. Cape Town: Juta. 11–25.

Van der Vyfer, J. 1992. 'State sovereignty and the environment in international law', *South African Law Journal* 109: 472.

Wallace, R. M. 1986. *International Law: a Student Introduction*. London: Sweet and Maxwell.

Index

Copyright acknowledgements

Figure 1.1: Peet, R. 1998. *Modern geographical thought*. Oxford: Blackwell, p. 10.
Figure 2.1: Goudie, A. S. 1993b. *The nature of the environment*. Oxford: Blackwell, Figure 2.17
Table 2.2: Kates, R. W., B. L. Turner, and W. C. Clark. 1991. 'The great transformation', in B. L. Turner (ed.), *The Earth as Transformed by Human Action*. Cambridge: Cambridge University Press, Table 1.3.
Table 2.3: reprinted from *Global Environmental Change*, vol.1, B. L. Turner, R. E. Kasperson, B. W. Meyer, K. M. Dow, D. Golding, J. X. Kasperson, R. C. Mitchell, and S. J. Rattick. 1990. 'Two types of global environmental change: definitional and spatial-scale issues in their human dimensions', pp. 14–22, Copyright 1990, with permission from Elsevier Science.
Figure 2.2: Summerfield, M. A. 1996. 'Tectonics, geology and long-term landscape development', in W. M. Adams, A. S. Goudie, and A. R. Orme (eds.), *The Physical Geography of Africa*, New York: Oxford University Press, 1–17.
Figure 2.3: reprinted from *Marine Geology*, vol. 56, R. V. Dingle and Q. B. Hendey, 'Mesozoic and Tertiary Sediment Supply to the Western Cape Basin and Palaeodrainage Systems in South-western Africa', pp. 13–26, Copyright 1984, with permission from Elsevier Science.
Figure 2.4: reprinted from *Marine Micropalaeontology*, vol. 4, W. L. Prell, W. H. Hutson, and D. F. Williams, 'The subtropical convergence and late Quaternary circulation in the southern Indian Ocean', pp. 225–34, Copyright 1979, with permission from Elsevier Science.
Figure 2.5: Partridge, T. C., D. M. Avery, G. A. Botha, J. S. Brink, J. Deacon, R. S. Herbert, R. R. Maud, A. Scholtz, L. Scott, A. S. Talma, and J. S. Vogel. 1990. 'Late Pleistocene and Holocene climate change in southern Africa', *South African Journal of Science* 86: 302–5.
Dollar, E. S. J. 1998. 'Palaeofluvial geomorphology in Southern Africa: A review', *Progress in Physical Geography* 22 (3): 325–49.
Figure 2.6: Acocks, J. P. H. 1953. *Veld Types of South Africa*. Memoirs of the Botanical Survey of South Africa, 28. Pretoria: Government Printer.
Figure 3.1: Thompson, L. 1990. *A History of South Africa*. New Haven: Yale University Press, p. 34.
Figures 3.2, 3.4, 3.7 & 3.8: Lester, A. 1996. *From colonization to democracy: A new historical geography of South Africa*. London: I. B. Tauris, pp. 18, 48–9, 119, 127.
Figure 3.3: Oliver, R., and G. N. Sanderson (eds.). 1985. *The Cambridge History of Africa, Volume 6*. Cambridge: Cambridge University Press, p. 482.
Figure 3.5: Parsons, N. 1993. *A new history of Southern Africa*. London: Macmillan, p. 150.
Figure 3.6: Christopher, A. J. 1984. *South Africa: The impact of past geographies*. Cape Town: Juta, p. 33.
Table 4.1: Platzky, L., and C. Walker. 1985. *The Surplus People: Forced Removals in South Africa*. Johannesburg: Ravan Press.
Figure 4.7: reprinted from *Political Geography,* 16 (7), M. Lupton and C. Mather, 'The anti-politics machine: GIS and the reconstruction of the Johannesburg local state', 565–80, copyright 1997, with permission from Elsevier Science.
Table 5.1: Awireng-Obeng, F., and P. J. McGowan. 1998. 'Partner or Hegemon? South Africa in Africa', *Journal of Contemporary African Studies* 16 (1): 5–38.
Economist Intelligence Unit. 1998b. *Country Report: South Africa: 1st Quarter 1998*. The Economist Intelligence Unit: London
World Bank. 1994. *World Development Report*. World Bank: Washington.
Table 5.2: Tomlinson, R., and M. Addelson. 1987. *Regional Restructuring Under Apartheid: Urban and Regional Policies in Contemporary South Africa*. Ravan: Johannesburg, p. 266.
Figure 6.1: part of Map 3 of the Gordon Collection, 1777–87, redrawn in 1952 by Marie Kenyon for reproduction in V. S. Forbes, 1965, *Pioneer Travellers in South Africa*. Cape Town: Balkema.
Table 7.1: Haldenwang, B. B. 1997. *A socio-demographic profile of the Southern African Development Community region*. Occasional Paper 36, Institute for Futures Research, University of Stellenbosch. Bellville.
Haub, C., and M. Yanagishita. 1996. *World Population Data Sheet 1996*. Washington: Population Reference Bureau.
United Nations Population Fund. 1996. *The state of world population (1996) – Changing places: population, development and the urban future*. New York: UNFPA.
Table 7.3: Van der Merwe, I. J., and L.O. van Niekerk. 1994. *Language in South Africa – Distribution and Change*. Stellenbosch: University of Stellenbosch.
Table 8.1: Potter, R. 1992. *Urbanisation in the third world*, p. 8. By permission of Oxford University Press.
United Nations. 1995. *World Urbanization Prospects: the 1994 Revision*. New York: Population Division, pp. 74–85.

Table 8.2: United Nations. 1995. *World Urbanization Prospects: the 1994 Revision*. New York: Population Division, pp. 74, 87, 95.
Figure 8.2: *Geojournal,* Supplementary Issue 2, 1981, pp. 59–72, 'The spatial formation of the South African city', Davies, R. J., with kind permission from Kluwer Academic Publishers.
Table 9.3: McGladdery, K. 1992. Political land division in South Africa in the context of agro-ecological zones. Unpublished Geography III project, Rhodes University, Grahamstown.
Pickles, J., and D. Weiner. 1991. 'Rural and regional restructuring of apartheid: ideology, development policy and the competition for space'. *Antipode* 23 (1): 2–32.
Table 9.4: Van Zyl, J., J. Kirsten, and H. P. Binswanger (eds.). 1996. *Agricultural Land Reform in South Africa*. Cape Town: Oxford University Press.
Table 10.1: Simon, D. 1996. *Transport and development in the Third World*. London: Routledge.
Figure 11.1: *Earth's dynamic systems* (6th ed.), by W. K. Hamblin, ©1992. Reprinted with permission of Prentice-Hall, Inc., Upper Saddle River, NJ.
Figure 11.2: Tarbuck, E. J., and F. K. Lutgens. 1987. *The Earth: an introduction to physical geology* (2nd ed.). Columbus, Ohio: Merrill, p. 403.
Figure 11.4: Pickering, K. T., and L. A. Owen. 1994. *An Introduction to global environmental issues*. London: Routledge.
Figure 11.5: Petters, S. W. 1991. *Regional geology of Africa*. Berlin: Springer-Verlag, p. 11.
Figure 11.6: McCourt, S. 1995. 'The crustal architecture of the Kaapvaal crustal block South Africa, between 3.5 and 2.0 Ga. A synopsis', *Mineralium Deposita* 30 (2): 89–97. Copyright Springer-Verlag.
Figure 11.7: Partridge, T. C., and R. R. Maud. 1987. 'Geomorphic evolution of southern Africa since the Mesozoic', *South African Journal of Geology* 90 (2): 179–208.
Figures 12.1, 12.2, 12.3, 12.4 & 12.5: Preston-Whyte, R. A., and P. D. Tyson. 1988. *Atmosphere and Weather of Southern Africa*. Oxford University Press, Cape Town, pp. 210, 166, 260, 204, 202.
Table 12.1: 'Human response to the environment', C. Vogel, in *People and the Environment,* T. Binns (ed.). Copyright John Wiley & Sons Limited. Reproduced with permission.
Spio, K., and L. Nchabeleng. 1997. Workshop on drought mitigation in the Northern Province of South Africa, Proceedings of the Drought Mitigation Workshop, 11–12 March, 1997, Pietersburg, held under the auspices of Disaster Mitigation for Sustainable Livelihoods Unit for Southern African Development education and Policy Research, University of the Western Cape.
Figure 13.2: Marker, M. E. 1995. 'Further data for a Pleistocene periglacial gradient in southern Africa', *Transactions of the Royal Society of South Africa* 50: 49–58.
Figure 13.4: Beckedahl, H. R., T. A. S. Bowyer-Bower, G. F. Dardis, and P. M. Hanvey. 1988. 'Geomorphic effects of soil erosion', in B. P. Moon and G. F. Dardis (eds.), *The Geomorphology of Southern Africa*. Johannesburg: Southern, 249–76.
Figure 14.6: Smith, H. J. 1997. Definition and demarcation of high potential land for rainfed annual crop production. Unpublished M. Inst. Agrar. dissertation, University of Pretoria.
Table 14.2: Oldeman, L. R. 1993. *Global extent of soil degradation*. ISRIC Bi-annual report 1991–92: 19–36. Wageningen: ISRIC.
Figure 15.1: Croall, S., and W. Rankin. 1992. *Ecology for Beginners*. Cambridge: Icon Books.
Figure 15.2: Bond, W. J. 1985. 'Canopy-stored seed reserves (serotiny) in Cape Proteaceae', *South African Journal of Botany* 51: 181–6.
Figures 15.3 & 15.4: Wilson, E. O. 1992. *The Diversity of Life*. Harvard University Press.
Figure 15.5 & Table 15.1: Cowling, R. M., and C. Hilton-Taylor. 1994. 'Patterns of plant diversity and endemism in southern Africa: an overview', in B. J. Huntley (ed.), *Botanical Diversity in Southern Africa*. Pretoria: National Botanical Institute.
Table 15.1: Cowling, R. M., G. E. Gibbs Russell, M. T. Hoffman, and C. Hilton-Taylor. 1989. 'Patterns of plant species diversity in southern Africa', in B. J. Huntley (ed.), *Biotic Diversity in Southern Africa: Concepts and Conservation*. Cape Town: Oxford University Press, pp. 19–50.
Figures 15.6 & 15.7: Archibold, O. W. 1995. *Ecology of World Vegetation*. London: Chapman and Hall.
Figure 16.8: Rutherford, M. J., and R. H. Westfall. 1994. 'Biomes of southern Africa: an objective categorization', *Memoirs of the Botanical Survey of South Africa* 63: 1–94.
Figure 15.9: Cowling, R. M., D. M. Richardson, and S. M. Pierce. 1997. *Vegetation of Southern Africa*. Cambridge: Cambridge University Press, p. 93.
Figure 15.10: O'Connor, T. G., and G. J. Bredenkamp. 1997. 'Grassland', in R. M. Cowling, D. M. Richardson, and S. M. Pierce. 1997. *Vegetation of Southern Africa*. Cambridge: Cambridge University Press, pp. 215–57.
Table 15.2: Cowling, R. M., D. M. Richardson, and P. J. Mustart. 1997. 'Fynbos', in R. M. Cowling, D. M. Richardson, and S. M. Pierce. 1997. *Vegetation of Southern Africa*. Cambridge: Cambridge University Press, p. 103.
Table 15.4: Rebelo, A. G. 1997: 'Conservation', in R. M. Cowling, D. M. Richardson, and S. M. Pierce. 1997. *Vegetation of Southern Africa*. Cambridge: Cambridge University Press, p. 574.
Table 16.1: Van der Leeden, F., F. L. Troise, and D. K. Todd. 1990. *The Water Encyclopaedia* (2nd ed.). Michigan: Lewis Publishers.
Figure 16.2: Midgley, D. C., W. V. Pitman, and B. J. Middleton. 1994. Surface Waters of South Africa 1990, Water Research Report no. 298.
Figure 16.3: Dunne, T., and L. B. Leopold. 1978. *Water in Environmental Planning*. San Francisco: Freeman.
Table 16.2: Alexander, W. J. R. 1990. *Flood Hydrology for southern Africa*. Pretoria: The Southern African National Committee on Large Dams.
Figure 17.2: Russwurm, L.H. and E. Sommerville, 1974, *Man's Natural Environment: A Systems Approach*, North Scituate, Mass: Duxbury Press, p. 5.